Felicien Machotte

The Automobile Magazine

1899-1900

Felicien Machotte

The Automobile Magazine
1899-1900

ISBN/EAN: 9783741113000

Manufactured in Europe, USA, Canada, Australia, Japa

Cover: Foto ©Andreas Hilbeck / pixelio.de

Manufactured and distributed by brebook publishing software
(www.brebook.com)

Felicien Machotte

The Automobile Magazine

THE

AUTOMOBILE MAGAZINE

ILLUSTRATED

October, 1899, to March, 1900

NEW YORK

THE UNITED STATES INDUSTRIAL PUBLISHING COMPANY

1900

Contents of Volume One

The Automobile Magazine

Contents of Volume One

95

The Automobile Magazine

THE AUTOMOBILE
MAGAZINE

Index for Volume I, October, 1899 to March, 1900

Illustrated articles are marked with an asterisk (*).   Reviews of books are marked with a dagger (†).

THE "LOCOMOBILE"

is equal to the high standard of American workmanship, and cannot be surpassed. There will be no better self-propelled vehicle made. It is the ideal for pleasure or business. The motive power is known and understood the world over—is no experiment. It is easily controlled, costs but ¼ cent per mile, and can be renewed at even the smallest country towns here or abroad. The "Locomobile" meets the approval of critical buyers. It is durable, handsome in appearance, and weighs but 400 pounds; is unequalled for speed, hill-climbing, or travelling over bad roads, and is absolutely reliable under all circumstances. It easily ascended Mt. Washington, altitude 6,300 feet. Noiseless, odorless.

$600

DELIVERY IN NINETY DAYS

SIMPLICITY

SPEED

LOW COST

The Automobile Magazine

VOL. I   OCTOBER 1 1899   No. 1

CONTENTS

AGENCY FOR FOREIGN SUBSCRIPTIONS:
INTERNATIONAL NEWS COMPANY
BREAM BUILDING, CHANCERY LANE   STEPHAN STRASSE, No. 18
LONDON   LEIPSIC

Price, 25 Cents a Number; $3.00 a Year

Newport Night Parade.

The Automobile
MAGAZINE

Vol. I   OCTOBER 1899   No. 1

Newport in the Lead
By Edwin Emerson, Jr.

JUST as the *pièce de resistance* comes after the *entrée* and the dainties follow the salad, so the Newport horse-show followed the yacht races, after which came the automobile festival. This long-expected parade of Newport's " upper ten " in their much-cherished rigs finished the summer season of America's " four hundred " in a blaze of glory. For Newport it was indeed the end, for that very night the last of the weekly Newport dances was held at the Casino, and next day the summer cottagers began to scatter in all directions, Mrs. O. H. P. Belmont, the leading spirit of the automobile festival, betaking herself to the Virginian Hot Springs, Mrs. William Astor to Europe, and a bevy of the others to Lenox or the Adirondacks.

The present craze for automobiles among the leaders of American society, so-called, began at Newport two years ago, when Mr. Oliver H. P. Belmont imported a French automobile carriage, and thus made the sport " good form." Harry Payne Whitney promptly stocked a whole stable with dashing automobiles, as he was wont to do with hunters, and William K. Vanderbilt, Jr., and his cousin Alfred took to driving automobiles in such reckless fashion that their mishaps rivalled the accidents of the hunting field and polo grounds. Thus on one occasion young William Vanderbilt, coasting down a steep hill at breakneck speed, by suddenly reversing the wheels of his new gasolene motor, made the automobile turn a complete double-back-action somersault. He escaped serious bodily injury only by his luck in taking a header into a grassy bank by the roadside.

This summer the women at Newport took up the craze, and insisted on running their own automobiles. The first to appear

The Automobile Magazine

in public was Miss Daisy Post, a niece of Mrs. Frederick Vanderbilt. This level-headed young *chauffeuse* took as much pride in twisting her swift electric Stanhope trap through the narrow turns of Thames Street or the mazes of Bellevue Avenue crowded with fashionable turnouts, as she has in lifting her hunter over a six-bar fence. After Miss Post's rig had become a familiar sight on the drives and avenues of Newport, Mrs. Herman Oelrichs and Mrs. William K. Vanderbilt fell victims to the craze early this summer, and thus it came about that when Lord

Miss Daisy Post on Bellevue Avenue.

Beresford visited Newport on his return from China, he received his first impressions of the latest product of Western civilization in that place.

Mrs. Stuyvesant Fish took her first automobile lesson early in July. It is safe to say that had an announcement of the time and place of her next lesson been made public a large crowd would have come to witness it. There may be some portions of the well-kept grounds surrounding her house that were not ploughed up by the wild rearing and charging of her automobile, but careful scrutiny on the morrow failed to locate them, and the appearance of the cross-walks, walls, and shrubberies led a few to suppose that some venturesome person had been trying to ride a bucking bronco about the place.

Mrs. Fish, with Miss Greta Pomeroy by her side, started to take the machine around the grounds next day. It was their intention to stick closely to the drives and crosswalks, but the automobile saw a stone wall and made for it, head on. The

6

Newport in the Lead

automobile won, and a large section of the stone wall fell with a thud. The motor shook itself free from the *débris*, and all went well for a while; but soon the automobile proceeded to lay low a clump of choice shrubbery on the lawn. For the next half hour Mrs. Fish had a more or less exciting time, and then came the climax. The carriage finally dashed against the steps of the villa, whereupon there was an awful, sudden stop, a crash and a snap, and the side of the automobile gave way, and the day's lesson necessarily came to an end.

Even with these mishaps Mrs. Fish enjoyed her experiences immensely. Miss Pomeroy is an experienced hunter, or she would surely not have been able to keep her seat during this up-to-date steeplechase. She was very much shaken up as it was, but she stuck to the machine during all its contortions, and thus successfully defied it to unseat her.

After this, at Newport, it became merely a question how fast

Peter Cooper Hewitt at Newport.

the manufacturers could supply the demand. Many of the ladies, like Mrs. Whitney, Mrs. Drexel, and Mrs. John Jacob Astor, preferred the old-fashioned style of comfortable victorias with electric running-gear, manipulated from the coachman's seat, but the men were apt to use more serviceable vehicles, such as the electric golf-break, with back-to-back seats for four, introduced by Mr. Alfred Vanderbilt, or the light gasolene phaeton in which Mr. Peter Cooper Hewitt and others of the younger set made their daily runs back and forth between the golf links and yacht landings.

The Automobile Magazine

Even Mrs. Astor's little seven-year-old boy was seen driving here and there in his father's automobile rig, nor would the little chap suffer any footman or driver to accompany him on these rides.

Then the ladies raised the question of costume. This all-important problem had been left utterly unsettled by the men, who wore all sorts of clothes, from pink hunting-coats and top-boots to reefers and yachting-caps, according to the exigencies of the moment. Of the ladies but a very few were guilty of such heresies.

To guide an automobile one must have free use of the limbs. To attain this end the Parisian modistes have designed a gown of silk jersey. This costume is very expensive, one dress costing no less than $500, but, needless to state, those fashionable women who are on automobile driving bent find it simply indispensable.

On the day of the parade, white gowns were in preponderance. Most of the men, similarly, appeared in light-colored

Waiting for its Master.

8

Newport in the Lead

Phœnix Ingraham and Miss Scott in Electric Golf Brake.

flannel suits with wide-brimmed hats of white felt or manila straw.

Inasmuch as the first automobile seen at Newport was the now discarded vehicle which Mr. Oliver H. P. Belmont brought home from Europe, it was but meet that the first automobile event of social importance should be held under the auspices of the Belmonts. As it turned out, it proved the greatest success of the season.

On the seventh day of September, the date set for the automobile parade, Indian summer had set in with its perfect weather, and the entire cottage colony, with the townspeople and a delegation of summer residents from Jamestown and Block Island, took advantage of the weather to appear in greater numbers than had gathered for either the yacht races or for the horse show just ended.

Early in the afternoon Bellevue Avenue was crowded with smart traps and victorias, the occupants of which craned their necks to catch a glimpse of the floral decorations of the different automobiles in gala attire, awaiting their owners under various portes cochères. On the front lawn of Newport's most fashionable florist stood a dense crowd watching a small army of foot-

9

men and white-aproned maids put the last finishing touches on
Mrs. Astor's automobile carriage, while Colonel Jack Astor
stood by lazily watching one of the Vanderbilt boys feverishly
adjusting the levers on his own flower-laden carriage. In the
meanwhile, all of Mrs. Belmont's household was busily engaged

First Prize for Speed.

William K. Vanderbilt, Jr., in his Steam Locomobile.

in transforming the smooth green lawn of Belcourt into an " obstacle park." Wooden horses attached to carriages, and sundry
dummy figures representing policemen, nursemaids. and loung-
ers, were scattered all over the place, after the manner of the

Newport in the Lead

French automobile driving academies, and among them all a devious course was staked out by means of golf flags. While this was going on, the end of Bellevue Avenue leading into Belcourt and the adjoining streets became choked with carriages, horsemen, and bicycles. The first of the prettily decorated automobiles that arrived upon the scene, found the crowd collected in front of the place a more serious obstacle to progress than any of the carefully devised pitfalls awaiting them on the other side of the hedge within the " obstacle park."

The " meet " was directly in front of the entrance to the grounds, and there the automobiles entering the competition stood quietly awaiting their turn, while excited horses and drivers thronged about them, pushing, backing, and turning around one

The Kissing Bug Turnout.

Mrs. O. H. P. Belmont and H. J. Gerard.

another in the crowd, so as to make pedestrians and wheelmen wish they could all be changed into automobiles. When a policeman had at last restored some semblance of order, and a way had been cleared, the competing automobiles, one by one, entered the field and did their " stunts " in what Mrs. Belmont was pleased to call her circus.

First came Mr. O. H. P. Belmont, driving with Mrs. Stuyvesant Fish in a runabout automobile decorated with yellow field-flowers surmounted by an arbor of cat-o'-nine tails bearing a stuffed eagle from whose beak ran blue and yellow streamers festooned to a floral pole, upon which were numerous sea-gulls. Pink hydrangeas with trailing vines and small colored electric lights ran along the arbor supports. The floral pole extended in front of Mr. Belmont's automobile by nearly twice its length, so that, when his turn came to wind in and out among the ob-

A Rough Rider Tamed.
Reginald Ronalds with Mrs. Drexel.

stacles, no amount of skilful steering could save Mr. Belmont from touching two of the dummy figures. Thus he lost his chance of receiving a prize.

Mrs. Belmont, with J. W. Gerard for a partner, rode in her golf rig. They sat in a bower of blue hydrangeas, with electric lights of various colors running along the supports. Clematis was clustered all over the carriage, as well as tastefully arranged bunches of daisies at the foot of the arbor supports. Large bows of wide pink satin ribbons were placed at advantageous places, and the wheels as well as the other parts of the carriage were covered with flowers in appropriate shades. Mrs. Belmont carried a whip made of white and blue hydrangeas, with daisies. On a pole rising above her fluttered two very large butterflies, nearly six feet apart. A horde of unabashed small boys, in Eton jackets, hailed these monstrous insects as " kissing bugs."

Mrs. Astor's automobile was decorated with large bunches of pink hydrangeas with a tree of growing American beauty roses and a floral umbrella of white and pink hydrangeas. Mrs. Astor carried a little dog on her lap which was decorated around the collar with blue hydrangeas. If there had been a booby prize her escort, Mr. H. Lehr, would have won it. By careful steering this gentleman managed to collide with every one of the obstacles, and succeeded at last in utterly demolishing one of the straw figures obstructing his course.

Colonel Jack Astor, driving with Mrs. Ladenburg, drove a small Stanhope carriage tastefully decorated with white and green clematis. He steered his automobile with the same coolheaded dash that distinguished him while serving under fire before Santiago. His example was emulated by Reginald Ronalds, one of Roosevelt's Rough Riders, now riding for new glories in the automobile of Mrs. John R. Drexel, who accompanied him through the contest.

Mrs. John Jacob Astor and Harry Lehr.

13

They were followed by Mrs. Herman Oelrichs and Mr. W. G. Max Muller, of the British Embassy. They sat enthroned in a mass of white and pink hydrangeas, beneath two crossed arches of daisies guarded by twelve white doves perched on broad yellow ribbons. From the bills of the doves on the dashboard to the doves on the arbor ran delicate-tinted satin streamers of white and pink.

Mrs. Oelrichs was the first woman automobilist to enter the contest. The success of her forerunners apparently spurred her on to do even better than they. So she boldly ran her trap into the park at a brisker rate than any. She weathered the nurse-girl obstacle, but in making a sharp turn just beyond she did not slacken her speed sufficiently. There was a snap, and down went the carriage with a broken axle. Mrs. Oelrichs was thrown from her seat, but escaped without any bodily injury. A friend gave her and Mr. Muller seats in his carriage, in which they continued the parade, consoled by the evident good impression made on the judges.

Mr. William K. Vanderbilt, Jr., with Mrs. Vanderbilt, came driving rapidly behind Mrs. Oelrichs in a pretty locomobile. It

The Prize Winners.
Mrs. Oelrichs and Mr. Max Muller.

Newport in the Lead

was decorated with immense bows of corn-colored ribbons interwoven with white mulle, on which were hung a number of canary birds with trailing vines and poppies. He steered his vehicle so cleverly around Mrs. Oelrichs' trap when the latter had come to grief that he should have earned another prize for that feat alone.

Mr. Shoemaker, who drove with Mrs. Burke-Roche, had a beautifully decorated automobile that was a favorite with the

H. R. Taylor with Miss Marie Winthrop.

spectators. It was covered with red poppies and corn-flowers, topped by an immense corn-flower parasol, which went by the board early after the last goal had been passed in the driving contest.

Miss Clapp, driving with the Marquis de San Vinto, had a pretty pink floral arbor of pink and white hydrangeas with small lanterns. The design was to convey the impression of a victoria.

The Automobile Magazine

An elaborately decorated carriage was that which bore Mr. Winthrop Rutherford and Miss Fifi Potter. It was a Columbia electric Stanhope, but not even an expert could have recognized it as such without being told, for, banked across the dashboard from the top of the carriage down to nearly a level with the ground, and likewise over the back of the carriage, were two solid screens of pink tipped hydrangeas, hiding everything from view but the heads of the occupants.

The last automobile, guided into the lists by Mrs. DeForest, proved to be the prettiest. It had a minimum of floral decorations, consisting of a simple wreath of blood-red salvia festooned across the dashboard and back of the light vehicle, the body of which were lacquered a deep automobile-red. This left the design of the light electric vehicle plain to view, and made it appear trim and graceful, in marked contrast to some of the other flower-laden affairs.

The judges having awarded the prizes for best driving and decorations to Mr. Stuyvesant Le Roy and Mrs. Oelrichs, respectively, the start was made for a short country run to Gray Craig, Mr. Belmont's farm in Middletown. Behind the score of fantastic appearing vehicles came an automobile ambulance, with expert mechanicians to repair any possible break-down along the road.

The run to Gray Craig took one hour. By this time darkness was already setting in, and little electric lights began to twinkle in the foliage of the trees and shrubbery. In the centre of the lawn, banked with flowers, was a pond and on its island was an illuminated floral automobile in miniature. Here the horseless company supped and made merry.

The end came when the automobile parade started for home at midnight. Every vehicle was brilliantly illuminated with countless little glow-lights interspersed among the floral wreaths, and each in turn let the rays of its headlight play on the many-colored decorations of its forerunner. Thus the procession of scintillating vehicles sped swiftly over the dark country roads, and at last glided into the sleeping old town of Newport, like a veritable pageant of fairy chariots.

A belated stage-driver, pulling up his horses on the post-road, rubbed his eyes in wonder. To him and to his snorting steeds it was like a whiff of the century to come.

Club Medal.

The Automobile Club of France

By Baudry de Saunier

IT was not in existence in 1895. In 1899 it is one of the first clubs of the whole world! The Automobile Club of France to-day, like those self-propelling vehicles which are worthy of their name, progresses, one may say, at the rate of sixty miles an hour. Its name is now inseparable from all progress in the new locomotion. This club of two thousand members is a considerable personality, counting as it does almost all that our country contains of intelligence, of wealth, and the friends of progress by way of backing. Why not write its history?

When was the club founded? The motor race of Paris-Rouen of 1894, from which dates the real public history of the automobile, shed a great light on many minds. The carriages that had travelled without obstacle from Paris to Normandy, at the modest speed of a dozen kilometres an hour, contained the gem of a great revolution of our customs, an unprecedented improvement of our roads, an expansion of industrial interests, and even of social influence until then unsuspected. A phalanx of scientists, of manufacturers, of rich philanthropists was attracted by the new light, and they began to flock about the strange mechanical carriage. They were allured by the enormous future of the idea. So, when the Count de Dion, that irresistible instigator of the most successful projects of our time,

The Automobile Club of France, Place de la Concorde, Paris.

The Automobile Club of France

proposed, in 1895, a second and more startling demonstration of the merits of the automobile, by a run of 1,200 kilometres without stop, between Paris, Bordeaux, and Paris, within a few days' time the subscriptions for this monster trial reached the sum of 100,000 francs.

Beside the names MM. de Zuylen and de Dion, were inscribed those of Menier, Bennett, Chauchard, Vanderbilt, Chasseloup-Loubat, Peugeot, Edeline, Levassor, and a score of others equally well-known in two hemispheres.

" Paris-Bordeaux-Paris " was run in June, 1895. The success was prodigious, the demonstration irrefutable. An automobile carriage, it was shown by Levassor, could, without stop, travel forty-eight hours continuously, at the average rate of twenty-seven kilometres an hour. Veritable armies of automobilists arose on all sides.

But, since they were not united, much good will was wasted, for the most ardent defenders of this revolution, as a rule, were ignorant of each other, or did not in the least understand how to combine forces for a great concerted movement.

It was the Count de Dion who, first of all, had enough faith and audacity to discover the common aim and end of these aspirations.

" Our run of last June," he often remarked, " demands a definite result. We have seen the enthusiasm which is aroused through all France by the idea of mechanical locomotion! We cannot let it languish after all this. The proprietors of automobile carriages should unite, should meet, for the purpose of discussing their common interests."

Our *confrère*, Paul Meyan, to whom, one day in October, 1895, the Count made this observation, cordially approved the idea.

" Moreover," he answered, " why should you not found, or why should *we* not found, a club of amateur engineers, an automobilists' society. . . ."

" *L'Automobile Club de France!* " interrupted the Count de Dion. " The name is good. It is capital! Let us go to work! "

Two days after, the Count de Dion brought together at luncheon, in the charming dining-room of his house, the Baron de Zuylen and M. Paul Meyan.

It was then, *inter pocula*, that the great club was born. Projects were mentioned, statistics discussed, committees planned of persons who should be called to furnish the first nucleus of founders. They prepared the outfit for the infant, and gathered around the cradle its illustrious god-fathers.

The Automobile Magazine

Le Figaro, always on the track of novelties, above all when they may supply public utility, had already got the news. On the second of October, 1895, it devoted a column to " The Automobil' Club."—(*sic*.)—After having enumerated the persons who had helped the automobile out of obscurity, it gave, in some sort, the programme of the incipient society. The first sentence of the article, as I recall it, indicates clearly the proposed aim of the founders from the very beginning.

" This will be not only a club, properly speaking, as its name seems to indicate, but rather a society for the encouragement of this new sport and industry, the centre of which will be in Paris, with branches and correspondents in the principal provincial towns, similar to our jockey and yacht clubs.

" In Paris we shall have a club-house where the members can gather, and there also shall be centralized all information concerning automobile locomotion, technical commissions, libraries, and so forth. There shall be held the reunions of various committees charged with representing the cause of automobilism before our people as well as to the world, to assume the establishment of special facilities in Paris and out of Paris, so that the members may find desirable accommodations wherever their machines take them.

" Will it not be necessary, furthermore, to facilitate for excursionists the means of sojourning in the large towns? It is in the great centres that the greatest difficulties present themselves for replenishing and storing petroleum or gasolene carriages. On account of the public markets, which are held every week, the hotels of small towns are fitted to receive numerous vehicles; but just ask accommodations at the hotels of Lyons, Marseilles, or Bordeaux! It is as much as they can do to house the omnibus which takes their passengers to the station! The Automobil' Club will guard against these inconveniences. It will have its engineers, its mechanicians, its lawyers. It will also aid with its advice and protection those municipalities which desire to establish a service for travellers, or transportation for merchandise; it will concern itself with collecting and placing before the public eye, through the press, all the progress realized in this new and interesting branch of industry; finally it will organize expositions, prize competitions, annual races, etc."

The *Figaro* it was again that put forth afterward the idea of organizing automobile trips in a tour around Paris, so that Parisians may avail themselves and enjoy the privileges. Later the provinces will be interested in this great movement whose origin and dedication depend on capital.

Club Roof Garden.

21

The Automobile Magazine

Little time elapsed before the feverish activity of the persons who had started the idea, and the powerful aid of the *Figaro*, placed the club on its feet.

On November 12, 1895, in the dwelling of the Count de Dion, Quay d'Orsay, were gathered together MM. Marcel Deprez, Georges Berger, Baron de Zuylen, Count de Dion, Ancoc, Ballif, Clément, Dufayel, Edeline, Gaudry, Pierre Giffard, Artigue, Jeantaud, H. de La Valette, R. Lebaudy, de Lucenski, Marc, P. Meyan, Mors, Pérignon, Pierson, Recopé, Sabatrin, and Varennes.

By acclamation MM. Marcel Deprez and Georges Berger were named presidents of honor, for they had lent the support of their names and authority to the brilliant manifestation of June, 1894. Then M. Meyan read the statutes which he had drawn up with the approval of the assemblage. Finally the Council of Administration was thus constituted: President, Baron de Zuylen; Vice-Presidents, MM. H. Menier and Count de Dion; General Secretary, M. Paul Meyan; Technical Adviser, Count de Valette; Treasurer, M. André Léchideux. A commission was named for the purpose of finding a place to install the club. After having visited the finest quarters of Paris, the committee selected a second floor in house No. 4, Place de l'Opéra. Three parlors, a dining-room, library, and secretary's office were secured, all for the sum of 25,000 francs.

At the end of two years this accommodation became insufficient. A rare chance offered itself immediately; the hotel of the Plessis-Bellièn, Place de la Concorde, was for sale. The adjudication was made in three days. . . . How should they find immediately the necessary million and a half?

M. Robert Lebaudy offered generously to furnish the funds, and he bought the hotel for 1,500,000 francs. A few days later he reconveyed it to the civil society which had been rapidly formed among several members of the club. The annual rent is 105,000 francs.

The Automobile Club of France appointed M. Gustave Rives, one of its most distinguished members, as its architect, and charged him to make a new palace out of an old mansion. With praiseworthy disinterestedness, M. Rives refused, in advance, all remuneration for the enormous task which he had undertaken, and he accomplished wonders worthy as well of the artist who had conceived them as of the dignity of the club which was to enjoy them.

A rapid survey of the interior of the club-house may serve to amuse the reader. The embellishments are not yet entirely

Club Balcony.

finished—for this 700,000 francs will be necessary—though this is told in confidence—and at the entrance there is still room for much architectural display.

The property of the Place de la Concorde comprises four stories. Entering by the *porte-cochère*, we leave our carriage at the rear of the court, in a vast court where a mechanician takes charge of the motor so that on our departure we shall have only to mount and sink into our seats. The ground-floor comprises besides, a coat-room, an ante-room, an office, toilet-rooms, and an inclined platform in case an exposition should be organized in the *entresol.*

Two large stairways lead above to the vestibule, salons, dining-rooms for visitors, shower-baths, toilets, barber-shop, open verandas, and the like.

The elevator takes us to the first floor, where are the reception-halls, foyer, halls for conference, banquets, fêtes, congresses, and general assemblages, buffets, and similar rooms.

The second story is set apart for the committee, containing as it does a general meeting-room, library, foyer, hall, offices, and small rooms.

On the third story we find a billiard hall, a dining-room, ladies' toilet-rooms, steward's offices, kitchen and linen-rooms.

Finally, above, there is the roof-garden, with its fountains and banks of flowers. Here the lucky club members smoke their cigars and sip a quiet glass of absinthe at a height where common mortals only have smoking chimneys! . . .

Such is the simple organization of our Society of Encouragement. It has never deviated from its purpose of enriching France with a brand-new industry, which little by little has come to enlighten the whole world.

AN AUTOMOBILIOUS UTTERANCE.

Before the Dreyfus " affaire,"
France led *in* civilization;
Now all that is left, over there,
Is *her* automobilization.

The Earliest Auto.

The Genesis of the Automobile

By John Grand Carteret

THE idea of a vehicle moved by an unseen power is as old as the hills. Among the Egyptian sculptures, on the banks of the Lower Nile, is a relief representing a royal chariot ascending to heaven, borne upon clouds. In Greek and Roman art, too, there are representations of mystic chariots drawn by invisible coursers. These were efforts of the imagination, like the fancy which pictures sea-shells or rose-leaf-carriages drawn by butterflies, swans, or cupids; but they are nevertheless indications of what must be regarded as a conscious straining for higher ideals of locomotion.

The first actual appearance of self-propelled vehicles was in Persia, in the day of Alexander the Great, when his phalanx had to sustain the onslaught of tremendous cars, armed with projecting spears and scythes, which his Persian foes rolled down the steep sides of the mountain-passes crossed by his army. These are the contrivances mentioned by Roger Bacon in a treatise on the secret forces of nature, entitled: " Epistola Fratris Rogerii Baconis de secretis operibus artis et naturæ et de nullitate magiæ." There he says:

" With the aid of science and art alone it is possible to make wagons roll in a fixed direction without the help of draught animals, as did the battle-cars of the ancients with their formidable wheels, armed with scythes and sickles."

The Automobile Magazine

Similar contrivances were the redoubtable sickle-wagons of the Swiss mountain peasants, which they used in their terrible battles against the feudal knights of Austria, early in the Middle Ages. If we are to believe the chronicles of the fifteenth century, the onslaught of these wagons, thundering down the mountain-side, propelled only by their own weight, proved overwhelming to the Austrian nobles.

In 1649, Johann Hantsch, of Nuremberg, astonished the thrifty burghers of that town by his invention of a carriage which was run by clock-work, and could be made to attain a speed of three leagues an hour. This machine achieved a great reputation throughout Germany, and was finally sold to Prince Charles Augustus of Sweden for three hundred thalers. From Nuremberg, too, came the invention of the wheel-chair for paralytics, not long afterward, which, in a modified form, is still in use among invalids. In 1660 an issue of the *Mercure de France* contains some comments on those carriages, or moving chairs, circulating in the streets of Paris, which are moved by a hidden mechanism, suggesting some trick of the devil, by the marvellous manner of their locomotion.

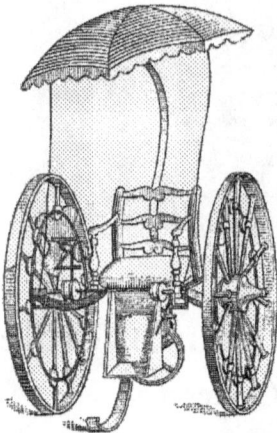

Nuremberg Automatic Wheel-chair.

In 1680 Sir Isaac Newton launched his project of a carriage propelled by steam, consisting practically of a spherical boiler mounted on four wheels. None of these machines achieved a signal success, and a lull of a hundred years followed in the continuous pursuit of the idea of automobilism, until the experiments with steam of Dr. Robinson, 1759, Dr. Erasmus Darwin, 1765, Matthew Boulton, the future associate of Watt, and young Edgeworth brought about a renewal of interest on the subject.

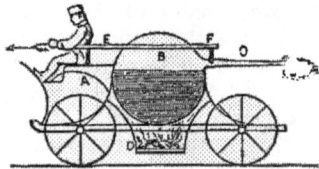

Newton's Steam Carriage.

Through three centuries man had been seeking steam,—three centuries, during which the inspirations of wise men led them to

26

The Genesis of the Automobile

make constant research, without success. And when the problem was clear the first practical step toward an invisible propelling power was assured. While awaiting the possibility of clearing the air, man now is eager to roll at full steam over the earth. Faster and faster, and even faster, does his desire of speed transcend the wonder of steam and cycle locomotion.

Cugnot's Second Steam Wagon, 1770.
Now in the Conservatoire des Arts et Métiers, at Paris.

The first steam-propelled vehicle was constructed in 1765 by a French officer, Nicholas-Joseph Cugnot, of mechanical tastes, encouraged by the patronage of Marechal Saxe. It was made for the purpose of transporting cannon, and, though making four kilometres an hour, frequent stops were necessary. In 1770, the partial success of this tentative having encouraged Cugnot, a new and more powerful machine was submitted by him to new experiments. This time it was a long carriage with three wheels, holding in front a boiler of conical form and a single motorwheel. This locomotive, though theoretically satisfactory, left much to be desired. The violence of its motions was such that it was necessary to cease its use. On one occasion, it is said, the headstrong device actually broke down a wall which stood in its way. From this singular machine, cabriole, steam-engine, flamboyant-car, or whatever fantastic term may be applied to the monster, was evolved our locomotive and the railroad. The carriage of Cugnot leads us to the steam-carriage, as Count de Valette, the President of the Automobile Club of France, prophetically declared a few years ago. Then he said:

" If the Cugnot carriage had been further perfected, its boiler more stanch, its fire of greater heat, and had attained a quarter of the speed which our automobile possesses, we should have had no need of rails, and should not have been obliged to invent them. If we had not launched the mechanical locomotive on rails, on account of industrial interests, we should not have seen

the interminable file of attached cars barring the roads, nor en-
dured the fatigue of promiscuous journeys; nor the monotony
of view in the depths of trenches and tunnels; nor the necessity
of being on the hour, which hinders those who are in haste, and
hurries those who are not; nor the round-about railroad junctions which compel the traveller, who would journey from
Nantes to Marseilles, to pass through Paris, nullifying the principle that the straight line is the shortest distance between two
points.   In fine, we would not have known railroad disaster.
For years we should have gone straight to our ends, wherever

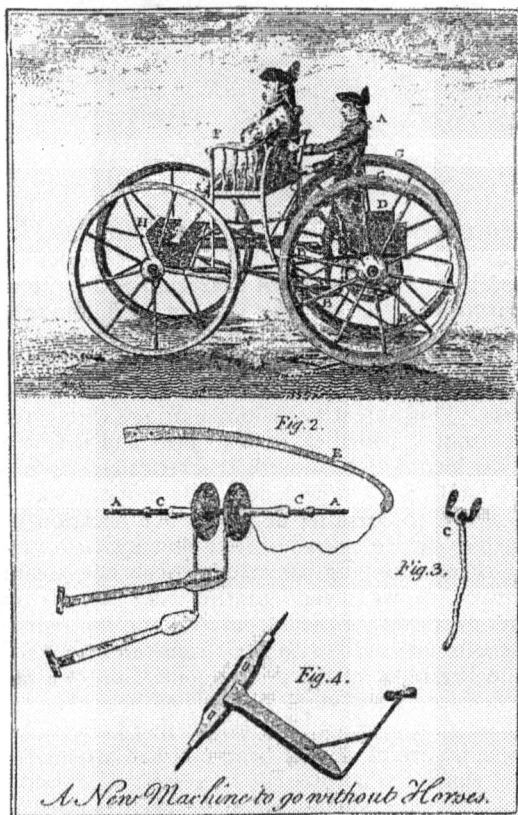

Old English Print, dating from 1780.

The Genesis of the Automobile

business or pleasure might lead. We should have had the automobile carriage, wheeled pavillion, or travelling dwelling-house,
for which we shall be obliged to wait even to the end of the
century."

After Cugnot, in 1780, Dallery constructed a steam-carriage

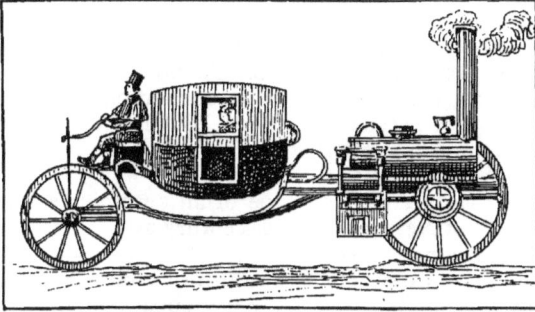

Symington's Steam Carriage, 1786.

with a tubular boiler, which circulated in the streets of Amiens,
and was afterward exhibited at Paris, in 1790, in the work-shop
of the celebrated mechanician, Brézin, where everyone could see
it. Charles Dallery and Mark Seguin were thus, if not the inventors, at least the perfectors of the locomotive, since the inventions made travelling twenty to twenty-five leagues an hour
a possibility. In the patent taken in 1784, by James Watt, the

Guerney's Steam Stage, 1825.

29

locomotive machine figures, and, in the same year, with the help of the great inventors, Murdoch constructed a model which was capable of great speed, giving birth, as a result, to a kind of model, called the " grasshopper machine."

In 1786, Oliver Evans, one of the most ingenious mechanicians that America has ever produced, demanded a patent in Pennsylvania, for the application of steam-power to mills and vehicles. His demand being rejected, toward 1800, he renewed his first idea, commencing the construction of a steam-carriage, propelled by a machine without condensation. Unfortunately he was unable to pursue his project. But he, too, prophesied, a few years later, the attainment of swift transportation between cities like Washington and New York.

Another early American type of an automobile was in use in the streets of New York a half-century ago. It was invented by Robert Dudgeon, who built the " steam road-wagon," as it was called, to carry him to and from business and to convey his family to church. It was a noisy apparatus and consumed two bushels of coal and nearly a hogshead of water on each trip. Mr. Dudgeon rode in it for ten years, when the city authorities forbade his further use of it. The machine still reposes in the inventor's Long Island home. A similar automobile, built by Mr. Dudgeon, was exhibited at the Crystal Palace, London.

In England, early in 1786, William Symington was constructing, with his father's help, a model of a steam-carriage, and in 1800, Nathan Read, who had done so much for the introduction of steam navigation, patented another model which he exhibited publicly, to the end of procuring resources necessary to its manufacture. Richard Trevithick, whose name is connected intimately with the creation of railroads, in 1802 constructed a new model, without result. The following year, assisted by Vivian, at Camborne, he finished the first model carriage that was ever seen in its actual size, and then exhibited it in London. " She made the distance, ninety miles, under the action of her machines," records Thurston in his " History of the Steam-Power Machine," " from Camborne to Plymouth, where she was embarked." A short time afterward, Trevithick, for some unknown reason, demolished the machine, sold separately the carriage and the motor, and began to construct a locomotive for a railroad. The same year that the first English steam-carriage made its appearance in London, Cugnot, the inventor of the first French carriage, was dying in his own country, in a state neighboring on misery.

The first quarter of a century seems to have been a period of

The Genesis of the Automobile

many fantasies of vehicle. A consciousness that some new form of locomotion was to be expected was in the air. The imagination of caricaturists and wags and other persons, with a taste for the startling, were busy, as well as the earnest minds of mechanicians and inventive geniuses. Mechanical contrivances, velocipedes, steam vehicles—(the eccentricities of locomotion)—if we may trust the caricatures, even included men in harness, or machine wagons with life-size toy animals of impossible form in front, or a *charivari* of steam, boat-shaped, and furnished with a huge bellows, doubtless, in the fancy of the artist, to be used for blowing the fire.

New devices for steam-propelled carriages to run on open roads followed thick and fast, but all were so heavy and clumsy in their construction that they practically compelled the use of

Alleged Steam Plow. (A German Print.)

steel rails, and thus came to be withdrawn in favor of equally ponderous railroad engines. Among these early automobiles should be mentioned Gurney's steam stage-coaches, which ran for three years, between Bank and Paddington, after their early trial trips in 1822; William H. James' steam-gas carriage launched at Winsom Green in 1825; Dr. Scarborough's steam phaeton, similar to a contemporaneous invention by Griffiths; William Denton's perambulating locomotive constructed in 1829; and a series of steam-carriages, modelled on Gurney's perfected machine, put forward between 1828 and 1833, by such inventors as Walter Hancock, Sir Charles Dance, Ogle and Summers, Joseph Gibb, Heaton, and Colonel Macerone.

31

The Automobile Magazine

One of the last inventions proposed during that period was a steam-plough, which, if one is to believe the picture of a German artist, may have been operated by a rustic gentleman in silk hose, smoking a long pipe, and taking his ease on a cushioned seat.

Finally the triumphs of railroading put an end to these early attempts at automobilism. It had been a period of storm and stress, during which the adherents of the automobile idea and those who pinned their faith to steam-engines, supported by rails, split into two factions, which roundly abused one another with scientific invective. Had the carriage-builders at that time grasped their opportunity as the bicycle companies of these days are doing now, by producing some form of vehicle light enough to render propulsion along ordinary road-beds an easy problem, the issue would have been otherwise, and railroad construction, as such, might have been indefinitely retarded.

After the triumph of the iron track, roads were no longer considered. It remained for the bicycle to reopen interest in the subject. The general introduction of the automobile must be followed by an unprecedented development of good roads. Locomotion can no longer be concentrated at certain privileged points, and the present luxury of travelling will no longer be limited to certain districts, but come within the grasp of all.

Thus it is not only conceivable, but more than likely, that the coming century will render the streets of all our great cities as devoid of horses as were the bridges and causeways of Venice in the days of old. Omnibuses and horse-cars will be superseded by self-propelled storage-battery cars; public electric-mobiles will drive out the present antiquated forms of horse-drawn hansoms and cabs; and jaunty automobiles, guided by their owners, will take the place of the present fashionable turnouts that throng the mazes of the Bois, Rotten Row. and Greater New York's Speedway. Even our bicycles. as propelled at present by enthusiastic men and women, it is fair to assume, will largely give way to the new forms of motor cycles that have become so popular in France and on our race-courses.

Who can doubt that such a marvellous transformation will mean an immense benefit to civilization? In the wake of such a reform, it must be remembered, would follow more far-reaching changes. For instance. what the late Colonel Waring could not do for New York will be brought about by the automobile. In short, we may hail in the automobile the latest and one of the greatest benefactors of mankind.

A Missing Link Vehicle

WHILE French and American motor manufacturers seem bent on eliminating all horse-drawn vehicles, a German inventor, Joseph Vollmer, has evolved a device destined to bridge the gap between the old era and the new. Already his invention has been applied to the postal delivery-wagons of the German Empire; and it is stated that the entire cost of this new government contract has been more than offset to the postal authorities by the great saving of money formerly spent on post-horses. In addition to this there has been a great saving of time in postal deliveries, the newly transformed wagons covering twice their former distances. Now, the device has been patented in this country, and negotiations are being made by several large carriage manufacturers to convert their entire stock of old-fashioned vehicles into automobiles, by the simple substitution of these detachable motors in place of the former front wheels and drivers' boxes. Thus any old stage-coach or horse-drawn four-wheeler can be converted into a motor vehicle, and one and the same motor can be used for different styles of vehicles, such as heavy trucks, delivery-wagons, or pleasure-carriages.

In Germany this new missing link motor is known as the Kuhlstein-Vollmer Motor Vorspann, a word for which there is no exact equivalent in English. Kuhlstein's motors, as used in Hamburg and Berlin, depend on gasoline for their driving-power, since electric charging stations are still too scarce in Germany to make electric motors readily available, but the American patents taken out fully provide for the employment of electric stor-

Vollmer Vorspann.

The Automobile Magazine

German Postal Wagon.

age-batteries in similarly constructed motors. The construction of these motors comprises the following elements:

1. A movable axle, which has the two-fold advantage and object of diminishing, as much as possible, the sliding motion necessary for the rotation of the axle, and of rendering the transmission of the motive force upon the axle independent of the various speeds of rotation of the wheel, and this without the use of intermediate gear which it has been necessary to employ hitherto.

2. A coupling mechanism, which effects the engagement of the several mechanisms from the driver's seat, notwithstanding their rotation during turning, as the latter operation is effected by the rotation of the forward section of the vehicle and the engine-box as an entirety.

The invention is illustrated in the accompanying drawings:

Fig. 1 is a side-elevation showing the general arrangement of the forward part of the vehicle, motor mechanism, and housing therefor. Fig. 2 is a cross-section through the same, taken upon the front axle.

Referring to the drawings it will be seen that the whole of the driving-gear is arranged in a rectangular box or housing K, Figs. 1 and 2, situated above the centre of the axle-wheel, the top plate P' of said housing being suspended from the pivot plate P' by means of pivot block S as shown. Said top plate P' forms an integral member with internal teeth, Z, as shown. The pivot block S is rotatably mounted in the hub N of the plate P',

34

A Missing Link Vehicle

and is retained by nut m. Upon the bearing plate P′ is rigidly attached the steering-rod L, served at its upper extremity with a hand-wheel, RX. Said rod terminates at its lower end in a crown wheel, whose teeth mesh with the internal teeth Z of the plate P′. Rollers or balls r are also provided on the bearing surfaces of plates PP′ for the purpose of diminishing friction. In order to decrease sliding motion the differences in the rotation of the wheels upon rounding curves are compensated by means of a differential gear, D, which is mounted directly upon the axle a and at one end of the same for economy of space. The differential gear D is situated directly between the two members, a and b, of the driving-axle, and, in order that the latter may form a rigid entirety, the shorter member a is sleeved within the member b as shown in Fig. 2. By this arrangement the axle bearings C are relieved from lateral strain, whilst the provision of an

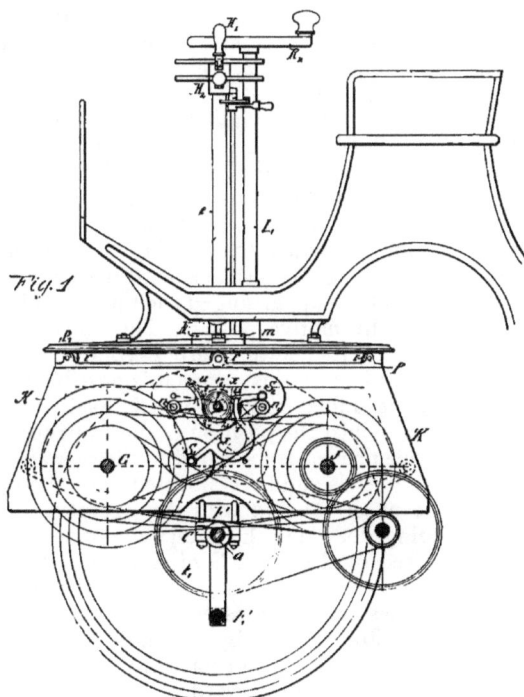

Fig. 1

Side Elevation.

intermediate bearing for the separable members a b, such as has hitherto been employed, is rendered unnecessary.

In order that both the springs F, which, it will be observed, carry the housing K, may support an equal load upon the two bearings, C, motion is imparted by means of the chain or gearwheels k1 k2 in such a manner that both act simultaneously upon the bevel pinions w w of differential gear D, which is effected by means of sleeve c, the length of axle b thereby still further obviating any necessity for the provision of an intermediate bearing.

The axle-bearings C are made in two parts, whereby the members a b can be readily inserted. The upper half of the bearings C, which form a strap or staple, are directly connected with the springs F, by which the entire engine-box and forward section of the vehicle are carried. The lower halves of the bearings are connected by means of a bent shaft, E, which imparts the necessary stability to the bearings, and maintains the distance between them always constant. Lubrication is effected by means of annular lubricators, o, served with the lubricant, through suitable ducts as shown.

The parts required for throwing the motor into gear, consisting of tension-pulleys and the requisite operating levers H' H2, are supported upon the top plate P of the engine-box. Through this plate P, at the point where it is pivoted, passes the hollow pivot block S, which extends into the pivot plate P' at its centre, in such a manner that during the turning of the vehicle it is capable of angular displacement, with respect to the plate P. Within said tube are arranged the upright tubes e e' e2 serving to regulate the motive power. The object of this arrangement is, that when the top plate P rotates, the vertical tubes may rotate with it. By this means it is possible to regulate the movement of the under frame, which is itself turning in a very simple manner, whilst the position of the front section, with respect to the vehicle proper, is indicated to the operator.

The vertical tube e' is rigidly connected with the operating lever H2, and the outer tube e2 with the operating lever H2, both said tubes being sleeved in the outer tube or sleeve e, which latter is rigidly fixed in the pivot S of top plate P, and at its upper portion carries the locking disc s of the operating levers. The lower extremities of the tubes ee' are connected with the bevel pinions r' r2, Fig. 2, thereby enabling the shafts v' v2, journaled in the top plate P, to be rotated by turning one of the operating levers H' H2 from o toward o', or from o toward o2, in a right or left-hand direction. The object of this arrange-

A Missing Link Vehicle

ment is, that a pair of tension-pulleys, S' S2, may be so operated that one of them only (S' for example) stretches the belt imparting motion to the vehicle, while the other tension-pulleys, S2, remain motionless. The operation and arrangement of the tension-pulley gear with a pair of tension-pulleys, S' S2, arranged upon the shaft v, are shown in various positions. The two tension pulleys, S' S2, which are arranged upon levers, pivoted upon opposite points n'n2 upon the top plate, can be disengaged, that is to say, in both cases the toothed pinions z' z2, mounted upon the shaft v2, are out of engagement with the toothed segments, u, of the tension-pulley levers. In addition to this lever the lower teeth i, of the toothed segments u, are raised a certain distance above the points of the teeth of the segments z' z2, because the noses x, upon the tension-pulley levers, rest upon the concentric portion of the cam-shaped hubs,

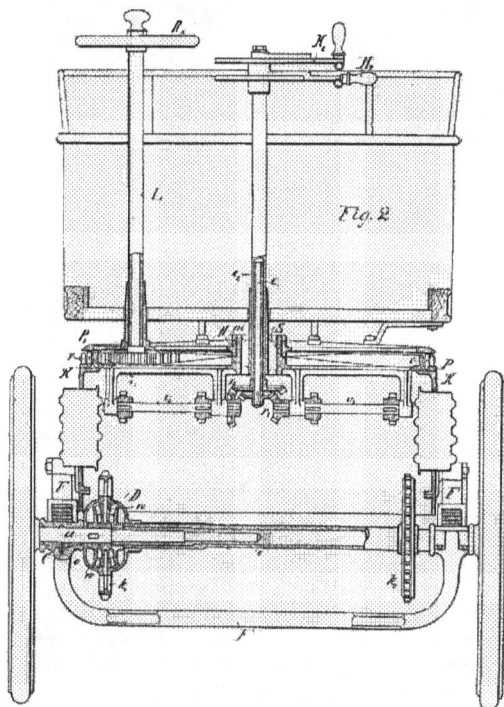

Cross Section.

37

y, of the toothed pinions z′ z2. If the shaft v′ is caused to rotate in a left-hand direction (for example) by means of the hand-lever H′, the nose x of the tension-pulley S2 falls along the reduced cam-shaped portion of the hub y, and the teeth i gear with teeth z2, thereby displacing the tension-pulley S′, which remains stationary, because the nose x continues to slide upon the concentric portion of the hub y and thus experiences no displacement. Only when the operating lever is drawn back does the tension-pulley resume the initial position, and upon continuing to rotate the lever, the tension of pulley S′ becomes operative whilst S2 remains stationary. The pair of tension-pulleys

Vollmer Cab.

of the hand-lever H2 act in a similar manner. Motion is effected by means of stepped pulleys G, and J. (Fig. 1.)

In the benzine-motors the vaporizing of the benzine into gas is automatic through a peculiarly constructed carbureter. The carriage can go a speed of twenty miles an hour. The advantage of its construction is that in the speed of the vehicle, the slowest movement can also be obtained, being regulated as desired. The use of fuel is just as high and the same as that of other construction, and the amount of horse-power per hour is 0.6 litre benzine. There is no reason, however, why any other driving-power should not be substituted, since the essentials of the device remain the same.

An American Exhibition for 1900

At the recent auto-truck and motor-wagon exhibition, of London, two Kuhlstein-Vollmer motors were exhibited, and were submitted to trial trips. One of these, as shown in the accompanying illustration, was an ordinary London cab changed into an automobile by the substitution of a Vollmer motor for its fore-wheels. This cab ran at the rate of fifteen miles an hour, and easily took the steep grades provided for these tests. One filling of benzine lasted for one hundred miles, at an expense of less than one cent per mile.

An American Exhibition for 1900

THE widespread popular interest in everything relating to the automobile, together with the extraordinary development in that direction within the past twelve months, shows that the time is ripe for the holding of a specific automobile show.

The recent Electrical Exhibition gave a remarkable demonstration of the extent to which the general public is interested in the automobile arts. The most attractive part of the show was furnished by the various electric automobiles that were exhibited by five different makers. It being an exclusively electrical exhibition, of course no automobiles propelled by any other motive power could find a place there. But in a distinct Automobile Show there would be brought before the public every kind of mechanical propulsion.

A factor that should appeal with special force to our growing automobile manufacturing interests lies in the opportunity for reaping the rich rewards that will be ready for them at the coming universal exposition in Paris. A most potent instrumentality to that end would be the holding of an Automobile Show. The supremacy of this country in the various engineering industries indicates that that supremacy will be extended over the manufacture of automobiles. For this industry the export trade will be of particular importance.

The organization of an Automobile Show in New York would be one of the best means to help our manufacturers to get ready for the Paris Exposition, and the most opportune time for holding it would seem to be in February, from the beginning to the middle of the month. This would permit the makers to ship their exhibits to Paris in due time, having already established for them a reputation which would not be without weight at Paris.

The New Pegasus

No more the bard, leaf-wreathed and solemn,
In syllables of airy gold,
Sings seated on the spinal column
Of Pegasus the steed of old,—
Because to-day old Pegasus
Has been unhorsed—he's now a 'bus!

No more o'er dizzy heights he'll hurtle,
As he has done since time began;
He'll soon be hymned as choice mock turtle—
A pearl imprisoned in a can.
The bard upon a new horse sings,
That goes on wheels and not on wings.

The bard whose rosy fancy rises
Into the heavens brightly starred,
Who art for art's sake idolizes,
Yet writes his sonnets by the yard,
Sings " Money makes, in weal and woe,
The horseless carriage horseless go."

In youth we learned with studious rapture
About the ancient Siege of Troy,
And how that great historic capture,
Which glads the heart of man and boy,
Was made by men hid head and heels
Within a horse whose hoofs were wheels.

Then let the horseless carriage gather
The laurel wreath on any course!
Or call it simply, if you'd rather—
'Tis quite correct—the horseless horse,
That glides along without a break,
And, though he eats not, takes the cake.

So, clear the track and let this peerless
Equine outfly the whirling wind,
On noiseless wings, in spirit fearless,
He'll leave all other shapes behind!
For Science we must say, perforce.
Has put the cart before the horse.

R. K. Munkittrick.

How the Horse Runs Amuck

By Sylvester Baxter

" Hosses don't know but dreadful little, really. Talk about hoss sense—wa'al the' aint no such thing." —DAVID HARUM.

THE most typical " horse " man in American fiction was characteristically right. A lover of children would not permit his affection to go so far as to give them their way in the world unrestrained. So those who have a genuine affection for the horse should have no disposition to restrain progress, to the end that that animal may continue to ride rough-shod over civilization. That is to say, it would be most irrational to hold back, or injuriously to restrict, the present great advance in mechanical locomotion that is about to complete the cycle begun by the railway and steam-engine. For thereby would be perpetuated the noise, filth, and unsanitary conditions that lead to many of the greatest evils and discomforts of city life, as well as the great wastefulness of money and energy caused by the construction of improper roads all over the country and the incessant destruction of the comparatively few good roads that we possess, simply to retain the horse in his old place in the world.

It is peculiarly appropriate that some cold facts about the horse be laid before the public at the present time. These will substantiate the assertion that the horse is an animal of extraordinary little sense—using the word as synonymous with judgment. He has a remarkably delicate perception, coupled with a very slight power of correlation. He is therefore subject to seiz-

The Automobile Magazine

ure at any moment with fits of the most violent insanity, induced at the slightest provocation. This, together with the enormous reserve strength of the animal, makes him an exceedingly dangerous engine to be practically given the freedom of the road in our populous communities. Only familiarity makes the peril seem endurable.

It is quite natural that the horse should have a nature so unbalanced mentally; evolved, as he is, from an ancestor who was one of the most timid of wild animals, possessing no weapons of offence or defence, and therefore finding his only safety in flight. He had ever to be on the alert, with his keen senses of perception ever tense; ready to urge him into a mad gallop at the slightest movement, or rustling of a leaf, which, perhaps, might betray the neighborhood of some lurking and terrible beast of prey about to spring upon him and tear his life out with lacerating claws or teeth. It is no wonder, therefore, that at any unaccustomed sight, noise, touch, or motion the horse of to-day, in spite of countless centuries of training in the service of man, under the ancestral impulse that dominates his most intensely nervous organization, should still be seized with an ungovernable terror that expresses itself in a mad onward rush whose frightful power is fraught with destruction for everything about him.

The horse, moreover, with all his intelligence in certain ways, and with many virtues that win him the love of those about him —for commonly he is docile, affectionate, patient, and kindly— has an exceptionally small brain capacity in comparison with his size, or even with the size of his head. He is therefore a huge and powerful animate machine, most difficult to control, under any circumstance out of the ordinary, and consequently as dangerous as any powerful mechanism would be if actuated by some form of motive power that easily became ungovernable.

One of the greatest manufacturing concerns in the country, being interested in motor-vehicle development, recently began to look into the facts concerning the horse so far as they related to his part in the civilized life of to-day. In connection therewith the accumulation of accounts of runaway accidents was begun, and newspaper clippings obtained from all parts of the country were arranged in scrap-book form. The record thus obtained is necessarily incomplete. Although the clippings represent accidents occurring in all parts of the United States and Canada, a very large number must go unrecorded, occurring as they do in parts where they do not get into the newspapers. If all accidents occurring had been reported, the list might perhaps be more than tripled.

42

How the Horse Runs Amuck

As it is, however, a record of about six weeks comprises 476 runaway accidents of various kinds. These involve, probably, at least 600 horses, together with five mules. The mules were all frightened by bicycles! A classification of the first 269 of these accidents shows that 34 of them were due to the breaking of harness. The breaking of the carriage caused 32. The dropping of the reins caused six. Fifty were occasioned by the sight of something that seemed unwonted. Forty-nine were caused by some sort of noise. The dropping of something upon the animal, or a sudden contact with some object, caused 14. Fright from some object that combined both sight and noise, like a railway train or electric car, was the occasion of 62. Of extraordinary causes—not including various preceding instances of some remarkable nature—together with collisions and other unclassified causes, there were 24. It is worthy of note that but two of the accidents were caused by the sight of an automobile.

The true inwardness of these two accidents may be judged from the following letter, which J. B. Hoecker, one of the members of the Brooklyn Automobile Club, sent to a friendly editor:

"I think owners of horseless carriages need protection from 'brainless' drivers and 'ownerless' horses. I made a short run yesterday to New Jersey with my gasolene-carriage, going by the Annex boat. As I was going up Mercer Street, in Jersey City, a horse, attached to a carpenter's wagon, took fright, but was stopped after running about half a block. Whoever had charge of this wagon was out of sight.

"Returning, we met a horse, attached to a wagon of some soap manufacturer, leisurely travelling down Mercer Street all by himself, there being no one in the wagon to guide the animal. Two blocks above the City Hall he started on a run. A police officer, who was on hand and could easily have checked his progress, made no effort to do so. After running two blocks the horse fell in front of the City Hall. I sincerely hope that the animal was not injured and that owners of horses will learn that it is not safe to leave them on the street unattended."

Many of the other causes for accidents were of a remarkable and curious nature. When it is stated that a horse, long accustomed to the sight of yellow trolley-cars and paying no attention to them, shied and ran away at his first sight of a red one, it may be inferred that no occurrence is too trivial to startle such an animal to flight.

A glance at some of the headings is not without interest: " Near to Death in a Runaway!" "Two Horses Dash through Big Windows!" "Is Horribly Maimed!" "Knocked Down a

The Automobile Magazine

Lady!" "Meets a Sudden Death!" "A Vicious Runaway!"
"Killed by Brewer's Runaway Horse!" "Dragged by a Run-
away Horse!" "Shied at a Roller!" "Horses Plunged on a
'Cycler!" "Runaways Scare another Team!" "Piece of Paper
Caused Death!" "Two Wheelmen Hurt!" "Frightened by
a Street-car!" "Smashed into a Car!" "Two Women Killed!"
"Horse's Wild Dash!" It all makes a pretty exciting succes-
sion of captions. Let us look at some of these runaways:

In Boston, one day last June, two horses went through two
plate-glass windows, within a few hours of each other. The first
was "Old Jim," a big bay horse, whose name was synonymous
with equine sober-sidedness. He was jogging down Tremont
Street, early in the morning, attached to a herdic cab. Something startled him and he ran until he struck a corner post that
tore off a wheel and sent the driver to the ground painfully hurt,
while his own head went through a florist's window, with resulting severe cuts. In the afternoon a horse, attached to a buggy,
took fright at the sparks from an electric car in Scollay Square.
He rushed onto the sidewalk, cleaning out a long and elaborately
laden fruitstand, scattering oranges, peaches, and apples to the
delight of bootblacks and newsboys, and pushed his head and
shoulders through the plate-glass window of a tea-store.

In Humboldt, Neb., a trotting-horse exercising on the driv-
ing-track, frightened at the explosion of one of the rubber tires
on the vehicle, kicked his owner from the seat and broke the
latter's knee.

In Burlington, Ia., a coachman was driving a pair of three-
year-olds down hill. They had been hitched to a carriage for the
first time. The neck-yoke broke, the horses ran, and the driver
was hurled over a bridge to fall fifteen or twenty feet. His left
leg was broken in three places, his left arm in two, and his right
leg at the ankle. In a preliminary dragging his face was fearfully
lacerated, half his moustache was scraped off, and an ear nearly
torn away.

In Waterloo, Ia., a lady was driving with her two daughters.
In turning, the buggy tipped, frightening the ladies. Their
screaming frightened the horse into running. Mrs. W—— tried
to jump out and was caught in the wheel. The carriage upset,
the daughters were thrown out and slightly injured, and the
mother was dragged a quarter of a mile. She was horribly mangled and lived but a few minutes.

In Baltimore a team of horses, frightened by pieces of paper
caught up by the wind, ran until they collided with a horse and
buggy, badly damaging the latter.

How the Horse Runs Amuck

At Oxford, Pa., an umbrella, carried by a little girl, frightened a horse into a runaway and an upset.

In West Springfield, Mass., a horse with a carriage was frightened by the snorting of a locomotive. He ran for five miles and rushed down a long hill like a rocket. A young man on a bicycle was going down the hill at the same time. He said that the horse chased him so that he could not dodge, so he had to put on full speed. At last the horse turned a corner and the wheelman got away; the former, after sundry adventures, wrenched himself from the carriage and finally brought up in the river.

In Woonsocket, R. I., a horse, standing with an express-wagon, was frightened into insane flight by the sight of a cow.

A horse in Gardner, Mass., left standing with a wagon, was slowly walking off when somebody called out to him to stop. That started him into running away, and the wagon was smashed.

At Williamstown, Mass., a party of eight hotel guests was ascending Greylock Mountain when a storm came up. The horses, frightened by hail, dashed down the steep mountain-side. The carriage was smashed to bits. All but two ladies managed to get out; the latter were upset with the carriage, but not seriously hurt.

In Chicago a lady and gentleman were out driving. The lady opened her parasol. This frightened the horse. A runaway, upset, and serious hurts followed.

In Little Rock, Ark., a prominent citizen was driving with his wife. The horse took fright at some passing bicycles, the lady was thrown out and killed by the crushing of her skull against an electric-light post.

In Findlay, O., a doctor's horse slipped and fell. A gentleman held its head for the doctor to unharness. The horse suddenly became uncontrollable, wrenched himself loose, and ran.

In Philadelphia a young man was driving a cart. A piece of paper blew beneath the horse's feet. The animal ran and the young man was thrown out and killed.

Near Milwaukee a young man and girl were driving in a buggy. The horse took fright at a speeding team, plunged across the road into a group of bicycle riders, seriously injuring two of them, and upset the carriage.

Near Centralia, Pa., an old man was driving down a steep hill. The harness broke. The horses ran down the mountain for two miles, finally upsetting the wagon and throwing the old man out to die with fractured skull.

At Titusville, Fla., a horse with a wagon, standing near the railway station, was taking his meal. Another horse came up quietly and gave him a kick. That frightened him into running, and the wagon was smashed.

At Maumee, O., the sting of a wasp made a horse run away, throwing out a man with his two little daughters.

At Paducah, Ky., a pair of mules, left standing with a wagon, were frightened by a bicycle. They ran, and five persons were seriously injured.

Near Easton, Pa., a man and his wife with their small son were out driving. The horse took fright at the sight of some colts in a field and ran. The man and woman were badly hurt.

At Cambridgeport, Vt., a horse ran away from fright caused by getting choked with grain while quietly eating from his feed-bag.

Here is a bit of runaway humor from Chicago, as set down in the Chronicle of that city; " Horse Levi was a guest at the Transit House yesterday noon, and the only thing he registered was several kicks. He sent these in all directions and came near working great havoc. Horse Levi was yesterday the property of Morris Levi, who took the animal to the stock-yards to sell him. In front of a Dexter Street commission house, Archie Anderson entered the road-cart attached to the horse, to drive to the market. The animal did not like this at all. He sprang suddenly forward, sending Anderson heels over head to the side-walk. Then he headed directly for the Transit House. The cart was quickly banged to pieces, and the horse was not encumbered with it when he dashed up the steps like a man on the last second of the dinner-hour, and entered rapidly without waiting for an invitation. A dozen guests were sitting in the lobby. As they heard the first steps they thought it was a Dakota stockman coming into the lobby. When they saw the form of a horse looming up they scattered in all directions. The beast surveyed the interior and then stepped around and aimed a kick at the clerk's shirt-stud. He missed, and several of the men having regained their courage tried to seize him by the head. He wheeled sharply and cast his heel about as free as a boy distributing handbills. The guests retreated. Presently he saw another horse at the other end of the lobby. He headed for it at lively gait. He was apparently surprised to see the other do the same thing. To avoid a collision he came to a sudden halt and Michael Hunter, a stockman, caught the excited animal just in time to avoid the smashing of a $500 mirror."

At Columbia, Pa., a runaway did great damage at an outdoor gospel meeting. The horse, with a covered wagon, was stand-

How the Horse Runs Amuck

ing at a hitching-post while the people were attending the meeting, held in a grove close by. The horse had invariably been gentle and frightened at nothing. Suddenly, apparently maddened at the sting of some insect, he ducked and violently raised his head, stripping off the bridle. Though headed up the street he wheeled and made straight for the assemblage. All attempts to turn him in another direction by a frantic waving of arms and hats failed. He made a bee-line for the meeting-place, and in an instant was leaping over the benches and people, dragging the heavy wagon after him and knocking persons right and left. Nobody was killed, but eleven persons were injured, some badly.

At the close of a wedding in Natick, Mass., a hack was waiting to receive the bridal pair. The bride was about to enter and the merry guests were coming after, laughing and showering rice. Some of the rice hit the horses and they ran away; the driver had a narrow escape, but nobody was hurt.

At Connersville, Ind., the turning on of the electric-lights frightened a horse, causing a smash-up with the injury of a man and the animal.

At Schenectady a gentleman and lady were out driving in the evening; behind them were two gentlemen in a carriage going in the same direction. Two drunken men came tearing along past them, likewise in the same direction, yelling and lashing their horse. One of the fellows lashed at the horse of the two gentlemen. The horse sprang forward and dashed into the carriage in front, wrecking it, but only slightly damaging the occupants. The carriage behind was little harmed, but its two occupants were seriously hurt.

At Cleveland the breaking of a rein caused a pair of horses, attached to a heavy truck, to run away. They dashed into the rear of a street-car, killing a boy and injuring three men.

At Woodstock, Mass., a horse took fright at boys playing ball in the road, and threw a man from his wagon, breaking his nose.

At Charleston, Ill., the sudden raising and lowering of a window-curtain frightened a passing horse, and a young man and young woman were thrown out of their buggy.

The noisy celebration of the Fourth of July was, as usual, responsible for numerous runaways, some of them fatal in their consequences. Probably the worst was that at Cynthiana, Ky., where the explosion of a giant cracker frightened a horse and caused the death of three of the four occupants of the carriage— a man, his wife, and their boy. Their infant child escaped unharmed.

47

The Automobile Magazine

Varied as they are, the comparatively few instances, here enumerated, by no means exhaust the list of distinctive causes of runaway accidents. Such things as a lump of ice left in the street, a baby-carriage, and the flapping of an awning were sufficient to startle horses into a runaway fright. Indeed, there appears to be no kind of a noise, sight, or movement of the most everyday occurrence that, at any moment, might not occasion a fit of equine insanity with disastrous potentialities. It is by no means an extravagant assertion to say that, had a steam-boiler been as liable to explode from a fraction of the number of causes that set in action the tremendous force for harm exerted by a runaway horse, the marvellous transformations of the industrial world and of civilized society that have characterized the nineteenth century would never have been known.

In the 476 runaways, described in the collection under consideration, 36 persons were killed, most of them outright, and 290 others were injured. Among the injured were not a few whose hurts were undoubtedly fatal. On the other hand, the many persons who were merely bruised or ordinarily hurt by cuts, scratches, etc., are not included in the category of the injured. It may be observed, however, that it not infrequently happens that the nervous shock from such accidents, though taken little account of at the moment, not infrequently so affects a person, apparently unhurt, that their consequences remain through life.

It is not to be wondered that already, for some time past, the nervous tension and the various dangers caused by the multiplication of bicycles and of electric-cars upon the highways, particularly in and about the cities, have been causing thousands to abandon the use of the horse for pleasure-purposes. This has been a very considerable cause contributing to the falling off in what may be called the equine population of the country, in addition to the loss from the substitution of mechanical for animal traction on street-railways, and the various ways in which the bicycle has taken the place of the horse. All this had been happening before the setting in of the automobile movement. And what reason will there be for keeping a horse in any part of the country where there are good roads when a good motor-vehicle can be had for the price of an ordinary horse and carriage with equipment, while the items of depreciation and of running expenses on the part of the automobile are trivial in comparison?

The Turgan-Foy Petroleum-Motor

M OST of the motors hitherto built have been motors with horizontal shafts. As a result of this construction, each stroke of the piston could be felt in the springs; and the carriage was subjected to vibrations far from pleasant to the occupants.

If, on the other hand, the motor-shaft were vertically arranged, if it were driven by two motors oppositely disposed and horizontally mounted, the cranks being set at 180°, the vibration, already considerably reduced by the arrangement of the motors, would act in a horizontal plane on the entire frame of the carriage. Such a construction forms one of the characteristic features of the petroleum-motor carriage of Messrs. Turgan & Foy.

Turgan-Foy's Petroleum-Motor Carriage

The frame of the carriage is made of steel tubing, and is mounted on springs fastened to two tubular axles carrying bicycle-wheels provided with pneumatic tires. The front wheels are swiveled in the usual manner; and the rear wheels act as the driving-wheels.

49

The motor with its carbureter is situated in the front part of the frame and comprises two cylinders, cooled by spiral flanges and mounted horizontally end to end, on a casing which serves as a support.

The piston-rods, by means of a finger, rotate two spur-wheels, engaging a third intermediate wheel of double the diameter of the other two; the two smaller spur-wheels thus act in conjunction; and the power of each of the cylinders is applied in unison.

The intermediate gear-wheel is keyed to a vertical shaft carrying the fly-wheel and the starting mechanism. The gear-wheel has but half the velocity of its engaging spur-wheels, owing to its greater size, and, through the medium of the shaft on which it is mounted, controls the exhaust-valves and the electric gas-exploding devices.

At each side of the cylinders the admission valves and the exhaust-valves for the escape of the exploded gases are arranged. The valves are therefore completely independent of each other; and there is no danger of their becoming overheated. The two vertical shafts of the motor pass through the centres of the two spur-wheels, which, as already mentioned, are directly driven by the piston-rods. The two motor-shafts, at their lower ends, are provided each with a horizontal pulley. The carbureter is secured to the rear face of the casing. The entire mechanism occupies but little space, and is enclosed in a light box in the forward portion of the carriage.

On the rear axle a casing inclosing the speed-changing gear and the differential gear is secured. On the frame of the latter two gear-wheels of different diameters are mounted. By means of a simple mechanism, operated from the front of the carriage, these gear-wheels can be alternately thrown into engagement with two small pinions mounted on a shaft projecting from both sides of the casing, and carrying two sets of two pulleys each; the diameters of the pulleys composing each set are different. The one pulley is loose, and the other keyed to the shaft.

Two belts transmit the movement of the working-pulleys of the motor-shafts to the rear pulleys. Let us assume that the gear-wheels on the frame of the differential gear are thrown out of operation, and the belts are running on either the loose or the fast pulleys. The motor under these circumstances being in operation, the carriage will be at rest.

Let us now assume that one of the gear-wheels on the frame of the differential gear is thrown into operation. and that one of the belts is left on the corresponding loose pulley, while the other

The Turgan-Foy Petroleum-Motor

belt drives the shaft. The carriage will then be started at a certain speed. A second speed can be obtained by throwing the pulley, previously idle, into operation, since the diameters of the two pulleys are unequal. The same position of the parts will give two other speeds if the second gear-wheel be employed. These four different speeds can be obtained by operating two levers, one of which controls the intermediate rear shaft, each of its two possible positions corresponding with two speeds. The other lever controls two forks, between which the belts pass. When the lever acts upon these forks, the belts on the loose pulleys are shifted on the fast pulleys. This shifting is very easily effected, since the two pulleys are both of the same diameter.

The second lever can also be moved to two different positions, and when used in conjunction with the first lever enables the automobilist to obtain any of the four different speeds desired.

A small handle within reach of the operator controls the electric gas-exploding devices, regulates the interval between the electric sparks, and hence the speed of the motor. The induction-coils and accumulators are located so as to be readily accessible.

The safety-appliances of the carriage comprise two pedal-brakes and an interrupter. One of the brakes acts on two drums, secured to each of the rear driving-wheels; the other is connected with the frame of the differential gear. The commutator enables the automobilist to break the circuit of the electric current, and thereby to bring the motor to a quick stop.

The steering mechanism consists essentially of a worm engaged by a pinion, operated by means of a small hand-wheel. The worm is connected with a lever, which, through the medium of links, acts upon the two rear driving-wheels. This steering-device has the merit of not being reversible, and therefore renders the steering of the carriage both trustworthy and easy.

ON THE LAST TACK.

First Horse.—Used up all those tacks?

Second Horse.—Neigh, neigh; I have some left.

First Horse.—Scatter a few close to the curb; here comes an automobile.

Liquefied Air for Automobiles

By J. Ravel

NEVER has any industry been submitted to so many revolutions as that of automobilism. During the five years that it has existed, every change of any kind has been qualified as a revolution. For many years we had gas-motors of which the cylinders were provided with ribs or flanges to prevent overheating; but such appendages, when applied to stationary motors, presented to us an aspect of conservative quietude. Placed upon the motors of automobiles, however, these same cooling devices were the signal of revolutions. Upon vertical stationary motors they were, at first, arranged longitudinally; then they were placed transversely, and that was the first revolution. Some, who were more audacious, placed them obliquely; and, finally, some anarchists came to the front and colored them black or white. There was here a Homeric contest; and, as has been said by a dithyrambic writer, the future of automobilism depends upon cooling-flanges.

Alas! My profession of consulting engineer has just put me upon the track of a conspiracy, in other respects more formidable; but, to begin with, there are no flanges in the matter.

In a letter that I have before my eyes, a banker proclaims to a capitalist that he no longer believes in flanges, and that the latter have carried him too far. The following is an extract from this letter: "A highly scientific man has just invented a liquefied gas-motor, which I am going to push. There is a great revolution organizing."

Nevertheless, I desire to reassure patent agencies, which will not remain idle on this account, since there will spring up a new crop of inventors, who will take upon themselves the mission of doing battle, not with heat, but rather with cold.

Therefore, make way for liquefied gas-motors!

In the first place, I suppose that it is not a question of hydrogen gas, since such an application would not even be new, and would not involve the recasting of the present motors, which would be called gas instead of gasoline-motors; but the distinctions without a difference would be innumerable. I believe, rather, that it is a question of *liquefied air*.

Let us, then, examine the properties of this new state of air.

For some years past, considerable attention has been paid,

Liquefied Air for Automobiles

especially in America, to the liquefaction of gases—a process of obtaining oxygen, hydrogen, etc., in a thousand fold smaller volume.

If, therefore, in the case of atmospheric air, we compress a cubic metre (35 cubic feet), at a pressure of a thousand atmospheres, such cubic metre will be reduced to the volume of one litre (by cubic inches). The air, however, will no longer be in a gaseous, but rather in a liquid state. In practice, instead of simply compressing the air, which would raise it to too high a temperature and prevent liquefaction from taking place; recourse is had to the following method of procedure: When the pressure reaches about two hundred atmospheres, a portion of the air is made to expand in a receptacle containing the coil in which the compressed air circulates. The partial expansion of this air lowers the temperature to 212° below zero. Then, under these two actions of compression and cooling, the air that passes into the coil becomes liquefied, and is finally collected in its new state.

This liquid air is of a light-blue color, and may be preserved for several days in glass cylinders with double sides, between which a vacuum has been formed; an absolute vacuum being a perfect calorific insulator. When exposed to the ordinary temperature it resumes its gaseous state.

A manufactory for the industrial production of liquid air has been established in New York City. The compressing machines of this utilize a total of about 160 horse-power, and the production is 4½ litres (1⅛ gallons) per minute. It is said that the net cost of the liquid is about fifteen cents per gallon.

Numerous industrial applications have already been made of liquid air, and it is employed also as a source of respirable air and as a powerful refrigerant.

If we introduce a little liquefied air into the cylinder of any kind of a motor, the piston of the latter will immediately be set in motion through the impulsion of the liquid, which, in resuming its gaseous state, exerts a pressure upon the lower surface. It will be seen from this that upon constructing a motor having a proper distribution, we shall obtain the same mechanical results as with steam or compressed air. In reality, the motor for this purpose will be sensibly the same as that employed by Mekarsky, for compressed air street-cars, with the difference, however, that instead of having large reservoirs containing air under a pressure of thirty atmospheres, we should have small receptacles containing liquid air that would be vaporized according to the requirements of the motor.

53

The Automobile Magazine

The consequence, then, would be a considerable diminution in the bulk and weight of the reservoirs, a reduction in the size of the motor, suppression of reducers and multipliers of speed, an instantaneous starting of the motor, and a running as easy as that of a locomotive; and all that without smoke, without fire, without electricity.

Alas! all this, at present, is again but a mirror for catching larks—that is to say, for catching capitalists. Let us look a little at the other side of this mirror.

The considerable amount of cooling that takes place in expansion air engines, and the consequent easy production of icicles, which interfere with the running of such motors, has prevented their prevalent application in the industries.

On street-cars run by compressed air it has been necessary, for heating the air, to employ boilers—indispensable adjuncts of some weight that occupy considerable space upon the platform.

In order to obtain a proper useful effect from liquefied air applied to automobilism, we should be obliged to have recourse to a high degree of expansion; but, in this case, we should have to contend with a greater degree of cold than in compressed air engines, since the air of the latter has a pressure of but thirty atmospheres, while liquid air is obtained at a pressure of two hundred atmospheres and a reduction of temperature of 212° below zero. Under such circumstances, then, in order to cause the liquid to resume its gaseous state, a gasoline-heated boiler of large dimensions would be required.

Having no very positive data as to the physical and mechanical laws that govern liquefied air, it would be idle to make calculations that would show with but little certainty the quantity of liquid air that would have to be stored up in order to run an automobile.

Nevertheless, I believe that I am not far from the truth in granting that it would require the vaporization of from 9 to 10 kilogrammes (20 to 22 pounds) of liquid air per horse-power. Allowing 9 kilogrammes, it would therefore require, for a four horse-power, 36 kilogrammes (79 pounds) of liquid per hour, say a supply of 216 kilogrammes (475 pounds) for a run of fifty-four hours. The cost of such a trip I pass over in silence. The present drivers are all millionaires, or nearly so.

This liquid air, then, shares with electricity the drawbacks of dead weight and the necessity of recharging. The same conditions exist also with regard to liquefied carbonic acid. Of this, it would require about 17 kilogrammes (37 pounds) per horse-

A Tight Finish

power, plus the gasoline to be burned in order to prevent icicles from impeding the action of the piston.

"*Qui ambulat in tenebris nescit quo vadit*," which may be translated thus: "The driver who travels about in darkness does not know whither he is going." Therefore, it is necessary to enlighten drivers, and that is why I must add the following postscript: In my brief statement of the industrial applications of liquefied air, I forgot *one* detail, which has its importance. *It may be, and already is employed as an explosive!* In fact, under the influence of a violent shock, it detonates forcibly, and instantaneously resumes its gaseous state, and its initial pressure is then about a thousand atmospheres. Powdered charcoal mixed with it and cotton soaked in it are used for making cartridges, which detonate under the action of a capsule of fulminate.

These last few lines make me suppose that many drivers will hesitate before laying in a supply of liquid air in order to make the "*Tour de France.*"

A Tight Finish

He longed a motor ride to take,
    To rumble all about—
Alas he went but twenty miles,
    The gasolene went out!

His pocket flask was with him, though,
    To aid him in his plight;
But whiskey couldn't make it go—
    It got the wheels so tight!

W. R. ROSE.

The Amphibious Automobile

By de Glatigny

IN this *fin de siècle* period, when all minds are turned toward automobilism, we have seen the cycle, thanks to its happy development, become what the horse used to be, the indispensable companion of every active young man or woman. We have seen the automobile reach such an important degree of improvement that all records relating to locomotive-speed have been equalled or broken by fearless *chauffeurs* like Girardot, the Count of Chasseloup Laubat, Genatzy, and many others. We have seen the submarine automobile, under the forms of the Goubet, Gymnote, and Gustave Zedé, recently make its trials and prove to the Tyrants of the Sea that their big iron-clads and unconquerable cruisers are already threatened by an enemy that cannot be seen or reached, foreboding that all the fleets of Great Britain may any day share the fate of the " invincible " Spanish Armada of Philip II. We have seen aeronauts like Santos-Dumont dashing victoriously to the conquest of the winds.

With these achievements we had thought that automobilism had covered all possible grounds, inasmuch as on land it had broken the records of the *Gazelle* and the swiftest racers, while under the sea it was becoming the competitor of the dumb inhabitants of the waves, and finally contesting with the birds the aerial domains. But its limits had not been reached; beside those creatures that fight for the possession of *terra firma*, the liquid element and atmospheric realms, nature has endowed others with special qualities—often monstrous beasts, like the hippopotami, that attacked the boats of heroic Marchand, or like those huge alligators which still swarm in the hot lands where civilization has not yet penetrated. These creatures, *Amphibia*, which can exist either in water or on dry land, had not yet been imitated by human beings; or, at most, the latter had made but a few timid attempts to imitate them, and had always failed.

The gap is now closed, and this time it is not from America that light has come to us, but from the North. A Swedish engineer, Mr. Magrelem, has lately built a tramway-boat that is really an amphibious means of locomotion, for it moves along quite as easily on the water as on a railway track. In fact, it might be said that automobilism, by enabling us to go up or

The Amphibious Automobile

down the steepest inclines and to turn the sharpest curves with the greatest ease, had already rendered useless the enormous engineering works, the embankments and tunnels that have cost so much money during the latter half of this century. But Mr. Magrelem's happy invention is going to render equally unnecessary the construction of bridges.

The tramway-boat was devised to ply across two lakes north of Copenhagen, divided by a neck of land, about three hundred meters wide. It had been nearly decided to dig a canal across the isthmus, in order to connect the two sheets of water by a continuous channel, when Mr. Magrelem offered his plan.

The amphibious steamboat is 15 meters long, by 4 meters beam; she weighs 11 tons when empty, and 15 tons with her maximum load. The engine is 25 horse-power, and its action is transmitted by means of a triple gear, either to the propeller or to the rimmed wheels. She can carry 70 passengers.

In passing from the liquid medium to the solid road, the boat enters a small canal having a slightly inclined bottom and two grooves, where the front wheels are engaged and then meet the rails with which they keep in contact by the action of some pins or spikes. As soon as contact is secured, the whole power of the engine is brought to bear on the front wheels, which easily pull the tramway-boat up the incline to the dry land and make her travel on the rails like an ordinary car. On reaching the other side of the isthmus she launches herself by the same process, inverted, but the brakes are then strongly applied to the wheels to prevent a too rapid descent. The trip is then continued on the water.

The changes of action from the propeller to the wheels, or from the latter to the former, are made instantly and without stopping the boat, whose speed is kept unchanged while moving on the water or on the track. Since this amphibious automobile was launched it has been doing service continuously, in all kinds of weather, and has carried over 20,000 passengers without any trouble or accidents whatever.

UP TO DATE

" That girl ran away and married her father's coachman."

" Oh, no; they have an automobile, and he was their electrician." —*Chicago Record.*

The Lepape Carbureter

THE Le Pape carbureter comprises a cylindrical body provided with a chamber, o, which can be closed both at top and bottom by valves a and b, the stems of which are respectively surrounded by coiled springs, X and z. The cylindrical body at its top is closed by a cap, g, through which passes an adjusting-screw engaging the stem of the valve a. In the lower portion of the cylinder a lamp is mounted, which is surrounded by wire gauze, P, through which the heated air passes. The liquid to be vaporized enters at m, beneath the valve b. A quantity of fresh air can be added, when desired, by turning the collar d. The explosive mixture finds its exit through the tube T.

In inoperative position the lower valve is slightly forced from its seat by the upper valve, the two valve-stems telescoping within each other. The movement of the stems is limited by stops, e. The valves being in this position the liquid will fill the chamber c. When the intake of the motor is open the resitance of the wire gauze will cause the cap g to be depressed, and, likewise, its adjusting-screw V. The upper valve-stem will be hence plunged into the chamber filled with liquid.

By this operation the lower valve will be closed, thus cutting off the communication between the supply reservoir and the chamber c. As it continues to fall, the cap will force the valve-stem f into the liquid contained in the chamber c, and will cause it to displace a volume of liquid equal to that of the immersed portion. The volume immersed and consequent displacement can be regulated to meet with the requirements of the motor, by means of the adjusting screw V of the cap g. By turning the collar d, a supply of fresh air can be admitted to diminish the vacuum produced by the intake, and consequently to regulate the quantity of liquid which falls on

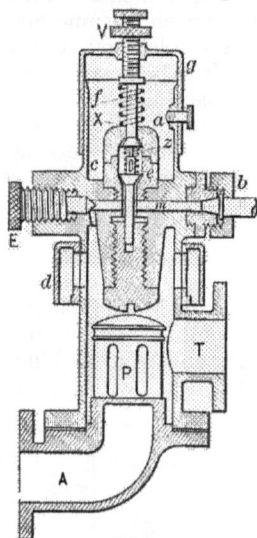

Lepape Carbureter

the wire gauze P, since this volume depends upon the degree of immersion of the stem. The liquid which falls upon the wire gauze is vaporized by the hot air and passes to the cylinder of the motor, mixed with air.

It therefore follows that the admission of a supply of cold air regulates the quantity of liquid which should pass to each cylinder, and the proportions of air and gas in the explosive mixture introduced within the cylinder of the motor. There is neither odor nor smoke; and the mixture is perfectly carbureted.

The screw E serves to force out the air from the liquid supply-tube, and to permit a small quantity of liquid to flow, in order to facilitate the starting of the motor.

An Improved Serpollet Engine

THE latest invention of M. Léon Serpollet, the inventor of the steam-carriage which bears his name, is an engine utilizing steam superheated to a temperature of 752° to 932° F.

In the accompanying diagrams, which represent the engine in partial side-elevation, end-elevation, and in plan, C and C′ are two cylinders, arranged end to end on a casing, B, constituting a supporting-frame in which the crank v revolves. To the crank, two connecting-rods, d, are coupled and pivoted to the pistons D. The connecting-rods are held on the crank by two sleeves, c and c′. The shaft A projects from both sides of the casing, and at one end carries the fly-wheel V.

Steam is admitted by two valves, S and S′, symmetrically arranged, and connected with the steam-pipes s and s′ and with the cylinders C and C′. The exhausted steam escapes through the pipes O and O′, when, by the action of the piston, they are opened, the cylinder being left full of steam at atmospheric pressure.

The steam is not actually exhausted; for the piston D compresses it on the return stroke; and if the pressure to which it is subjected is greater than that in the boiler it will return through the pipe s and valve S. The admission of steam on the forward stroke of the piston is not affected by this operation. Moreover, the compression prevents condensation in the cylinder and does away with the injurious effects of the clearance. The two valves

The Automobile Magazine

S and S' are operated by an auxiliary shaft, a, above the casing B, and parallel with the main shaft A. The motion of the shaft A is communicated to the shaft a by gearing, E, the speed of the two shafts being the same. The auxiliary shaft is longitudinally slidable on a long sleeve, I, provided with cams, act-

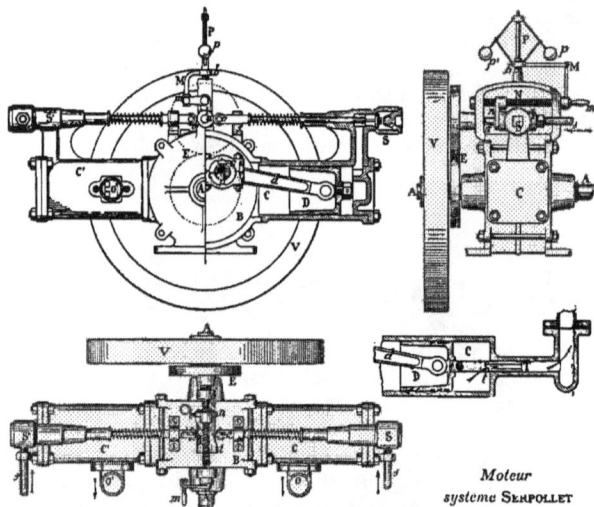

Moteur
système SERPOLLET

uated by an arm, u, which is screwed on and operated by a screw-shaft, N. The screw-shaft N can be turned by means of a crank-handle, m, automatically operated by the centrifugal governor P of the engine. When the balls pp' recede from or approach each other, the collar b respectively rises and falls, and thereby turns the crank m and the screw-shaft N, through the medium of an arm, M, and varies the position of the cam i, on the sleeve I. As a result, the friction-rollers oo', journalled in the ends of the stems of the valves SS', will be more or less advanced, so that the valves will be opened at the proper intervals to produce a variable cut-off.

The cam also serves to reverse the engine. In this case the position of the cam with regard to the rollers is changed by turning the crank-handle m manually.

Frederick Gougy, C.E.

The "Cyclone Motor"

THE cyclone motor is a two-cylinder petroleum engine cooled by flanges, and making 800 revolutions at the rate of 400 explosions in the cylinder. It is, therefore, not so easily heated as the one-cylinder motors of the same form making 1,200 revolutions and more per minute. Of the two sizes in which the motor is made, the one develops 135 kilogram-meters (975.105 foot pounds) of energy, and weighs 32 kilograms (2.2 pounds to the kilogram) with the fly-wheel; and the other develops 200 kilogram-meters (1,446.60 foot pounds) of energy at 800 revolutions, and weighs but 50 kilograms approximately.

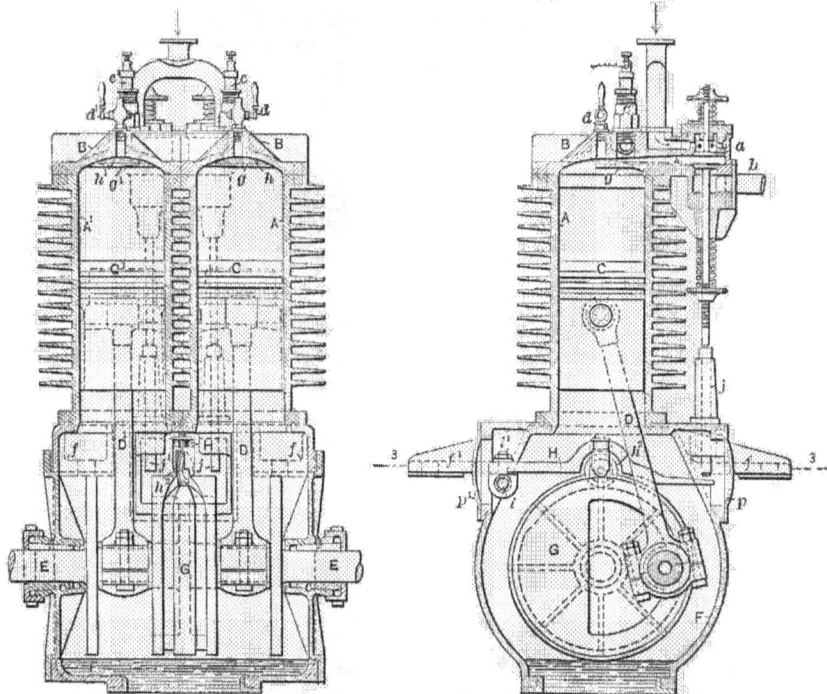

FIG. 1   Section of "Cyclone Motor"   FIG. 2

In Figure 1, A and A' are two vertical cylinders arranged side by side and provided with heads B and B', each having an automatic admission valve a and an escape valve d, controlled by

61

The Automobile Magazine

means which will be later described. The igniters are represented by c, c', and the cocks controlling the compression by d d'.
The pistons are of the single-acting type, and are connected by
rods with the shaft E. The lower portion of the operating mechanism is inclosed in a casing F, on which the cylinders are secured. The motor is fastened in position by the shoes ff.

At the upper portion of each cylinder is a partition g, which
serves to deflect the fresh gases admitted through the valve a
and prevents mingling with residual gases near the igniters.

One of the novel features of this motor is the method employed for controlling the escape valves b b'. A cam G is keyed
on the crank of the shaft E (between the two piston rods D D')
and turns therewith. The periphery of this cam is formed with
two grooves intersecting at x (Figure 3).

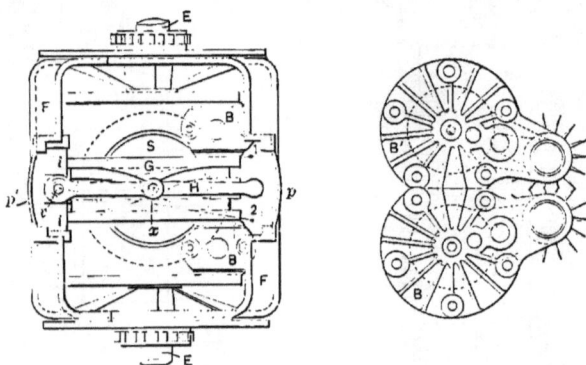

FIG. 3   Exhaust Valve Governor   FIG. 4

At the upper part of this cam is a roller h engaging the groove
and journalled in a lever H (Figure 2), fulcrumed at i i.

The lever has both vertical and horizontal movement, the
latter being due to the peculiar 8-shape of the groove. When
the lever is shifted horizontally, its flattened free end alternately
moves to the extreme points indicated by the numerals 1 and 2
(Figure 3). In these two positions the lever is given a vertical
movement by the cam by which the stems j j' of the escape valves
b b are alternately elevated. The reseating of the valves b b is effected by a spring, the tension of which can be regulated by
means of a collar screwing upon the stem. The two pistons, in
response to the movements of the single crank of the shaft E
will always act together. As the cycles alternate, the motor
shaft will be driven at a uniform rate of speed.

The Grosse-Boubault Motor-Tricycle

THE Grosse-Boubault tricycle, exhibited at the Automobile Exposition in the Tuileries, was one of the most interesting things to be seen there.

The tricycle is remarkable chiefly for its motor, which, although its weight and size are small, develops two and one-half horse-power. The exhaust-muffle, which, viewed from the exterior, is of considerable size, contains a chamber for the detention of the gases, which are allowed to escape through the top. This arrangement, in addition to the spiral flange, running from the base to the top of the cylinder, completely prevents overheating.

The expansion-chamber is integrally formed with the compression-chamber, the two being separated by a partition, which provides a seat for the exhaust-valve. Oxidation of the valve-stem is prevented by a metal tube, which operates in conjunction with the valve so as to prevent a binding of the stem and to insure perfect operation. A forced circulation of air about the helicoidal flange of the cylinder is obtained by the employment of a casing, so arranged as to utilize the current of air generated when the vehicle is in motion.

The efficiency of the motor's operation is still further increased by the employment of a carbureter, also patented by Messrs. Grosse and Boubault. The carbureter automatically feeds the motive agent in such a manner as to save 50 per cent. thereof, and to insure a perfectly uniform operation of the motor, irrespective of the character of the road over which the vehicle may be running.

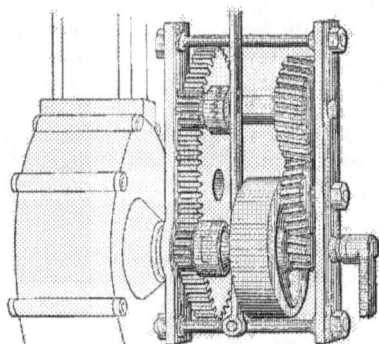

Speed-changing Gear

The speed-changing gear used can be applied to all forms of tricycles. The mechanism employed comprises an automatic friction-clutch, by means of which the tricycle can be gently started, and a crank-operated driving-gear. A patented ratchet-connection has been provided, whereby the feet can be maintained at the same height when the tricycle is driven by the motor alone.

The speed of these tricycles, as manufactured by the Grosse-Boubault Company, has been demonstrated on several cycling race-courses in France, where they have been used as pacers.

During the recent interesting handicap races in France of pedestrians, runners, horseback riders, wheelmen, motor-cyclists, and an automobile, one of these motor-tricycles showed its speed by passing all its competitors excepting only the specially designed racing automobile, which won the race by a narrow margin.

As a roadster, too, this new motor-tricycle is bound to win popular favor, since its construction, as described above, shows it to be peculiarly fitted for continued long runs where hard strains are apt to be encountered.

AN AUTOCAR PARADOX

When a motor-man tackles a very steep hill
(He can't always run on the level),
And back starts his car with the threat of a spill—
Why, then his good angel's the " devil " !

Dust-proof Speed-changing Gear

WHEN a layman for the first time looks at the driving-mechanism of a motor-carriage, he exclaims: " That is not a carriage; that is a travelling machine-shop." The prospective purchaser of an automobile is dismayed chiefly by the apparent complexity of the power-transmitting mechanism, a complexity often further increased by the labyrinth of pipes which connect the motor with the carbureter and petroleum-tank, which circulate the water for cooling the cylinder, and which conduct oil to the various parts to be lubricated. There is also a maze of connecting-rods, crank-shafts, and levers, by means of which the driver controls the various operating parts, the air and petroleum admission valves, the gas-exploding devices, and the brakes. The unmechanical purchaser immediately concludes that such a carriage is not practical.

This appearance of complexity has been avoided to a large extent by the new gear-casing which Messrs. Montauban & Marchandier recently exhibited at the Tuileries, in Paris. Such a casing completely conceals the mechanism, with its accompanying shafts and gearing, the whole arrangement being compact and not readily put out of order.

All the gears, shafts, and their controlling devices are inclosed in an envelope, or casing, containing oil. Their wear is hence reduced to a minimum; and their mechanical efficiency increased to a maximum. They require but little care, since they are perfectly protected from dust and foreign bodies. Cleaning and oiling, two operations particularly distateful to the automobilist, are rendered unnecessary.

The usual levers controlling the speed-changing gear have been done away with. A single operating-wheel, to the right of the automobilist, and a bicycle-chain, take the place of the levers. With this simple contrivance the carriage can be started and its speed changed to four different rates; or it can be stopped, or caused to run backwardly at the lowest speed.

The Montauban-Marchandier speed-changing gear is inclosed in an aluminum casing, the two sections of which open like a shell. These sections are hinged together and rendered perfectly tight by a packing of leather.

Three shafts are mounted in this casing. The first, A, is merely an extension of the motor-shaft, with which it can be thrown into engagement by means of a friction-clutch, in order to start the carriage gently.

The second shaft, B, parallel to the axle of the driving-wheels, is provided with two bevel-gears, *a* and *b*, engaging a third on the motor-shaft. These three gears, when the friction-clutch is thrown into engagement with the motor-shaft, are rotated. Between the two bevel-gears a clutch is mounted having teeth, adapted to interlock with teeth on the gears. The carriage can be run forward or backward by throwing the clutch into engagement with the proper gear.

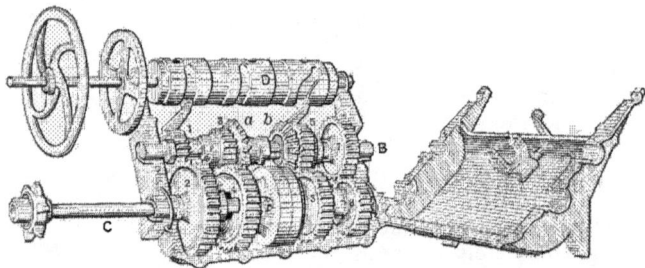

Dust-proof Speed-changing Gear

On both sides of the bevel-gears *a* and *b*, and on the same shaft, are mounted the gear-wheels 1, 3, 5, 7, meshing respectively with the gear-wheels 2, 4, 6, 8, and giving four different changes of speed.

Between each pair of gear-wheels 2-4 and 6-8, on the lowermost shaft C, is another clutch, which when thrown to the right or to the left engages one of the pairs, 2-4 and 6-8.

In order to have but one set of gear-wheels in operation at a time, the lowermost shaft is made in two sections, projecting from both sides of the casing and carrying at each end a small sprocket or gear-wheel, which transmits the movement to the rear driving-wheels. In order to throw each pair of the gear-wheels, producing the different changes of speed, successively into operation, each clutch is thrown by means of a fork, carried by a rock-shaft and operated by a crank-arm. The crank-arms are engaged by cam-grooves on the drum D. By rotating the drum, the pairs of gear-wheels can be successively thrown into or out of mesh with the corresponding pair. Less than one revolution suffices to pass from a complete rest to full speed, or to stop and reverse. The drum is operated from the carriage-seat by a hand-wheel, two small sprockets, and a bicycle-chain.

The advantages of employing this incased gear are too obvious to be dilated upon.

Calculating Horse-power and Speed

SOME time ago two Parisian engineers, Messrs. Boramé and Julien, prepared a table and diagram in which were given the force exerted tangentially to the driving wheels of automobiles, the factors considered being the load (including the weight of the vehicle), grade, and speed. This table and diagram were designed to assist automobile builders in reckoning the dimensions of the various parts, by showing what forces these parts had to resist.

The first edition of this table and diagram having been quickly exhausted, the authors have issued a second edition, in which the work performed in horse-power during a run is recorded, the factors considered being again the load (including the weight of the carriage), the grade, and the speed.

The exigencies of space will hardly permit us to reproduce the numerical portion of the table; and we shall therefore content ourselves with giving that part of the diagram in which the work performed in horse-powers for different loads on a level road is indicated.

The numerical and graphic results of the authors are based upon the general formula:

$$\text{Nch} = \frac{eP\,(0.025 + 0.007\ v + p) + sv^2 \times 0.0048}{75}.$$

in which Nch = the expression of the work in horse-powers, e = the distance, in metres, covered by the carriage in one second, $P(0.025 + 0.007v + p) + sv^2 \times 0.0048$ = the force exerted tangentially to the driving wheel, in which expression P is the total weight in kilograms (2.2 pounds to the kilogram) of the carriage and the load; 0.025 the coefficient of the resistance encountered on an up-grade for 0.80 m. wheels provided with rubber tires; 0.007v the term proportioned to the speed and relative to the resistance due to jars occasioned by inequalities in the road; p, the grade in fractions of a metre, per metre; s. the square metres of surface acted upon by the resistance of the air; and $sv^2 \times 0.0048$, the resistance of the air.

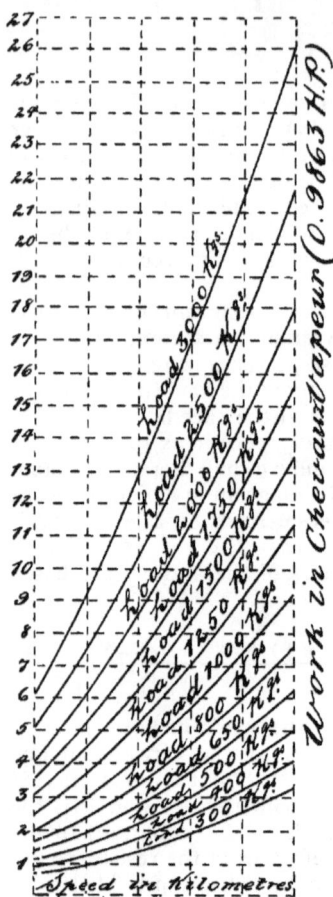

The diagram reproduced herewith indicates at a glance the power Nch of a motor necessary to drive on a level road a motor-carriage which has a total weight P, and a speed v. By a simple algebraic operation the approximate speed v can be obtained at which the motor having a power Nch can climb a grade p. The auxiliary formula employed is the following:

$$v' = \frac{Nch \times 75 \times v}{Nch \times 75 + \dfrac{Pp \times v}{3600}}$$

in which v' and v designate in metres the distance covered per hour.

Example: Let us assume that a carriage weighing 1,000 kilos (including the load) is to be driven at a speed of 30 kilometres per hour on a level road. The diagram shows that in order to attain this speed a motor of about 5.6 H. P. will be required.

If it be desired to ascertain at what speed v' this carriage with its 5.6 H. P. motor could climb a grade of 0.07 metre to the metre, the formula given above is applied:

$$v' = \frac{5.6 \times 75 \times 30000}{5.6 \times 75 + \dfrac{1000 \times 0.07 \times 30000}{3600}} = 12562 \text{ metres.}$$

It is evident that this diagram and the auxiliary formula which accompanies it are of no little value in computing automobile speeds and horse-powers.

Spring Suspension for Motor Vehicles

ALL moving bodies are controlled by the law of gravitation. Then we must overcome that law. " Ah," I hear you say, " it cannot be done! " Well, as Galileo told the Inquisition, " It does move," so I say the law of gravitation can be, and is being, overcome every day. " How?" you say. " Very simply "—but we will come to this later.

The baby, when he first attempts to walk, finds his equilibrium very difficult. But when he does walk, he is as safe as his father; for the same law keeps them both centred to the earth. Were they to fall, they would both drop down; were they to jump up, they would both arrive where they jumped from. Nature has provided man with legs, and it only requires a little experience for him to discover that they will do more than provide locomotion. Should he jump from a safe distance, he finds upon landing, that his knees will absorb the upward thrust his body must meet, when his feet strike the earth. This is one method of overcoming gravitation, and it is a feat every man has accomplished at some time in his life. It will give the reader the key to the principle that I shall unfold as we get further into the subject.

The man weighing one hundred and eighty pounds, and the baby weighing thirty pounds, must, when they jump, be prepared to have their knees absorb the blow, and they do land more easily for this very knee action. Were they to have kept their knees stiff, the thrust or shock would have been severe. You can have a better idea of this by springing up from the floor, say twelve inches, and landing with your legs stiff. There will be a decided rattle, especially in the jaws, as they are the bones having the most delicate adjustment.

The question of the weight of the man and baby does not enter into the argument, provided they absorb the thrust with the knee action. But if each were to keep his legs stiff, he would find the jar proportionate to his weight; therefore we find the weight the springs of the vehicle carry has nothing to do with

The Automobile Magazine

the absorption of the thrusts, providing the springs are proportioned to the load.

In reading a paper, recently written by A. Von Borries, upon "Spring Action," we were greatly surprised that he should consider the load carried as the great factor, whereas the load has nothing to do whatever with the riding. The axles and wheels are to carry the load.

The springs are to absorb the upward and downward thrusts of the wheels as they rise and fall. Due to the uneven plane they have to roll over, these thrusts vary in degree, being governed by the inequalities and the momentum.

In looking over the drawings of the earliest steam-carriages, we find the half-leaf spring was used then as now. Owing, no doubt, to a lack of understanding of its nature, and its limited possibilities, many combinations of springs and equalizers have been produced. For proof of this, one can study the spring suspensions of the most modern motor-vehicles that have been built during the past year. The writer has failed to find any suspensions of different builders showing the same methods of spring equalization for the same type of carriage.

There seems to be considerable uncertainty among motor-carriage builders as to what the elliptical and half-leaf spring is capable of doing in the way of absorbing thrusts. As for the spiral, it does not seem to them to be worthy of any more consideration than a block of iron or wood—in fact, if rubber was cheap, they would maintain it was far away a better thrust absorber than a spiral. To fully appreciate this only requires a trip abroad among some of them, and when you mention springs you will hear all kinds of heresy.

There are really two kinds only—leaf and spiral. The nature of the full elliptic spring is nearly identical with that of the spiral.

If you load it beyond its sensitive point, it will, when forced by a jolt or jar, close up and rebound, or open out considerably more than the normal position when either without load or fully loaded.

Broken leaves in this type of spring attest the truth of this statement, for to pull an elliptic spring apart a given distance will invariably wreck some of the leaves. Another feature of this type of spring is the tendency to keep the load—if the load happens to be a little below the sensitive point of the spring—jouncing up and down, a motion frequently observed in passenger-cars where the track is rough, and in buggies on a rough pavement. To say the elliptic spring is suitable for heavily loaded bodies, moving at high speed over an uneven plane,

Spring Suspension for Motor Vehicles

would be to subject myself to criticism, and I will therefore dismiss the elliptic as entirely unsuitable for the service.

The half-leaf spring presents a basis for the ideal method of supporting all moving bodies that are carried on wheels over an uneven plane. There is no rebound to this type of spring, you may underload it far beyond the level of the top of the band, and it will, if built of good material, come back to its original set; this allows for a wide variation in the load.

In the electric-car we have the widest range of constantly varying load: here the car leaves the end of the line empty, and before going a block 25 persons embark, another block and 25 more, another block and 10 more, and so on until the car is carrying, in the seats and on the running-board, 135 passengers, and at an average of 150 pounds each, this would mean over 10 tons, and this load will be dropped in a distance of a mile or two, and probably leave one passenger on the car. With the idea on springs of the average motor-carriage designer under this car, this poor, lone passenger would think he was on the "Rocky Road to Dublin," so stiff and hard would the car ride, and some of them do; but, as a rule, they do not, because the street railways cater to the great army of cranks, and the president who would allow many rough-riding cars on his line would have his life made miserable by the letters he would receive.

Therefore we have the two extremes of loading, the motor-carriage, the ideal; the electric-car, the opposite of this ideal. But why is it there are so many rough-riding motor-carriages? Because the matter of the springs and spring suspension receives little or no attention; in most cases the springs are made by outside spring-makers under contract, at so much per pound; the contractor, being the designer of the spring, invariably makes a spring heavy enough to carry three or four times the load, because he has to renew all the broken ones, and as he knows they break, he takes no chances, erring always on the side of safety, which in his case means money. Therefore, he adds extra leaves entirely regardless of the duties the springs are to accomplish.

A RURAL RELAPSE.

" John is so absent-minded."
" What's the matter now? "
" He bought a load of hay for our automobile."

—*Chicago Record.*

A Compressed-Air Fire Engine

THE time is comparatively near at hand when every city will have its public supply of air power, and when such a time arrives the utilization of compressed air will enter into almost every nook and corner of industrial, municipal, and domestic enterprise. Its usefulness, according to *Compressed Air*, is so comprehensive that a central compressed-air power plant might in these days look for a large and prosperous patronage.

One of the most forcible arguments for introducing a public supply of compressed air will be the degree of economy and the vastly increased protection which would be afforded by compressed air in the matter of fires. It is entirely feasible to propel a fire-engine to a fire without horses, and when it reaches the water supply to operate it by compressed air.

Protection against fires is one of the most important municipal questions, and the expense to the public of handling a modern fire department requires skilful financial treatment to keep it within bounds and at the same time have an efficient department.

A public supply of compressed air would certainly reduce the expense of a fire department to within a reasonable figure, and the proposition to introduce automobile compressed-air fire engines will in time bring forth potent reasons for the adoption of this system. The installation of a compressed-air supply would not have to be borne by the fire department, because private enterprise will be glad to install central compressed-air power plants, relying on the patronage of the general public in the matter of power; but when the advantage of such a system is well known any proposition to lay pipes in the streets for this purpose will receive the moral support of every fire board and a majority of the citizens, all of whom are interested in fire protection.

The plan to be followed is similar to that followed by the installation of gas and water plants. Mains would be laid in all streets, and air hydrants would be close to the water hydrants.

An automobile carriage with a suitable fire-pump would constitute a fire-engine, to be propelled by compressed air, electricity, gasoline, or other motive power. With such an apparatus, the time used in reaching a fire would be reduced to a

A Compressed-Air Fire Engine

minimum, and the operation of applying the air power for pumping the water would take but a few seconds.

The advantage of this system, which does away with cumbersome boilers and dangerous progress through crowded streets with wildly excited horses, will be apparent to everyone. If necessary a speed of thirty miles an hour may be obtained in going to a fire and without any undue commotion. The weight of the vehicle itself would be very much reduced. An ordinary engine to be drawn by horses weighs about 8,500 pounds, and an equally efficient automobile engine will not weigh more than 5,000 pounds, or possibly less.

As an illustration of possible economy in a city, we will give the statistics of the fire department of Newark, N. J.

The manual force of this department is 202 men, with the following apparatus: Fourteen steam fire-engines, 4 hook-and-ladder trucks, 1 aerial truck, 1 chemical engine, 4 engineers' wagons, 2 wagons for superintendent of fire-alarm telegraph. In reserve are 2 steam fire-engines, 1 hook-and-ladder truck, 1 chemical engine, and 1 engineer's wagon; and the cost of operating the department for one year is $251,000. Eighty-two horses are kept in this department, aggregating the value of upwards of $30,000, and the mortality among horses in fire departments is necessarily high; consequently there is a constant expense from this cause.

Automobile fire-engines should not cost more than ordinary fire-engines. The number of men employed for taking care of horses would be dispensed with. No coal or other fuel need be used.

The item of keeping the above number of horses would amount to $19,680 a year, counting the expense of keeping one horse at $20 per month. The total saving in a department of the size of Newark, N. J., would probably reach $30,000 a year, or about 7½ per cent. of the total cost of running the department.

In the foregoing estimate no account is taken of the cost to the department for the use of the compressed air for operating engines which would be used during fires, and the reason for omitting it is because any company putting in a compressed-air plant would readily concede the free use of power for such purposes in part compensation for the franchise privileges. But even if it had to be paid for at the regular rates the cost would not amount to much.

The development of such a scheme is within practical limits, and it only needs discussion and intelligent demonstration.

Daimler's Gasoline Motor

W E illustrate herewith a gasoline motor of the Daimler & Daimler-Phenix type, manufactured by MM. Panhard & Levassor, of Paris.

This motor, which has been long and favorably known, is distinguished from all others of the kind that have been constructed up to the present by the following important features: (1) The use of a purer and more powerful gaseous mixture; and (2) a greater rotary velocity. Consequently, with the same weight, its power is much greater than that of any other engine of the same class, or, to put it in other words, with the same power, the weight is very much less. This peculiarity of the Daimler motor is very important, since, on account of it, it may be employed as a stationary engine in places where space is very limited, and also be used for actuating all kinds of vehicles.

The following are some of the more important details of construction of this motor: The cylinder piston is provided with a valve for controlling the admission of the explosive mixtures. In the upper end of the cylinder, above the piston, is located the explosion chamber, with which are connected the inlet and exhaust valves. Every other stroke of the piston is a working one.

The gasoline designed to furnish the explosive mixture is contained in a carburetter placed at the side of the motor, and which may be seen to the left of the cylinder in the accompanying illustration.

The velocity of the motor is controlled by a very sensitive governor which acts in such a way as to cut off the supply

Panhard & Levassor Petroleum Motor Carriage

74

Daimler's Gasoline Motor

of gasoline that may be excessive, and regulates the number of the explosions according to the power to be obtained.

The ignition is effected, pneumatically, through a rapid compression of the gaseous mixture in the interior of a capsule raised to a red heat by an external flame. Therefore, there is no need of a valve nor of electricity.

The cooling of the chamber in which the explosion takes place, is effected by a circulation of water in the jacket in which it is encased. The water for this purpose may be obtained either from a conduit or from a reservoir of small capacity. In the latter case the liquid may be used over and over again. The lubrication of the principal parts of the engine, which are enclosed in the frame, is performed simply by lubricators placed upon the cylinders. The saving in this respect is very appreciable, since not a drop of oil, so to speak, is lost.

The motor is started by acting upon a small winch placed at one of the extremities of the fly-wheel shaft. After the first explosion has taken place, the winch frees itself and the engine begins running. The Daimler motor, which is made with from one to four cylinders, from one-half to twenty horse power, is well adapted for use in all kinds of industries, and is capable of actuating all sorts of machines, such as pumps, printing presses, machines for working wood and metals, hoists, weighing apparatus, etc. Its high speed renders it serviceable for driving dynamos, while its extreme lightness and compactness well adapt it for the running of automobiles, small street cars, etc., and for the propulsion of boats.

To sum up, the motor under consideration recommends itself by the following advantages: (1) extreme lightness and great simplicity of its parts; (2) a compactness that permits of its being installed almost anywhere; (3) a high rotary speed that allows of reducing the extent of the intermediate transmissions, and in some cases of suppressing them; (4) ease and cheapness of installation; (5) a feeble consumption of gasoline and lubricating oil; (6) absence of smoke and odor; and (7) low price.

It may be added that the Daimler motor may be run with gas also. In this case, the carburetter is done away with and replaced by a special regulating cock.

MM. Panhard & Levassor are likewise manufacturing a full line of motor carriages of styles and sizes capable of accommodating from two to fifteen persons. These vehicles are actuated, according to size, by a two or four-cylinder Daimler motor, which is usually placed in front, and is easily accessible from all sides. A reservoir, which likewise is placed in front, contains a sufficient supply of gasoline for a run of forty-eight miles; but, by carrying a supplementary reservoir 150 miles or more may be made. The steering is done through the intermedium of either a lever or a hand-wheel, and, by reason of the peculiar combination employed, gives rise to no fatigue. The variations of speed, from high to low, are obtained, as usual, by acting upon a lever. The low speed is employed upon heavy grades and bad roads. On a road that is dry and in good order, grades of one inch or more to the foot may be ascended with ease. These carriages are provided with powerful brakes that allow them to be stopped almost instantaneously. Fuller particulars in regard to these vehicles may be found in an illustrated catalogue, which the firm will be pleased to mail, postage paid, upon the receipt of two francs fifty cents.

Daimler-Phenix Petroleum Motor

The Jeantaud Steering-Gear

THE steering-gear is one of the most vital portions of a motor carriage, and not a day passes but some improvement is made which renders it safer in operation. The most efficient of the many gears is undoubtedly the irreversible, in which the mechanism can be operated from one end only. The simplest gear of this type is the endless worm; but its simplicity is offset by certain disadvantages, among which may be mentioned the number of revolutions of the steering-wheel required to turn the carriage wheels.

A steering-gear which is free from these defects has been invented and patented by M. Jeantaud, and is illustrated herewith. Figure 1 is a vertical section through the axis of the gear; and Figure 2 is a plan taken on the section A, B, C, D of Figure 1.

On the steering-shaft a the wheel b is loosely mounted. Secured to the wheel b is an arm c, provided with dogs d and d'. To the shaft a an arm e e is keyed, to which are rigidly fastened two plates f f' by two adjustable connecting bolts g g'. The arm

The Jeantaud Steering-Gear

e e is also provided with dogs *m m'*. The steering-shaft *a* is revolvubly mounted in a socket on which is secured a casing *i* inclosing all the parts. Surrounding this casing are two spring-brake bands *k* and *k'* covered with leather.

The spring-band *k* is formed with two notches, one adapted to engage the dog *d*, the other the dog *m*. On the spring-band *k* there is also a cam surface corresponding in angle with an inclined portion *n n* of the plate *f*. The spring *k'* is similarly formed.

When, under the action of the steering-wheel, the arm *c* is made to turn, one of the dogs, *d* for example, enters the notch of the spring-band *k* and forces the spring away from the casing *i*.

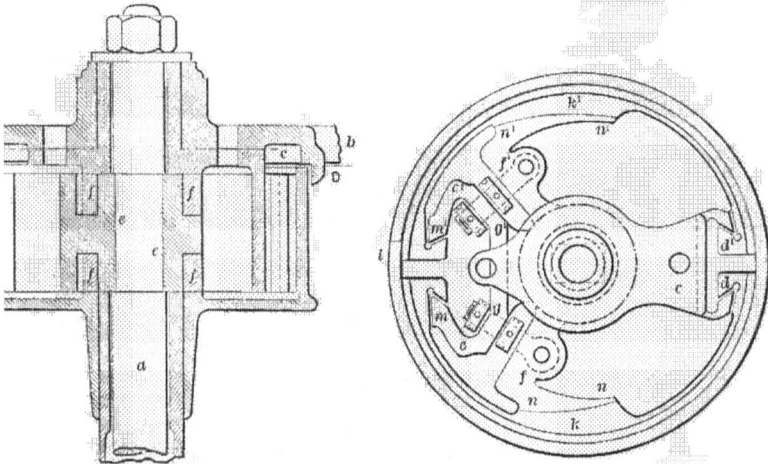

Jeantaud Steering-Gear.

The driving force is transmitted to the dog *m* of the arm *e* by the corresponding notch of the same spring *k*, and hence to the shaft *a* rigidly secured to *e*. If the steering-wheel be turned in the opposite direction the shaft *a* will be actuated by the spring-band *k'*. It is therefore evident that the movement of the wheel is integrally imparted to the shaft *a*. Let us now assume the shaft *a* to be turned by the carriage-wheels. The arm *e* is then displaced and one of the dogs, *m'* for example, enters its notch. But, by reason of the cam surface *n'*, the spring *k'* is bound against the casing *i* and cannot transmit the driving force to the arm *c*. The arm cannot therefore be turned by the wheels. The steering-wheel is consequently not displaced; and the front wheels cannot turn independently of the operator.   R. Denham.

77

A Novel Steering-Gear

WITHIN the last few years many motor carriages have been invented, provided with steering-gears the operation of which, besides being trustworthy, did not fatigue the operator too greatly; but, despite the efficiency of these gears, there is still room for many improvements.

Among these improvements may be mentioned the invention of W. D. Priestman, S. Priestman, and T. Wright, of Kingston-upon-Hull, which is shown in plan in the engraving annexed. Instead of a steering-wheel, a handle-bar is used secured to a vertical steering-shaft mounted in a hollow post fastened to the bottom of the carriage in front of the driver. At its lower extremity this shaft carries a pinion meshing with a segment-gear, connected with a vertical shaft, the upper end of which is seen at *G*. At this upper end *G* the shaft carries a double, eccentric segment-gear *H*, which, when viewed from above, resembles a butterfly with outstretched wings. These two segment-gears in turn mesh with two other segment-gears *I*, which are likewise eccentric and which are formed with arms pivoted to links *K*. Each of these links is operatively connected with a lever secured to the fork *M*, between the two members of which the front wheel is mounted.

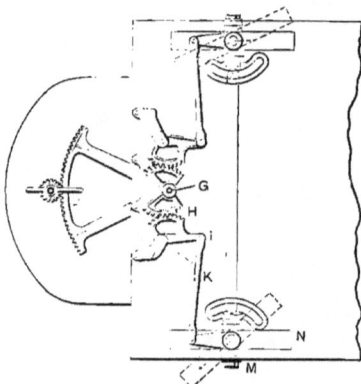

Priestmann & Wright's Steering-Gear.

If the handle-bar be turned to the left of the carriage, it is evident that the one wheel will be turned through a greater angle than the other, as shown by dotted lines in the figure. This result is due to the eccentric form of the segment-gears *H* and *I*.

It should be mentioned that the centre of the pair of segment-gears *H*, as well as the centres of the gears *I*, are revolubly mounted in a casting bolted to the carriage-body. The exact shape of the segment-gears is carefully determined, so that they are always in mesh, the one with the other, and that the variant movement described may be obtained.

A Successful Storage-Battery

EVER since electricity was first used as a source of motive power for automobiles, the storage battery has been so vastly improved that its general efficiency has been considerably increased. The accumulator is without a doubt the most vital part in an electrically driven carriage; for upon its ability to withstand mechanical disintegration after continued use or during rapid and heavy discharges, upon the relation of its active surface to the weight of its elements, upon the good contact between the active material and the grid, the general excellence of the carriage depends. The theoretical and practical considerations affecting the storage of electricity for automobile use have long held out great promise to the inventor; and it is therefore not astonishing that many accumulators have of late been devised which are particularly adapted for carriages. Among these batteries may be mentioned an accumulator which has been used with great success by the American Electric Vehicle Company.

The storage battery, which is of special design built under patents of the company, gives the greatest amount of output with the least weight, and combines low cost of production with high efficiency. A battery ordinarily lasts a vehicle from one and a half to two years, and the cost of replacing the positive plates at the end of that time is far less than the sum which would have been required to keep a horse shod.

The lead grids for the positive plates are stamped in presses into a network of metal adapted to hold the active material and at the same time to furnish a connection to all parts of it. The plate is formed from a thin sheet of rolled lead. Rows of small square holes are punched into the lead; diagonal cuts are then made from the corners of the holes, and the flap thus formed is turned at right angles to the surface of the plate, thus forming a series of pockets for containing the active material. The thickness of the completed plate is thus equal to the sum of the thickness of the lead and the pockets formed by the portions turned at right angles. The opposite faces of the upturned portions form the side walls of adjacent pockets, while the greater portions of the plate thus not in contact being the side edges of the sections. This insures a maximum amount of surface to the active material, therefore resulting in a maximum rate of discharge for a given weight. After the grids are stamped,

79

The Automobile Magazine

a mixture of finely divided lead and lead oxide is placed in them under heavy pressure. They are then formed. The positive plates are separated from the negative plate, in making up the cells, by thin pieces of wood, sawed in cross directions so as to form small gratings. These wooden insulators are treated chemi-

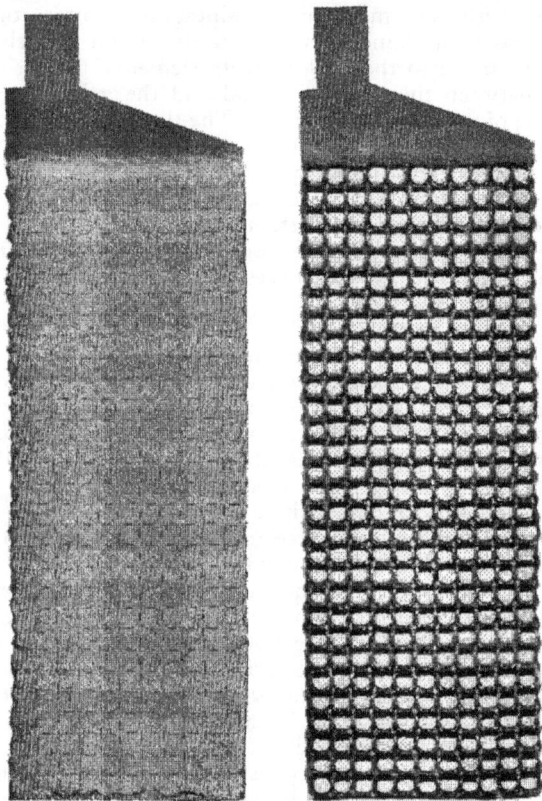

Battery Plates.

cally and made water- and acid-proof, so that they closely resemble vulcanite. The completed cells are held together with rubber clamps and put in hard-rubber jars closed at the top except for small openings, with an electrolyte of dilute sulphuric acid.

The Automobile
MAGAZINE
An *ILLUSTRATED* Monthly

Vol I No 1   NEW YORK OCTOBER 1899   Price 25 Cents

The Automobile Magazine is published monthly by the United States Industrial Publishing Company, at 31 State Street, New York. Cable Address: Induscode, New York. Subscription price, $3.00 a year, or, in foreign countries within the Postal Union, $4.00 (gold) in advance. Advertising rates may be had on application.

Editorial Comment

THE NEED OF ORGANIZED ACTION

THERE are certain significant indications that, on the part of automobile interests, every possible effort, in the way of organization and of energetic action for the influencing of public sentiment in the right direction, will be necessary to overcome the opposition that in various forms is sure to arise. Otherwise there is danger of the interposition of obstacles which, though they cannot remain insuperable, may nevertheless retard progress to a serious degree. A phenomenal activity characterizes the early stages of the movement in the United States. All manufacturers are overwhelmed with business; one concern alone having orders for nearly five thousand vehicles! Everybody is talking about the coming form of locomotion, everything written on the subject is read with avidity, and the community is apparently impatient for the day to arrive when the new vehicle shall become as common as bicycles on the streets. It would therefore seem to not a few sanguine minds that one great automobile wave is about to sweep over the country, carrying everything before it, and even drowning the horse in its flood, leaving the bones of the animal that has borne man through long stages of civilization to lie beside those of his antediluvian progenitor, the eohippus, to be recovered from the fossil remains of the present period of the earth's

The Automobile Magazine

history by the palæontologist of future ages! The figure of a wave is, indeed, peculiarly applicable to the automobile movement. The forces and impulses of modern invention have been assembling their energies for a new advance in the mechanical progress that signalizes the present age above all others, until, with conditions made favorable, the flood moves forward with irresistible momentum, changing the face of things in its course. Like other great achievements in the field of invention it will doubtless appear that the automobile has come in the very nick of time, the ground having been so prepared for its reception that it would seem as if the world could not possibly have got along without it any longer. Such, for instance, was the case with the telephone. In a moment, as it were, that instrument at once became indispensable, so that now the transaction of business without it is well nigh inconceivable.

RESTRICTIVE AND PROHIBITORY LEGISLATION

Strong efforts, however, will be made to dam the flood, to stay the advance of the wave, and even to turn it back in its course. In the nature of things these efforts must be unavailing. But if they are allowed to proceed unstayed or unopposed, they will create a deal of useless friction and involve the making of much unnecessary trouble. The history of the bicycle shows what a tremendous hostility had to be overcome to secure for that most useful instrumentality its right place in the civilized world. The intelligent steps taken by organized wheelmen, the sagacious and far-seeing action of leading interests involved, remain an example to be followed under the new and not dissimilar exigencies. The opposition of prejudice, of ignorance, of cupidity and corruption, of interests that feel themselves based on vested rights, has again to be overcome. In the case of the bicycle the horse was menaced by a powerful rival, whose limitations, however, were such that the attainment of only an equal place by his side was sought. Now, however, in the eyes of many, it has become a struggle of life and death for the horse. It is true that many thousands of owners of horses stand ready to welcome an instrumentality that will rid them of the cost, the inconvenience, the trouble, and the risk of maintaining an equine equipment. But the sentimental antagonism of "friends of the horse" is something by no means to be despised, while there is to be expected a most strenuous opposition from the side of dozens of different horse interests, short-sighted though it all is. It may therefore be expected that every automobile accident

82

Editorial

that occurs, near or far, will be dwelt upon and enlarged upon, and that every effort will be made to turn the tide of public sentiment against an engine " so dangerous and destructive." This has been the case with every new step in locomotion, from the railway and steamship to the bicycle and the electric car, and it will, of course, be the case with the automobile to at least an equal extent.

The Bar Harbor Attitude

More serious is the menace in the way of restrictive and even prohibitory regulation on the part of legally constituted authorities, both local and State, in this or that part of the country. The action of the town authorities at Mount Desert and of the Boston Board of Aldermen furnish typical instances. Mount Desert is so rugged in its mountainous character that there is but little field for driving except under the roughest conditions. But there are a few good roads, made possible by the enormous values that real estate has attained through the influx of summer residents. A number of these summer residents anticipated this year a deal of pleasure from the use of their automobiles over the limited courses thus available. But early in the season, at a town-meeting held in Bar Harbor, measures were adopted in the guise of rules regulating the use of automobiles, which,. to all intents and purposes, practically prohibited their use altogether. The regular residents of Bar Harbor are people of a semi-maritime, semi-bucolic character, who would not unnaturally be prejudiced against such an innovation as the automobile. While the field at Mount Desert is so limited as to make little appreciable difference in the movement, yet this action indicates a tendency that may possibly find expression at other summer resorts where it would be a more serious matter. There is a good field for automobile missionary work in such directions.

The Attempt of the Boston Aldermen

Of a different type is the form of opposition represented by the action of the Boston Board of Aldermen early in the past summer. That body, in exercise of the power conferred by the statutes to make certain rules and regulations relative to the use of vehicles in the public streets, passed an amendment to a certain section of its regulations so as to make it provide that no person should " in any street use any vehicle, other than a railroad or railway vehicle, or a vehicle of the fire-department, or a vehicle drawn or pushed by an animal, or a vehicle of a con-

struction approved by the Board of Aldermen as not endangering
the life or property of others." The Board may not impossibly
have been actuated by high motives of solicitude for the public
welfare. But since the Boston Aldermen have long allowed certain streets in the business section of the city, thronged with
passers throughout the day, to be used as pacing and exhibition
courses for the profit and pleasure of the proprietors of sales-
stables abutting thereon, greatly to the annoyance and peril of
the public, it may be inferred that considerations of averting
danger to life and property were not their controlling motive.
The Boston Board of Aldermen, in common with similar bodies
in other great American cities, has not, as a rule, been of a
character to induce admiration for the motives governing
its actions. A not uncommon form of blackmail is that which
actuates the members of legislative bodies, who "are there for
what they can get out of it," to introduce measures affecting
detrimentally this or that interest in prospect of being
persuaded to change their views in return for substantial consideration. The hypothesis that the Boston Board of Aldermen may not have been oblivious of such contingencies is not
at all strained by the coincidence that certain very strong and
immensely wealthy interests had only recently made arrangements to introduce automobile vehicles on a most extensive
scale in that city. What better opportunity for thrusting a very
prehensile finger into a very luscious pie could the Aldermen
have had than that presented under the power of deciding
whether certain vehicles that were about to enter into enormous
popularity, backed by capital representing a vast field for boodle-
exploitation, should meet with their approval as "not endangering the life or property of others"?

Fortunately the authority in question is exercised by the Aldermen only conjointly with the Mayor of the city, and Mayor
Quincy very promptly vetoed the measure. The Aldermen appear to have estimated their own masticating capacity very
highly, for they certainly bit off a very large mouthful when they
made their regulation so sweeping in form that even bicycles
were brought under its provisions and were thereby excluded
from the streets unless formally approved—as the Mayor pointed
out would be the case. A curious circumstance was that this
vicious attempt at wanton interference with a great interest was
made by amending a certain section of the regulations for the
use of vehicles, which solely provided that the speed at which an

Editorial

animal was allowed to draw a vehicle over a drawbridge should be limited to a walk. And the amendment was so drawn as to abolish that very useful provision altogether and substitute for it the proposed measure.

THE PRIVILEGE OF THE PARKS

The restrictions upon the use of motor-vehicles in public parks, in various cities of the country, are a matter of great importance. The Park Commissioners of New York, of Chicago, and of other cities have adopted rules that prohibit their use in the parks, while the Park Commissioners of Boston forbid them after ten o'clock in the morning, except upon the traffic roads that traverse certain parks and pleasure grounds. The bicycle was similarly treated in its earlier years. It was absolutely forbidden in the parks of New York and of various other cities. The Boston Park Commission, however, always pursued a liberal course toward the bicycle, and from the start its use in the public parks of the city has been encouraged. It may well be asked if a public park is intended for the recreation of the public in the various ways for which it was designed, including the use of its roads for driving, riding, bicycling, etc., why should the users of motor-vehicles find themselves barred out, when they, as taxpayers, contribute their share toward the establishment and maintenance of such pleasure-grounds?

The Boston Park Commissioners base their restrictions upon motives other than that of hostility to the form of locomotion. They are by no means opposed to the automobile, they say. On the contrary, they are disposed to welcome it as a beneficent factor in modern civilization. But they hold that it is their duty to make the parks attractive and safe for the people who use them. At present the people who frequent the park roads, beside bicycle riders, are chiefly those who resort thither in horse-drawn vehicles. Horses are not yet used to the sight of carriages moving without their aid and are liable to take fright at the mysterious spectacle. While the horse has got to become used to the sight, as he has had to become used to many other innovations shocking to his nerves, the commissioners say that he will gain that experience in the public streets, and until he has reached such familiarity they feel it to be their duty not to subject the people resorting to the parks to the dangers arising from the startling of horses by objects so unwonted. Since, with the rapid increase of motor-vehicles in the thoroughfares of our

85

The Automobile Magazine

cities, horses will very soon be led to view them with the same indifference that they show toward electric cars, bicycles, and even steam fire-engines, it cannot be long before the automobile will be placed on a par with horse-drawn carriages in the particularly beautiful and extensive parks of Boston. And there can be no valid reason why they should long continue to be excluded from public parks in New York and elsewhere.

There is all the more reason why motor-vehicles should be welcomed by the governing authorities of public parks since the great cost of maintaining park roads in proper condition is due to the action of horses' feet, whose steel-clad hoofs exert an incessant chopping action upon the ground until finally the road-bed is fairly "chewed" to pieces. To take care of the litter caused by horse-droppings is also a matter of very considerable expense. If it is not removed it is ground finer and finer, and reduced to a pulp by the action of the weather and of passing wheels. It is thus turned into a mud that spoils the road, entering into the road-bed, clogging the macadam, and stopping proper drainage. With the replacement of the horse by the automobile these detrimental effects would disappear. The cost of road maintenance in parks and elsewhere would be reduced to a minimum, with the action of the elements as the only cause of "wear," while the "tear," which proceeds entirely from the impact of horses' feet and the cutting of metal-tired carriage wheels would be entirely done away with.

As to the subjection of motor-vehicles frequenting public parks to well considered regulation, there can be, of course, no reasonable objection. Rules, for instance, excluding forms that make disturbing noises or produce offensive odors would be quite proper, as well as requirements limiting speed to a very low rate. In the latter regard, it should be borne in mind that the excellent roads of a public park are not designed for speeding purposes, but to enable the leisurely and peaceful enjoyment of the landscape to which they give access. Such enjoyment cannot be had, whether by bicycles or any form of carriage, while dashing along at high speed, and there should be ample opportunity for speeding outside of the public parks. But even were there no such opportunities there is no reason why speeding should be allowed within park limits, for there it is foreign to the purpose of a park as well as a menace to public safety.

86

Editorial

The Sentimental Plea for the Horse

In the case of " Horse *versus* Automobile," now before the court of public opinion, the sentimental aspect of the plaintiff's side has probably been no better set forth than by that most charming writer, the Listener, of the Boston *Transcript*. The Listener is a very humane and genial person, fondly attached to the horse as a pet, as he is to all domestic animals—so that even the annual pig-killing, or the more frequent sacrifice of a fowl —he lives in the country—must bring a pang to his kindly heart. Therefore, with the conservative tendency that leads him to speak strenuously at first against all mechanical innovations of the century and finally to fall in with the procession—he now even rides a wheel—the automobile yet finds little favor in his eyes. He says: " It is hard enough to live or work in a city even with the daily sight of strong, sweet, sensible, beautiful, patient horses all about one. Without them it would be simply impossible—not for a moment to be thought of. Neither wild horses nor any sort of self-propelling vehicle or locomotive could drag me into a horseless city." The appositeness of the word " sweet " in connection with a horse will be seriously questioned by most persons who ever smelt a stable or anything else intimately associated with the animal. One must infer that the Listener has never been in Venice, the most famous of horseless cities, but from his poetic temperament it would be safe to wager that, having once been dragged into the lovely place whose streets are paved with water it would be much against his will that he would be dragged out of it, even by a wild gondola. Its dream-like atmosphere, its freedom from bustle and roar, make it the ideal of what a city should be in respect to the noises that grate most harshly upon the human system and are responsible for a large proportion of the nervous maladies to which town-dwellers are subject. A city without noise is the ideal of civilization, and it is toward that ideal that the automobile is bringing us. The horse is now one of the greatest drags upon the wheels of progress. He is the chief maker of the vast and overwhelming din that envelops the modern city. It is for the horse that the stone pavements are laid, to keep his feet from slipping. And over every separate stone, set to that end, he drags bumpingly, poundingly, crashingly, with unceasing rattlety-bang, the endless procession of clamoring carriages, wagons, carts, and drays. Then at the end of the day every frequented street that meanwhile has not been constantly cleaned is literally carpeted with a warm,

87

The Automobile Magazine

brown matting of comminuted horse-dropping, smelling to
heaven and destined in no inconsiderable part to be scattered in
fine dust in all directions, laden with countless millions of dis-
ease-bearing germs. Then what words can characterize the
nuisance for hundreds of city neighborhoods caused by the stables that house the " sweet " animal, ruining the property of the
unfortunate persons near whom their erection chances to be permitted by the perversity or cupidity of municipal authorities?
It has been shown that without the horse the cost of street-cleaning would be so reduced, that with the amounts thereby saved—
not to mention the tremendous economy in maintenance cost—
a city could very soon be paved throughout with smooth and
noiseless sanitary pavement, to the unspeakable advancement of
health and comfort. Horse-excrement constitutes the chief
breeding-ground for flies; and a horseless city means also, practically, a flyless city, and likewise a city much less infested by
disease. For disease germs are not only very extensively generated by the filth of horses, but are very extensively distributed
by those winged messengers, the flies. A horseless city means
a clean city, a quiet city, a wholesome city, a more odorless and
beautiful city. And we are sure that our friend the Listener
would love such a city.

PUBLICITY VS. SECRECY

In the French automobile press it is significant that there is
now a great abundance of detailed description of inventions,
where formerly a mysterious reserve was maintained. Manufacturers were then endeavoring to hide every mechanical part.
The inventor apparently believed that somebody was lying in
wait to spy upon and copy his work, and many were in the habit
of imploring the press not to publish anything about their devices. Of late all this has changed in France. Fundamental
patents in automobile practice do not exist. Patents can be
claimed only on special parts, and every patentee can protect
himself fully so far as legal procedure can go. Realizing this,
French manufacturers and inventors have become eager to take
the public into confidence through the automobile press, fully
explaining their inventions and the methods of application. The
very best French makers describe and illustrate their automobiles in every part, in the form of printed lessons showing " How
to handle and take care of our vehicles," and those manufacturers
sell the most.

An American manufacturer recently wrote to this magazine

Editorial

as follows: "I send you, enclosed, the photo of my motor-car. Please publish the same. I do not wish to give you any details of the mechanical parts, although covered by patents, as I do not want to be copied—but you may say that it is the best and the most economical automobile in the world."

Had this manufacturer not stated that his invention was covered by patents, the photograph of his vehicle would gladly have been reproduced, under the impression that his patents had not yet been issued and that the publication of the picture might possibly influence capital in its favor. But after investigation it was learned that this particular vehicle had never yet been subjected to a public trial.

The public wants to know—before buying—and rightly. Fear of imitation cannot be pleaded, since a few hundred dollars could buy such a vehicle and let out the cat! Fortunately these short-sighted manufacturers are few, but this magazine will not encourage the public to buy vehicles whereof nothing can be known.

The Problem of Safety

Automobile engineering is an art in itself. As the new industry grows, the proper authorities will, in all probability, take the necessary steps to subject vehicles to a technical inspection before they are delivered to the public. While most of the recent automobile accidents that have occurred in France were due to the recklessness of the users, there have been some that have demonstrated the desirability of so modifying the construction of automobile motor-vehicles as to increase their safety. According to Mr. René Varennes, theoretical safety in a motor-carriage can be secured:

1. By providing a steering-gear which is perfectly trustworthy and does not exceed a safe angle in turning. The steering-wheel, whether horizontal, inclined, or vertical, is no safer than the ordinary handle-bar if the connection between the wheel and the front axle has not been properly effected, as is too often the case.

2. By considerably lowering the centre of gravity. In many automobiles this is placed far too high for the speed exacted, and the occupants are in constant danger of an upset.

3. By providing brakes controlled by a swingle-tree or similar steering-device, so that they can be applied both right and left.

4. By not using for the wheels any ball-bearings in which tempered steel works against tempered steel. Such bearings,

without warning, may suddenly bind, as recently happened in the accident to M. Duchemin. The bearings should be of tempered steel, working on bronze, so that if a binding of the balls should occur it would be gradual instead of instantaneous.

Practical safety, however, depends upon a vast number of circumstances which can be understood only after an experience so long that one instinctively anticipates and meets the trouble. If the ground be wet and slippery, if it be sandy and friable, as on the edge of certain roads, if the grades be such that the additional speed acquired from gravity retards the action of the brakes, a velocity of twelve miles an hour may be dangerous even for a carriage which, on a dry, firm road-bed, could safely make twenty or twenty-five miles an hour.

While American-made electric vehicles have been thoroughly tested here as well as in Europe, vehicles made here in which gasoline, steam, or compressed air provide the motive-power will have to undergo the most searching trials, and the public should only accept such makes as have been thoroughly tested as to safety, durability and speed.

We shall hear in the very near future that the American rights for certain world-famed French motors have been secured by such-and-such a company. But if the vehicle itself, be it a bicycle, tricycle, quadrucycle, carriage, delivery-wagon or truck, be not built on safe lines—all the mechanism properly distributed, the suspension secured according to the use of the vehicle— it does not necessarily follow that the American make should deserve the same fame as the French product has won. In connection with a badly engineered vehicle a very good motor will give but poor satisfaction.

Public Tests

While it is wise and business-like on the part of the manufacturers to subject their vehicles to conclusive tests, as some of them have done, either by long-distance rides or by hill-climbing tests, it must not be forgotten that such tests can only be considered as private trials. The public expects that such tests be made under the auspices and control of some reputable club and the press combined, as is done in Europe. Good makers can only be benefited thereby, while those eager rather to sell the stock of their companies than safe and practical vehicles, will be weeded out.

Editorial

A PICTURESQUE ANNOUNCEMENT

If anything were needed to confirm the prevailing belief in an enormous development of the automobile industry in the near future, it would be the announcement of a new publication called *The Automobile*, at 112 West Thirty-second Street, this city. The edition for October is to be 100,000 copies; November, 200,000; December, 500,000; and the Christmas number a round million, or rather more than the combined circulation of *Harper's Magazine*, the *Century*, *Scribner's*, *McClure's*, and *Munsey's*. The monotony of this arithmetical progression in circulation, month by month, will be relieved by the "marvels of pictorial beauty, chaste in tone and unique and original in conception," which *The Automobile* will present to its readers, according to the prospectus. Surely nothing could be more marvellous in pictorial beauty than the figures enumerated—especially those for the Christmas number.

THE QUESTION OF TERMINOLOGY

The question of nomenclature for the new vehicle seems to be exercising the minds of many people, and the public prints are teeming with all sorts of propositions and suggestions. There is, however, no occasion for any vexation of spirit in this regard, for, so far as immediate purposes are concerned, the question is well settled by the designations now in common use. For generic purposes the word "automobile" could hardly be improved upon. Objection is made that it is French. It is true that the name originated in France, where it has the official sanction of so scholarly and conservative a body as the French Academy, the supreme authority in French orthography. But while it is proper that the country where the automobile was born should give a universal name to its child, it happens that "automobile" is also good and intelligible English, both of its component parts being in common use in our language. There might seem to be more validity to the objection that one of the parts is Greek and the other Latin in origin, and that therefore a combination of the two is a violation of philological propriety. But where the procedure is sanctioned by an authority so regardful of such proprieties as the French Academy, it would appear a bit over-nice to find it repugnant to the spirit of a language so flexible and eclectic as the English. Where root words have been well incorporated into English from any foreign source it would seem

The Automobile Magazine

quite in order to combine them in any manner that convenience may dictate, irrespective of considerations of linguistic uniformity or consistency, so long as the demands of verbal symmetry are observed. Such combinations have long been in order. There should be no more objection to the conjunction of Greek and Latin forms in our language than to the combination of Anglo-Saxon and Latin forms. For example, the use of the Anglo-Saxon terminal " ness," signifying quality or state, is as common in connection with Latin as with Anglo-Saxon derivatives, " comprehensiveness " being as good a word as " newness."

Greater objection is made to the word " automobilism," which in English has a somewhat unwonted look. The word, however, is well accepted in French, as are similar forms of it in other Latin languages; as for instance, the Spanish " automobilismo." Its strangeness in English comes from the fact that the terminal " ism " has been practically confined to words having intellectual abstract, or theoretical significance, such as " catholicism," " sophism," " republicanism," " absolutism," etc., while it has not been applied to more concrete terms. According to dictionary authority, however, the termination denotes " the theory, doctrine, spirit, or abstract idea of that signified by the word to which it is appended." So Webster says. " Automobilism," therefore, should be a good and admissible word, denoting the theory of the automobile.

The words having " motor " for a common compound have become well established in automobile phraseology. They are good and most convenient. " Motor-carriage " is the best, as it is the most familiar designation for the thing that it represents, conveying the idea perfectly. Excellent, likewise, are the terms " motor-vehicle," " motor-cab," " motor-wagon," " motor-cycle " (motocycle in French), etc. Terms representing the form of motive power are also in common use, like " electric-carriage," " steam-carriage," " gasolene-carriage," etc. " Hydrocarbon motor " is a good term to cover the various forms comprised under the explosion type of engine, using some grade of petroleum, commonly gasolene for fuel. In France the popularity and almost universal use of the automobile has developed a very extensive nomenclature, with special names for every form of vehicle. Most of these are so peculiarly French that they are not likely to be adopted into English. We shall, therefore, be most fortunately spared the infliction of such words as " petrolette," one of the common of French names for a gasolene motorcycle.

Editorial

There has been a most industrious search for some brief word that shall popularly designate the automobile. Some most extraordinary verbal freaks have been evolved to that end, but all have proved abortions so far. One ingeniously simple word which failed to catch on was the contraction " mobe." " To mobe, or not to mobe, that is the question," said one of the newspapers that passed it along. Such terms, as a rule, are not a result of deliberate effort; they come of themselves, nobody knows how. The word " bike," that most harsh and unlovely abbreviation for " bicycle," with its wholly arbitrary and illogical substitution of the hard *c* for the soft, suddenly leaped into usage in some such fashion. If a short name is ever evolved for the motor-vehicle, let us hope it will be something more euphonious than that. Meanwhile we shall get along very comfortably with our present stock.

In China.

What is amusing these Chinamen so ?
What is that queer thing a-wheel ?
There's only one Cochin China you know,
And only one automobile.

Carolyn Wells.

The Automobile Magazine

AN ECHO OF THE PAST.

He.—" How do you like your new automobile? Is it hard to manage? "

She.—" Oh, no. The only trouble I have is that somehow I can't break myself of saying, ' Get up!' when I want to start the thing, and ' Whoa!' when I want the machinery to stop."

An Autolift.

AN AUTOGO.

It was an automobile,
  Began to balk and rant,
And when 'twas told to move on, said,
  " I auto, but I shan't!"

—From *Harper's Bazar.*

Book Reviews

E. Bernard & Co. (Paris) have just issued a new book, by Henry de Graffigny, on " Light Motors " (*Les Moteurs Légers*). The volume contains 336 pages, large octavo, and is well illustrated by many engravings. It has been the object of the author to discuss the various attempts made by inventors and mechanics to reduce the weight of automobile engines, generators, and operative mechanism. In several chapters he has carefully analyzed and studied the motors which are rivalling one another for future supremacy. After having stated the theory of operation of the vapor, petroleum, and electric motor, and recapitulated the principles upon which this operation is based, he describes the most interesting among the various methods of producing power for automobile propulsion. True it is that certain omissions are noticeable; but in a second edition, which is soon to be published, those motors which have been apparently forgotten or involuntarily omitted will doubtless find a place. Taken as a whole, the work describes, in a compact and convenient form, the principal motors used on French carriages, and is, therefore, to be recommended to the automobile maker as well as to the automobile user.

" Traité théorique et pratique des Moteurs à Gaz et à Pétrole et des Voitures automobiles." Par Aimé Witz, Ingénieur des Arts et Manufactures, Professeur à la Faculté libre des Sciences de Lille.

The third volume of Witz's *Traité* appeared toward the end of 1898, and its success was fully equal to that of the preceding volumes. In the present volume, the third chapter, in which the various gases and vapors used in thermic motors are treated, is undeniably the most interesting, and certainly one of the most valuable. Although somewhat cautious on the use of acetylene, Professor Witz nevertheless admits that the great explosive power of the gas particularly adapts it for use in engines. He is even inclined to believe that when its employment is attended with less danger it may become the motive power of the gas-engine of the future. The alcohol-motor has also been discussed with no little care, and perhaps with perfect justice, since the use of alcohol, in the words of Professor Witz, is of great " economical importance." The conclusions drawn regarding the utility of the alcohol-motor are based largely on the experiments made by Pétréano. The concluding chapter of the volume

The Automobile Magazine

is devoted to a description of automobiles. After a brief historical review, Professor Witz discusses the principal types of explosion-engines used in automobiles. Perhaps he is right in not burdening his work with motors that differ from one another in unimportant details; for an endless array of engines which are practically similar is not only wearisome but also useless.

" L'Annuaire Générale de l'Automobile et des Industries qui s'y rattachent," published by Messrs. Thévin et Oury, is a veritable cyclopedia in its way. It contains everything that can possibly be of interest to the " chauffeur " or the manufacturer. In a work so encyclopedic in character it might be inferred that the enormous mass of information would render it difficult to look up a subject upon which one desires to be enlightened. But the material has been so methodically classified that the information sought is almost immediately found. It is to be hoped that the *Annuaire* for 1899 will meet with even greater favor than that accorded to the *Annuaire* of 1898.

" Connaissances pratiques pour conduire les Automobiles à Pétrole et Electriques, par Félicien Michotte."

This work on the management of petroleum and electric automobiles is based upon a course of lectures delivered by Professor Michotte at the Association Polytechnique. As its title indicates, the book is written entirely from the practical point of view. There are no historical chapters, no profound expositions of the theory of the motor, no innumerable formulæ and mathematical operations such as are found in almost all similar works. Whatever there may be of the theoretical or mathematical has been inserted to simplify the explanations and render the descriptions more intelligible. Professor Michotte has neither enumerated nor discussed all the various kinds of motors in use; such descriptions he has left to the catalogues of the various manufacturers. Although the innumerable types of motors now in use have found no place in his book, Professor Michotte has by no means neglected to describe the various methods of propelling an automobile. He carefully names all the parts of a motor-carriage, shows what the function of each of these parts is and how it is removed and replaced, gives rules for the care of and management of an automobile and for the prevention of accidents. In a word, he has told everything about an automobile worth while knowing, and that in a manner so attractive that his book can be recommended to all " chauffeurs."

Book Reviews

"Les Secrets de Fabrication des Moteurs à Essence, pour Motocycles et Automobiles," is a new work by Georgia Knap, in which the rapid progress made in recent years in the building of high-speed automobiles is admirably discussed. It has been the author's object to produce a book which should aid manufacturers in overcoming the obstacles which they encounter when they first attempt to build automobiles. Expansion, oxidation, condensation, lubrication, distortion of overheated parts, binding of rods and gears, and a host of similar difficulties besiege the inexperienced builder of motors and engines. It is only after an experience gained by long years of practical work that a manufacturer can hope to produce a motor which shall hold its own in the automobile market. In order to build a motor which shall meet all the requirements of an automobile, the manufacturer should be something of an electrician and chemist. He must also be familiar with the theory of explosives in order that he may know what effect the discharge of a carbureted gas will have upon his engine.

All this and more is described by M. Knap in his work. Without the use of difficult mathematical operations, or of pretentious formulæ, he has clearly explained how a motor should be built, how its parts should be proportioned to obtain the greatest possible efficiency. The exhaustiveness and patient research which characterize his work, and the fruits of the long practical experience which he offers to his readers, should earn for his book a place in the technical library of every automobile manufacturer.

"Manuale dell' Automobilista: Guida del Conduttore D'Automobili." Del Dottor G. Pedretti. Ulrico Hœpli: Milano, 1899. Lire 5.50.

Although this work can hardly be considered as a very valuable addition to automobile literature, its author nevertheless deserves no little credit for having been probably the first in Italy to publish a book on motor-carriages. So far as technical description is concerned, the work certainly leaves much to be desired, its defects being particularly noticeable in the chapters devoted to electric-carriages. The illustrations could also be considerably improved upon, both in the way of size and draftsmanship. The motor-carriages described comprise most well-known automobiles, such as the Panhard-Levassor, Beutz, Gautier-Wehrlé, Fisson, Daimler, Roger, Mors, Duryea, Klaus, de Riancey, de Dion-Bouton, as well as various vehicles made by Italian manufacturers. A preface to the work has been written

The Automobile Magazine

by Count Roberto Biscaretti di Ruffia, President of the Automobile Club of Italy.

" Carnet du Chauffeur," par M. le Comte de la Valette.

Count de la Valette's hand-book is a veritable *vade mecum* for the " chauffeur;" he has condensed within comparatively few pages much indispensable information regarding the care and management of horseless vehicles.

The book will be of no little service to the travelling " chauffeur," for here are recorded most of the formalities through which one must go when journeying in France, such as the paying of tolls and duties. Portions of the book have been reserved for memoranda as may be of interest.

The descriptions and the operation of the best automobiles will be of assistance to the " chauffeur " when his carriage, for some cause unknown to him, suddenly stops on the road. Indeed, it may perhaps be not too much to say that these simple technical explanations would grace a work more pretentious in character.

The fourth volume of the Traité de la Construction, de la Conduite, et de l'Entretien des Voitures Automobiles (Treatise on the Construction, Management, and Care of Automobiles), compiled under the editorship of Charles Vigreux, by Messrs. Milandre and Bouquet, has just appeared.

This fourth volume is entitled " Voitures Automobiles Electriques " (Electric Automobiles), and is more exhaustive and more complete than the volume on petroleum-motor carriages. It is evident that the authors are more favorably disposed to electricity as a motive power for automobiles than to petroleum or its derivatives.

The first part of the work is devoted to a description of the most important types of accumulators, such as the Fulmen, B. G. S., Faure-Selon-Wolckmar, Pulvis, Phœbus, Pisca, Blot-Fulmen.

In the second part electric-motors for automobiles, the method of regulating the speed of electric-carriages, power-transmitting mechanism, and auxiliary appliances are discussed.

The third portion treats of the cost of running automobiles, both in public and private service.

In the fourth part will be found a description of the best known types of motor-carriages. Among the automobiles described we find the Jeantaud and the Kriéger, the vehicles of the

The Foreign Automobile Press

Cie Française des Voitures Electromobiles, New York's Electric-cabs, the B. G. S., Patin, and Mildé carriages.

At the end of the volume is an appendix comprising a few notes.

Appearing at a time when electric-carriages are becoming more popular, this fourth volume of the " Traité " should be very favorably received.

The publishers of the work, it should be mentioned, are E. Bernard et Cie, 53ter Quai des Grands Augustins, Paris, France.

" Notice de Route sur la Conduite et Entretien des Tricycles avec Moteur de Dion-Bouton," par J. Wolff. 75 cents.

The great merit of this work is found chiefly in the fact that the author, in writing his book, has consulted only his own experience in running the de Dion-Bouton mototricycle. His descriptive matter is characterized by a praiseworthy clearness of style and trenchancy of expression, and is devoted largely to the prevention of accidents in the motor described and to the overcoming of such difficulties as are now and then met with in the operation of the tricycle.

The Foreign Automobile Press

AUSTRIA

Club Organ des Oesterr. Automobile Club, Vienna.

BELGIUM

L'Automobile illustré, Brussels.
L'Autocar de Belgique, Brussels.

FRANCE

Dailies

Le Velo, Paris.
Le Journal des Sports, Paris.

Weeklies

La Locomotion Automobile, Paris.
La France Automobile, Paris.
Les Petites Annales de l'Automobile, Paris.
L'Auto Cycle, Paris.
Cycle et Automobile, Paris.

Tous les Sports, Paris.
Le Vie au grand air, Paris.

Biweekly

Le Chauffeur, Paris.

Monthlies

L'Industrie Vélocipédique et Automobile, Paris.
L'Avenir de l'Automobile, Paris.

GERMANY

Das Journal für Wagenbaukunst, Berlin.
Der Motorwagen, Berlin.

GREAT BRITAIN

The Autocar, Coventry.
The Automotor and Horseless Vehicle Journal, London.
The Motor Car, London.

The Automobile Index

Everything of permanent value published in the technical press of the world devoted to any branch of automobile industry will be found indexed in this department. Whenever it is possible a descriptive summary indicating the character and purpose of the leading articles of current automobile literature will be given, with the titles and dates of the publications.

ACCUMULATORS—
The Crowdus Accumulator. Illustrates and describes a cell which has been claimed to be the most suitable yet designed for automobile purposes. Automotor Journal, London, April, 1899.

Electric Automobiles in Bremerhaven. Dr. Sieg. Discussing mainly the use of storage batteries for automobile service. Elektrotechnische Zeitschrift, Berlin, May 11, 1899.

Crowdus Electric Accumulator for Automobile and Launch Motors, described and illustrated. The Autocar, May 27, 1899.

Paper on Accumulators. E. C. Rimington. Automotor Journal, July 15, 1899.

Successful Accumulators of the American Vehicle Company of Chicago described. Automobile Magazine, New York, October, 1899.

AIR RESISTANCE—
Air Resistance to Motor Cars. Comments on the little known of this subject, and suggests methods of making experiments. Automotor Journal, London, February, 1899.

AIR MOTOR—
The Hoadley Knight Compressed Air Motor. An illustrated description of the motor and its operation. Scientific American, New York, February 4, 1899.

ALCOHOL AS FUEL FOR EXPLOSION MOTORS—
Discussion about Alcohol as Motor Generator, after W. Petreano's researches. La Locomotion Automobile, Paris, January 5 and 12, 1899.

Communication of Mr. Lucien Perissé to the Société of Civil Engineers, France, on the use of alcohol in explosion motors. La Locomotion Automobile, Paris, July 27, 1899.

APPLICATION OF OIL ENGINES TO AUTOMOBILES.—
A serial article. The Autocar, London, September 2, 1899.

AUTOMOBILE ASSOCIATION—
A National Automobile Association. An editorial calling attention to features of the situation which make the organization of such a body desirable. Electrical Review, New York, July 19, 1899.

AUTOMOBILE BRAKES—
Emergency Brake. System Cuénod, described and illustrated. L'Industrie Automobile, Paris, July 25, 1899.

Gros & Pichard Brake, described and illustrated. La France Automobile, Paris, February 26, 1899.

AUTOMOBILE CLUBS—
Automobile Club of France. Illustrated article by Baudry de Saunier. Automobile Magazine, New York, October, 1899.

AUTOMOBILE DESIGN—
Some New Features in Motor-Vehicle Design. Thomas H. Parker. Read before the Liverpool Self-propelled Traffic Association. Describes the work of the writer in this field and the success attained. Industries and Iron, London, March 3, 1899.

AUTOMOBILE FREIGHT WAGONS
Automobile Freight Wagons in England. Stating the desire for self-propelled freight road wagons in Great Britain and the success thus far attained. Invites Ameri-

The Automobile Index

can competition. U. S. Consular
Reports, No. 338, January 31, 1898.

Motor vs. Horse Haulage. An
Account of Our Nine Months' Experience. S. H. Sparkes. Read
before the Liverpool Self-Propelled Traffic Association. Gives
particulars of the service of a
motor-van used by Messrs. Fox
Brothers & Co., owners of large
woollen mills in England. Industries and Iron, London, January
20, 1899.

AUTOMOBILE GUN—
System Simms-Maxim. Mitrail-
leuse gun on a quadricycle illustrated. La Locomotion Automobile, Paris, July 20, 1899.

AUTOMOBILE HISTORY—
A Short Historical Account of
Motor Vehicles. Sketches of the
most prominent motor plants of
this country. The Hub, New
York, May, 1899.

Condensed History of Automobilism in Paris since 1890. Scientific American, New York, May 13,
1899.

GENESIS OF THE AUTOMOBILE—
Concise Illustrated History of
Horseless Traction, by John Grand
Carteret. Automobile Magazine,
New York, October, 1899.

**AUTOMOBILE MOTOR EXAMINA-
TIONS—**
The technical publication, La
Locomotion Automobile, is organizing a competitive motor examination in order to standardize the
correct horse-power of motors.
La Locomotion Automobile, July
27, 1899.

AUTOMOBILE MOTORS—
The Different Motors—Vapor,
Electricity, Steam, Compressed
Air—and their Comparative Mer-
its. Scientific American, New
York, August 5, 1899.

**AUTOMOBILE PLATFORM SIDE-
WALK—**
Description of Moconible Auto-
mobile Platform Sidewalk. La
France Automobile, February 5,
1899.

AUTOMOBILE SITUATION—
A descriptive article, with illustrations, by Mr. Hiram Percy

Maxim. Cassier's Magazine, New
York, September, 1899.

AUTOMOBILE WHEELS—
Swerdna Automobile Wheel, described and illustrated. The Autocar, London, April 22, 1899.

AUTOMOBILES IN EUROPE—
Impressions of an American Elec-
trical Engineer. By M. K. Eyre.
Electrical Review, New York,
July 26, 1899.

CARBURETERS—
Petreano Carbureter, described
and illustrated. Le Chauffeur,
Paris, January 11, 1899.

Raymond & Oury Carbureter,
described and illustrated. La
France Automobile, Paris, February 19, 1899.

Jupiter Carbureter, described
and illustrated. La Locomotion
Automobile, Paris, March 2, 1899.

Automatic Regulating Carbureter, Crouan Patent, described and
illustrated. La France Automobile, Paris, March 12, 1899.

Riotte Carbureter, described and
illustrated. La Locomotion Automobile, Paris, April 6, 1899.

Sangster System, described and
illustrated. La Locomotion Automobile, Paris, July 13, 1899.

Lepape Carbureter, described
and illustrated. The Automobile
Magazine, New York, October,
1899.

CHAINS STANDARD—
Proposition of the Touring Club,
France, to Standardize Automobile
Chains. La France Automobile,
February 19, 1899.

La Locomotion Automobile,
Paris, February 23, 1899.

Le Chauffeur, Paris, March 11,
1899.

**COMBUSTION CHAMBER OF HY-
DROCARBON MOTOR—**
Improved Combustion Chamber,
A. Eldin System, described and il-
lustrated. La Locomotion Auto-
mobile, Paris, February 23, 1899.

COMMERCIAL OPERATION—
The Motor Vehicle in Commercial Operation. G. Herbert Con-

The Automobile Magazine

dict. Considers the electric motor the form of power presenting the least objectionable features, discusses the station, wheels, etc. Electricity, New York, March 1, 1899.

COMPARISON OF CARRIAGES AND DRAYS—
Article with Statistics, by M. Forestier, C.E. La Locomotion Automobile, Paris, August 3, 1899.

COMPARISON OF COST—
A general review of the applicability of electric driving for automobile vehicles, and a comparison of costs with animal power. L'Electricien, Paris, May 6, 27; June 3, 1899.

A paper by Professor G. F. Sever and Mr. R. A. Fliess, read at the general meeting of the American Institute of Electrical Engineers, gives valuable information regarding the relative economy of horse and electric delivery wagons. They have found that the average delivery wagon covers a distance of about 11,268 miles per year, at a cost of 13.86 cents per mile for a two-horse vehicle. An automobile will cover the same distance for 2.65 per mile less, equivalent to $298.60 per year. A further advantage is the higher speed of the electric vehicles, enabling the deliveries to be accomplished with greater celerity. Electrical Review, New York, July 19, 1899.

COMPETITION—
The Second Competition of Automobile Cabs (Deuxième Concours des Fiacres Automobiles). With descriptions of the competing vehicles, and a summary of the results of the trials. Génie Civil, June 24, 1899.

COMPRESSED AIR AUTOMOBILES—
Hoadley & Knight Compressed Air Yard Truck, described and illustrated. The Autocar, London, February 11, 1899.

COMPRESSED AIR FIRE ENGINE—
Advantages of Compressed Air Fire-Engines over the Ordinary Horse-drawn Engine, in Weight, Speed, and Economy, explained and illustrated. Compressed Air,

August, 1899. Also, The Automobile Magazine, New York, October, 1899.

COMPARISON OF ELECTRIC AND GASOLINE TRACTION—
A Comparative Study of the Effects of Electricity and Gasoline as Applied to Automobile Traction. La France Automobile, Paris, April 2, 9, 23, 1899.

CONTESTS AND TRIALS—
The Boston (U. S. A.) Moto-Vehicle Contest. An account of the entries and the points considered, with extracts from the judges' report. Automotor Journal, London, February 15, 1899.

The Royal Agricultural Society's Trials of Moto-Vehicles. The judges' report, description of vehicles, summary of times and speeds, consumption of fuel, and cost of running, with conclusions. Automotor Journal, London, October 15, 1898.

CONTINUOUS COMBUSTION GENERATOR—
Woillard System, described and illustrated. Scientific American, New York, September 16, 1899.

DIFFERENTIAL GEAR—
De Dion-Bouton Tricycle Differential Gear, described and illustrated. La France Automobile, Paris, March 5, 1899. Also, Automotor Journal, London, April 9, 1899.

DRIVING WHEELS—
The Design of Driving-Wheels. A discussion of needed improvements in design to satisfy the requirements of safety and comfort. Suggested by the Harrow accident. Automotor Journal, London, March, 1899.

ELECTRIC AUTOMOBILES—
The Milde Electric Voiturette. A brief illustrated description of a novel automobile. Electrical Engineer, New York, January 26, 1899.

Joel Electric Motor Carriage. Illustrates and describes a new form of motor carriage, with little vibration. Electrical Review, London, April 7, 1899.

The Automobile Index

American types illustrated. Electrical Review, New York, April 19, 1899.

Electric Road Vehicles. With numerous illustrations of automobile omnibuses, coaches, cabs, etc. Elektrotechnische Zeitschrift, Berlin, May 25, 1899.

Waverly Electric Runabout and Columbia Electric Emergency Wagon. Illustrated descriptions of two electric vehicles, exhibited at the exhibition in Madison Square Garden, New York. Scientific American, New York, June 10, 1899.

Patin's Electric Phaeton, of One Ton, 1,000-pound Accumulators, Good for a 60 Mile Run, described and illustrated. Le Chauffeur, Paris, June 25, 1899.

Draullette's Electric Autocab, described and illustrated. L'Industrie Automobile, Paris, July 25, 1899.

Jeantaud's Electric Cab (first prize in the Motor Cab Trials, Paris, 1899), described. Electricity, New York, July 26, 1899.

Summer Types Exhibited in Paris, illustrated. Electric Review, New York, August 2, 1899.

Waverly Delivery Wagon, described and illustrated. American Machinist, New York, August 3, 1899.

Monnard Electric Phaeton, described and illustrated. The Automotor and Horseless Vehicle Journal, London, August 15, 1899.

ELECTRIC BRAKES—
An Independent Motor to Utilize the Brake Energy. Merrill E. Clark. Abstract of a lecture describing a system of independent motor and a means of utilizing the brake energy in connection with street railways and other traction, Compressed Air, New York, April, 1899.
The Value of Electric Brakes as Recuperating Devices for Automobiles. F. B. Booth. Considers that the value rests wholly with the conditions under which the vehicle is used, and that it is not likely to be used on vehicles built for city service. Electrical World and Electrical Engineer, New York, April 8, 1899.

ELECTRIC DELIVERY WAGONS—
Advantages of Using Electric Delivery Wagons. Engineering Magazine, September, 1899.

ELECTRIC IGNITER- -
Houpied System, described and illustrated. La Locomotion Automobile, Paris, August 17, 1899.

ELECTRIC MOTOR—
Thompson Electric Motor for Vehicles, described and illustrated. The Autocar, London, January 7, 1899.

ELECTRIC MOTOR TRIALS—
Speed Trials in Paris. La Locomotion Automobile, Paris, August 3, 1899.

Competitive Trials of Steam Heavy Trucks in Liverpool. La Locomotion Automobile, Paris, August 17, 1899.

ELECTRIC STATIONS—
Electric Vehicles and their Relations to Central Stations. Abstract of paper read before the Northwestern Electrical Association. Considers the application of these vehicles to practical purposes and the advisability of stations providing their plants with the apparatus necessary to charge them. Electrical Engineer, New York, January 26, 1899.

Electromobile Station in Paris, described. Electrical Review, New York, July 19, 1899.

ELECTRIC TRICYCLE—
Paquin & Requillart. Storage Battery of 350 Pounds, for 24 Elements—able to cover 30 miles. Description and illustration. La Locomotion Automobile, Paris, July 27, 1899.

Waltham Electric Tricycle, described and illustrated. Carriage Monthly, Philadelphia, September, 1899.

ELECTRIC VEHICLES
Electric Automobile Fore-carriage, Speiser Patent, described and illustrated. La France Automobile, Paris, March 26, 1899.

The Automobile Magazine

Vedovelli & Priestley Electric Automobile, described and illustrated. La Locomotion Automobile, Paris, August 17, 1899.

Electric Automobiles. Elmer A. Sperry. Read at Boston meeting of the American Institute of Electrical Engineers. Discusses types of batteries, tires, problems to be solved by designers, etc. Electricity, New York, July 19, 1899.

Riker's Electric Wagons. Hugh Dolnar. Working drawings showing the principal mechanical inventions in motor wagons made by this establishment, with explanation of methods. American Machinist, July 13, 1899.

An Electric Fire Wagon. Illustrates and describes a vehicle with electrical equipment, designed for the Paris Fire Brigade, and reports highly encouraging experimental trials. Engineer, London, July 7, 1899.

ELECTRIC VEHICLE DESIGNS—
Need of Æsthetic Reforms. Engineering Magazine, New York, August 9, 1899.

EVOLUTION OF AUTOMOBILE CARRIAGES—
The Evolution of the Motor-Car. E. Shrapnell Smith. Excerpt from a paper read before the University College Students' Engineering Society. Thinks the problems most pressing are to combine strength, lightness, and efficiency, and to secure sufficient adhesion on greasy surfaces. Automotor Journal, London, February, 1899.

Evolution of the Motor Vehicle as Shown by Patents. A serial article, illustrated, by Mr. Leonard Huntress Dyer. Horseless Age, New York, September 6, 1899.

EVOLUTION OF HYDRO-CARBON MOTORS—
Evolution of Benzine and Petrol Engines, by G. Lieckfeld. Engineering Magazine, New York, August, 1899.

FRENCH AUTOMOBILES—
Description of Types Exhibited in Paris. Electrical Review, New York, August 2, 1899.

GASOLINE FORE-CARRIAGE—
Amiot & Péneau's Fore-Carriage, described and illustrated. La Locomotion Automobile, Paris, March 16, 1899.

Vollmer Fore-Carriage, described and illustrated. The Automobile Magazine, New York, October, 1899.

GASOLINE RESERVOIR—
Desponts & Godefroy Gasoline Reservoir for Motocycles, described and illustrated. La France Automobile, Paris, May 7, 1899.

GASOLINE SUPPLY—
The Supply of Moto-Car Spirit or Petrol. George Herbert Little. Notes that a great majority of these vehicles use products of petroleum and discusses the facilities for a proper supply and distribution of the oil known as moto-car spirit and petrol. Automotor Journal, London, November 15, 1898.

The steady increase in the demand for gasoline with the increase of hydrocarbon automobiles raises the question of using oil instead of gasoline. La France Automobile, Paris, January 29, 1899.

GEARS—
A Novel Dust-Proof Gear, illustrated and described. Automobile Magazine, New York, October, 1899.

HUB—
Perry's Hub, described and illustrated. Le Chauffeur, Paris, January 11, 1899.

HYDRO-CARBON AUTOMOBILES—
Gobron & Brillé's Vehicle and Motor, described and illustrated. La Locomotion Automobile, Paris, January 5, 1899.

Th. Cambier's (Lille, France), Vehicles, Motor, Truck, described and illustrated. La Locomotion Automobile, Paris, January 19, 1899.

English Daimler Vehicle, description and illustrations. La Locomotion Automobile, Paris, January 26, 1899.

Automobile "Cyrano," System Popp, truck, motor, gearing, and carriage described and illustrated.

The Automobile Index

La Locomotion Automobile, Paris, February 9, 1899. Also, La France Automobile, Paris, February 12, 1899.

Panhard & Levassor Light Carriage, described and illustrated. La France Automobile, Paris, February 19 and 26, 1899.

The Improved Daimler Motor Vehicles. With numerous illustrations of the application of this compact and convenient petroleum motor to launches, carriages, and heavy vehicles. Glaser's Annalen, Berlin, March 1, 1899.

Automobiles of Brouhot & Cie, described and illustrated. Simplicity in all machinery parts as well as in construction. La Locomotion Automobile, Paris, March 2, 1899.

Barisien Double Motor two-seat Carriage, described and illustrated. La France Automobile, Paris, April 30, 1899.

" Bolide " Carriage, described and illustrated. La France Automobile, Paris, May 14, 21, and 28, 1899.

Light Swiss Automobile, described and illustrated. The Autocar, London, May 27, 1899.

Brothier & Paugnaud Gasoline Carriage, described and illustrated. La Locomotion Automobile, Paris, June 1, 1899.

De Dietrich's Vehicles, truck, motor, carbureter described and illustrated. La Locomotion Automobile, Paris, June 29, 1899.

Henriod Vehicles, technical description and illustration of motor, regulator, gearing, and steering apparatus and truck. La Locomotion Automobile, Paris, July 20, 1899.

Leon Bollié's two Motors Voiturettes, described and illustrated. La Locomotion Automobile, Paris, July 27, 1899.

Voiturette Barisien—weight 450 pounds, 2 de Dion Motors—described and illustrated. La Locomotion Automobile, Paris, August 10, 1899.

Voiturette Renault frères Vehicle, motor, speed-changing gears, and steering apparatus described and illustrated. La Locomotion Automobile, Paris, August 10, 1899.

Peugeot Carriage, described and illustrated. La France Automobile, Paris, June 4, 11, and 18, August 13, 20, 27.

André Py Carriage, described and illustrated. Le Chauffeur, Paris, August 11, 1899.

Voiturette Underberg, described and illustrated. La Locomotion Automobile, Paris, August 17, 1899.

Henriod Two-seat Automobile, described and illustrated. Le Chauffeur, Paris, August 25, 1899.

The Turgan-Foy Motor Carriage. Illustrated article. Automobile Magazine, New York, October, 1899.

Daimler Automobiles, described and illustrated. The Automobile Magazine, New York, October, 1899.

HYDRO-CARBON BICYCLES—

Ridel Light Gasoline Motor Bicycle, described and illustrated. La France Automobile, Paris, April 23, 1899.

System Lamandière and Labre Motor and Bicycle, described and illustrated. La Locomotion Automobile, Paris, July 6, 1899.

System Labre, described and illustrated. L'Industrie Automobile, Paris, July 25, 1899.

HYDRO-CARBON MOTORS—

Motor Malesieux, described and illustrated. Le Chauffeur, Paris, January 11, 1899.

Henriod Motor, described and illustrated. La France Automobile, Paris, January 29, 1899.

Two-Cycle Motor " Le Gaulois," described and illustrated. La Locomotion Automobile, Paris, February 9, 1899.

Mors Motor, described and illustrated. La France Automobile, Paris, February 12, 1899.

System Bounard de Bouvand Rotary Motor, described and illustrated. Le Chauffeur, Paris, March 11, 1899.

System Alten and de Platen. Le Chauffeur, Paris, March 11, 1899.

Henri Chaudun Rotary Motor, described and illustrated. Le Chauffeur, Paris, April 25, 1899.

Gaillardet Motor, described and illustrated. La Locomotion Automobile, Paris, May 4, 1899.

Wiedknecht Explosion Motor Improvements, described and illustrated. La France Automobile, Paris, May 7, 1899.

Daniel Auge " Cyclope " Motor, described and illustrated. Le Chauffeur, Paris, May 11, 1899.

Pichard Gasoline Motor, described and illustrated. La France Automobile, Paris, May 28, 1899.

Nunn & White Gasoline Motor, described and illustrated. The Autocar, London, June 10, 1899.

Two-Cylinder Motor Sphynx, illustrated and described. La Locomotion Automobile, Paris, July 6, 1899.

Hire & Horn two-cycle explosion Motor, described and illustrated. La France Automobile, Paris, July 9, 1899.

Fritscher & Houdry Gasoline Motor, described and illustrated. Le Chauffeur, Paris, August 11, 1899.

Daimler's Motor, illustrated and described. Automobile Magazine, New York, October, 1899.

" Cyclone " Motor, described and illustrated. The Automobile Magazine, New York, October, 1899.

HYDRO-CARBON TRICYCLES—
" Ariel " Gasoline Motor Tricycle, described and illustrated. The Autocar, London, February 25, 1899.

Bluhm Motocycle, described and illustrated. La France Automobile, Paris, April 16, 1899.

Grosse & Boubault Tricycle, described and illustrated. The Automobile Magazine, New York, October, 1899.

IGNITING CANDLE—
Reclus candle described. La Locomotion Automobile, Paris, June 29, 1899.

KEROSENE OIL AUTOMOBILE—
Koch Kerosene Oil Automobile, described and illustrated. La Locomotion Automobile, May 4, 1899.
La France Automobile, July 16 and 23, 1899.

LAW—
The Law Relating to Automobiles in France. Rules for traffic on the public roads, with editorial. Automotor Journal, London, April, 1899.

LIQUEFIED AIR—
Critique by J. Ravel. Automobile Magazine, New York, October, 1899.

MILITARY AUTOMOBILES—
Automotors and Military Transport in Campaigns. M. le Capitaine Bardonnant in La Revue du Cercle Militaire. Abstract and criticism of an article by Major Mirandoli, published in the Rivista di Artigliera e Genio. Automotor Journal, March 15, 1899.

MOTOR TESTS—
Rules and Regulations for Competitive Motor Tests. La Locomotion Automobile, August 10, 1899.

MOTOR VEHICLE GEARING—
Chain and Sprocket and Spur Gear Compared. The Hub, New York, July, 1899.

NEW YORK AUTOMOBILE SHOW—
Advocacy for an Automobile exposition to be held in New York. The Automobile Magazine, New York, October, 1899.

OILERS—
Bonnefis Oiler, described and illustrated. La Locomotion Automobile, Paris, January 26, 1899.

OIL MOTOR—
The Koch Heavy Oil Motor. Illustrates this motor, which is considered the most successful of heavy oil motors which has ap-

The Automobile Index

peared, and calls attention to some features. Engineer, London, February 3, 1899.

OLD HORSELESS CARRIAGE—
Patent for a Horseless Carriage in Louis XIV.'s Time. Inventor, Jean Théson. La France Automobile, Paris, March 5, 1899.

OPERATING COST—
Operating Cost of Horse and Electric Delivery Wagons in New York City. Abstract of a paper read at the meeting of the American Institute of Electrical Engineers. Comparison of cost of operating in favor of the electric vehicles, with statement of advantages. Electrical World and Electrical Engineer, July 8, 1899.

PARIS AUTOMOBILE SHOW 1899—
Automobiles in Paris. Comments on the vehicles at the late exhibition, the lack of novelty, but evidences of improvement, the power used, etc. Engineer, London, July 14, 1899.

PARIS TRIALS—
Motor Cab Trials in Paris. An account of the trials, with description of the vehicles entered. Engineer, London, June 16, 1899.

POWER CALCULATING METHOD—
Power Calculating Method given. The Automobile Magazine, New York, October, 1899.

PRESSURE GAUGE FOR EXPLOSION MOTORS—
Duflos Clairdent System, described and illustrated. La Locomotion Automobile, Paris, February 9, 1899.

RESISTANCE FROM UNEVEN ROADS –
Illustrated description of a German solution of the problem of diminishing resistance caused by uneven roads. Automotor Journal, London, March 15, 1899.

ROAD INNS OF THE FUTURE—
Competition for Plans of a Modern Road Inn, with Repair Shop for Cycles, Automobiles, Charging Station, etc. Le Chauffeur, Paris.

ROAD VEHICLES–
A Study of Mechanical Traction on Roads. A discussion of the conditions of heavy haulage on roads and in city streets, with reference to the design of motor trucks for such service. Le Génie Civil, Paris, May 27, 1899.

ROADS—
Good Roads. Discussion of effect of Automobiles on Roads. Sports Afield, July, 1899.

ROLLER BEARING—
Connolly Roller Bearing, described and illustrated. The Autocar, London, May 13, 1899.

SCOTT SYSTEM " FIELD " BOILER—
Scott System Traction Boiler, described and illustrated. La Locomotion Automobile, Paris, March 30, 1899.

SPEED CALCULATING METHOD—
Speed Calculating Method given. The Automobile Magazine, New York, October, 1899.

SPEED REDUCER—
Angular Speed Reducer, Humpage System, described and illustrated. La Locomotion Automobile, Paris, July 6, 1899.

SPEED VARYING GEARS—
Conget's Speed Changing Apparatus for Motocycles, described and illustrated. La Locomotion Automobile, Paris, January 19, 1899.

Louis Brun's gears, described and illustrated. La Locomotion Automobile, Paris, January 26, 1899.

J. Didier Speed-Changing Gear, described and illustrated. La Locomotion Automobile, Paris, February 2, 1899. Also, L'Industrie Automobile, Paris, July 25, 1899.

Automobile Tricycle with Variable Speed Gear. Describing a practical form of variable gear to be used in connection with a petroleum motor, enabling a smaller motor to be used, and permitting low gears to be obtained for ascending grades. Revue Technique, Paris, April 10, 1899.

Diligeon & E. Mathieu speed-varying gear, described and illustrated. La France Automobile, Paris, April 30, 1899.

The Automobile Magazine

Petitjean & Sevette speed-varying gear, described and illustrated. La France Automobile, Paris, July 30, 1899.

Device for regulating speed, by Messrs. Roe and Knight. La Locomotion Automobile, Paris. August 3, 1899.

E. Buchet System, described and illustrated. La Locomotion Automobile, Paris, August 10, 1899.

Cherrier speed-varying gear, described and illustrated. L'Industrie Automobile, Paris, August 25, 1899.

Hugot speed-varying gear, described and illustrated. L'Industrie Automobile, Paris, August 25, 1899.

Montauban & Marchandier System, described and illustrated. The Automobile Magazine, New York, October, 1899.

SPRING SUSPENSIONS—
Spring suspensions discussed. Automobile Magazine, New York, October, 1899.

STEAM AUTOMOBILE ENGINEER-ING—
Excerpt from a paper read before the Liverpool Self-Propelled Traffic Association, by W. Norris. Discusses the writer's view of the correct lines on which a successful steam motor-driven vehicle should be designed. Automotor Journal, London, February, 1899.

STEAM AUTOMOBILES—
Steam Automobile of la Société Européene d'Automobiles — two-seat runabout, weighing 700 pounds; dogcart, four seats, 900 pounds—Patent Tatin & Tanière, described and illustrated. Le Chauffeur, Paris, February 11, 1899.

Piat Heavy Steam Truck, described and illustrated. La France Automobile, Paris, April 30, 1899.

Coulthard Heavy Steam Truck, described and illustrated. The Autocar, London, June 10, 1899.

A Ride in Professor Elihu Thomson's Steammobile. Electrical Review. New York, July 19, 1899.

Leyland Steam Break, described and illustrated. Le Génie Civil, Paris, August 5, 1899.

Serpollet's Steam Omnibus, steam being generated by crude petroleum, described and illustrated. La Locomotion Automobile, Paris, February 23, 1899.
Automobile Magazine, New York, October, 1899.
Automotor Journal, London, April, 1899.
Le Génie Civil, Paris, August 5, 1899.

Thornycroft Steam Wagon, described and illustrated. The Automotor Journal, London, August 15, 1899.

Some American Steam-Driven Motor Vehicles. Horace L. Arnold. Illustrates and describes four vehicles made in or near Boston, which collectively exhibit all the features of an ideal automobile. Industries and Iron, London, June 30, 1899.

Early Motor Vehicles and Modern Practice. Illustrated notes of early steam carriages. Automotor Journal, June, 1899.

STEAM BOILERS FOR AUTOMO-BILES—
A Light Steam Boiler for Automobiles, Tangye System, described and illustrated. La Locomotion Automobile, Paris, February 9, 1899.

STEAM ENGINE—
New Traction Engine for Overland Freighting, described and illustrated. Horseless Age, New York, September 13, 1899.

STEAM GENERATOR—
Rousseau Non-Explosive Steam Generator, described and illustrated. Horseless Age, New York, September 13, 1899.

STEAM MOTOR—
Steam Motor Serpollet, described and illustrated. The Automobile Magazine, New York, October. 1899.

STEERING AXLE—
Gobiet & Mersier Steering Axle, described and illustrated. La France Automobile, Paris. May 14, 1899.

The Automobile Index

STEERING GEAR—

Conink Epicycloidal Movement Steering Gear, described and illustrated. La France Automobile, Paris, February 5, 1899.

Jeantaud's Steering Gear, described and illustrated. The Automobile Magazine, New York, October, 1899.

STREET ELECTRIC SWEEPER AND SPRINKLER

Amiot-Peneau System, described and illustrated. La Locomotion Automobile, Paris, August 24, 1899.

STREET PAVING—

Motor Carriages and Street Paving. Thomas Conyngton. On the progress and outlook for these vehicles and the effect on the life in cities that would result from their general use. Engineering, Indianapolis, April, 1899.

TIRES—

" Teuf-Teuf " Tire. A combination of a pneumatic and a full tire, described and illustrated. La France Automobile, Paris, January 12, 1899.

Means for Preventing the Rapid Destruction of Rubber Tires by the Application of Metallic Pieces. La France Automobile, Paris, January 29, 1899.

Automobiles and the Tire Problem. Opinions of prominent men in the trade concerning the effect of the automobile industry on the demand for rubber. India Rubber World, New York, February 1, 1899.

Pneumatic Tires for Moto-Vehicles. Excerpt from a paper read before the Automobile Club. Considers cost, advantages, and disadvantages. Automotor Journal, London, March 15, 1899.

The General Relation of Tires to Motor Vehicle Construction. Considers the relation of the parts, and the effect on the wear of tires.

Horseless Age, New York, April 5, 1899.

Rolling Resistance to India Rubber Tires. The present article deals with the elasticity of the air in pneumatic tires, and discusses how far it is perfect or imperfect. Automotor Journal, London, April, 1899.

General Notes Respecting India Rubber Tires for Carriages and Moto-Vehicles. Illustrates and describes the chief existing types of tires, gives suggestions relating to the choice of tires, and advice as to their maintenance and preservation. Automotor Journal, London, April, 1899.

Simms Compound Pneumatic Tire. Illustrated. The Autocar, London, June 24, 1899. Also, La Locomotion Automobile, Paris, August 10, 1899.

TIRE TRIAL—

Chameroy Tires tried on bad roads, after a run of 4,500 miles, found in perfect condition. La Locomotion Automobile, Paris, August 17, 1899.

TRANSMISSION GEAR—

Transmission for Automobiles, Crouant Patent, described and illustrated. La France Automobile, Paris, February 26, 1899.

Spooner Transmission Gear, described and illustrated. The Autocar, London, May 20, 1899.

TRUCKS—

Henriod Truck, described and illustrated. La France Automobile, Paris, January 29, 1899.

WARNING SIGNALS—

Device of Mrs. Sophie Straus. La Locomotion Automobile, Paris, August 3, 1899.

WATER COOLER—

Benoit-Julien Water Cooler, described and illustrated. The Automobile Magazine, New York, October, 1899.

The Automobile Magazine

VOL. I   NOVEMBER 1899   No. 2

CONTENTS

AGENCY FOR FOREIGN SUBSCRIPTIONS:
INTERNATIONAL NEWS COMPANY
BREAMS BUILDINGS, CHANCERY LANE   STEPHAN STRASSE, No. 18
LONDON, E. C.   LEIPSIC

Price, 25 Cents a Number; $3.00 a Year

Motor Scouts in Action

The Automobile
MAGAZINE

Vol. I   NOVEMBER 1899   No. 2

The Automobile in War

By Edwin Emerson, Jr.

First U. S. V. Cavalry

"Napoléon s'est moqué de Fulton et de ses bateaux à vapeur ; nos gouvern-ants devraient tâcher d'éviter qu'on ne leur reprochât un jour d'avoir méprisé la voiture automobile."—Jean Baptiste Jobard's *Observations Recueillies*, 1835.

THE God of Battles, according to a cynical utterance of Napoleon, is apt to favor the side that musters the greatest numbers. Bismarck amended this to the effect that the God of Battles smiles upon those who manage to get there first with the greatest numbers. Admiral Dewey and Colonel Roosevelt, in the face of heavy odds, put their trust in "good guns and straight shooting," and such seems likewise to be the last article of faith with Oom Paul and his pious Boers.

England, in her present championship of modern civilization in South Africa, puts her reliance on a general superiority of force, both as to actual numbers and more effective engines of war. The sharp-shooting Boers are met by machine guns belching forth cartridges at the rate of a thousand shots a minute, and the nimble skirmishing of the Veldt rough riders is offset by the operations of a balloon corps, bicycle division, armored cars, and quick-firing automobile batteries, with an advance guard of motor scouts.

In the face of such radically different methods of warfare pitted against one another the world can foresee but one final outcome—the same which resulted when the frenzied spears-men of the Khalifa hurled themselves upon the serried ranks of Lord Kitchener's Maxim guns and magazine rifles at Om-durman.

To military men, no matter on what side their sympathies

may be enlisted, this *fin de siècle* campaign in a far-away corner of the earth is bound to prove of transcendental interest, with its opportunities for testing the latest important product of modern invention. Accordingly military attachés from all the armies of Europe, as well as from our own, have seized this chance to see service in South Africa.

The automobile gun which has found most favor with the British war office is the so-called "motor scout," invented by Frederick G. Simms, and first exhibited by him at Richmond. It is a motor quadricycle, fitted with a Maxim quick-firing machine gun. The frame is of stiff steel tubing, well braced together, and having a standard on the front upon which is

Old German Print of Cugnot's Steam Cannon (1765)

mounted the gun and a light steel shield. The ammunition is carried in boxes placed on the framing in the front. If desired, the gun can be dismounted and a seat substituted for it. A tripod mounting can also be carried if required. The wheels are made according to Messrs. Simms's design, and are fitted with pneumatic tires, the principal feature of which is the thick india-rubber tread; this is so made as to be detachable while worn, when it can be replaced by a new one. The rim of the wheel is made in three parts, the metal rim to which the rubber tire is attached, and the two side-flanges which keep the latter in place. In the case of damage, these side-flanges are easily replaced. In the event of a serious puncture, the tire collapses inside the rim, which then takes the wear. Side-flanges to the rim are shaped so as to permit a rim-brake being used with them.

The motor scout is fitted with a $1\frac{1}{2}$ horse-power petroleum

The Automobile in War

motor, which can propel it at any speed up to about eighteen miles an hour for a distance of 120 miles, or further with a reserve supply of petroleum. It is convertible, carrying either two persons or one person, and a light Maxim gun. In the latter case the gun is mounted in front over the leading-wheels and so arranged that it can be fired either to the right or left or straight ahead, with the vehicle going at full speed, while in a tray below it there is room to stow 1,000 rounds of ammunition within easy reach of the rider.

Another type of "war motor car" designed by Mr. Simms, is much larger and heavier, is armor-plated all over, and has a ram both in front and behind. The armament consists of two

Pennington's Autoram and Battery (1899)

quick-firing Maxim guns carried in two revolving turrets. The steering is managed by means of mirrors, and it is claimed that the crew need never expose themselves outside the protection of the armor. An electric searchlight is provided, the dynamo being worked from the main engine, which is a four-cylinder Daimler motor developing sixteen horse-power nominal, and at close quarters the electrical equipment can be further utilized to give a shock to persons touching the outside of the car. The tires with which the wheels are fitted are such as will enable the vehicle to travel over very rough ground.

A third type designed by Mr. Simms is a military railway inspection car, also armor-plated, and carrying a Maxim gun.

Another similar war engine has been designed and patented by Mr. E. J. Pennington, the well-known inventor, who has made

The Automobile Magazine

such great strides in the autocar and motor-cycle movement.
The equipment consists of two rapid-firing guns, mounted on

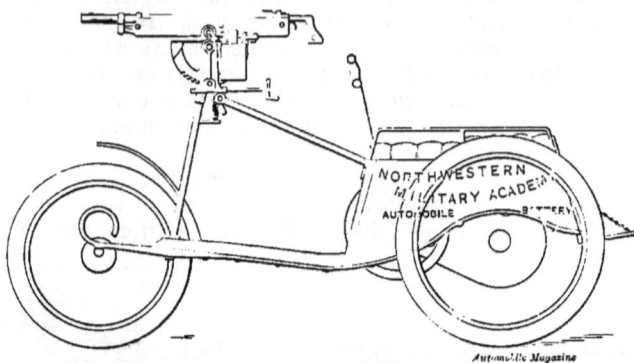

Major Davidson's Automobile Colt Gun

an autocar. The autocar is driven by a sixteen horse-power engine. the wheels having four-inch solid rubber tires. The guns
are on a swivel. and are arranged to discharge at various angles,
and they rotate automatically while they are being fired, if desired. The firing is controlled entirely by the engine which
drives the autocar, and the guns may be fired while the autocar
is in motion or when standing still. The ammunition is fired at
the rate of from fifty to seven hundred rounds a minute. Each
gun is supplied with five thousand rounds of ammunition. The
shields round the autocar are arranged so that an ordinary bullet
from a rifle would not penetrate, as they are on an angle. and
when these machines are in motion it will be very difficult for
a large field-piece to hit them.

The great trouble in the past with rapid-firing guns. according to English military authorities. has been that it is not possible to fire them at varied speeds. and sometimes the entire
charge is lost owing to this fact. With a gun of the above type,
it is claimed, even after the operators are shot down. the engines
would continue running, and would fire the entire charge of
ammunition. A rapid-firing gun is supposed to be a very good
thing, provided you can come in contact with your enemy, but
where you have a gun mounted on a fast-running machine.
which is capable of making forty-five miles an hour, it should
be an easy matter to reach close quarters. One can readily
imagine what would be the effect if an appreciable number of

The Automobile in War

these engines of war were placed under one command and charged into a large body of troops.

The first constructor of automobiles for war purposes, if we leave out of account Cugnot's attempts in the direction of steam cannons during the last century, was M. Serpollet, the well-known French inventor. In 1893 he built a steam artillery and ammunition wagon, which gained the warm approval of French military authorities.

The problem set by the Ministry of War was as follows: To render self-moving a wagon of large dimensions capable of containing a load of ammunition or cannon of a weight of 3,300 pounds. With this load it was to be capable of running a distance of twenty-four miles without stopping, at a mean speed of 4.8 miles per hour, whatever be the state of the roads, gradients included. This same wagon was to be able to haul a second one in its train, loaded with 6,600 pounds of useful weight, and,

Section and Plan of Serpollet's Artillery Autocar

a, motor; *b*, generator; *c*, water tank; *d*, steering apparatus; *e*, lever of the starting pump; *f*, reversing lever; *g*, starting pump; *h*, coal bunkers; O, swivel gun plate.

121

with this new charge, the speed to be, on an average, only 2.4 miles per hour, to utilize for this purpose an already existing wagon without making any change in its arrangement, the ordinary system of fore-carriage being preserved and the pole being replaced by another controlling piece.

The accompanying diagram shows the suspended artillery wagon and the arrangements of the mechanical adaptation that have converted it into a self-propelling steam wagon. The motor is placed in the rear under the floor of the vehicle. It has two cylinders and a change of direction by the Stephenson slide. The diameter of the pistons is five inches and their stroke also is five inches. The motor actuates the two driving-wheels through chains of a special system connected with the driving-wheels by an intermediate shaft with two rates of velocity. This shaft carries the differential movement.

Gradients of thirteen per cent. have been ascended with this machine without any difficulty, the wagon then having a weight of 9,460 pounds. These figures naturally lead to a comparison with traction by horses. The expense of fuel for a trip of twenty-four miles, counting coal $8 per ton, is eighty cents, adding ten cents for lubrication, it costs but ninety cents to carry 3,300 pounds to a distance of twenty-four miles. Let us compare this expense with that which would accompany eight spans of horses necessary to make the same distance in the same time, and we shall see that the first is not a third of the second. Besides, when the wagon under consideration has travelled twenty-four miles, it suffices to replenish its supply of water and fuel in order that it may immediately start on a similar journey. It would be out of the question to call upon horses to repeat such work.

The success of the tests to which M. Serpollet's invention was subjected led such progressive military men in France as General Le Marquis de Gallifet, General Zédé, and Colonel Picquart, all three of whom were supposed to be prejudiced in favor of the dashing tactics of cavalry and horse artillery, to recommend the adoption of an automobile corps to be operated in conjunction with a flying detachment of military cyclists.

Germany and Italy followed suit by introducing automobile and cycle tactics into the general field manœuvres of the last few years. At the German Kaiser manœuvres this year, it was generally commented upon as a significant thing that a military bicycle corps, supplemented by a wagon train of automobile delivery wagons, during a difficult movement over rough country, succeeded in outstripping the cavalry and in making so

The Automobile in War

decisive an onslaught upon the cavalry of the supposed enemy that the horsemen were ordered to beat a retreat.

In this country the foremost military champion of the Automobile is Adjutant-General Avery D. Andrews, of New York, the President of the Automobile Club of America, who first introduced the use of the bicycle as a police machine.

On the part of the Regular Army the utility of the automobile as a means of transporting the heavy paraphernalia of the United States Signal Corps has recently been tested under the personal supervision of General Miles. Some months ago the signal corps entered into a contract with the Fischer Equipment Company of Chicago to furnish two heavy delivery wagons and a light carriage, to be propelled by electricity from storage batteries. They were to be used in connection with the balloon service and experiments in wireless telegraphy. The heavy wagons were to be capable of carrying not less than eight hundred pounds, besides the driver, and must run for twelve miles on one charging. They may be converted into a signal corps station, and a switch-board is provided, by means of which the entire output of the battery at 55 volts may be available for general service.

These vehicles have lately been received in Washington after their efficiency had been tested at Fort Meyer. All three of the wagons are painted a pale sage-green to make them harmonize with the landscape and prevent their discovery by the enemy. The equipment of the delivery wagons is sufficient to run them thirty miles without recharging, carrying a load of 1,500 pounds. Independent motors are attached to each of the driving-wheels and are arranged to develop 6-horse-power each on grades, or 12-horse-power for the entire wagon. The average speed is ten miles an hour. All the wagons are fitted with attachments for the use of horses, and that motive power can be used at any time. The batteries in such case are taken out, leaving additional freight-room, and can then be used in connection with the field telegraph instruments, utilizing every part of the vehicles at all times. The wheels are provided with ball bearings and a rim of solid rubber makes travel smooth. Electric side lamps are on all the vehicles, and there is an electric light in the interior of the carriage. The vehicles are to be equipped with search-lights when placed in field service.

More recently still, an automobile gun-carriage constructed with a view to service in the Philippines has been thoroughly tested by Major Davidson and Sergeant Rice, of the Northwestern Military Academy.

The carriage employs the Duryea system of propulsion, which has been in use several years on carriages. It mounts a Colt automatic rapid-fire gun, firing about five hundred shots per minute. The cartridges are seven millimetre U. M. C., with smokeless powder and nickel-jacketed bullets, and have a velocity of two thousand feet per second. There is provision for carrying twenty-five hundred rounds of ammunition, a few accessories, and fuel. The large fuel tank for gasoline is placed under the forward floor, so as to be well protected from possible danger, and fuel for two hundred miles can be carried. This tank is of heavy seamless sheet-iron, and is practically bullet-proof.

The range of the gun which has a seven millimetre calibre is about two thousand yards. The gun points forward, and is ready for use at any time; it is mounted on a swivel and can be swung around, up or down, to cover any object, its sweep being that of a full half circle with a sighting range of 2,000 yards, and a firing range, when elevated, of 4,000 yards, or more than two miles.

The carriage weighs about nine hundred pounds, of which

United States Army Wagon

The Automobile in War

quantity the gun accounts for eighty. The running gear is made very strong, to withstand the rough usage to which the carriage may be subjected.

United States Signal Service Brake

Besides this it will carry an apparatus for pulling itself out of ditches and over embankments. It consists of a windlass and tackle. It thus has the power of pulling itself up by its own boot-straps, quite a remarkable feat for a piece of artillery.

The advantages which such a cannon has are many. It can travel thirty-five miles an hour, while horses do well to cover ten miles an hour in an emergency spurt when lashed to their fullest speed. When they have travelled at this rate for two hours they usually founder and fall dead.

What an immense superiority such a piece of artillery has over the kind now in use can be imagined, when it is remembered that a horse gun carriage has to be swung completely around, stopped and unlimbered before a shot can be fired.

In the face of such startling successes some people still insist on clinging to hoary traditions of warfare dating from the days of chivalry. Thus the editor of a leading American newspaper recently expressed himself on the subject in this wise:

The Automobile Magazine

"The automobile may be crowding the horse, but it will be a long day before it dethrones the army mule. Representatives of the English army have been buying 1,000 mules in the

Major Davidson and His Gun

South for use in the pending Transvaal campaign. The South African problem being a stubborn one, the army mule will be right in his element there."

On the day after this was printed, appeared a despatch from Durban relating the experiences of a cattle ship sent to South Africa laden with horses and mules.

In the midst of a severe storm her steering-gear broke, and the ship fell off in the trough of the sea. Being light, she tossed about very violently, and the horses being thrown against their poorly constructed stalls, broke them down faster than they could be repaired. In a few hours all the stalls were demolished, and their splintered stanchions and side boards, with projecting spikes, were mixed up with a moving mass of the half-dead, killed, and mutilated bodies of the horses, all shifting together with every toss of the ship. Finally the whole quivering, putrifying mass settled on one side, and would have caused the ship to founder had not the crew by herculean efforts managed to throw the horses overboard, killing a few that gave trouble. All were lost.

Some time previous to this General Otis, at Manila, protested against the shipment of further horses or mules, on the

The Automobile in War

ground that they were impracticable for operations in the Philippines and cost too much for subsistence.

His protest was emphasized shortly after, a despatch announcing the loss of 300 valuable mules on the U. S. transport Siam. The ship, while nearing the end of its journey, encountered a severe typhoon. Most of the forage which was on deck was swept overboard; all the stalls for the animals were smashed, and the mules were hurled from side to side, and frightfully mangled. Their legs and necks were broken, and the wretched animals fell in such a confused mass that the attendants were unable to relieve them. All but nineteen of the animals were lost.

Such are some of the problems presented by the use of animals for war purposes. As a matter of fact, a well settled opinion

F. G. Simms on His Motor Scout

is gaining ground among the best military authorities that the horse, considered as an instrument of war, is all but obsolete.

The first power to recognize this, characteristically enough, is Japan, which has lately withdrawn all its cavalry from Formosa.

after a series of disastrous experiences in its mounted operations against the rebels in the field.

In our own war with Spain, our Government was put to the expense of more than $100,000 in purchasing cavalry horses, but a couple of hundred of which could be used for service against the enemy. Of the few horses that were used in Cuba, the majority perished under miserable circumstances, and no end of trouble was caused by their loss.

No one, surely, who ever witnessed such pathetic scenes as those pictured in Frederick Remington's famous drawing, " How the Horses Died at Santiago," would care to subject his own mount to such tortures. The suffering of the poor animals on such occasions seem the worse for serving no practical end. Thus the writer well remembers how all artillery operations came to a standstill before Santiago because the fagged-out horses could not draw the cannon over the heavy roads.

Theodore Roosevelt's experiences with the unbroken horses of the Rough Riders and his own two mounts are a case in point. After no end of trouble with the wild bronchos that were impressed into service, all had to be left behind, to be cared for through many dreary weeks and months by the unfortunate Rough Riders who had to stay behind with them. Colonel Roosevelt in his book wrote this about it :

" As it turned out, we were not used mounted at all, so that our preparations on this point came to nothing. In a way, I have always regretted this. We thought we should at least be employed as cavalry in the great campaign against Havana in the fall; and from the beginning I began to train my men in shock tactics for use against hostile cavalry. My belief was that the horse was really the weapon with which to strike the first blow. I felt that if my men could be trained to hit their adversaries with their horses, it was a matter of small amount whether, at the moment when the onset occurred, sabres, lances, or revolvers were used; while in the subsequent mêlée I believed the revolver would outclass cold steel as a weapon. But this is all guesswork, for we never had occasion to try the experiment."

Of his own two horses that were taken to Cuba, the first, named " Rain-in-the-Face," was miserably drowned while disembarking at Daiquiri; the second, " Little Texas." to be sure participated in the charge up San Juan hill. but had to drop out at the first obstruction—a barbed wire fence. All the other officers and men fought on foot.

The Automobile in War

Among the general comments upon the uselessness of horses in that part of the country the writer likewise recalls how General Wheeler, cavalry leader as he was par excellence, gave it as his personal experience that cavalry attacks in modern warfare were suicidal, and should be undertaken only as " a last hope."

This coincides with Moltke's dictum after the terrible losses of the French squadrons at Wörth, Vionville, Beaumont, and Gravelotte, that the use of cavalry for general attacks could only be justified as an " ultima ratio regis."

At this latter-day stage of scientific warfare such an ultra feudal form of reasoning can no longer prevail. Any king who would send his horses against self-propelled batteries of quick firing magazine guns might as well quit the game, as did Napoleon after he had sent the flower of his cavalry into the sunken road at Waterloo.

Considered as a military machine the horse is done for.

" Done For "
(*From a photograph taken at the front by the author*)

"The Winner"

An American Auto Race

A UNIQUE feature of the recent Carnival of Sports held at Galesburg, in Illinois, was the first speed contest in that section of two American-built automobile carriages. The race, which was to be over a fifty mile course for a $2,000 sweepstake, widely heralded throughout the West, was run immediately after an automobile floral parade, similar to the one first held at Newport. Those who participated in the parade, and the large crowd that had gathered, went in a body to witness the race, thus adding largely to the gayety of the scene.

Shortly after three in the afternoon the two machines appeared on the track, one machine containing E. B. Snow, of Wyoming, the challenger, and an expert machinist, C. L. Turner, of Peoria. The other machine, owned by Dr. Morris, the defender, carried Albert S. Gale, of Galesburg, and Mr. Collins, an expert from Cleveland, O.

Mr. Snow's automobile was manufactured in Springfield, Mass., and was four years old, having come from the factory in

An American Auto Race

1895. It weighed 1,200 pounds. Dr. Morris's automobile came from the Winton factory at Cleveland, O., and was purchased this year. It weighed 1,500 pounds.

The judges were all from Galesburg, to wit: Col. J. T. Long, Bert Chappell, and John T. Piatt.

Before the starting of the race the following agreement was entered into by Mr. Morris and Mr. Snow, and the judges so notified: That no time would be allowed for breakages. If they could be repaired on the track, the repairs to be made, but no time for the same to be deducted. If any damage done was not repairable on the track, then the principal governing that machine was to forfeit the race.

The start was made at 3.29, and was a good one. Snow immediately took the lead and held it throughout the first mile, which was travelled in three minutes, and when the tape was crossed at the end of the first mile Snow was two carriage lengths ahead.

The second quarter of the second mile witnessed a change in the Morris machine forging ahead, and when three lengths in advance, taking the pole. The second mile was completed in 2.39, with Morris one length ahead at the crossing of the tape.

Morris maintained the lead through the third mile, which was done in 2.42, and at the completion he was one length in advance.

Snow spurted during the fourth mile, at the half, and took the lead, holding it then successfully for four miles, at varying distances. The fourth mile was done in 2.43, with Snow half a length ahead.

The fifth was completed in 2.40, and at this juncture Snow crossed the tape four lengths ahead of Morris. The five miles had been run in 13.44.

At the end of the sixth mile Snow had gained, and was now six lengths ahead, with the mile done in 2.38.

The seventh mile was made in 2.39, and the distance between the machines was the same.

In the eighth mile, when nearing the first quarter, the Morris machine made a pretty spurt, overtaking the Snow carriage and passing it. The lead here attained was never relinquished, and from that time until the completion of the race Morris retained the lead. The eighth mile was done in 2.36, with Morris three lengths ahead.

In the ninth mile Morris gained more ground, and at the completion of the mile, in 2.32, a sixteenth of a mile separated the two carriages.

The tenth mile was made in 2.45, and Morris was one-eighth of a mile in the lead at the crossing of the tape. The ten miles had been made in 26.54, the latter five in 13.10.

During the eleventh mile the Snow machine made slight gains, but this was due to Morris stopping for a can of water for the cylinder. The mile was made in 2.54, with Morris a sixteenth of a mile in advance.

It was during the twelfth mile that trouble came to the Snow machine, and when near the half-mile post it had to stop. Examination proved that one of the sparkers in the cylinder had given out, which caused the lessening of speed and which left them with but one engine to complete the race, with which it was impossible to do so.

During this time the Morris machine had been continuing its course, for Snow had not yet forfeited. The twelfth mile was made in 2.44, the thirteenth in 3.25, the fourteenth in 2.50, and the fifteenth in 3.08.

Mr. Snow had now reached the judges' stand, and forfeited the race, and the Morris machine was stopped. The fifteen miles had been done in 43.54.

Shortly after the forfeiting of the race Mr. Snow gave authority for the announcement that he would immediately telegraph for a new Duryea machine, one of the same kind he used, and would challenge Dr. Morris for a race, to be held over the same race-track, the purse to be the gate receipts.

A NEW SPEED RECORD.

The world's record of speed for automobiles has been officially established here this week, said a Paris despatch, dated October 21st.

During the Bordeaux-Biarritz race, two weeks ago, a speed of forty miles an hour was maintained for five hours.

Yesterday, under test conditions, a representative of an automobile company averaged forty-six and one-third miles an hour during a run of six hours, without a single stop, over ordinary country roads.

Miss Betsy Carr in Control

America's First Automobile

By Lewis C. Strang

FRANK CURTIS, of Newburyport, Mass., does not claim to be the original automobilist, but he does modestly acknowledge that he was probably the first man in the New World to run an automobile. Only he does not call it by any such hifalutin name, preferring to designate it as a steam-carriage.

It was just thirty-three years ago that Mr. Curtis first showed his machine in public, to the astonishment of the citizens and the horses of Newburyport. At the same time the rash inventor incurred the openly expressed disapproval of the inhabitants of that quiet burg, who, indeed, were never brought to countenance such an uncanny thing as a wagon that went without horse-power. Mr. Curtis, indeed, was sadly in advance of his times.

This steam carriage, be it known, was not Mr. Curtis's first experience with steam-propelled vehicles. Back in the early sixties he was engineer in the local fire department, and he had charge of the first steam fire engine that the town owned.

" It seemed to me a shame to half kill two fine horses every time there was a fire," he says, in telling the story, " and that, too, with plenty of good steam going to waste. So I fixed up the old engine so that she would go herself. Oh, yes, she made a big stir, and Bean and Damrell came out from Boston to ride on her. She went first rate, too, only they wouldn't let me put any steering apparatus on her for fear she would run away— or for some other reason just as silly. So two men had to go on ahead and steer the thing by means of the pole.

" The only times I ever had any trouble with the fire engine was in the winter, when there was snow on the ground. That made hard going, and she used up a pile of water. I remember once we were going to a fire some distance away, and she worked so hard that she got out of water. The only hydrant that we could get to was frozen up, and, consequently, we never reached the blaze. Another time she broke through the ice into a big puddle in the street, and there the wheels stuck while the body kept going. We hoisted her back on the trucks, however, and went on to the fire. It was pretty late when we got there, though."

Mr. Curtis thus relates how he failed to make a fortune:

" I was the first one to use a donkey engine in the engine house to keep the steamer's boiler hot. I applied for patents for my propelling apparatus and for the heating idea. The authorities would not give me one for the propeller, but they said I could have one for the heater. That made me mad, and I made up my mind if I couldn't have a patent for the whole business, I wouldn't have one for part of it. A few years after a man offered me $20,000 for the patent on the heating idea, but, of course, I didn't have anything to sell. Then another fellow got it patented and sold it to New York City for a big price."

The steam carriage was built to order by Mr. Curtis for a Boston man—a gay Lothario, Mr. Curtis calls him. It was put together in a little shop in the old gold foundry on State Street, opposite the old Public Library, in the spring and summer of 1867, and was first used in August of that year.

" It cost me $1,000," Mr. Curtis complains, " and I never got a cent for it. The chap I made it for promised to pay half down when it was finished, and the rest if it proved satisfactory after trial. He used it a few times, but I never saw the color of his money, so I made up my mind that I had better keep it myself.

"Oh, it went all right; it was good for twenty-five miles

America's First Automobile

an hour any day in the week. I had it in running order for eleven years, and then I sold it to save my credit. The man who bought it let it go into a decline, and what is left of it is stored in a barn near my place now."

This is how Mr. Curtis thinks he saved his credit:

"You see, folks around here got to looking upon me as a sort of crank, and when I sold the steam carriage I suppose they thought I was getting sensible. At any rate, I could borrow more money after I got rid of it than I ever could before. Perhaps they only thought that I stood a better show of not getting killed, and so being able to settle up with them."

The first person to brave the dangers of an automobile ride with Mr. Curtis was a clergyman—the Rev. Dr. Seymour. The old inventor thus records his passenger's impressions:

"The parson came to me one day and asked me if I knew Betsey Carr. "'Yes, I know her,' I said; 'but she isn't Betsey Carr any longer; she's married to my uncle, Mr. Cheney.' The parson knew that, and he wanted to see her. She lived some little distance outside the town, and he didn't know how to get there.

"'I'll fix that all right,' I said. It was in the afternoon, and he had an evening service.

"'Can you get me back in time?' he asked.

"I told him I'd guarantee that, and I got out the steam carriage and we started. You never saw a man so tickled as the parson. He wanted to know if I could stop her as easily as I could steer her. So when we got to my uncle's place I ran her up between two posts and halted her as pretty as you please. Then I fetched him home and brought him up to his house with a whirl. He was immensely taken with the machine, and that evening, at the North Church service, after they had had some preliminary singing and praying, the parson got up to say something, and he talked about nothing but that steam carriage for an hour."

The following are the inventor's own specifications as verbally communicated to the writer: "She was run by a five horsepower engine. She had a vertical boiler, which weighed 109 prounds and cost $109. It was made by the Whittier Machine Works. The boiler was just in front of the back seat on which the engineer and pilot sat. She was steered by a lever, attached to the front wheels in such a way that you never felt any jolt when the wheels struck a stone or a hump in the road. The running gear was attached to the rear wheels. In front of the boiler I could rig up a dinky seat so that she would carry four persons.

The Automobile Magazine

The carriage and machinery weighed 618 pounds, and in a box in the rear I stowed away 80 pounds of coal. Under the seat there was a tank that contained 20 gallons of water. So the whole outfit, with four passengers, weighed over 1,100 pounds, and even then I could make 25 miles an hour.

"She never broke down with anyone except myself, and that was when I tried some experiments; and she never got stalled except once, on Plum Island sands, and she would have gone through them with the proper kind of wheels. There were no macadamized roads in those days, and she'd climb hills and never turn a hair, though you could hear her puff half a mile away. She had a blast like a locomotive, with a vent under the seat. That, of course, made a noise, and the last thing I did for the machine was to have a 'quieter' made, but I never put it on. I carried 40 to 45 pounds of steam, usually, and she'd blow off at 80 pounds. Going up a hill she would run up to the limit in a minute and a half. Going down hill she'd fall off below normal.

"With 80 pounds of coal the steam carriage would run about 30 miles, so she wasn't particularly economical. We had to stop for water every six or eight miles, though, and that was a nuisance. The longest stretch I ever ran without stopping was nine miles, and I did it in 26 minutes. Another time, when a man kept the fire going for me, I did six miles in 18 minutes. I used to wear gloves and put the coal in with my hands. The fire-box came just where the door could be opened with the toe. I remember once I came down a hill and along a road parallel to the railroad track. A train was going the same way we were, and we made as good time as it did. The passengers opened the windows and waved their handkerchiefs and cheered us.

"The only real trouble I ever had was with horses, but even then it all depended on the driver. If he wasn't afraid, nine times out of ten the horse wasn't either. Once I was out and I met two ladies in a carriage. I turned out on the side of the road and waited for them to pass.

"'Why did you do that?' asked the one that was holding the lines. 'We wanted to see you go.'

"'All right,' I said, 'if your horse doesn't mind,' and off I went. Then I came back and circled around them, just to show off a bit. I never ran over but one person, and that was a boy who persisted in trying to get across the road in front of me. I went over his ankle, but a stick of candy fixed him all right. He's living yet.

America's First Automobile

" It was the horses that finally drove me off the streets, though," Mr. Curtis added. " A wealthy woman in town had a coachman who was frightened to death of me, and he made so much talk that the woman finally complained to the Selectmen. They sent the sheriff after me, with order to keep me off the streets, or else bring me and the carriage into court. I was in the steam carriage down in Market Square when the sheriff drove up.

" ' I want you,' he shouted.

" ' What for? ' I asked. He went on to tell me, and I suggested that he'd better read his warrant. So he sat in his carriage and I in mine while he read it. When he got to the part about delivering me over to the court, I just pulled open the throttle, and away I went, with the sheriff driving like mad after me. He might as well have been on foot. He couldn't keep anywhere near me. I ran over to the next town, and stored the steam carriage in a barn until the excitement wore away. Then I brought her home and sold her."

ELECTRIC CABS IN GERMANY.

According to the *Elektrische Zeitung*, electricity seems to be the only available power for public automobile cabs in Germany, as the use of explosive motors is forbidden by the authorities. As the cabs must have certain measurements, prescribed by the authorities, old carriages formerly used with horses were changed into electric cabs, the Hellmann and Vollmer-Kuhlstein systems being used. In Berlin the two motors are placed under the seat of the driver; each weighs 110 pounds and drives one of the rear wheels by means of a chain and a flexible shaft. Each motor alone can drive the cab at reduced speed. At 85 volts the speed is 1,100 revolutions per minute, each thus giving two horsepower. The accumulators are placed under the cab in a special case, which can be replaced by a new one in two or three minutes. The battery of four cells has a capacity of 60 to 70 ampère-hours and is sufficient for a trip of 19 to 26 miles. The braking can be accomplished electrically as well as mechanically. The cab has a weight of 2,750 pounds without passengers; it will seat five persons besides the driver. In Hamburg, one motor, known as a Vorspann, is placed under the driver's seat, and thus the front body of the vehicle cab serves to propel the rest of the transformed cab.

The Oldest Chauffeuse

By Suzette

M RS. SARAH TERRY, of Philadelphia, having arrived at years of discretion—she has celebrated her one hundred and eighth birthday—the other day went out to see the town in an automobile. The ride was a mutual pleasure and surprise all around, for the town seemed as much in-

Mrs. Terry Takes an Outing

138

The Oldest Chauffeuse

terested in Mrs. Sarah Terry as Mrs. Sarah Terry was in the town.

This remarkable old lady in her old age has had but two desires, one of which has now been fulfilled. One is to receive a pension from the government because she is a daughter of the Revolution in fact as well as in name, and the other was to ride in an automobile.

To celebrate the event of her one hundred and eighth birthday she was taken out in a hansom belonging to the Pennsylvania Electric Vehicle Company, of Philadelphia.

The old lady was in a flutter of excitement all the morning before her prospective ride. She insisted upon dressing and being arrayed in her best bonnet and choicest shawl at least two hours before the vehicle was ordered.

The neighbors came to their doors and windows, and the passers-by paused on the pavement to see the extraordinary spectacle of an old lady, one hundred and eight years of age, walk unassisted down a long flight of marble steps and get airily into an automobile.

In spite of her years Mrs. Terry is a very charming woman, retaining the dainty manners of the last century that made her a favorite at the Danish court many years ago. Her face is framed in snow-white hair, which she wears in the quaint puffs of our great-grandmothers. There is always handsome black lace in the neck and sleeves of her gown, and her black bonnet ties in a broad bow under her chin.

The driver of the automobile took unusual interest in his passenger, and started slowly that she might not be startled.

"Doesn't it seem queer!" exclaimed the old lady, in high glee, as the automobile rolled noiselessly over the paved streets. Something of the old-time sparkle came into her eyes, and she grew animated in the bracing fall air.

When the automobile swung swiftly around a corner she opened her eyes a little wider and clung to the seat, but she gave no evidences of fear.

Once on Broad Street, the wide boulevard which is the city's pride, the quaint little figure in the horseless carriage attracted endless attention. Busy men stopped to look, people driving in their carriages stared in open-eyed amazement, the policemen grinned, and "all the town wondered."

The wind blew the old lady's bonnet strings till they fluttered out behind, but she sat up like a major, enjoying the novelty to the fullest extent. The faster the automobile went the better she liked it. The evident amazement of other old ladies she passed afforded her the greatest amusement.

The Automobile Magazine

" I rode in about the first steam-cars that were used," she said, " and I thought they were wonderful; but I never dreamed of riding in anything like this. Just fancy, no horses, no nothing; it all seems like magic. I should think the horses would be very jealous to see their occupation being taken from them in this fashion. When I was young, the only horseless carriages we knew were sedan chairs. It's so exhilarating, I don't see why people use horses when they can go like this. I am not nearly so uneasy as behind a skittish pair of horses."

The old lady was very anxious about her bonnet.

" I must keep it on straight, for I don't want to look rakish," she said, coquettishly.

She insisted upon the automobile stopping in front of the houses of her friends that they might see her, and so she went from one end of the city to the other, commenting brightly upon the changes that had been wrought within her recollection.

She was out two hours, and could scarcely be persuaded to go home.

" I am not a bit tired," she insisted; " it did not shake me up as ordinary driving does, and I do not see why I cannot stay out the rest of the afternoon."

In spite of her protestations she was somewhat flustered from the excitement, but she thanked the driver for the care he had shown and said it was one of the nicest birthdays she had ever had.

"And next year, when I am one hundred and nine," said the old lady, cheerfully, " I am going out again."

GROWTH OF THE NEW INDUSTRY

There are now over 7,000 owners of automobiles in Europe, and the number of vehicles is, perhaps, 10,000, and of this number, says the Annuaire Générale de l'Automobile, 5,606 are in France. There are 619 manufacturers in France, not including the makers of parts; 998 dealers in them, and 1,095 repair shops. For the remainder of Europe the figures are not very complete. There are 268 owners of automobiles in Germany, 90 in Austro-Hungary, 90 in Belgium, 44 in Spain, 304 in Great Britain, 111 in Italy, 68 in Holland, 114 in Switzerland. It is impossible to state at the present time how many automobiles are in this country, since so many new concerns are preparing to turn out motor carriages of all kinds in large quantities.

How to Buy an Automobile

By H. de Graffigny, C.E.

SO numerous are the types of motor carriages now in common use, that one who has but a superficial knowledge of mechanics hardly knows which form to select for his own purposes. For this reason I shall endeavor clearly to state those considerations which should govern the unmechanical layman in the choice of an automobile.

The use to which the vehicle is to be put is a question of paramount importance. It should be clearly determined beforehand whether the carriage is to be employed for long-distance traveling or for pleasure-riding within the limits of a city. In the former instance a petroleum-motor carriage will prove the most serviceable; while in the latter the electric carriage will answer all requirements, since at the end of each day it can be returned to its charging-station.

This first point having been disposed of, the next question to be considered is the price to be paid; and here a few remarks will not be out of place.

As in all industries which are still in their infancy, the hand-labor required in the building of automobiles is exceedingly costly. Good workmen are still difficult to obtain. Often the engineers are themselves compelled to assemble the various parts; for in this kind of work it is not a mere question of mounting an ordinary motor or building a common carriage. The one must be modified to conform with the requirements of the other. It is for this reason that the prominent manufacturers exact such high prices for their carriages; and since automobiles are rather fashionable, they receive orders for a year in advance. In spite of the prevailing high prices it is better to purchase a carriage from one of the large makers, since the more insignificant builders, not having the facilities of their better-equipped rivals, can make only inferior vehicles, and are thereby enabled to sell at low prices.

The automobile-industry at present is in very much the same state as was the bicycle-industry in 1889. At that time a good wheel was worth $150. Now a bicycle, better constructed and more durable, can be bought for one-third the money. The prospective automobile-buyer should therefore confine himself only to well-known firms, and should not hesitate to pay $200 or $300 more for a carriage that will perform all the services which he may exact of it.

When the amateur visits the various shops and compares different models, he should not allow himself to be influenced by the exterior of the carriage or by the more or less happy arrangement of parts; for external appearance has nothing to do with mechanical construction, and can be made to suit the taste of the purchaser. What is of greater moment is the motor and co-acting mechanism. The motor constitutes the essential part of the carriage; everything else is merely of secondary importance. The purchaser should carefully examine the carriage which he has in mind and ascertain whether its motor and power-transmitting mechanism are good and adapted to the needs of the carriage, whether the framework is strong and capable of yielding to strains and shocks, and whether the controlling and steering mechanisms are easily operated. The prime consideration, as I have already stated, is a good motor, simple in construction and operation—one, the parts of which are readily accessible and which is easily controlled.

The "voiturettes," or lighter carriages, of which there are several kinds, have the advantage of being cheaper than the larger automobiles. They are extremely serviceable, not only in cities, but also in long-distance traveling and touring. They are, however, not so comfortable; and, the motor being less powerful, their speed is not so high.

Motocycles of the de Dion, Gladiator, and Clément type have been hailed with delight by many wheelmen. In starting and in ascending steep grades the motocycle must be driven by the feet as well as by the motor—a feature which is considered a fault by some and an admirable quality by the majority. The few rare opponents of this system claimed that they would prefer to ride an ordinary bicycle; but the motocyclists reply that a wheel is always fatiguing and an automobile enervating, while the motocycle combines the merits of both, in so far as it requires an amount of exercise which can be varied at pleasure.

By reason of its small height from the ground the motocycle, like the light automobile or " voiturette," is not the most pleasant of vehicles in rainy weather or on muddy roads. Unless his wheels be provided with guards, the cyclist will soon be covered with mud; and on slippery pavements he may even meet with accidents. These are the disadvantages of the motocycle compared with its advantages. As to steam motocycles, they present so many inconveniences that they are hardly to be recommended for general use.

A Floating Steam-Carriage

ON March 29, 1809, Charles Dallery filed an application in France for a patent on a steam-carriage. The patent was issued on October 2d, numbered 6,776, and entitled *Système mobile perfectionné appliqué aux transports par terre et par mer* (An improved motive system applied to land and water transportation). The patent describes an unsubmergible boat and a carriage, both driven by the same means. We shall concern ourselves only with the carriage.

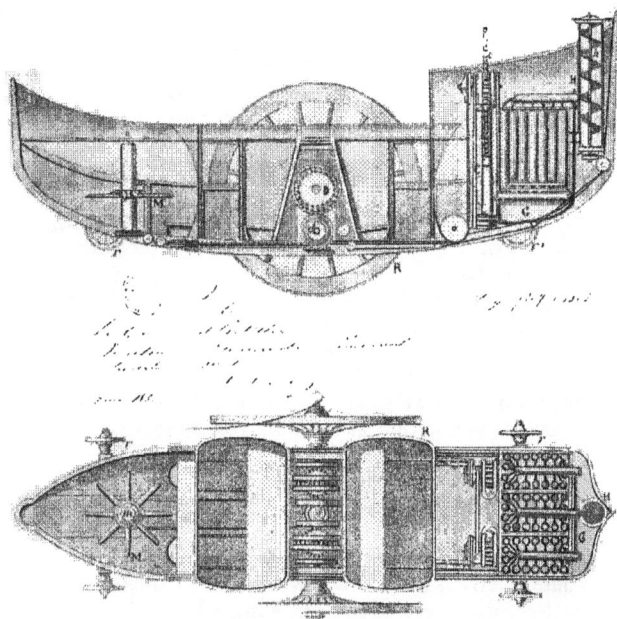

Dallery's "Boat on Wheels"

In his very obscure description of his invention Dallery states: "The carriage has the form of a boat, from the centre of which rises a casing divided into two equal portions by two back-boards (*dossiers*) separated by a space of eighteen inches."

"The carriage is traversed by an axle divided into two equal parts, each extremity of which is provided with a wheel (R) of

143

the ordinary size; the axle is carried by a movable frame which breaks the force of shocks received."

" The fire-engine (*machine à feu*) has in the present form but three furnaces; its proportions correspond with the size and weight of the carriage. The furnaces are located toward the rear end of the boat (G), at a distance from the casing sufficient to provide room for the man who attends to the fire."

[The smoke escapes through a stack H, the draft being produced by a screw *h*, driven by the motor.]

" Two steam-cylinders (CC′) have pistons provided with two racks (*cc′*) which turn two cog-wheels (*ee′*). The shafts of these two wheels transmit their movement by means of two ropes to two ratchets secured on a shaft (*d*), which is held by the frame. The two ratchets transmit the movement by means of a pawl (*cliquet*) having two wheels secured to the said shaft. These wheels transmit the movement to two drums (D), the said drums turning about the centre of each portion of the axle."

" Up to this point the movement has not been communicated to the axle; in order to attain this end, other intermediary means must be employed. Each portion of the axle, toward one side of the hub of the wheels (R), carries a vertical gear-wheel, the teeth of which are turned toward the centre. From this point to that of their reunion, the bearings of the axle are round; at this latter point and on each inner side are two other movable, double-toothed vertical gear-wheels, the inner teeth of which engage a cog-wheel acting as a pinion to connect the two wheels (R). The cog-wheel is operated by the wheel (M) mounted on the vertical shaft (*m*), and serves as a point of support; it operates to turn the wheels (R) parallel with each other or in a curved line, so that one turns more rapidly than the other."

" The drums inclose a wheel which meshes with the teeth of the wheel fixed on the shaft, toward the hub of the wheels (R), and with the outer teeth of each moving wheel, also mounted on the axle, at the point where the two portions meet. By reason of this engagement, it follows that the impulse is communicated to the opposite vertical gear-wheels in order to drive the wheels (R) forming part of the respective portions of the axle.

Four small wheels (*r* and *r′*) maintained the stability of the vehicle when running on up or down grades.

From this obscure description of Dallery's it is evident that a vehicle mounted on two large wheels, with small supporting wheels at the ends of the vehicle, is no new idea, although such contrivances have been patented in our own day.

The First Motor Ride

By John Grand Carteret

SIR GOLDSWORTH GURNEY, the well-known English inventor, early in the century, succeeded in perfecting a motor carriage, run by steam, to the point of exciting serious attention and comment. Even the *Britannic Review*, in one of its issues of 1829, acknowledged a discreet possibility in the new idea, in the following measured terms:

"This carriage, now perfected in all its parts, has been examined by a large number of enlightened persons. It has been put in movement in their presence. The simplicity of construction, its rapidity, the facility with which it is guided, and, above all, the evident security, have all been noted. The result of this experiment has convinced the most incredulous that the new invention will obtain the favor of the public; and that the application of the principle on which the vehicle rests will soon extend to all sorts of carriages, and thus become a universal usage."

But in 1830 the same *Review* questioned in less general terms the particular difficulties of the scheme. "The great obstacle to the introduction of steam carriages was the weight of the machinery and the resistance opposed thereto by the inequalities in the surface of roadways. Our best roads present a resistance to the progress of these vehicles, because at each rise in the ground a double or triple expenditure of force is necessary to preserve headway. We should have machines of great energy which do not create a resistance equal to their weight. . . ."

An account of a ride in the Gurney stage coach was published by *The United Service Journal* (1829). The impressions of one of those who took part in this unique journey are set down as one might who made his first voyage in an air-ship or submarine boat at the present time. Here is the thrilling story:

"We numbered four in a coach attached to the steam carriage, and we had travelled without experiencing any difficulty or mishap as far as Longford, where they were repairing the bridge built over the Cambria.

"On this bridge was a large pile of bricks, so high as to conceal what was happening on the other side. Precisely at the moment we began to cross the bridge the mail-coach from Bath arrived at a brisk trot on the other end. As soon as we perceived it we shouted to the driver to take care; but, as he was not aware of the extraordinary vehicle he was going to meet, he paid no attention to our warning and did not slacken speed. To avoid a collision, Mr. Gurney guided our steam carriage into the pile of bricks. Some damage to our apparatus resulted, but was repaired in less than a quarter of an hour. As to the horses

Old English Print

of the coach, they had taken the bit between their teeth and had to be cut loose.

"When we entered Reading it was 8.20, and we remained two hours to repair one of our wheels. Mr. Gurney had noticed that a certain small chain of the action was broken; this accident was, without doubt, the result of what had happened on the bridge at Longford.

"We left Reading only at 10.30, and arrived at Melksham toward eight o'clock in the evening. We had made about six miles an hour, including stops.

"It must be remembered that our principal object had been to avoid mishap; and to that end we had provided in advance an abundant water supply. In order to incur no risk, we had made a rule to run no more than four miles without taking water. Thus we stopped whenever we noticed water facilities on the route, often when we had made but two or three miles, in the fear of missing it farther on. We were altogether eight travellers, and as many engineers and workmen, for we were followed by a wagon which carried our fuel.

" No smoke was visible when we burned coke, though coal, which otherwise made a good fire, gave it out in a large amount. At Devizes the coke which we procured proved to be of so bad a quality that we were obliged to use coal. Smoke began to pour out, and, as night was falling on our arrival at Melksham, bright sparks were seen flying from the smokestack. We were agreed that these sparks might be of danger along the way if we should meet a load of straw or a hay-wagon, but we were able to avoid danger by returning to the use of coke.

" Upon our arrival at Melksham, we found that there was a fair in progress, and the streets were full of people. Mr. Gur-

Old English Print

ney, who combines with his inventive genius the most amiable qualities of heart, made the carriages travel as slowly as possible, in order to injure no one. Unfortunately, in that town the lower classes are strongly opposed to the new method of transportation. Excited by the postilions, who imagined that the adoption of Mr. Gurney's steam carriage would compromise their means of livelihood, the multitude that encumbered the streets arose against us, heaped us with insults, and attacked us with stones. The chief engineer and another man were seriously injured in the head. Mr. Gurney feared we could not pursue our journey, as two of his best mechanics had need of surgical aid. He turned the carriage into the court of a brewer named Ales, and during the night it was guarded by constables with the authorization of magistrates. The next day we resumed the journey to Bath under escort.

" In Mr. Gurney's opinion, the machine was in better condition and made more headway on our return journey, which is proved by the fact that we made it in four hours' less time. The nearer we approached to the end of our journey, the more rapid became our speed. The road was, however, exceedingly muddy from a heavy rainfall."

A few notes taken on the return journey show us to what subterfuges the inventor was driven to avoid a recurrence of the disagreeable events of the first day:

" We were six miles from Devizes at three o'clock in the morning. . . . Mr. Gurney had passed through the town of Melksham in the night with horses, so as to give the roughs no excuse for rioting. The machinery was set to work as soon as we were well on the outskirts. At Devizes Mr. Gurney put on all his speed to escape the popular aggression of the manufacturing districts. Our course was so rapid that the horses of the post carriage which accompanied us were quite winded.

" At the foot of the hill of Devizes we met the mail-coach and another carriage, which stopped to see us ascend the steep height. We mounted rapidly. The travellers in the coach, delighted at the unusual spectacle, encouraged us with shouts of applause."

To have assisted at the experiment of Gurney's steam carriage was, in those days, almost a title to glory. These carriages became speedily one of the curiosities of London. Foreign travellers who crossed the Channel and printed accounts of their journeys, did not fail to devote a chapter to the new means of locomotion. Jean Baptiste Jobard, Belgian savant and economist, was of the number, and so were Mr. Cuchetet, St. Germain Leduc, and C. G. Simon, three prominent scientific writers of that time.

M. Jobard's impressions, as noted down at the time, are worthy of record:

" My first visit in England was to the starting station of Sir Goldsworth Gurney's steam omnibus, which, during the last fortnight, has been running between London and Bath. This carriage, which can accommodate thirteen persons, does not differ materially from other stage-coaches, nor has it had any serious mishap as yet. For my benefit it manœuvred back and forth over the street pavement and later on the smooth macadam of the highway, without any apparent difficulties of guiding. The drivers of other stage-coaches are agreed that the

The Progress of Science—Whitechapel Road in 1830.

(Caricature of Aiken, published in "Modern Philosophy," in 1828.)

thing is a success, and that before long it will do them much harm. After the coach had ceased running for the day, the engineer, Mr. Gordon, showed me the mechanism in detail. When he showed the engine and boiler I saw that they had been injured by the excessive heat of the fire and that the front wheels had suffered so much from the same cause that it was necessary to make repairs.

"A common reproach against these carriages is that they frighten horses, but a parliamentary inquiry has proved that this reproach is not well founded and that most accidents were due to the carelessness of the coachmen. A curious thing is that the Society for the Prevention of Cruelty to Animals has espoused the cause of this new system of locomotion, for the purpose, so it is said, of protecting horses against the great strain that has been put upon them by the increased speed demanded from the stage-coaches of to-day."

The concluding words of this author are as effective to-day as they were then:

"Napoleon made fun of Fulton and his steam-ships. Our present rulers would do well to avoid falling into a similar contempt of the modern steam-carriage."

Steel Tubing for Automobiles

By Charles G. Canfield

SCIENTIFICALLY, seamless cold-drawn steel tubing is the best known form in which a given amount of like material can be used to resist the greatest torsion or support the maximum weight. Just when, in the world's history, this important fact was discovered and used is difficult to ascertain. It is sufficient to state that the tube is the best form of strength to resist torsion and bending.

Observe the universal appearance of the tube in vegetable life, and the most familiar example is the growth of the straws of grain. What a marvel of power to resist torsion and bending is the rye straw, which frequently rises five feet above the ground and supports the head of grain four or six inches in length and twenty-five times its weight. The bamboo of India is another example of vegetable fibre formed in a tube, of such

Steel Tubing for Automobiles

strength and lightness that it is common to a thousand uses in the arts. Every one of those trees is a tube; in some the body and limbs are solid, but the form is tubular; and in many tropical species the trunks are hollow.

In animal life the same thing appears. The bones of most animals are round, excepting where strength is sacrificed to form. The ribs that contain the vitals in the chest inclosure are made strong by being formed of flattened tubes. This is a mere economy of space. Even the vertebræ of the fishes are tubular, and the form is so common in animal life that there is no need of specifying.

Passing from the study of the tube in nature, the most wonderful example of its manipulation and adaptation by man is the bicycle. He sought to do in mechanics precisely what is found in nature—to produce the strongest frame with the least possible material in quantity and weight. In this particular he succeeded beyond the wildest dreams of the early makers. If the reader will reflect how crude, clumsy, and heavy the original frame of the velocipede was when it appeared as a pacing machine, built with a wood frame; how the bicycle followed, with a heavy iron frame, and, lastly, the safety, with its thin steel tubing—a marvel of lightness and strength—he will realize how slow has been the evolution of the bicycle.

At first thought it seems fortunate that the bicycle builder has preceded the engineer in his work of perfecting the automobile, but experience does not justify this conclusion. With the splendid achievements of the safety bicycle before him, the designer of the new vehicle has largely disregarded the lesson so plainly taught him in bicycle construction, and almost wholly omitted from the automobile the use of the tube. This is a singular error.

At the present time—the end of this century—the leading manufacturers of automobiles in the United States are imitating horse vehicles, and build on the old-fashioned carriage-makers' lines of design and method. It is the purpose of the writer to point out how this mistake is made, and suggest the remedy, having a faint hope that it is possible to save time and money in the evolution of the ideal automobile.

It seems perfectly natural that the automobile constructor should imitate the carriage-builder—first, because his work was similar, and, second, his desire to please the public, which had no other standard of taste than the horse carriage. The first thought was to make a four-wheeled vehicle to carry two or more people comfortably, gracefully, and steadily, and having

The Automobile Magazine

the mysterious power to propel itself. This labor was chiefly mechanical in its inception, for a machine with sufficient power and under perfect control must first be produced. Greatly to his credit the engineer solved this problem, but so absorbed was he in his mechanical work that he seems to have overlooked the important conditions of weight and material, and overbuilt the vehicle in all its essential parts. In the end this mistake will prove beneficial, because the engine required to drive an overbuilt machine will exert more power than required to propel one of less weight.

In this connection let us consider a single example. The body and running-gear of its electric cabs, as now constructed in America, aside from machinery and storage battery, weigh 2,200 pounds, and the load to be carried is two persons and one operator, estimated at 600 pounds. To this add 1,600 pounds of battery and 500 pounds of machinery, making a total of 4,900 pounds. From this it is plain that seven times the weight of passengers to be transported is employed in the weight of the vehicle.

With tubular construction of frame and body, at least 1,200 pounds can be eliminated in weight, together with 800 pounds of battery and 200 pounds of machinery, making a total reduction of 2,200 pounds—a little more than one-half the weight of the entire vehicle. We are aware that improvements in batteries as well as motors have been made. These reduce the weight, but the principle remains the same, for whatever reduction is made in the weight of the vehicle is a saving as well in the weight of motor power. If tubular construction be generally used in the body and running-gear of the automobile vehicle, the greatest strength and least weight will be realized.

The second cause which led to heavy-weight automobile construction was the natural desire of the maker to please the people, and he constructed the vehicle to resemble the horse carriage. This, to some extent, is a pardonable offence. The designers of horse vehicles have been employed for centuries, educating and cultivating the taste of the people to their standard, and in this they have succeeded to fix and prejudice our judgment.

It is but natural that the automobile designer should shrink from changing these conditions, and give to his patrons, as near as possible, the same thing. But the time has arrived when originality must take the place of imitation and copy, and the ultimate creation must be a distinctive automobile vehicle.

Why Not Send the Horse to School?

By Sylvester Baxter

T would be greatly to the advantage of both the horse and the automobile if training - schools for horses were established in every centre of population. The greatest drawback to a much more rapid introduction of the latter for domestic purposes, both of pleasure and general utility, is the fear of frightening horses.

" I should get me a motor-carriage at once; it would be more convenient in every way, and save me a deal of expense and trouble. But I think I shall wait awhile until horses get used to them. I do not want to be continually haunted by the fear of frightening them and causing runaways, smash-ups, serious injuries, and perhaps loss of life. After a while horses will take them as a matter of course, just as they do bicycles, and then I will lose no time in getting me an automobile."

One very frequently hears remarks of the foregoing tenor. Public sentiment is almost universally favorable to the automobile. Its advent is sure of a hearty welcome all along the line, with the exception of the comparatively insignificant factors, even though large and extensive in themselves, devoted to maintaining the old order of things. The uncomfortable obstacle is as aforesaid, lying in the difficulty of reconciling a very sensitive animal to the presence of his mechanical rival on the thoroughfares. If this obstacle could be overcome it would benefit the horse and his owners as well as it would the automobile interests, for thousands of persons are abandoning the use of the horse, even before they are ready to adopt his substitute, simply for fear of his taking fright at any unlooked-for moment under the ever-increasing occupation of the highways by other forms of traction.

This process of abandonment was well under way even before the coming of the automobile, largely in consequence of the multiplication of bicycles, until they have to an enormous extent attained the majority among vehicles on the public road, and likewise because of the extension of electric-car routes throughout all the populous sections of the country. The necessity of constantly encountering these things, and consequently of being ever on the lookout for careless wheelmen riding with heads down, or on the wrong side of the road, or without lanterns at night; or for the coming of electric cars at crossings or curves— all these things are apt to induce such a condition of the nerves as to make even " pleasure-driving " anything but a pleasure for the one who drives. Therefore a large proportion of the thousands who have found delight in visiting pleasant rural scenes, or the beautiful parks of the large cities, and who have depended upon horse and carriage to get there, have come to deprive themselves of a form of daily or regular recreation that was once one of the greatest of their sources of enjoyment—simply for the reason that conditions of fear have grown to outweigh the conditions of pleasure.

And now, with the coming of the automobile, this unpleasant condition of things will naturally be greatly augmented unless something is done to mitigate the situation very essentially. So why not send the horse to school? The education that he gets at present he picks up at random, for the most part, much like the street-urchin of the city, learning whatever happens to come in his way, but omitting to learn many desirable things that would stand him in good stead on occasions. All horses come to town from the country, and it is commonly in the country that they receive the schooling for the road known as " breaking in." And the thousands that come to live in the city adapt themselves to the irritating conditions of urban life mainly by the experience acquired as they are driven about the streets. Not infrequently, however, something happens that their experience has not prepared them for, with dire consequences to those about them, if not to themselves.

The horse's training in the country nowadays includes, of course, familiarity with the bicycle, and to a considerable extent with the trolley-car as well as the steam-cars. But even now, when bicycles are everywhere thick as blackberries, so to speak, at times we hear of a runaway caused by fright at a bicycle. It seems as if almost anything—perhaps even a bucketful of oats! —would scare a horse out of his senses should he happen to see it under circumstances at all unusual.

Although the horse has a very small brain capacity for a

Why Not Send the Horse to School?

creature of his size, he seems to make up in acuteness of perception for his deficiency in judgment. He is quick at recognition, and has a retentive memory. He is therefore amenable to instruction, as a rule. The accomplishments of the circus-horse are good evidence of this. And though congenitally one of the most timid of beasts, the horse can so readily be adapted to extraordinary conditions that he can be made to achieve at least the semblance of valor. He advances fearlessly into battle, and delights in the rush of conflict, the clamor of war—unheeding frightful sounds in vast accumulations, any one of which would be sufficient to strike terror to his heart if heard under ordinary circumstances. All this is done by schooling. So the horses of a city fire department likewise seem to take huge pleasure in the excitement of alarm, the intense rush to the scene of action, the tremendous clatter and clanging of the apparatus, the streams

of flying smoke, and the confusing steam. It is the unwonted that terrifies the horse, and if pains should be systematically taken to accustom him to all sorts of unwonted things and circumstances, the danger of his taking fright would be very much diminished.

It has been observed that in the first stages of an innovation in the vehicle line there is commonly an epidemic of horse-frights and runaways. But after a while the horse seems to accept the situation, and go his way unconcernedly in presence of the strange thing. Even green horses from the country, on beholding one of these city spectacles for the first time, are not unlikely to take the experience quite tranquilly. Perhaps they notice that other horses accept the thing as a matter of course, and so they conclude to do the same. Horses, in common with other animals apparently, have some means of intelligent communication with each other, and so when a new thing comes up, like the bicycle, the trolley-car, or the automobile, those who have got used to it pass the word around to all of their kind not to mind the thing; it won't hurt them one bit! Some hold that this is done by telepathy. And since scientific investigation appears to have established the fact of thought-transference, there seems to be no reason why it should not occur in some degree among the lower animals as well as among human beings. But after all, it seems best that we should make sure of the training of the horse in respect to the things that are liable to make him lose his head rather than leave him to pick up his knowledge from the fellows of his species.

There are certain features about the motor-vehicles that make it perhaps more difficult for a horse to get used to them than to the other things that successively have been trying his nerves in the course of the present century. In the first place, the automobile is a carriage, and the horse perceives the fact. Therefore, he cannot understand how it is that a carriage should move along without a horse to pull it. He is alarmed by the noise made by some; and the clouds of steam that proceed from others apparently make him think that a locomotive has escaped from the rails and taken to the highway. And the electric-cab, in its earlier shape at least, in its bulkiness of aspect reminds him of the hated steam-roller. Motor-bicycles and even tricycles are least likely to give him a fright, unless they make too great a noise, for they look like more familiar things.

In the large cities, where the automobile is coming into use, a training-school for those who are to run them is one of the essentials. In this sort of a school, besides practice under skilled teachers in ascending and descending grades, making curves, turning corners, etc., the avoidance of all kinds of obstacles is taught. To this end the way, here and there, is obstructed by all sorts of things that one may at times encounter in the highway—sticks and stones, brickbats, bicycles, baby-carriages, wheel-barrows, and dummy figures representing nursery maids. old women, drunken men, policemen, etc. Among all these things the driver of an automobile is taught to steer his way with care and precision.

Now, why should not this institution also be utilized as a school for the training of horses in paths of familiarity with things that are liable to cause them fright? If this were done, it would be equally useful on both sides. The user of the automobile would have practice in the most important matter of knowing what to do in the presence of horses; how to pass them on the road face to face, or coming up from behind, and how to act with presence of mind when a horse gives the slightest indication of showing fright. On the other hand, horses would have the very

Why Not Send the Horse to School?

best of opportunities to get used to the various classes of auto-mobile. Not only this, but the course of training for the latter should comprise all possible things under all practicable circumstances in the way of startling sights and sounds: the flapping of awnings; the flaunting of banners; newspapers, etc., blown about the road, and under a horse's feet; all sorts of strange things lying about the highway, of which the obstacles placed for automobile practice would in themselves furnish a goodly variety; the explosion of fire-crackers and torpedoes, the firing of pistols, rifles and cannon, " German bands," and Salvation Army bass-drums!

A certificate of thorough instruction, with lessons well learned, given by such a training-school would be appreciated by purchasers of horses, and would soon be likely to be demanded by them. It would even be well if it should be required by law that no horse should be sold without the giving of such a certificate, for the very good reason that a horse not so trained is an animal too dangerous to be allowed the freedom of the highway in these days.

A GRADUATE

Gallery of American Automobiles

American Electric Vehicle Co.—Stanhope.

American Electric Vehicle Co.—Brake.

(*This Department will be a regular feature of the* AUTOMOBILE MAGAZINE.)

Gallery of American Automobiles

The Riker—Brake.

The Riker—Coupé.

(*This Department will be a regular feature of the* AUTOMOBILE MAGAZINE.)

Automobile Power-Transmitter and Speed-Changer Combined

By E. E. Schwarzkopf

IN the operation of automobiles as constructed at present, the greatest drawback to their successful operation is derived from the loss of power in sudden starts and in climbing heavy grades.

In all the existing designs of electric automobiles the motor is positively connected to the axles or the wheels of the vehicle, with the result that when it is required to start the vehicle from a static condition, the electric motor must also be in a state of rest. In starting, the motor, being at its least efficient condition, meets its severest strain, and the storage battery thus receives excessive shocks, which must necessarily shorten its life. It must be remembered that the vehicle has to be gradually started from rest to a required speed, during which time it is necessary to use resistance, causing a great loss of energy. Resistance must also be used to permit the vehicle to trail wagons in the street, the motor at this time running at a low efficiency. When heavy grades are encountered the severe shock to the storage battery is too well known to require comment. These conditions are the principal causes of the rapid disintegration of the storage battery.

The same wasteful condition of energy exists in automobiles when driven by either a gasoline or steam engine.

A new mechanical invention has recently been put forward which is bound to prove an important factor for service in this connection. It is designed to overcome the various drawbacks we have enumerated as existing to-day. It will be clear, from the accompanying illustrations, that many combinations of this transmitter can be made other than those shown in the illustrations. It will be noticed that in all the illustrations the same basic principle is used in transmission, namely, a new system of differential gears; that the prime and last movements are in one straight line; and further, that the strains imposed on the transmitter are all of a tortional character. The details of this transmitter are briefly as follows:

On the prime moving shaft are fastened two eccentrics of equal diameter, equal face, and equal throw; the throws of ec-

Power-Transmitter and Speed-Changer

centrics are set diametrically opposite each other; thus the eccentrics are in equipoise. On these eccentrics are loosely mounted two discs, having fastened to them on both of their faces either internal or external gears, and meshing with these gears are respectively external or internal gears. One of these latter gears is fastened to the last mover, the balance being fastened to flanges operated by the brake mechanism. The last mentioned gears act as fulcrums in transmission, while the gears attached to the last mover and those in mesh with same act as levers in this combination. In order to transmit motion to the last mover in this transmitter, it is necessary to hold, through the action of the brakes, one of the fulcrum gears. These brakes are operated either by hand or electrically, and it will be plain to those skilled in the art that this transmitter, driven by one motor, will have, in addition to the function of differential speed of the last mover, with a proportionate increase of torque for overcoming the inertia of the vehicle and its load in starting for varying its speed, owing to the conditions imposed by street traffic or for hill-climbing, also the function, without further complications, of permitting the vehicle automatically to turn curves. In order to make this clear to the lay reader, we will refer to the illustrations showing this invention. In Fig. 1 this transmitter is shown with its electric motor mounted on the axle of the vehicle; on each side of the motor are shown duplicate transmitters connected on each side to a carriage wheel, both transmitters possessing differentialities of speed from that of the motor in ratios of 35 to 1, 13 to 1, and $7\frac{1}{2}$ to 1. All the brakes are electrically operated. To explain this brake action a rheostat is set close to the hand of the driver, having three points of contact respectively connected to the brakes on both sides, engaging the three differentialities enumerated. The wires from the rheostat lead first to the steering handle of the vehicle, and thence to the brakes. If the driver wishes to turn to the right, then the electric current which actuates the brakes operating the left-hand transmitter would be cut out, and the left-hand wheel would run idly, but the transmitter connected with the right-hand wheel would positively drive the vehicle around the curve. Referring to Fig. 1, and in order to clearly show the functions of this invention, we will assume that we start the vehicle. At this time the motor would be stationary. The operation would be as follows: the driver would switch the current to motor, when the motor would idly revolve to its highest efficient speed, having no connection to the axle of vehicle, until the driver engages the brakes through the rheostat. He first engages the brake which gives a differentiality from the speed of the motor of 35 to 1 with its proportionate torque, and thus overcomes the inertia of the vehicle. He then moves the rheostat to engage the brake giving the differentiality of 13 to 1, and in turn moves the rheostat on to engage the ratio of $7\frac{1}{2}$ to 1, driving by the last combination the carriage to its full speed.

It must be clear that in the operation of the cycle we have given, no excessive draught on the storage battery has been made, the vehicle having moved from a static division to, say, fifteen miles per hour. To stop a vehicle operated by this transmitter the driver would, by the aid of the rheostat, reverse the cycle before given, and the vehicle, by a cushioned effect, would pass through the several differentialities of speed governed by the transmitter. By placing the

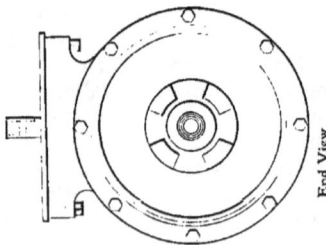

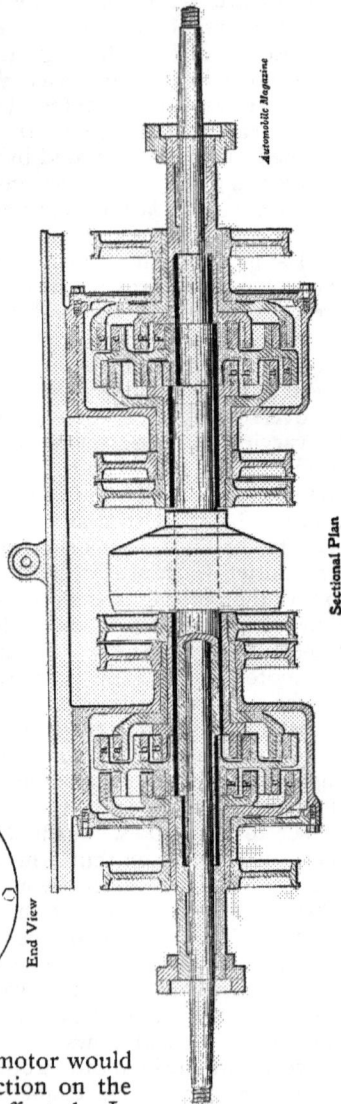

rheostat on the dead point, the motor would revolve idly, and, having no action on the vehicle, a quick stop would be effected. In

162

all known types of automobiles, when a stop is desired, the prime mover by its own momentum still drives the vehicle against the action of the brakes, notwithstanding that the current or power has been cut off. This is a potent reason for the many collisions by automobiles. Assuming that the three differentialities shown in Fig.

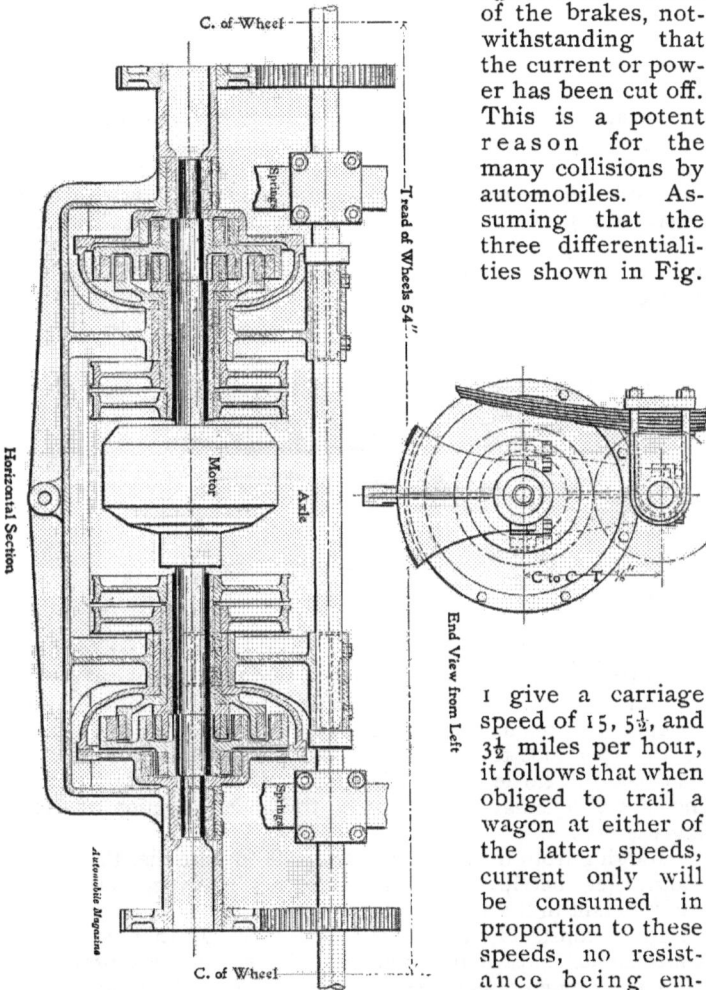

1 give a carriage speed of 15, 5½, and 3½ miles per hour, it follows that when obliged to trail a wagon at either of the latter speeds, current only will be consumed in proportion to these speeds, no resistance being employed; and when climbing excessive grades the leverage on the carriage axle would be increased in direct proportion to the higher ratio of the

differentialities employed. In Fig. 2 is shown another combination of this transmitter, the last movers being a gear in mesh with a gear fastened to the carriage axle. The alignment of the transmitters with the axle is positively secured by their being yoked to the same, the frame of transmitter on the opposite side being fastened to the carriage body by bolt. By this arrangement of suspension the alignment of the transmitter to the carriage axle is always preserved irrelative to the position of the carriage body, owing to the contraction or expansion of the carriage springs. In this combination there are two differentialities on each side, the same automatic arrangement as shown in combination Fig. 1 for turning curves being retained. The one differentiality would be used for the full carriage speed, the other for overcoming the inertia of the carriage or as a hill-climbing device.

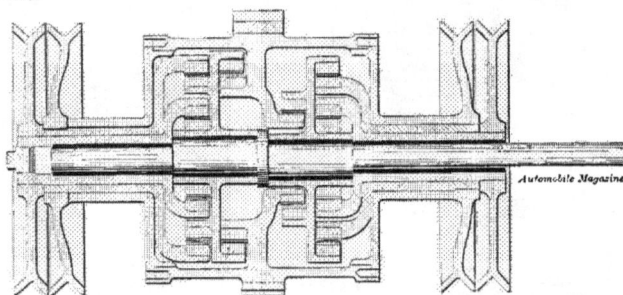

Automobile Magazine

In Fig. 3 is shown a combination of this transmitter for driving heavy delivery wagons. In it is shown three go-ahead speeds and one backing speed. We believe that in this class of automobiles the possibilities of this invention will be clearly shown, as all experience to date has proven the absolute necessity on this class of vehicle of variable speeds, with their proportionate increase of leverage in addition to the added function of backing the vehicle.

Judging from the above, it seems safe to state that the Birrell transmitter, herein described, overcomes many serious drawbacks extant in the automobile of to-day. It is simple in design, cheap in construction, self-contained in its arrangement. As a mechanical construction it is said to have won the approbation of many competent engineers in this country and abroad.

The New Sport Abroad

(By Our Own Correspondent)

THESE late autumn days in Paris have been delightful for lovers of automobile sport, who are taking ample advantage of the charming weather, filling the roads of the Bois with a kaleidoscopic variety of their vehicles, and presenting a spectacle which for a stranger affords a marvellous testimonial to the universality of the new form of locomotion. Even the Parisian never tires of its fascinations, and finds one of his greatest delights in watching the movement that shows the supremacy of his beloved capital in the latest contribution to modern mechanical achievement.

One of the most interesting things for us lies naturally in the fine additions made to the magnificent house of the Automobile Club. These include new dining-rooms, salons, and a beautiful little theatre, which will make a capital place of assemblage for the international automobile congress, which will probably be a feature of the Exposition next year.

Every new interest, of course, develops its freaks, cranks, and 'phobiacs. The worst instance of automobiliphobia I have heard of is that of a citizen of Cobourg, a village near the mouth of the Seine. This man, who lives in a villa with grounds bordering on a frequented highway, has somehow conceived an insane hatred for the automobile, and has devised a fiendish sort of contrivance to carry his malice into effect. He has cut off one of his trees at the ground, and converted it into an automobile-trap by hinging the trunk to the stump. He holds the thing in place by a kind of tackle, by which it may be dropped across the road when an automobile draws near, just in time to assure a collision. Either a lunatic asylum or a prison should be the destination of this iconoclast.

French Tree Trap.

The Automobile Magazine

There is a deal of talk about making long-distance tours into remote parts of the world—but until the era of universal good roads arrives these must necessarily remain in the air, the projects of uninformed enthusiasts. The ways are so bad in these remote lands that one might as well attempt to run an automobile down the rocky bed of an Alpine torrent as try to go over them. Thus there was much talk about a race from Paris to St. Petersburg, but even our national affection for the land of the bear would hardly be sufficient to induce the most intrepid of our *chauffeurs* to venture upon the journey after the report of the committee that undertook to investigate the subject. M. Thevin, who went over the route as a delegate for the committee, or attempted to go over it, speaks in high praise of the German part of the journey, for there he found the roads perfect, but of the Russian portion he said: " The roads only exist on the maps; the pools of mud and the wretched swamps are not to be dignified by the name of roads, and they would wreck the stoutest vehicle driven at more than six miles an hour." There are also long stretches of uninhabited wastes that would be most difficult to traverse.

Another chimerical-sounding project is that of an Englishman who proposes an automobile trip from Hongkong to Paris and London, starting next February. Dr. Lehwess, of London, is the man. He proposes to use a stout carriage with a motor burning ordinary kerosene. He thinks he can make the trip in three months! For the difficult stretch between Pekin and Khiachta, on the Russian frontier, a distance of 800 miles, he proposes to carry a kind of anchor, or grapnel, with a length of wire rope, to haul the car over otherwise impassable places. The men who made the various transcontinental bicycle tours found the obstacles well-nigh insurmountable, and it is a very different proposition for an automobile. Andree had a simple task in comparison. One might as well attempt to navigate an ocean liner overland. Fish should stick to the water, and automobiles to good roads.

In a recent trip to the other side of the Channel I was much impressed with the progress made in England. They have at last got rid of the great obstacles imposed by antiquated legal restrictions, and the Automobile Club of London is to celebrate the passage of the " Light Locomotives on Highways Act " by a great banquet.

There is a regulation of the British Local Government Board which provides that " light locomotives," as they say in England, shall stop at the request of any police constable, or of any person

A French Floral Prototype.

having charge of a restive horse. As this regulation is often being made use of simply for the annoyance of automobilists, the Automobile Club has issued a notice to drivers of motor-vehicles to the effect that only in certain instances, which are specified, does the right hold good to stop a "light locomotive," and that a person not the driver of a restive horse, or the driver of a horse that is not restive, has no right to require it to stop.

Newcastle-on-Tyne is to be the centre of an important "auto-motor service." A new company has been organized, called the North Yorkshire and South Durham Syndicate, to institute motor-omnibus lines along the banks of the Tees. Orders have been placed for two carriages to carry a score of passengers each, and six to carry ten each. These are to run between Stockton, Eaglescliffe, Yarn, Middlesbrough, and Marton Lane End, the service later to be extended to various other places.

Whoever has had the fortune to enjoy the hospitality of the Travellers' Club in London cannot easily forget the interesting persons from all parts of the world he has met there, as well as residents who have been in all parts of the world, and the many fascinating experiences and adventures there recounted. With the extension of the automobile to remote parts of the earth, we shall be likely to find our automobile clubs becoming centres just as entertaining in a similar way! We may even enjoy meeting some automobile lion-hunter, some Selous-on-wheels, so to speak, or listen to the thrilling account of grizzly-shooting by an automobilist among the Rocky Mountains! When your Rocky Mountain roads become equal to those of the Alps or the Black Forest—and they will be, sooner or later, for now we are going to have good roads everywhere—this will not be an impossibility, any more than it was for that talented countryman of yours to make those wonderful instantaneous photographs of wild beasts in action, in the heart of the wilderness. As it is. I remember that my learned Dutch friend, Dr. ten Kate, who was with Cushing when he made those extraordinary discoveries in your arid Southwest, told me that in Arizona it would be quite practicable to shoot rattlesnakes by the dozen from an automobile, the deserts there are so hard, smooth, and level.

But the nearest we get to such narratives at present is in tales of Alpine adventure from daring *chauffeurs*, who have made their way across the snowy ranges on the way to Italy or Austria.

A distinguished party has recently left Paris for an automobile tour of several months in Algiers and Kabylia. The party consists of Count Bosson de Perigord, who is the second son of

The New Sport Abroad

German Floral Show (Berlin).

the Duke de Talleyrand and Sagan, Count de Crisnoy, Count de Moustiers-Méronville, Count Marius de Gallifet, and M. de Lazarches d'Azay. As our French civilization means good roads everywhere it is carried, they will doubtless find smooth ways, and correspondingly pleasant experiences, for the greater part of their sojourn in African wilds. And perhaps we shall hear of the first automobile lion-hunt from them!

Last July Baron Duquesne gave us a pretty thrilling account of Alpine adventure on his automobile journey through Switzerland to Austria—an account calculated to discourage the most intrepid *chauffeur* from venturing into Alpine wilds. But perhaps the valiant Baron wanted to keep the field to himself! For now my friend, M. Baudry de Saunier, has lately had a letter from a

The Automobile Magazine

friend of his telling about a round trip just made from Vienna to Paris by the direct route, returning by way of Switzerland and the Tyrol—all without the slightest misadventure, and surmounting long and difficult twenty-per-cent. grades without any trouble. The heavy mountain work was pretty hard for the motor, he said, but his Dietrich vehicle sturdily overcame every obstacle without the least injury. He enjoyed the novel pleasures of a pioneer, for in crossing the Alps he went by roads the greater part of which were traversed by an automobile for the first time; the Simplon, at an altitude of 2,009 metres above the sea, followed by the Maloja at 1,811 metres, the Bernina at 2,330 metres, and finally the highest point with the Stelvic, at 2,814 metres.

For the automobilists of Munich, where motor-vehicles have achieved a remarkable vogue of late, the passage of the Brenner has become a favorite diversion. This famous historical highway crosses the Alps at an altitude of 2,034 metres. It leads all the way through delightful scenery, separates the Zillerthal Alps of the Tyrol from the Stubay-Otzthaler group. For time immemorial this pass has been the main highway between Germany and Italy. In the middle ages it was called the *Kaiserstrasse*, the Emperor's way, and up to the present time it has been the most frequented of all trans-Alpine highways. In 1772 the road was thoroughly rebuilt under Maria Teresa, and is now a fine example of mountain highway engineering. I once went over the route from Munich on a bicycle, and the memory of the glorious scenery is with me yet, although many years have passed since then. I well remember how pleasantly the highway impressed me as it wound along within sight of the train. Our brethren of Munich are to be congratulated on their fortune in possessing such a noble automobile route from their very doors into enchanting Italy. I hope to have the pleasure of retracing my trip some day on an automobile. The journey is practicable, not only for strong and heavy motor-carriages of eight or ten horse-power, but for the light vehicles of three horse-power or so. Baron von Franchetti, a devoted automobilist and musical composer, recently made the trip in a " Velo-Comfortable." Previous to the opening of the railroad in 1867, as many as 25,000 vehicles of all descriptions used to pass over the Brennerstrasse, as the Germans call the road, every year, but since then comparatively few have used it. Now, with the development of automobilism it will become again a very popular pleasure route. The road leaves the valley of the Inn at Innsbruck, and ascends through the Wippthal to the lake called the Brennersee, which is famous for its good fishing. The posthaus at the summit is not far from the celebrated health resorts, the Brennerbad and the Wildbad, with warm springs, the water at the latter having a temperature of 21° centigrade. The descent is through the Eisackthal by way of Hohensass, Sterzing, Brixen, and Botzen into the Etschthal. At Botzen there is a wonderful view looking back towards the Dolomites. A slight detour takes the tourist to the famously beautiful Lago di Garda, which lies partly in Austria and partly in Italy.

A section of this route was the scene of a trip of the Bavarian Automobile Club, of Munich, the past summer. It was intended to make the entire trip from Botzen to Munich, but at the last moment difficulties in the way of stations for replenishing the gasolene supply presented themselves, and so the programme was shortened, making Innsbruck the starting point. There were twelve participators. The start was made at five o'clock on Sunday morning, July 23. Three violent tempests tested the qualities of the tourists and their vehicles. Baron von Dietrich-Luneville, in a handsome carriage of his own invention and construction, was the first to reach the goal, terminating the stretch of 173 kilometres at 10.55 o'clock—or in five hours and forty-five minutes running-time. His motor was of sixteen horse-power. This was the first mountain trip of an automobile club in Germany. Next year the club will doubtless make the entire distance across the Brenner.

A notable automobile journey was that made by a Berlin journalist and his wife in July, from Berlin to Paris and back. The trip was made by means of a motor-tricycle, with a two-seated voiture attached. An expert driver rode the tricycle. Four weeks were taken by the journey, which covered a distance of at least 2,400 kilometres, and crossed the Harz and Vosges mountains. There were no mishaps whatever, and the tourists returned to the German capital enthusiastic over their delightful experiences. The weight of the three persons and their baggage was about 545 pounds, the latter weighing about 150 pounds. Only about ninety litres of gasolene were used, at a cost of nine dollars. In the dry battery used with the motor, the same element sufficed for the entire trip. At starting it showed an energy of a little less than six volts; on the end of the journey it gave out four and a half volts.

An event of the season in Vienna was an improvised outdoor exhibition and parade, to which the owners of all motor-vehicles at the time in the city were invited by the Austrian Automobile Club. The response was general, and a great variety of vehicles took part. The meet took place on the old Exerzier platz in front of the Franz Josef barracks. Among the participators were distinguished personalities like the Count and Countess Kielmansegg, Baron Springer, Count Pötting, Baroness Haas-Wächter, Consul-General Singer, and Dr. Suchanek. Several excellent photographs of the company were taken, and then there was a parade through the principal streets.

The Rheinischer Automobile Club is the latest of the organizations that testify to the rapid growth of motor-vehicle inter-

Group of Vienna Exhibitors.

ests in Germany, a country that now stands second only to France in the advances made. The new club makes its headquarters in Mannheim, the pioneer city in automobile manufacturing in Germany, and has for its field the Rhine countries of Baden, Hessen, and the Palatine.

The enormous growth of the automobile interests in Germany is shown by the important general exhibition opened in Berlin on September 5, with 120 exhibitors—of whom eighty-two were from Germany, thirteen from France, four from Belgium, and two from Switzerland. These makers were represented by 140 different vehicles. The national government testified its interest

Viennese Ladies in their "Schnauferl."

by sending the State Secretary of the Post-office, General von Pobbielski, to open the exhibition. The progressive Post-office Department of Germany will naturally be quick to avail itself of the advantages for facilitating its work presented by the automobile. The opening of the exhibition was also celebrated very appropriately by the Berlin Omnibus Company, in beginning its service of electric omnibuses, running at a maximum speed of thirteen kilometres, or about eight miles, an hour.

In the Low Countries there are some curious restrictions placed upon automobiles. In Amsterdam, for instance, they are forbidden upon a street paved with asphalt! And in Belgian towns the speed-limit is placed at ten kilometres an hour. There are reports of internal troubles in the Automobile Club of Belgium. Some of the more energetic members are talking of a new club with more " go " to it.

A comical instance of rustic avarice comes from Normandy. A peasant made complaint against an automobilist who ran over a duck and a drake belonging to him, and made out a bill of damages for thirty francs, charging seven and a half francs apiece for each of these items: (1) The drake; (2) the duck; (3) the eggs that the duck would have laid, and (4) the ducklings that would have been hatched from the eggs! And the court made the automobilist pay! Since the Dreyfus affair anything seems to pass for justice.

Petroleum Motors

THE high economy which may be obtained by the complete combustion of liquid fuel in an internal-combustion motor, according to the *Engineering Magazine*, is now generally conceded, and as a result there have been numerous attempts to design motors which shall prove acceptable for general use. The Diesel motor has been fully noticed in the columns of the American technical press at various times, and now we have the Dopp motor, which was discussed at a recent meeting of the Verein deutscher Maschinen-Ingenieure.

Dopp maintains that the high compression advocated and used by Diesel is not necessary to the attainment of superior thermal economy, and claims that equally good results can be secured by the use of vaporized petroleum, drawn into the cylinder with the proper proportion of air, and burned under practically the same conditions as obtain when gas is used in a well-designed gas-engine.

Petroleum Motors

It seems to be generally admitted that the main element in the economy of a petroleum motor lies in the complete combustion of the fuel. While this is secured by providing a compressed atmosphere, it does not appear that it is necessary to use a compression materially greater than is now employed in the gas-engine. The Dopp motor does not differ in general construction from an ordinary gas-engine, except that the petroleum fuel is gasified by the heat of a lamp before it is drawn into the cylinder, and the excellent economy which appears in the regular service is claimed to be due only to the completeness of the combustion, attained by a thorough mixture of the fuel with the proper quantity of air.

Herr Dopp gives figures from a number of his motors in daily use which show a consumption of 0.197 to 0.240 kilogramme of petroleum per horse-power-hour, the lower result being obtained with a ten horse-power motor after it had been in practical service for more than eleven months. This result is better than was attained by the Diesel motor of twenty horse-power tested by Professor Schroter, although under less favorable conditions.

Herr Dopp maintains, as has been claimed by others, that the Diesel motor by no means realizes in practice the theory enunciated by its designer, and shows that some of the fundamental points, which, according to the theory, are essential to the highest economy, are distinctly controverted in the working of the motor. From this he deduces that the high economy of the Diesel motor shows that the theory is not sustained.

An important feature of the motors constructed by Herr Dopp lies in the fact that they can be constructed and operated in a satisfactory manner for small powers, good results being obtained with motors of two to five horse-power, while the construction of the Diesel motor is such that it does not appear advisable to make them for less than twenty horse-power.

Regardless of the theoretical questions at issue, there seems to be little doubt that very simple, efficient, and convenient petroleum motors can be made upon the same general design as that already in public use for gas, and that care in design and in the correct proportion of air to fuel supply can, with a moderate degree of compression, insure such a complete combustion as to leave little or no trace of soot either in the cylinder or in the exhaust gases. Under such circumstances there can be little doubt that the petroleum motor has a most useful future before it, especially for small powers.

Axle Bearings

By Myron and Frederick B. Hill

SOME of the advantages and disadvantages of a few of the bearings which have been employed at various times in the endeavor to eliminate the factors of friction and wear form an interesting study.

The plain axle bearing, often called the parallel bearing, was naturally the first and simplest method of mounting a wheel upon an axle. As long as we could count on draught animals to overcome the friction in vehicles, the need of better bearings was not so obvious. But with the advent of the bicycle, when the burden fell upon man himself, came the demand for a bearing involving less friction in the running parts. The wear and the necessity of frequent lubrication were so troublesome that some more convenient and economical bearing was demanded. First, cone bearings were tried, and then the well-known ball bearing. And now in more recent times the development of motor vehicles has created the demand for a new bearing which will sustain greater weight than the ball bearing, and still eliminate as much friction and trouble as possible. It is this age of automobiles which has brought into use the roller bearing.

Let us first see what has been accomplished with parallel bearings. In laboratory experiments it has been found possible to obtain a coefficient of friction as low as .001, but this was under conditions theoretically perfect and such as could never be secured on the road. It was with a bearing that was highly polished, but not worn, and with an expensive lubricant lavishly applied. Moreover, this low coefficient was maintained for a short period only, and was soon increased with running and the consequent wear.

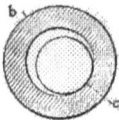

Fig. 1

Experiments with good parallel bearings, such as are used on vehicles, have shown fair results. At a moderately high speed. such a bearing, if supplied with sufficient and well-distributed lubricant, will have a coefficient as low as .07 or .08. This is, of course, when the bearing is new and free from wear.

In Fig. 1 is represented a section of a parallel bearing, showing in an exaggerated way how the axle rests within the hub, touching on the side *a* only. The weight upon the axle, combined with the pull or forward motion, causes the contact at this place. In action, the hub slowly wears away, and, as it is soft metal, there is more aggregate wear than upon the axle, although it is distributed over its surface. But the wear upon the axle comes all on the under side, and if the hub is .003 inch loose, not much more than that amount of wear on the axle causes it to assume a form somewhat like that shown in Fig. 2, with the surface of contract extending from *b* to *c*. This causes the surfaces to bind on each other, and, as they have the same curvatures, the distribution of lubricant becomes more difficult. The extent of this surface of contact depends upon the comparative wearing qualities of the hub and axle.

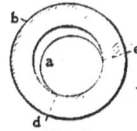

Fig. 2

This wearing away of the axle and the resulting deterioration of the lubricant largely increase the coefficient of friction, so that it runs up sometimes to .20, or even .30, and the loss of power becomes a material factor. The average coefficient of parallel bearings running at low speeds, fifteen miles an hour or less, is variable and difficult to determine, but probably lies somewhere between .15 and .30, dependent of course upon the speed, the weight carried, the character of the lubricant, and the amount of wear received by the bearing. The factor of end-thrust also adds materially to the loss of power.

To reduce the wear on the axle, which is expensive to re-place, the bearings in heavy automobiles have sometimes been made of soft metal, but it has been proven that the life of such bearings is short, lasting but a few months only under ordinary conditions. Another factor which affects the wear and life of parallel bearings is the mixing of the worn particles of metal with the lubricant.

The chief disadvantages, then, of the parallel bearing are the loss of power, the requirements of constant renewal of lubri-cant, the occasional renewal of the bearings, and once in a while a new axle. Then there are the dangers of hot boxes and abra-sions, due to poor distribution of lubricant or carelessness in applying it, or to the presence of sand or dirt. Under the weight of a heavy automobile, the pressure and binding are at times se-vere and the distribution of lubricant so poor that there is con-siderable abrasion; that is, the softer metal of the bearing is torn away, thus ruining it.

The difficulties and disadvantages of parallel bearings proved so great with the early form of velocipede or bicycle that some other bearing was sought which would have, not only a greater efficiency, but also a long life and little need of attention. Cone bearings were then resorted to, and they proved to be superior to the parallel bearing in many respects. The best feature of the cone bearing, of course, was its adjustability to take up wear, and the fact that it did not bind so much after a slight amount of wear. It was quickly superseded, however, in bicycles by the well-known ball bearing, which has been found admirably suited to this class of work. The suggestion has been made that the origin of the ball bearing is found in the difficulty of making roller bearings work. The first experiments in the line of roller bearings were undoubtedly with cylindrical rollers. These, however, when not supplied with guides, had such a tendency to twist and bind in the bearings, causing them to heat up or smash, thus ruining them, that they were found impracticable, and the inventor of the ball bearing probably made up his mind that it was necessary to have a roller that could twist and turn at will without interfering with the ease of running. A spherical roller was the outcome.

For the light work of the bicycle, ball bearings are undoubtedly superior to any old-fashioned roller bearing, for in the latter the rollers rub against each other along the *line* of contact between them or against the cage confining them, whereas in the ball bearing the balls rub against each other on *points* only. For heavy vehicles, however, the demand for a greater rolling surface to sustain the increased pressure prevents the use of ball bearings from becoming general.

The original form of ball bearing, in which the balls rest upon flat cones, presents many difficulties for heavy work; the weight upon the balls is sustained upon points, and as the balls roll, the wear comes upon a line around each ball and upon a single line about the cone. With a light weight the wear may be slight upon well-hardened tool-steel balls and cones, but with the heavy automobiles of to-day, weighing several hundred and even several thousand pounds, the wearing action upon the balls and cones is very great. Under such conditions the ball flakes and grinds away, and the cones wear correspondingly. Before long, unless the cones are constantly readjusted to distribute the wear, the wheel wabbles. In practice the cones are apt to be left until the wear on one side becomes noticeable, and then adjustability is lost, for if the cones are screwed up to keep the ball tight on the side not worn, the balls will be loose on the side that is worn, and then the end thrust on the wheel causes it to rock. On the driving-wheel of a motor vehicle this rocking action twists the teeth on the pinion and gear at angles with each other and injures them.

Axle Bearings

A ball bearing quickly develops these conditions, and the fault is due to a slight rolling surface, scarcely more than a line. The cones and balls wear off, the cone becoming grooved, and the balls flattening and losing their spherical shape. Before long, if the adjustment is close enough, they are apt to twist, bind, and break. This is one of the frequent causes of broken ball bearings encountered by the bicycle-repair man.

Many attempts have been made to minimize this trouble with ball bearings. The most praiseworthy ideas, apparently, are the concaved cones and the " staggered " balls.

In the concaved cone we have a surface which fits more or less closely over the surface of the ball, as shown in Fig. 3. In this form of ball bearing the balls may have the same curvature as the cones, or a smaller curvature. If the same, the zones of the ball on either side of the equatorial line m rub on the cones; the circle n, for example, is smaller than the equatorial circle m, yet it runs upon larger diameters of the cup and cone. It is dragged to make up the difference, so that an efficient lubricant is essential. The wearing of the balls, the deterioration of the lubricant due to the presence of worn steel particles therein, and the friction of the dragging surfaces are the evil features of such ball bearings. The loss of power due to this dragging action is a particularly objectionable feature.

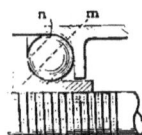

Fig. 3

If the curvatures of the cup and cone are slightly greater than that of the balls, the rolling action comes upon a line which crystallizes and wears away, so that considerable surfaces of the balls come in contact with the cup and cone, and the resultant bearing is similar to the one just described. Such a bearing is adjustable only within slight limits, and this also shortens its life.

In the " staggered " ball bearing the balls are held in a cage, so that each ball runs on a different line from its predecessor. This increases the rolling surface upon the cone, and undoubtedly increases the wearing quality of the cone, yet, as each ball has to wear its own groove in the cone, the wear upon each ball is more rapid than it would be if they all ran in a single line about the cone.

From this consideration of ball bearings it is apparent that some bearing is essential which has a greater rolling surface.

The automobile has thus introduced a new problem. The friction, wear and tear, and attention which are incidents of the parallel bearing, and the smaller rolling surface and conse-

179

quent wear of the ball bearing, make them unsatisfactory for the increased weights and strains of the motor vehicle.

Some ask, " Why not use the parallel bearing in spite of these disadvantages? " They argue that it has been adopted universally by road vehicles. The answer is this: There are two great differences between motor vehicles and horse-drawn carriages. In the first place, a horse-drawn vehicle has no complex mechanism to be attended to. The mechanism of the automobile, from its very nature, requires so much attention that the demand is strenuous for a vehicle requiring the least possible overhauling. Parallel bearings require constant lubrication. It is necessary for bearings and all running parts to be of such nature that they will require the least amount of attention. The difference of a few dollars in the first cost of the bearing parts will save constant expenditure of time and money afterwards. And, secondly, in the horse-drawn vehicle the difference of 10 or 20 per cent. in the power required to haul is lost sight of, for the horse is depended upon to overcome all friction. So here the loss of power through friction is not an obvious matter of expense.

In livery stables, of course, the overhauling of the bearings has been a material factor of expense, and many efforts have been made in past years to devise a successful roller bearing. All such efforts have had such dismal results as to give roller bearings a setback, from which they are but just recovering. The faults in the past have been in the failure of inventors to understand the necessities of a roller bearing. In a road vehicle a bearing is subjected to twists, strains, and end thrusts, which have to be provided for. Rollers have been used without any guiding or controlling device in some classes of work, but in vehicles they veer around and break sooner or later, the pieces going end over end, and the bearing is rendered worthless.

Some controlling device is, therefore, essential for a successful roller bearing. And the best roller bearing is the one having the best controlling device. This device must guide the rollers and keep them parallel with the axle, for, if they twist, the hub side of the bearing will rest upon the ends of the rollers, and the axle side upon the middle of the rollers, and the consequent binding of the bearing often causes the roller to break. This binding will happen sooner or later with the most perfect vehicle roller bearing which is not provided with some guide for the rollers. Moreover, when not guided or separated, the rollers rub against each other in opposite directions on their lines of contact, causing much rubbing wear and resistance against the operation of the bearing; and when rollers vary in size, from wear or other causes, they roll at different rates of speed and grind upon each other, and the wear between them is more rapid.

Various devices have been resorted to for guiding rollers. Rollers have been mounted in slots, in brass or iron cages, or upon pivots in cages. The pivots and slots are so constructed as to hold the rollers in a correct position on the axle. A great many forms of such cages have been devised, some of them so successfully as to last a considerable number of months and stand the wear and tear of several thousand miles. The great fault in such bearings is the wear upon the cages. Sooner or later the rollers will drill through the cages or wear away the slots or pivots diagonally, owing to their tendency to twist and their different rates of speed, and after a while the cage also gives way or the roller twists around so far as to bind and break. When this happens, the bearing is not only disabled, but sometimes the axle and hub are so injured as to require replacement or, at least, repair. A second evil is the rubbing of the rollers on the cages, causing a loss of power, and, to prevent the consequent squeaking and heating, proper lubrication is necessary.

Attempts have been made to secure an adjustable roller bearing. For this purpose rollers have been made in the form of cones, as shown in Fig. 4.

It will be noticed that the axes of the rollers, *o*, and the lines of the cup, *p*, and cone, *q*, forming the outer and inner races of the bearing, all converge to a single point, *r*. This secures the necessary result, namely, that the speed of the large ends of the rollers on the large circumference of the race shall be equal to the speed of the small ends of the rollers on the small circumference of the race. Such a bearing is difficult to construct accurately in large quantities, as is evident, and is therefore expensive. The cone upon which the rollers run is intended to be advanced into the bearing as it wears away or as the rollers wear.

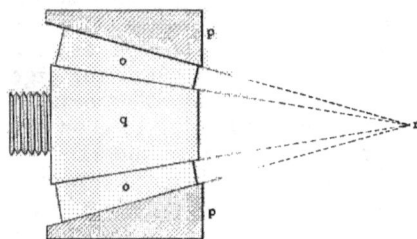

Fig. 4

In Fig. 5 is shown such a bearing, in which the cone is supposed to be considerably worn; the adjustment has forced the rollers partially upon a fresh surface, and they do not rest upon their flat sides as intended. In this position the roller would soon become ruined. Moreover, the lines of the rollers and of the races have lost their converging features, and one end travels more rapidly than the other, causing it to veer around and to wear upon its supports or guides. The end of a bearing so adjusted would be but a matter of a short time if the wear on the conical race was sufficient to demand adjustment.

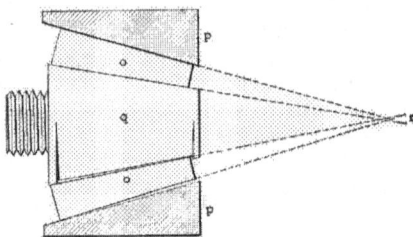

Fig. 5

For this reason the adjustable feature in roller bearings is believed to be of no value. If the bearing does not wear appreciably, no adjustment is necessary, and if it does wear, the adjustment does not remedy it, but, on the other hand, makes it worse.

As a matter of fact, the greatest wear upon a roller bearing of this kind is not upon the cone or rollers, for the wear of several thousand miles has been found so small as to be immeasurable. There is a material wear, however, upon the cage that holds them, and long before the rollers require any such adjustment as above noted, the cage has been disabled, and the rollers are no longer properly guided. So in this case, the adjustment theory has no chance to be favorably considered.

There is another serious obstacle in the way of using an adjustment for a roller bearing. While rollers, for all practical purposes, can be kept in their proper positions with a new cage, yet the slightest amount of loose motion in the cage, and the consequent slight displacement of the rollers which the cage allows, render it impossible to secure a tight roller bearing. There must be sufficient looseness in the bearing to allow the rollers to assume a slightly angular position on the axis. A thousandth of an inch is sufficient. The rollers, in assuming this slightly angular position with the axle, will require a larger diameter of hub than as if they were parallel with the axis at all times; hence in all roller bearings the race upon the hub should be slightly larger than required by the size of axle and rollers in a position parallel thereto.

The wearing action which is inseparable from rubbing surfaces, and which is rapid in roller cages, has caused the develop-

Axle Bearings

ment of a new controlling or guiding principle for the rollers. If rubbing can be eliminated from the controlling device, the wear due to rubbing will also be eliminated, and the bearing, if well constructed, becomes almost indestructible. Several attempts have been made to secure such a bearing.

The only bearing in which all rubbing has been successfully eliminated from the controlling device, as far as the writers

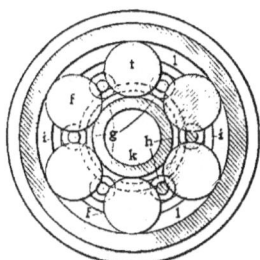

Fig. 6

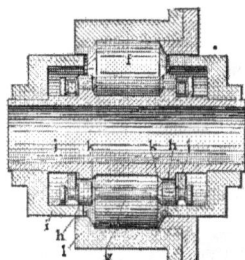

Fig. 7

know, is the "A. R. B." bearing, shown in the cuts, Figs. 6 and 7.

In this bearing the large rollers are separated by smaller separating rollers, g, mounted between the centres of the main rollers to prevent them from grinding on each other. The separators are so mounted at their ends that they *roll* upon their supports, i, j, without any rubbing or dragging; they have enlarged ends, h, overlapping the ends of the large rollers, as shown in Fig. 7, preventing them from twisting. These separating rollers cannot assume the position shown in Fig. 8 because of these enlarged ends, but are held in perfect alignment. All the rolling parts of this bearing are of hardened steel, and this, combined with the fact that there is no rubbing or dragging action, and that the rolling surface of a roller is equal to that on scores of balls, practically eliminates the element of wear. The absence of rubbing and dragging also renders the use of a lubricant unnecessary, except to prevent rust.

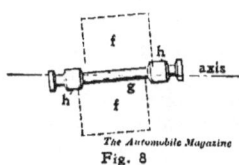

Fig. 8

In the ball bearing, the balls sustain weight upon points, and the lack of sufficient rolling surface causes them to wear away quickly; in the caged roller bearing, the rollers, while able to withstand the pressure upon their sides, rub against their cages and soon wear them

183

out, so that before long the rollers are no longer sufficiently confined, and soon bind and give way. In the "A. R. B." just considered, the rollers roll upon their separators, which in turn roll upon their supports, doing away with all rubbing, and hence eliminating all the consequent wear. The hardened steel separators and supports are simple and economical in construction, and are practically indestructible.

To secure an efficient rolling surface rollers should have generous diameters. The use of roller bearings will probably bring about larger hubs for a given weight than are now used. Road vehicles are subjected sometimes to such sharp blows on the wheel rims against a curbstone, a car rail, or stones, and often without the intermediation of the elastic tire, that the end-thrust bearing must be a stout one. The strains are momentary but effective in crushing a weak bearing.

In the "A. R. B.," above referred to, the end thrust is resisted by a bevel, against which the ends of the rollers roll. This is the principle of the bevelled car-wheel flange rolling against the rail, and of course its success is not open to discussion. The separating rollers are held in position and the main rollers held from twisting by similar bevels on the enlarged ends of the separators. Without rub, and its necessary friction and wear, without adjustment, with oiling only once in several months, and that to prevent rust, the "A. R. B." seems to have solved the bearing problem.

Electric Ignition by Dry Batteries

EVERY day electric ignition for gas-engines is gaining ground, and the time is not far distant when it will be generally employed, to the exclusion of ignition by burners. It presents so many advantages, offers so many facilities for quick starting, conduces so much to the easy controlling and running of the motor, that manufacturers who at first were its bitterest opponents are now adopting it one after another.

But these makers of motor carriages are confronted with an important question: What source of electricity should be employed—Should a generating armature rotating in the sphere of a magnet be used? or an accumulator? or a primary battery?

Experiments with the magneto-generator have not as yet yielded results of practical value. Hence there remain only the accumulators and the primary batteries to be considered;

Electric Ignition by Dry Batteries

and among the hundred of types of batteries which have been invented during the past century, the dry battery is the one which deserves the most attention.

Without detracting from the merits of the accumulator, for it undeniably possesses certain advantages so far as ignition is concerned, it would seem, from the automobilist's standpoint, rather than from the automobile-builder's, which is upon the whole the most rational view, that the dry battery is preferable to the accumulator. Those who have used accumulators know the inconveniences arising from their use. The plates are so exceedingly heavy, the internal short-circuitings cause such frequent stoppages, both at starting and on the road, the recharging of the battery is so tedious, and the acids will ooze out despite the care taken to seal the accumulators hermetically, that every automobilist has prayed for the introduction of some battery in which these faults were absent.

The dry battery has none of these disadvantages. If it be properly employed under normal conditions, if it be well made, if it be well arranged, if the material of which it is composed be carefully selected, the automobilist has at his command a source of electrical energy upon which he can rely with absolute certainty.

The only drawback to the use of dry batteries for ignition is the internal resistance of the cells which energize the primary of the induction coil (a resistance very great in comparison with that of an accumulator battery performing the same functions) and which has an influence on the character of the spark. The heat of the spark is not less intense; but for the same consumption of electrical energy in the primary coil it is smaller when dry batteries are employed than when accumulators are used.

The number of its points of contact with the explosive mixture is smaller, and the ignition is slower. That all this has very little effect on the operation of the motor is true enough; but that little is nevertheless perceptible. The influence of this internal resistance is greater as the consumption of electrical energy by the coil is less; and ignition by dry batteries necessitates the use of coils consuming as little energy as possible.

Many makers of cells have given their attention to this matter. Among them may be mentioned La Société Le Carbonne. Two years of experiment and constant trials have been crowned with success, and the battery now made by the Société is considerably more efficient than that made two years ago.

The Electric Automobile

By Prof. Félicien Michotte

THE electrically-driven carriage is generally known by one of two names—"electric carriage," by analogy with "steam-carriage," and "petroleum-carriage," or "electromobile," by analogy with automobile. Unfortunately, some electrician, with an incontrollable desire to display his cleverness, coined the word "accumobile," the meaning of which could be understood only if the steam-carriage were termed "boiler-carriage," and the petroleum-carriage "carbureter-carriage." Certain manufacturers have not stopped here, but have christened their vehicles "electrolabe," "electrobate," "velectra," etc. All this, no doubt, is well enough in its way; but it has the fault of being unintelligible and of leading the unwary to believe that there is something new where there is no novelty whatever.

THE OPERATION OF A CARRIAGE

Every electric carriage includes in its construction:
1. An electric motor;
2. An accumulator;
3. A transmission-gear;
4. A compound switch or circuit-controller.

The accumulator stores up electricity and supplies current to the electric motor through the medium of a distributer or switch. Under the influence of the current supplied the motor is made to turn at a great speed, developing power proportional to the electric current which it receives.

By adding a transmission gear, whereby the power developed by the motor is imparted to the wheels, we have a complete, operative, electric carriage.

Of all automobiles, the electric carriage is the simplest; for it does away with speed-changing gears, devices for throwing the parts into and out of gear, and requires only a mechanism for transmitting power to the wheels.

The electric motor can be driven at any desired speed; it is by far the cleanest of motors, evolves neither smoke nor smell, and is well-nigh noiseless in operation. The ease with which it can be controlled, coupled with the small amount of

The Electric Automobile

care which it requires, leads us to the conclusion that it is the ideal motor for the automobile.

Disadvantages: Accompanying the merits which have been mentioned is a slight inconvenience—the sphere of action of

Electric Carriage, Showing Storage Battery

the electric carriage is limited; at the end of its run it must seek some source of electricity and recharge its accumulators. This disadvantage is minimized in a city where a carriage covers hardly more than forty miles a day, a distance which can easily be traversed with good accumulators of the Planté type.

MOTORS

The number of electric motors is large and is increasing every day. They differ chiefly in the form of the armature, in the number and arrangement of the field-magnets, and in the winding of the wire.

Since these matters concern chiefly the engineer, and since the *chauffeur* and automobile buyer can learn but little from a profound scientific explanation, I shall confine myself to the briefest possible description. The electric carriage constitutes a whole in itself, and its owner buys and values it on the strength of its running qualities. Each element, taken separately, tells nothing.

187

The Automobile Magazine

On the subject of electric motors I have only one statement to make, and that is on the weight of a motor. Often we hear a *chauffeur* remark: " My motor weighs only so many pounds." The weight of a motor is an entirely relative question; for, if the total weight of a carriage be three or four thousand pounds, it matters little whether a motor weigh twenty or even eighty pounds, more or less.

Certain motors which have been proposed make as many as 3,000 revolutions. Such high speeds are detrimental to the elasticity of the motor, render the transmission-gears more complicated, increase the weight, and tend to overheat the parts.

The tension of the motor-current is usually 80 volts, this potential being taken to facilitate the charging of the accumulators on 90 to 110 volt circuits.

The switch or circuit-controller varies both in construction and arrangement, but its purpose in all cases is the same. It serves to supply the motor with a definite amount of current in order to obtain the speed desired. To this device all the accumulator-wires run; and by its means the cells can be coupled in parallel or in series and made to supply the motor with all or only part of the current.

To diminish the current supplied to the motor, there are usually placed in the circuit resistance-coils, which consist of nickel wire, and which, by reason of their low conductivity, resist the current, absorb it, and convert it into heat. By means of the switch, one or more coils can be switched into the circuit, so as to decrease the amount of current supplied to the motor.

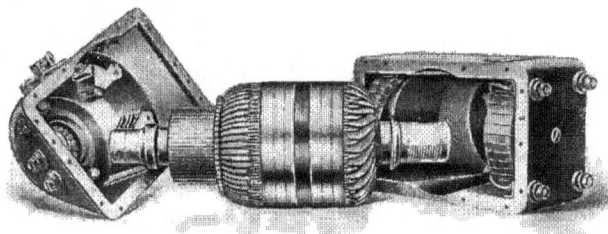

Motor with Parts Separated

Every carriage should be provided with a voltmeter and an ammeter. The latter may, at times, be dispensed with; but the former is necessary in regulating the charging of the accumulators.

The Electric Automobile

TYPES OF ACCUMULATORS

Many are the kinds of accumulators in use; and their number, like that of the motors, is constantly increasing. Most of these varieties are largely modifications of previously existing types; and the rest, batteries discarded fifteen years ago and now brought forth again under new names or as new foreign inventors.

All accumulators of the pasted oxid type are practically useless in automobilism, and all attempts at adopting them for use on motor carriages are vain. Celluloid and glass may be placed between the plates to retain the active material, but the active material refuses to be retained, and thereby impairs the perfect operation of the accumulator, shortens its life, and aids in its rapid deterioration.

Lead accumulators of the Planté type are the only batteries which should be employed; but here again discrimination must be used. Certain accumulators are admirable for the purposes of buildings, but, owing to the nature of their construction and their small capacity, cannot be used in automobiles. The entire problem of devising a practical form of electric traction depends for its solution upon the invention of a practical accumulator.

To be practical, one should be economical; and a pasted oxid battery, at best, is serviceable only for 2,500 to 3,000 miles, assuming that the active material is retained and that there will be no short-circuiting. Even at the maximum figure, 3,000 miles, which in practice frequently becomes reduced to 600, there is no economy in using a pasted accumulator. A battery costs $400; and the running expenses would therefore be thirteen cents per mile, without counting the cost of maintenance. A private carriage covering twenty miles per day or a public cab covering 40 miles would respectively cost $2.60 and $5.20 per day. Certain makers claim that the running expenses of their accumulators will not exceed 75 cents or $1 per day; but even at these figures I doubt whether the system be of any practical value for automobiles.

If, on the other hand, we consider the Planté battery, we shall see how great is its advantage over the pasted accumulator. A good Planté battery, when used every day, will last from three to five years. The distance covered for the three years at the rate of thirty miles per day will be approximately

$$350 \times 3 \times 30 = 31500 \text{ miles.}$$

189

The battery costs less than its pasted rival; but even at the price above given ($400) the cost per mile will be only

$$\frac{400}{31500} = \$0.01.$$

If the life of the accumulator be double that assumed, the cost per mile would be only one-half a cent. At the rate of one cent per mile, the daily running expenses in the examples pre-

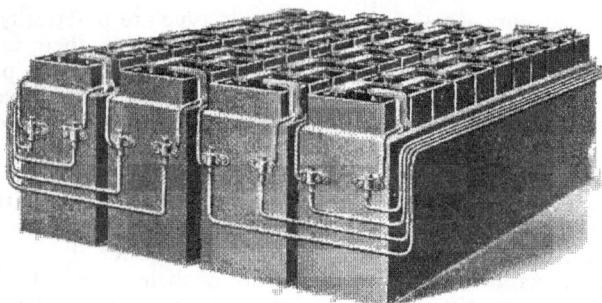

A Storage Battery

viously given would be twenty cents for twenty miles in the case of the private carriage, and forty cents for forty miles in that of the public cab. These calculations certainly speak for themselves.

ACCUMULATOR BATTERIES

The accumulator battery is composed of a sufficient number of elements to correspond with the voltage of the motor, each element being equivalent to two volts, and the total number of elements being increased by 4. If the voltage of a motor be 60, then

$$\frac{60}{2} = 30$$

plus the 4 additional elements, or 34 elements will be required. For 80 volts, 44 elements will be necessary.

The battery is inclosed in a casing, the interior of which has been tarred. The elements should be separated from one another at least 3 to 5 mm. (.12 to .20 in.); and between adjacent elements insulating rods should be placed.

The Electric Automobile

The accumulators of electric vehicles should always be enclosed in ebonite casings; celluloid should never be used, despite its lower cost and lighter weight. There is not a single maker of celluloid casings who has not one or two accidents on his conscience. Even cars are sometimes set on fire, proving that they, too, are endangered by the use of celluloid in cells. These accidents are due to short-circuiting.

The casings are usually closed either by an ebonite cover or by an insulating layer of some paraffin compound.

THE CHARGING OF ACCUMULATORS

Accumulators are charged at a constant potential, constant intensity, or constant wattage. The battery is charged at a constant potential when a constant voltage is maintained during the charging. The voltage can be read on the scale of the voltmeter, and by turning the handle of the switch in the proper direction a suitable number of resistance-coils can be thrown into and out of the circuit to secure the constancy desired. The voltage of the current should be equal to $2\frac{1}{2}$ times the number of elements. Thus, for 44 elements there will be

$$44 \times 2.5 = 110 \text{ volts.}$$

The intensity is constant when the ampèrage is constant. The power is constant when the number of watts is constant.

In charging accumulators a number of special devices are used, which are usually mounted on a single board. These devices comprise:

A voltmeter;

An ammeter;

A resistance-box or rheostat;

A switch by means of which the resistance can be varied;

A circuit-breaker;

Automatic cut-outs.

When charging at a constant wattage another special apparatus—a wattmeter—is added. These various appliances should be mounted preferably on marble or slate; or else the rheostat should be set up away from the board.

The rheostat serves to regulate the current in order to obtain the intensity desired.

The circuit-breaker is designed automatically to break the circuit, when, owing to a diminution in the intensity of the charging current, the accumulators will tend to discharge into the source of electricity.

The Automobile Magazine

The cut-out is composed of leaden wires, which will melt and break the circuit when the intensity of the current passes a certain maximum point.

The voltmeter and ammeter are instruments of electrical measurement.

CHARGING PRECAUTIONS

In charging accumulators, whether they remain in the carriage or whether they be removed, care should be taken that they be placed where air can circulate through the cells so as to avoid the corrosive action of the vapors arising from the acid solution as the charging nears its end.

If the accumulators be charged while still on the carriage, the casing of the battery should be entirely opened.

The density of the electrolyte, measured by an acidimeter, should remain constant, and the liquid should be tested from time to time. If the liquid be too concentrated, water should be poured in *after charging;* if the density be too small, acid should be added.*

The accumulators are charged by a current, the voltage of which is slightly higher than that of the battery. For a battery of 44 elements the voltage should be

$$44 \times 2.5 = 110 \text{ volts.}$$

If the voltage on the circuit used be too high, the resistance-coils will bring it down to the desired figure.

The time required for charging varies with the intensity of the charging-current.

With pasted oxid accumulators, the time is limited to a number of ampères per kilo of the accumulators (about one ampère); with lead accumulators three or four ampères per kilo is the ratio, the time for charging being thereby reduced to two or three hours for 10 and 12 kilos. As a general rule, it is preferable not to force the charging for the Planté battery, but to proceed slowly at a rate of two to three ampères.

* The acid used is sulfuric acid (20° Beaumé), commonly known as vitriol. It is extremely dangerous and should be handled with the utmost care, since it burns the flesh and the clothes. Should the acid come in contact with the fingers, the hands should be immediately plunged in water. The presence of the acid can be detected by its oiliness, and by the tingling, burning sensation which it produces. Acid-stained or burnt clothes should be neutralized by ammonia. The acid should be kept only in carefully-labelled glass receptacles. Never pour the water into the acid; but pour the acid into the water in a thin stream. The water used, it should be remarked, must be pure. Distilled water should be employed, if possible; but if it cannot be readily obtained, recourse must be had to spring or rain water.

The Electric Automobile

The end of the charging process can be determined by:
1. The voltage.
2. " Boiling."

VOLTAGE.—When the voltage is equal to the number of elements multiplied by 2.4, the charging is nearing its end. For 44 elements we should therefore obtain

$$44 \times 2.4 = 105 \text{ volts.}$$

" BOILING."—The term " boiling " does not indicate a rise in temperature of a battery, but rather the great evolution of gas which occurs when a cell is nearly charged, numerous small bubbles rising and congregating at the surface. As the charging progresses, the boiling increases in violence, until finally the liquid is covered with a layer of small bubbles.

When a solid layer of bubbles, as it were, extends over the liquid in the cells, it is impossible to perceive the boiling. But the sound produced can be distinctly heard. After a little experience it can readily be determined, when the charging has been completed, if a voltmeter be not at hand, merely by the boiling.

In order to drive a carriage, the cells need not necessarily be charged to their full capacity. If the automobilist be pressed for time, he can charge his accumulators sufficiently to cover the distance still to be run. A battery can be slightly charged in ten, fifteen, or thirty minutes. If the time for charging the battery completely be known, and likewise the distance which can be covered at that charge, then the time necessary for charging the cells with an amount of current equivalent to the distance still to be run can be easily calculated, if the intensity of the charging current be known.

(*To be concluded in next issue.*)

AUTOMOBILE STREET-CARS

The Siemens & Halske Company, of Chicago, has announced that it will introduce the Berlin system of autocabs in that city. These cars carry about twenty-five passengers, and can run on either street-car tracks or a paved roadway. They will be operated by storage batteries. C. S. Knight, vice-president of the company, states that the plant in Chicago will be enlarged for the construction of the busses and the work begun as soon as the great demand for private automobiles slackens.

The Fritscher-Houdry Petroleum Motor

O NE of the most remarkable engines of its kind—remarkable chiefly for its ability to withstand the severest tests to which it can be subjected—is the four-cycle petroleum-motor invented by Fritscher and Houdry.

At the Exposition des Tuileries, the motor ran continuously from nine o'clock until twelve, and from two o'clock until six, without the slightest noise, and without any sign of being overheated.

In our illustrations representing the motor in side and end elevation and in perspective, the cylinder C is surrounded by cooling-flanges, and incloses the piston D, to which is pivoted the piston-rod d, connected by a crank with the shaft A, provided with the fly-wheel V.

The motive agent, after having been carburetted in a vaporizing-chamber of a peculiar construction devised by Fritscher and Houdry, is injected in proper quantities into the cylinder, after each stroke of the piston. The motor starts almost immediately after the shaft has been turned a few times.

The explosive mixture is admitted by the automatic valve s,

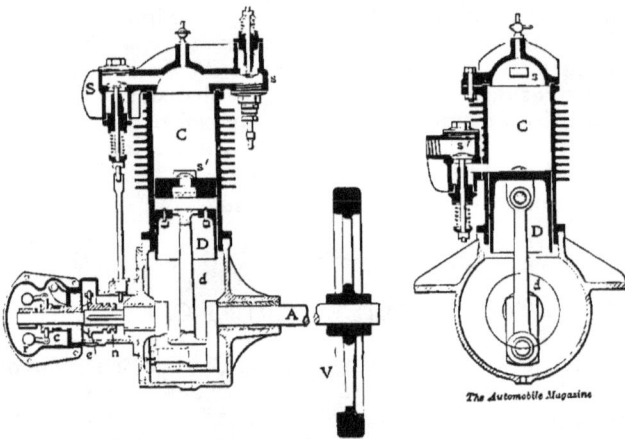

Side and End Elevations

The Fritscher-Houdry Petroleum Motor

and is ignited by an electric spark. The exploded gases are discharged through two openings:

1. An automatic valve s', located at the end of the down-stroke of the piston.

2. A mechanically operated valve S, through which the residual gases are discharged on the return stroke of the piston.

By reason of this arrangement the hottest gases are exhausted first, at the end of the third period; while the gases discharged through the upper valve S are considerably cooled, from which it follows that the valve cannot become overheated.

The Fritscher-Houdry—Perspective View

The valve S is operated by the rotation of a double-grooved sleeve n, which turns with the shaft A. If the speed of the shaft A be excessive, the sleeve and its auxiliary collar c will be displaced from their normal position by the action of the governor r, as its balls are thrown outward; and the grooves being out of engagement with the valve-stem, the valve S will be closed. The electric spark is produced by means of the cam c; the ignition can take place at any desired interval.

The one and two horse-power motors have their fly-wheels mounted within the casing, and are not provided with governors. The three horse-power motors (weight 210 lbs., for light carriages) are furnished with a governor and a fly-wheel on the exterior.

A New Sight-Feed Lubricator

WHEN the oil-reservoir of a motor contains a pump for injecting the oil into the cylinder or other parts, the feeding of the oil can be effected without soiling the clothes and without loss of time. You pull up the piston, you push it down again, and the thing is done. It is quick—but it is not certain.

When you have pulled up the piston to suck oil into the pump, how do you know that the liquid has actually risen? Nine times out of ten you have sucked up mostly air; so that, instead of supplying a full measure of oil to the part to be lubricated, you have furnished it with only half the desired quantity.

To remedy this defect there is but one way—to render the quantity of oil which is fed visible. Messrs. Desponts and Godefroy have solved this problem very simply indeed. In their system of lubrication the pump is discarded, and the oil flows directly by gravity from the reservoir in which it is contained to the lubricator gage. This lubricator is made of thick glass, and contains exactly one measure of oil; the automobilist can, therefore, readily see the amount of oil fed.

Sight-Feed Lubricator

The lubricator is so constructed that the same operation which permits the oil to flow closes the passage which leads from the reservoir to the liquid, and vice versa.

This light and neat little device, which has the great merit of dispensing with the untrustworthy oil-pump, seems one of the most practical forms of lubricators yet invented for use on automobiles.

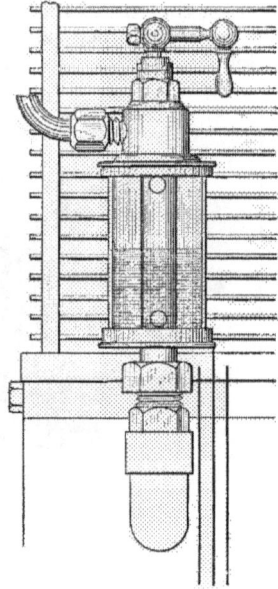

A Simple Water-Circulating Pump

IN order that the cylinder of a motor may be effectively cooled, the water must evidently be circulated rapidly and trustworthily. The thermo-siphon system long used in many stationary engines has been found inadequate for automobile motors; and it was therefore necessary to employ circulating tubes or pipes, which connected the motor with the cooler, a small pump driven by the motor being used to force the water constantly through the pipes.

In an automobile, the parts of which are in themselves already complicated, it was, of course, impossible to employ a reciprocating pump. For the use of such a pump with its valves and joints would have added another source of accident. The centrifugal pump, on the other hand, has no valves, and its joints can be hermetically closed.

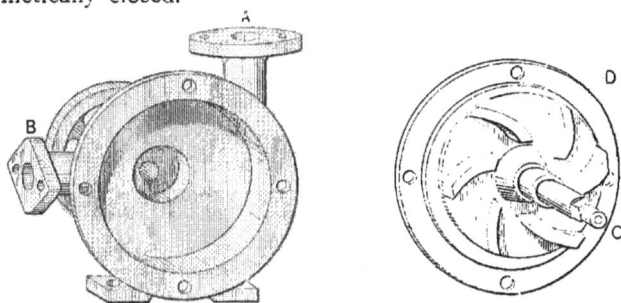

Water-Circulating Pump

We present herewith an illustration of a simple centrifugal pump made by Messrs. Dalifol and Thomas, which has been especially devised for use upon automobiles. Our engraving shows the pump with its parts separated. It is evident that the construction is the simplest conceivable. In the body of the pump the disk D is mounted, which disk is provided with radial curved blades, by which the water is forced out. This disk D, which is essentially a turbine-wheel, is given a rotary movement by means of the pulley B driven by a belt or by friction with the fly-wheel of the motor.

The chief merit possessed by this pump is the facility with which it performs its functions without in any way affecting the pipes which project from the pump-body at A and B. The disk C can be removed by unscrewing the four bolts by which it is held in place.

The Daimler Carriage

By D. Farman

A S a result of the collaboration of Europe's two most famous
automobile inventors, E. Levassor and G. Daimler, a
new motor for horseless carriages has been devised, which
is a modification of the celebrated Daimler motor.   Although
they succeeded in producing the new " Phœnix Daimler " motor, the two inventors could not agree on the method of transmitting the power to the driving-wheels.   Levassor adopted the

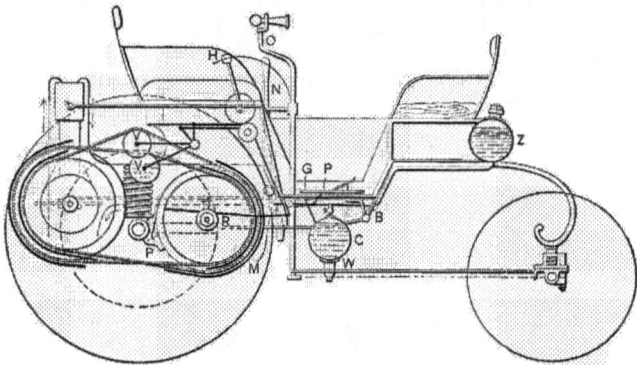

Side Elevation

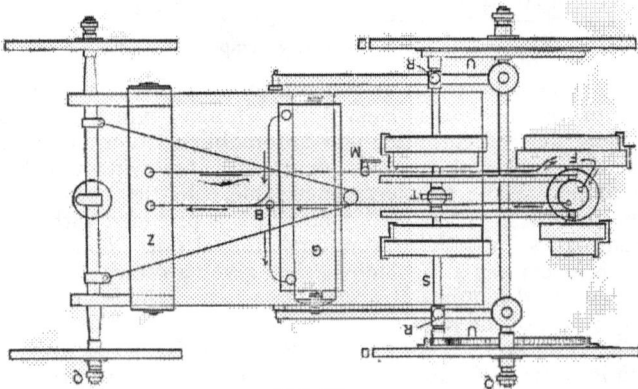

Plain View

sprocket and chain; but Daimler insisted upon the use of pulleys and belts in the speed-changing gear, and of gearing in the transmission of power. It was, however, agreed that the system which gave the best results was finally to be adopted.

But the death of Levassor cut short further experiment; and Daimler retained the belt-and-pulley system which he has used with such gratifying success on his own carriages.

In the above illustrations we have represented in side elevation and plan the motor and operative mechanism of the improved Daimler carriage.

The " Phœnix " motor is arranged in the rear portion of the carriage, and is provided at each side with two pulleys, which also serve as fly-wheels. Four other pulleys are keyed on the shaft, S, of the differential gear, which is connected with the rear wheels of the carriage by means of two pinions, R, and gear-wheels, U. The driving-mechanism and the speed-changing gear consist essentially of belts running over the four pulleys of the motor and the differential gear. By manipulating the lever, H, controlling the tightening-pulleys, V, the desired speed is obtained. By the employment of belts running over the tightening-pulleys the carriage can be gently started and silently driven.

The motor and the burners are supplied from a common petroleum-reservoir, Z.

The front axle is swivelled at its centre and is controlled by means of a chain-gear operated by the crank-handle, O. The lever, N, commands the brake-shoes of the wheels, and the pedal, P, controls the brakes of the differential gear.

The Doré Motor

A MOTOR which was recently exhibited by Doré & Co., at the Tuileries, in Paris, shows not a little progress in construction. Its chief merit is its simplicity; since, but a single-admission valve and a single-exhaust valve have been provided. There is no back thrust, no complicated gear to operate the exhaust-valve, no cooling of the motor by circulating water.

Another advantage obtained by the novel construction is lightness; for a five-horse-power motor weighs but 92 pounds without the fly-wheels, and 165 pounds with the fly-wheels.

One of our illustrations pictures the carbureter which is used

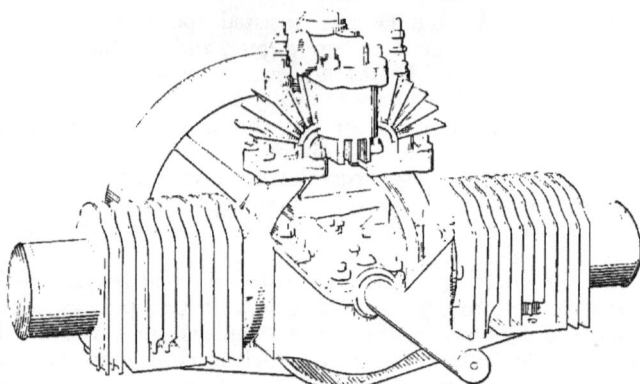

in connection with the motor.   Besides its efficiency of operation, the carbureter possesses the additional merits of being simple in construction, small in size and weight, and is applicable to the motor of any motocycle, tricycle, or motor-carriage.   Under all circumstances it will operate uniformly in conjunction with the motor; and its form is such that it may be mounted in any place that may be convenient.

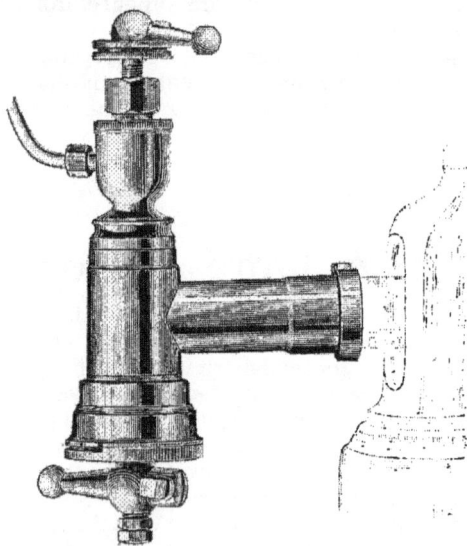

The Py Motor-Carriage

A MONG the carriages which attracted much attention at the last French automobile exposition may be mentioned a vehicle made by the Compagnie des Automobiles du Sud-Ouest, and invented by M. André Py, the manager of the company.

The Py carriage, which is represented in the accompanying illustrations in perspective, longitudinal section, and plan, is noteworthy for its handsome appearance and roominess, and for the simplicity of its mechanism.

The motor is a one-cylinder, four-cycle engine horizontally mounted at C, somewhat to the rear of the vehicle and to one side, so that its piston-rod can readily drive the crank v. The crank is secured to the end of the piston-rod, which is made in

The André Py Carriage—Perspective

two pieces held together by a steel collar. The explosive mixture of vapor and air passes from the carbureter c, through the automatic admission-valve s, and is discharged either by an ignition-tube or by an electric spark. At the end of the cycle, the exploded gases are discharged through the exhaust-muffle, situated in the rear of the carriage to one side.

The exhaust-valve S for the exploded gases is operated by a small shaft o, driven by the worm-wheel d, which receives its motion from the shaft A. The ball-governor *r* connected with this shaft is an exceedingly sensitive device which controls the valve *S*.

The fly-wheel V, secured to the shaft A, transmits the motive power to the driving mechanism by means of a friction-clutch G. This mechanism comprises two trains of gear-wheels so connected that the wheels and pinions, suitably secured to the motor-shaft A and to the auxiliary shaft *a*, are thrown into engagement

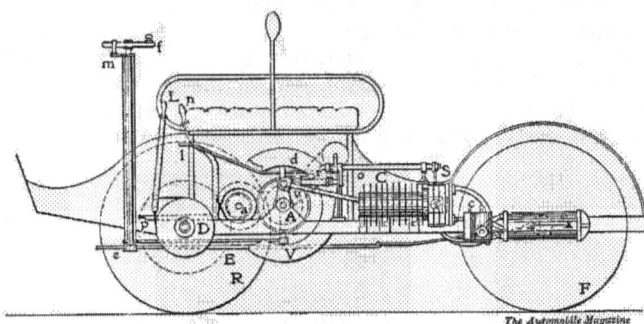

Side Elevation

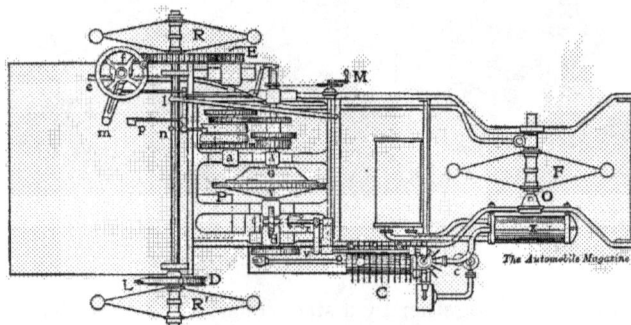

Carriage and Frame

to obtain three different speeds—five, eleven, and eighteen miles per hour. This last speed can be increased to twenty-two miles by driving the motor at its maximum speed. The carriage can be driven backwards by manipulating the reversing-lever *l*.

Power is directly transmitted to the axle of the front driving-wheels R and R' by a gear-wheel E on the right-hand side. A differential gear A, provided with a friction-brake, is located at D, and is controlled by means of a lever L.

The Py Motor-Carriage

All this mechanism is inclosed in a box or casing beneath the double seat of the carriage. The two seats being hinged, the various parts can be readily inspected, adjusted, and lubricated. This box also contains the petroleum-reservoir P, with a capacity of nearly three gallons—a quantity of petroleum which will enable the carriage to run 112 miles on an ordinary road.

The crank-wheel M, by means of which the motor is started, is one of the most efficient devices of its kind. A single turn of the wheel will set the engine in motion, the same movement simultaneously operating the escape-valve. By grasping the handle M firmly, the entire wheel can be immediately removed.

The carriage is steered by the rear wheel F, the steering mechanism being operated by the hand-wheel f, connected with the rack e, through the medium of which the axle of the wheel F is turned about its pivot O. By operating the brake of the differential gear D, the carriage can be turned within a circle the diameter of which is sixteen feet.

M. André Py has practically built a two-wheeled carriage, since the wheel F serves merely as a support and steering-wheel. The motor develops exactly 3.6 horse-power, the efficiency being 80 per cent. at the wheel-rim. With a load of over 600 pounds, grades of 10 and 12 per cent. have been ascended with a feeble motor.

The total length of the carriage is nine feet; the width between the wheels R and R', 4.65 feet. The seats are placed not quite four feet from the ground, by reason of which position, and of the arrangement of the motor and operating mechanism, the centre of gravity is very low, thereby rendering the carriage exceedingly stable. By the addition of a top, the appearance of the carriage can be still further improved.

AUTO-CAR TIRES

The fact that many owners of private motor-carriages in England still use solid rubber tires is a source of continual wonderment to Continental *chauffeurs*. The advantages, as regards comfort, absence of noise and jarring, safety in case of running over an obstacle, and the absorption of vibration which would otherwise be conveyed to the engine, are so great as to entirely outweigh any objection on the ground of puncture. Moreover, punctures are comparatively few in the case of carriages fitted with large tires of at least four or five inches diameter.—*London Daily Mail.*

A New Tire for Motor-Carriages

THE provision of a tire which would meet all the requirements of a motor-carriage has long engaged the attention of automobile makers. The problem, which at first sight seems one of no great moment, is worthy of no little consideration; for, if the tires be good, the motor will run more easily, and the life of the mechanism will be greatly prolonged.

Owing to the inadequacy of springs, the wheels have been provided with elastic tires, which were at first of solid rubber, and, later, pneumatic. When it was found that the solid rubber tire did not meet all requirements, the manufacturers of automobiles had recourse to the pneumatic tire, which, although undeniably superior to solid rubber, has many disadvantages, chief among which may be mentioned the danger of its being punctured. It has therefore been the aim of many inventors to devise a tire which should possess all the merits of the pneumatic tire without its faults. The tire which forms the subject of the accompanying engraving has been designed with this end in view.

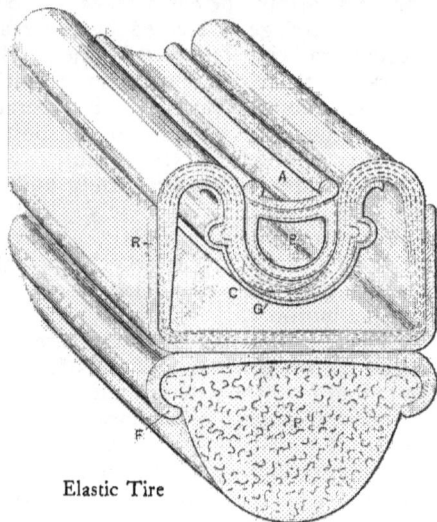

Elastic Tire

It will be observed that the tread of the wheel is composed of solid rubber *P*, held in a rim *F*. But this rim *F* is connected with the spokes by means of a removable pneumatic tire. The outer wall of the air-chamber is protected by an auxiliary rim *R*, and its extremities are overlapped at *C* and held in place by a half-tube *G* of steel. The extremities *C* are hermetically forced into contact with each other and with the tube *G* by means of a small elastic tube *D* secured to the rim *A* of the spokes. This tire evidently has the advantage of not being readily punctured. Its weight is probably considerable, for there are no less than four steel rims included in its construction.

Benoit-Julien Water-Cooler

EVERY automobilist knows what a difficult problem it is to cool the water which has circulated in the jacket of the motor for preventing the overheating of the explosion-chamber, and properly to arrange the pump which produces this circulation.

The problem seems to have been solved in a system devised by Messrs. Benoit & Julien, in which, as may be seen from the accompanying diagram, the principal elements are:

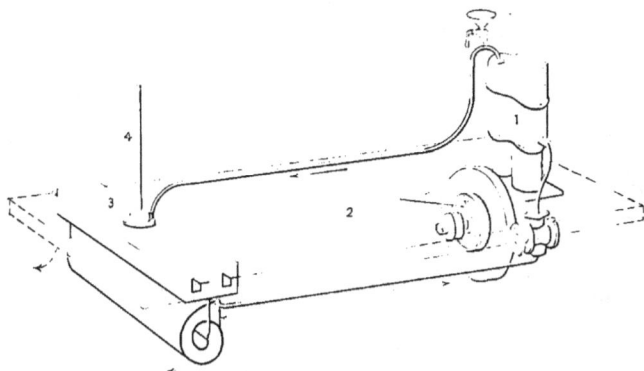

Truck Benoit-Julien

1. A motor having a double jacket.
2. A friction-clutch of the Julien type.
3. A rotary Benoit-Julien pump.
4. A Julien water-cooler surmounted by a water-tank.

The Benoit-Julien pump comprises a cylindrical body, closed at each end by two heads. Within the cylinder a screw is mounted, having a right and left thread, separated by a straight partition on the shaft. At a speed of 2,000 to 4,000 revolutions per minute, the pump can discharge 110 to 132 gallons per hour, the difference of level being 3.28 feet. With this pump water can be easily raised from a depth of two feet.

The two annexed views represent the Julien water-cooler in perspective and in cross-section.

The water-cooler is essentially a spiral tube in which the water to be cooled circulates in the form of a sheet 0.12 inch

The Automobile Magazine

thick and 4.10 square feet in area. When it is considered how thin is the surrounding envelope (it is only 0.02 inch thick), and

how large the cooling surface, it is evident that the water can be very quickly cooled.

The effectiveness with which this apparatus serves its purpose enables a water-tank of small capacity to be employed. The tank could even be entirely dispensed with, were it not necessary to keep it in reserve for losses occasioned by leakage at the pipe-joints and other places.

The Julien water-cooler can be secured by means of its inlet and outlet pipes under the water-tank, either in the front or rear of the carriage, with its mouth directed to the front. When the carriage is in motion the air is received by the mouth of the cooler and, after having circulated through the interior, is allowed to escape through lateral openings. The exterior surface is similarly acted upon by the air. The cooler can also be hung from iron supports secured to the heads A.

Mud does not readily cling to the cooler; but should it do so, it can be readily washed away, the water running off through small holes in the lower portion of the heads.

DOCTORS AND AUTOMOBILES

Take the case of a doctor in the country with a large and scattered practice. Four to six horses, a couple of men, stabling, shoeing, forage, wages, repairs, vets' bills, etc., will cost him at least £300 a year. A motor car, on the other hand, after the first initial expenses (they vary in price from 150 to 300 guineas) only requires the services of a lad to keep it clean, and it will cost considerably under £100 a year, including wages, repairs, and the fuel or motor power required.—*British Medical Journal.*

The Automobile
MAGAZINE
An *ILLUSTRATED* Monthly

Vol I No 2   NEW YORK NOVEMBER 1899   Price 25 Cents

The Automobile Magazine is published monthly by the United States Industrial Publishing Company, at 31 State Street, New York. Cable Address: Induscode, New York. Subscription price, $3.00 a year, or, in foreign countries within the Postal Union, $4.00 (gold) in advance. Advertising rates may be had on application.

Editorial Comment
THE TRANSITION

IT is just twenty-seven years since there swept around the world a mysterious malady that universally attacked the equine species. Horses everywhere were sick and unable to work. The disease seemed to sap their strength completely. Very few horses were exempt from the attack. For the time that it lasted business was almost paralyzed. In city and country, wherever horses were depended upon, transportation was almost entirely suspended. Ox teams were brought into the towns, and their services commanded great prices. Horse-car and omnibus traffic was at a standstill. In some cities the strange spectacle would be seen of a horse-car passing at rare intervals and hauled by men through the streets, carrying passengers to and from suburban places at twenty-five cents a fare; the employees of the company taking advantage of the opportunity to earn something, at least, in the period of their compulsory idleness. They dragged the cars by ropes, just as men used to drag the hand engines to a fire in the old days, and they made something of a lark of the event. The silence of the streets was that of rustic highways; it seemed oppressive, coming as an interlude in the customary Babel of urban noise; but, could our cities become permanently like that, the almost heavenly peace would be unspeakably welcome. A disastrous result of the epidemic was the great fire in Boston that destroyed the greater part of the business section, for the fire-department horses were sick, and there was a fatal delay in getting the engines to the spot.

It would have required but a comparatively slight intensification of the malady to make it fatal, sweeping every horse from the face of the planet. In all probability, not a few of the strange species that existed in prehistoric ages, as witnessed by their fossil remains scattered throughout the bosom of the earth, were annihilated by similar causes. We are not aware if the epizootic, as the epidemic of 1872 was called, ever before raged through the world. But there is no reason why it should not again occur. Suppose that it should? The industrial harm would now be far greater than then, for civilization has extended itself enormously in the past quarter of a century, opening up vast new regions to occupancy in all parts of the world. Its recurrence would unquestionably give a great impetus to the automobile movement, for it would impress men's minds with the desirability of taking quick advantage of the substitute at hand. But suppose that an intensification of the malady should occur this time! Suppose that, in the course of one short month, every horse in the world should pass out of existence! Not even the most impassioned friend of the automobile could wish for such a consummation. For it would mean a world-wide catastrophe of a most appalling nature. Doubtless human ingenuity would speedily devise a way out, and civilization would adapt itself to the new conditions. New methods of transportation would be hastily improvised, not only for those parts of the world where there are passable roads, but for the roughest and wildest of regions. An enormous leap forward in new inventions to these ends would be made. Practically all the energies of civilized society would be bent to the one great task of installing new forms of burden-bearing mechanism. But meanwhile there would be famine and pestilence to an unheard-of extent. In consequence there might ensue such an enfeebling of the race, such a sapping of the resources of civilization, that recovery would be exceedingly gradual. Therefore the improvement in transit methods would be obtained at a fearful price.

It should be remembered that animate instrumentalities, guided as they are by inherent intelligence trained to certain definite ends, are so flexible, so adaptable to widely varied uses, that to expect to replace entirely by mechanical substitutes a creature like the horse would be much like expecting to replace

Editorial

the human hand by mechanical devices. Hence, while efficiency is enormously increased in those functions where machinery replaces animate instrumentalities, there must remain a large class of operations where, for intricacy and delicacy of manipulation, the former must be retained. Therefore for a long time to come the horse will remain indispensable for a variety of uses. And even should he disappear, there would remain other beasts of burden whose use might be extended and developed to good advantage. For example, there is the trusty, though slow-going, ox; the plodding and patient donkey, cousin to the horse, so universal in Mexico and other Spanish-speaking lands; his other cousin, the zebra, the elephant and the camel; the ruminant llama of the Andes, industrious little brother to the camel; and even the humble dog, usefully earning his living in many European countries as well as in the polar and Arctic regions, where the reindeer likewise serves in harness. The utilization of many of these animals might greatly be extended, if demanded, and since most of them are without such a sensitive organization as the horse, with his tendency to insane panic, they could better serve side by side with the automobile, on more friendly terms with the latter. Man, too—the more's the pity—acts as a very efficient burden-bearing animal in many parts of the world.

The transition to the era of mechanical traction, now well under way, is hardly likely, however, to be accelerated by any sudden or dramatic happening. The change, though it will proceed with ever-increasing momentum, will nevertheless be gradual, until we find the horse replaced in the main fields of his activity throughout the populous parts of the civilized world. The process will probably be very similar in its course to that which has taken place within the past decade in the substitution of electric for animal traction on our street railways. Ten years ago electric traction had just begun to pass out of its experimental stages. It was tried here and there in minor localities, and then the great company that had acquired the street-railway systems of Boston began with changing over two of its lines. Trial on such a scale gradually showed various obstacles to be overcome and various improvements to be made; the old plant was made available for awhile in many respects, but very soon the tracks had to be entirely relaid with heavy rails, the small and light cars had to be replaced with larger and stronger ones, and it was not long before the entire plant had been replaced in almost every part by a new and far more costly one, with great additions and extensions. In all the other great cities of the country the same thing soon took place. The advantages of the new methods also caused an enormous extension of electric street railways throughout rural parts, affecting the life and the habits of a great portion of the country's population to a profound degree. So, in a very few years, we have seen this movement in a single phase of the transit problem effect a most significant revolution, building up great industries and creating new sources and forms of wealth that mount into the millions and millions in the aggregate, with far-reaching social and industrial consequences.

We are now witnessing the beginnings of a far greater change in transit instrumentalities—a change that is destined to bring about not a few remarkable innovations. Already most powerful combinations of capital are engaged in the work, but the industrial activity which the movement will cause will make present operations, great as they are, seem slight in comparison. Great manufacturing establishments are straining all their facilities to keep up with the demand, but soon we shall see not only practically every carriage-making concern in the country devoting its main energies to automobile work, but additions made to existing plants that greatly increase their capacity, and many extensive new factories built for the purpose. The bicycle manufacturing industry, extensive as it has become, is small in comparison to the industrial operations that the development of the automobile will entail.

The organization of cab services in the large cities is one of the most important aspects to occupy the energies of those engaged in the field. Side by side with this work will be that of supplying the large demand for domestic and pleasure purposes, in the way of light vehicles, ranging from victorias, buggies, runabouts, etc., down to motor cycles. The establishment of omnibus lines in the cities, after the example set by the electric omnibus service soon to be installed in Fifth Avenue, New York, will naturally follow the motor-cab movement, and the principle will soon be applied to all forms of cartage—trucks, wagons, etc.—until the middle of the first decade of the twentieth century will be likely to see the vehicular traffic of our cities almost entirely transformed, with the movement also ascendant throughout the rural parts of the country, wherever the condition of the roads will permit it.

Just as we have seen the street railway tracks reconstructed to meet the demands of electric traction, so we shall see a very general reconstruction of the highways, both city streets and country roads, to meet the changed conditions imposed by au-

Editorial

tomobile requirements. New features will doubtless be introduced of which we at present have scarcely an inkling, and the department of road-engineering will become one of the most important and most extensively practised branches of the engineering profession. Good roads will become universal in parts of the country where good roads are wholly unknown at present; and the revival of country inns to cater to the renewed use of the highways for travel, already extensively promoted by the popularity of the bicycle, will take place on every hand. The effect upon life in the country will be as marked and as beneficent as upon life in the city, and the social and industrial changes cannot fail to be very extensive throughout the world. It is a good time to live in!

Municipal Automobiles

The city of Boston has taken an advanced stand in relation to the use of the automobile for municipal purposes. The fire department has two automobile steam fire-engines in regular use. One of these has been doing excellent service for something like two years or so. It gave such thorough satisfaction that a second one was soon ordered. It is a thrilling spectacle to see these powerful machines rushing through the streets in response to an alarm of fire. An ordinary horse-drawn steam fire-engine is, however, sufficiently terror-inspiring in the elements of noise, so that the advent of the horseless machines does not seem to have contributed in any special degree to the agitation of the equine population of the city. In Mayor Quincy Boston has a chief magistrate who, like his two ancestral namesakes in the same office, takes a modern view of public affairs, and seeks the aid of all available instrumentalities of an improved nature in carrying on the city work. Owing to his initiative, two motor vehicles were ordered early in the present year for the use of officials in two of the administrative departments in going about their work in various parts of the city. One of these, a light steam-carriage, with seats for two persons, has been in use since the beginning of August, and proves to be equal to the work of at least three horses. It was intended for the use of the chief of the Repairs Division in the Public Buildings Department, who in the course of his duties has to make many trips to all parts of the city. But since, unlike a horse, it does not get tired, the automobile is available for other business of the department when not in use by the particular official in whose service it stands, and in this way a considerable economy is effected. This steam-carriage often does a turn of forty or

The Automobile Magazine

fifty miles a day within the limits of the regular working hours. It stands ready for use at any moment. It is proposed to try the various forms of motive power in the city work of Boston. The second automobile ordered is a gasoline-motor carriage, which, oddly enough, is assigned to the use of the superintendent of the Electrical Construction Department, instead of an electric carriage. As a very large number of horses are used by the various officials in going about their work in different parts of the city, the substitution of motor vehicles means a very considerable saving in expenditure and also a materially increased efficiency. There is a wide field for the automobile in municipal work, and should cities take a leading part in its introduction, as Boston appears to be doing, they will not only directly profit by the heightened efficiency in the operation of the various branches of municipal activity, but the example thus set will accelerate its general adoption. Therefore the consequent saving in the wear and tear of the highways, and the reduction in the cost of street-cleaning to a fractional part of existing outlays for the purpose, will effect an enormous diminution in heavy items of expenditure. This would set free large sums to be employed in reconstructing the streets with the most improved forms of pavement, smooth and noiseless, whose durability, when freed from the destructive action of horses' feet, would be vastly extended. Moreover, the saving would also enable cities to give more attention to those features of public utility which, although commonly considered as luxuries, are nevertheless indispensable to the equipment of a growing modern municipality—features such as parks, art galleries, tasteful public buildings, public baths, and other desirable improvements—many of which are considered beyond the means of the average municipality, sorely taxed as it is likely to be, but which would be made available in consequence of such economies. In the broad sense of the word, the municipal field for the application of the automobile transcends its use merely for transit purposes. The adoption of the automobile principle for steam fire-engines has been noted. The pioneer application of the principle, of course, was in the steam-roller, whose use has long been universal. There are other fields of mechanical work where its adoption will prove equally efficacious. One of these will be its application to street-sweeping purposes, and another in which it cannot fail to effect very great economies will be its employment for cartage purposes. We can look to inventive ingenuity to devise other forms of use in which the automobile will speedily take the place of animal traction in many other departments of municipal activity, greatly

Editorial

to the benefit of all concerned. And not the least benefit will be that derived from the incalculable sanitary improvement that will come with the disappearance of stables from our cities and of horse excrement from our streets; as well as from the reduction of nervous maladies that will attend the abatement of street noises.

AMERICAN AUTOMOBILE CLUB

As a worthy successor to the famous Automobile Club of France and its offspring the Automobile Club of Great Britain and Ireland, an Automobile Club of America has now entered the lists of militant organizations that are speeding the world along its accelerated course toward the new era.

The club was definitely organized at the Waldorf-Astoria Hotel, in New York, last month, with a hundred charter members, including the thirty-five original founders. General Avery D. Andrews, of New York, accepted the presidency, to be aided by George F. Chamberlain, first vice-president; V. Everit Macy, treasurer, and Capt. Hedge, secretary.

The Automobile Club of America has been officially recognized as such by the Automobile Clubs of France and Great Britain. According to present plans the organization of the club will prove of immediate benefit to American automobilists, for steps have already been taken to conduct test runs and inaugurate an automobile exposition. This show will be the first of its kind in the Western Hemisphere, and is expected to surpass the horse show as a social event. A majority of the intending exhibitors will take their vehicles to the Paris Exposition.

OBITUARY

We regret to record the death of M. Louis-Victor Lockert, E.C.P., editor-in-chief of the Paris *Chauffeur*, a charter-member of the Automobile Club of France, as well as of the Touring Club of France, and the author of a series of well-known monographs on automobilism. M. Lockert, before his lamented death at the age of fifty-six, achieved a national reputation as one of the foremost expounders of modern engineering problems. He represented his country as Chief French Commissioner at the International Exposition at Moscow in 1872, and as commissioner of the national engineering exhibit at the Paris International Exposition of 1878. In recognition of his good services he was appointed a Chevalier of the Order of Saint Anna of Russia, and bore the decoration of Emperor Francis Joseph of Austria. His burial in the grave-yard of Montmartre, last month, was attended by a number of the most prominent auto-mobilists of Europe.

Electric Automobiles

OUR contemporary, *The Horseless Age*, appears to consider that it holds a brief against the electric automobile. A few weeks ago we pointed out the worthlessness of some " evidence " it had offered in support of its position that " electricity cannot compete with horses in public cab service." The current issue of that journal contains another attack on electric automobiles, this time based specifically upon the assumed worthlessness of the storage battery upon which they depend for motive power. In this instance again, there is an extraordinary jumble of inaccuracy and misinformation. For example, an early experiment with storage battery traction is referred to as follows: " At Washington, D. C., *a few years ago* most extensive and disastrous experiments were conducted on the Metropolitan Street Railroad, ending in the usual way with the resumption of horse-power." The italics are ours. For more than four years the road mentioned has been electrically equipped. The purely experimental Julien storage-battery car, run ten years or more ago in New York, is also referred to, but, of course, no mention is made of the Chicago & Englewood storage-battery road which has been in successful operation for several years, nor of the successful storage-battery lines in Paris and elsewhere abroad—foreign reference being confined to a single storage-battery car, said to be now running in England, whose operation is stated to cost 36 cents per car-mile!

Among the specific charges against the storage battery are, that it has electrical and mechanical limitations; increased efficiency means greater weight, and decrease in weight lessens the efficiency and weakens the plates; the sulphuric acid in the cells must be kept at a uniform specific gravity; too rapid charging and discharging results in damage; and so on in a vein that might be similarly worked in an arraignment of the steam-engine or any of the imperfect product of the hand and mind of man. Again, we are told that the storage battery is a delicate subject even in stationary work, and that for cab and delivery service it is as often in the hospital as it is in service—for which reckless statement no proof is offered: that in traversing roads

214

Electric Automobiles

that are rough and intersected by street-car tracks, the whole overloaded structure is racked, gears are thrown out of line, wheels are strained, and the power of the battery wasted. Strangely, there are a few words of favor for the electric automobile, it being said that its high cost, simplicity of control, and undeniable æsthetic features, commend it to the rich, especially to ladies, for urban use, and that it will undoubtedly find other incidental uses where the demands made upon it are light. Some may question the ethics of selecting ladies and the rich to be the victims of such an abomination as the electric automobile is portrayed.

As our contemporary does not assume to have any technical knowledge of the storage battery, and apparently quotes merely at random in its phrases of condemnation, the technical argument advanced cannot be taken seriously. The more general charge concerning the limitations of the storage battery due to its very nature, may, however, be noticed. As to these limitations, none probably is more aware of them than advocates of the electric vehicle. But their point of view is that if the limitations do not apply to its profitable use in such work, they have not much more than academic interest—a stand, it may be remarked, which does not differ from that taken with respect to any other matter in this world. That the storage battery can be profitably applied to electric automobiles is no longer an open question in view of the already extended commercial experience had with it. The matter is thus essentially a commercial one, and is not affected by arguments based upon limitations of the storage battery. That the battery is more delicate than the generator of a steam automobile is a matter of indifference, if it commercially answers the purposes for which it is used. The position of our contemporary is, in fact, what might have been the position of the clock trade had the watch been suddenly introduced and considered to menace the clock industry. Experience long since would have shown that the more delicate nature of the watch mechanism could not militate against its use, any more than such arguments as our contemporary offers can influence adversely the employment of the storage battery for automobile work. During the past six months the demand for electric automobiles in this country would, we are in a position to state, have required 40,000 vehicles to satisfy it. Such a condition as this implies should warn our contemporary that, unless it recognizes the absurdity of its attitude, it may soon be an anachronism in a horseless age.—*Electrical World and Engineer*, October 14, 1899.

A PUNCTUAL GUEST

Never mind about instructions.

I can run the thing myself.

Whoa! Hold on there!

Will she never stop?

I can see my finish.

Am I in time?

(*Xaudaro in Blanco y Negro.*)

A Practical Demonstration

A PRACTICAL DEMONSTRATION

So you don't know what an automobile looks like?

I'll manufacture one right before your eyes.

First you put a motor into the body of the carriage.

If it's a petroleum motor you then light the wick.

And the automobile does the rest.

(Benjamin Rabier in Le Journal Amusant.)

A PRACTICAL POINT

Scorcher: How would you punctuate, " Look at that pretty girl, in her automobile, spinning down the avenue "?

Putter: That's easy: Comma after " pretty girl " and after " automobile."

Scorcher: I'd rather make a dash after that pretty girl.

"WE SEES OUR FINISH"

IN DIXIE

The old colored man was watching the new-fangled machine spin down the street, and remarked: " I reckens dat mus' be one er—one er dem autographs? "

" It's an automobile."

" Dat's it. I allus was gittin' dem two towns mixed, but I knowed it war some'ere in Alabama."

The New Era

THE NEW ERA

Pedestrian:—I was run down by him as I was crossing the street.

Bike Cop:—Never mind about that. First show me your permit for walking on foot.

(*Benjamin Rabier, in Le Journal Amusant.*)

The Automobile Index

Everything of permanent value published in the technical press of the world devoted to any branch of automobile industry will be found indexed in this department. Whenever it is possible a descriptive summary indicating the character and purpose of the leading articles of current automobile literature will be given, with the titles and dates of the publications.

ACCUMULATORS—
Electrical Laws and Units, by E. C. Rimington. Automotor Journal, London, September 15, 1899.

A Paper by James K. Pumpelly on Storage Batteries. The Autobain, Chicago, September 13, 1899.

Charging and Care of Automobile Batteries. An article on this subject by Mr. Theodore D. Bunce. Scientific American Supplement, New York, October 14, 1899.

ACCUMULATOR TESTS—
A Vibrating Table for Testing Accumulators Described. La Locomotion Automobile, Paris, September 7, 1899.

ACCUMULATORS WEIGHT FOR TRACTION—
A Theoretical Article, illustrated by diagrams, calculating formula for weight of accumulators. By Rosset. L'Industrie Electrique, Paris, September 10, 1899.

ALUMINUM—
Paper read by P. W. Northey at the British Carriage Builders' Convention. "Possibilities and Uses of Aluminum as Applied to Carriage Building." The Autocar, Coventry, September 16, 1899.

An article by Mr. Leon Auscher on the actual uses of aluminum, with various references to its applications into the construction of carriages. Le Chauffeur, Paris, September 25, 1899.

AUTOMOBILE CLUBS—
Organization of American Club commented upon. Automobile Magazine, New York, November, 1899.

AUTOMOBILE COMPETITION—
The International Competition of Heavy Weights (Concours International des Poids Lourds). The official report of the commission upon the exhibits and trials at the competition of the Automobile Club of France in October, 1898; with many illustrations. Three articles, one plate. Génie Civil—July 29; August 5, 12, 1899.

AUTOMOBILE CONTROL—
Description of a method of Control for Automobiles, as per English patent, issued to Henry Leitner, of London. Electrical World and Engineer, New York, September 23, 1899.

AUTOMOBILE EXHIBITION—
Automobile Exhibition at Paris. Editorial on the second exhibition of automobiles held under the auspices of the Automobile Club of France, calling attention to tendencies in the evolution. Engineering, London, July 21, 1899.

AUTOMOBILE FIRE-ENGINE—
A New Automobile Fire-Engine. From La Nature. Illustrates and describes a fire-engine invented by M. Porteu, which has been found to operate most successfully. Scientific American Supplement—August 5, 18, 1899.

AUTOMOBILE FIRE EQUIPMENT—
A short illustrated description of this automobile which has been on trial for four months with good success. The battery has a weight of 520-kg., which is only a fifth of the total, 2,500-kg., including the crew, and a capacity or 150 ampère hours at the rate of 35 ampères. If discharged in 17 hours 48 minutes, or in 7 hours 24 minutes, or in 3 hours 52 minutes, the accumu-

The Automobile Index

lators give 26.7, or 22.2, or 19.35 ampère hours per kg. of plates respectively, at 1.5, or 3, or 5 ampères respectively, per kg. of plates. It will run 4 to 5 hours at speeds of 15- to 22-km. per hour. —L'Industrie Electrique, August 10, 1899.

AUTOMOBILE FRAMES—
Canello Durkopp's Automobile Frame, described and illustrated. La Locomotion Automobile, Paris, September 7, 1899.

Henriod System, described and illustrated. The Automotor Journal, London, September 15, 1899.

AUTOMOBILE HANSOM—
New Type of French Automobile Hansom, described and illustrated. Electrical World and Engineer, New York, September 16, 1899.

AUTOMOBILE PHAETONS—
Phaetons at the Exposition of the Automobile Club. The first of a series of articles discussing the vehicles exhibited at the recent exposition in Paris. Scientific American Supplement. August 26, 1899.

AUTOMOBILE RACES—
An American Auto Race. Illustrated description of Galesburg race. Automobile Magazine, New York, November, 1899.

AUTOMOBILE TRICYCLE—
Detailed Description, with illustrations, of Sangster's English Patent. The Cycling Gazette, Cleveland, O., September 21, 1899.

BEARINGS—
Axle Bearings. Described and illustrated by M. F. Hill. Automobile Magazine, New York, November, 1899.

BRAKE—
Illustrated Description of A. Bollée's Band Brake. Le Chauffeur, Paris, September 25, 1899.

CARBURETERS—
Starr & Gogswell Closing Carbureter, described and illustrated. La Locomotion Automobile, Paris, September 7, 1899.

L. Aster's Carbureter, with an automatic regulator, described and illustrated. La France Automobile, Paris, September 10, 1899.

Carbureter de Sales, described and illustrated. Le Chauffeur, Paris, September 11, 1899.

De Sales & Braly's System, described and illustrated. Automotor Journal, London, September 15, 1899.

Kellett's System, described and illustrated. The Automotor Journal, London, September 15, 1899.

CHARGING HYDRANT—
Suggestions for a Charging Hydrant, with illustrations. Motor Vehicle Review, Cleveland, October 3, 1899.

COMPARISON OF AUTOMOBILES—
An article, with many illustrations of American, French, and English automobiles, discussing the relative advantages of electricity, gasoline, and steam-power for automobiles. The electrical system is claimed to be the best on account of the simplicity of operation and its availability for private carriages and delivery wagons, provided the distances are not more than thirty miles over ordinary grades at an average speed of eleven miles an hour, which represents one discharge of the battery, and provided the load to be carried by delivery wagons is under 1,000 lbs. The other motive powers are better able to fulfil the requirements of greater distances and heavier loads. Cassier's Magazine, New York, September, 1899.

COMPETITION, RULES AND REGU-LATIONS—
Rules and Regulations Governing Trial Tests for Heavy-Weight Automobile Vehicles, organized by the Automobile Club of France. La Locomotion Automobile, Paris, August 31, 1899.

CONTROL OF ELECTRIC VEHI-CLES—
A Paper by E. H. Cozens Hardy, indicating the methods employed to start, run, and stop electric automobiles; serial. The Automotor Journal, London, July 15, and September 15, 1899.

DOUBLE MOTOR AUTOMOBILE—
Double Motor Automobile, described and illustrated. Le Chauffeur, September 11, 1899.

DISTRIBUTOR—
Henriod Distributor, replacing the carbureter, described and illustrated. The Automotor Journal, London, September 15, 1899.

EARLY AUTOMOBILES—
First American Automobile, described and illustrated. Automobile Magazine, New York, November, 1899.

The First Motor Ride, Sir G. Gurney's Steam Stage-Coach. Described and illustrated by J. G. Carteret. Automobile Magazine, New York, November, 1899.

ELECTRICAL MEASUREMENT—
Fundamental Facts in Connection with the Problem of How Electrical Quantities are Measured. The Automotor Journal, London, September 15, 1899.

ELECTRIC AUTOMOBILES—
The Monnard Electromobile. Translation of an article by M. Delaselle in La Locomotion Automobile. Description, with several illustrations. Automotor Journal, August, 1899.

A Paper on Electric Automobiles and their Limitations. The Horseless Age, New York, September 27, 1899.

The Chapman Electromobile, described and illustrated. Scientific American, New York, September 30, 1899.

Riker's Electric Truck, described and illustrated. Electrical World and Engineer, New York, October 7, 1899.

Automobile Street Sweeper and Sprinkler, described and illustrated. Scientific American, New York, October 7, 1899.

Electric Break of the American Electric Vehicle Company described and illustrated. Electrical World and Engineer, New York, October 7, 1899.

Riker's Electric Demi-Coach, described and illustrated. Scientific American, New York, October 14, 1899.

Polemic Discussion. Electrical World and Engineer, October 14, 1899.

Monograph by Professor Felix Michotte. Automobile Magazine, New York, November, 1899.

ELECTRIC IGNITER—
Ignition by Dry Batteries. Automobile Magazine, New York, November, 1899.

FIRST AUTOMOBILE—
L'Ouest Sportif claims that the first automobile was constructed in Nantes in 1864. La Locomotion Automobile, Paris, September 7, 1899.

FRENCH REGULATIONS—
Automobile Regulations in France. A review of the new French regulations governing the use and sale of self-propelled vehicles of every type. Engineering Record, August 12, 1899.

HUB—
A Rubber-packed Hub for Automobile Wheels described. La Locomotion Automobile, Paris, August 31, 1899.

HYDRO-CARBON AUTOMOBILES—
Vehicles of the Société Bourguignonne, described and illustrated. La Locomotion Automobile, Paris, August 31, 1899.

Canello Dürkopp's Phaeton, described and illustrated. La Locomotion Automobile, Paris, September 7, 1899.

Calloch System, described and illustrated. La France Automobile, Paris, September 10, 1899.

Léon Bollée's Voiturette, described and illustrated. Le Chauffeur, Paris, September 11, 1899.

Léon Bollée's Vehicle, described and illustrated. Le Chauffeur, Paris, September 11, 1899.

Vallée's New Vehicle, described and illustrated. La Locomotion Automobile, Paris, September 14, 1899.

Description and Illustration of a Daimler Vehicle built for the Duke of Westminster. The Auto-

The Automobile Index

motor Journal, London, September 15, 1899.

The De Py System, described and illustrated. Automobile Magazine, New York, November, 1899.

Fritcher-Houdry System, described and illustrated. Automobile Magazine, New York, November, 1899.

HYDRO-CARBON MOTOR BICYCLE—

I. Hughes' & W. Owens' Patent, described and illustrated. The Automotor Journal, London, September 15, 1899.

HYDRO-CARBON MOTOR REGULATOR—

Canello Dürkopp's Motor Regulator, described and illustrated. La Locomotion Automobile, Paris, September 7, 1899.

HYDRO-CARBON MOTORS—

Trotler's Two-Cycle Motor, described and illustrated. La Locomotion Automobile, Paris, August 31, 1899.

The Bolide Motor and Mechanism. Illustrated description. The Automotor Journal, London, August, 1899.

Canello Dürkopp Motor, described and illustrated. La Locomotion Automobile, Paris, September 7, 1899.

Description of the Calloch Motor. La France Automobile, Paris, September 10, 1899.

Marius Marmonier's New Motor, described and illustrated. La Locomotion Automobile, Paris, September 14, 1899, also L'Industrie Automobile, Paris, September 25, 1899.

Henriod Motor, described and illustrated. The Automotor Journal, London, September 15, 1899.

Doré's Motor, described and illustrated. Automobile Magazine, New York, November, 1899.

Daimler's System, described and illustrated. Automobile Magazine, New York, November, 1899.

HYDRO-CARBON TRICYCLES—

Simms Tricycle, with two front driving-wheels, described and illustrated. La Locomotion Automobile, Paris, September 28, 1899.

LIQUID AIR—

An article by Frank Richards, on possible methods of applying liquid air to automobiles. Compressed Air, New York, October, 1899.

LIQUID FUEL BURNER—

Musker's Automatic Liquid Fuel Burner for Heavy Motor Vehicles, described and illustrated. The Automotor Journal, London, September 15, 1899.

LUBRICATORS—

Despont & Godfroy's Sight Feed Lubricator, described and illustrated. Automobile Magazine, New York, November, 1899.

MILITARY AUTOMOBILES—

Illustrated Description of the Function of the Automobile in War, by Edwin Emerson, Jr. Automobile Magazine, New York, November, 1899.

MOTOR-COOLER—

Automizer used as a Motor-Cooling Device, described and illustrated. La Locomotion Automobile, Paris, August 31, 1899.

Water Coolers, Julian System, described and illustrated. Automobile Magazine, New York, November, 1899.

POWER TRANSMISSION—

A New Power Transmitter and Speed Changer, described and illustrated. Automobile Magazine, New York, November, 1899.

PUMPS—

Dalligol & Thomas Water Circulating Pump Described. Automobile Magazine, New York, November, 1899.

RECONDENSATION OF STEAM IN STEAM MOTOR VEHICLES—

An article by Mr. Hudson Maxim on his invention for the recondensation of steam. Horseless Age, New York, October 11, 1899.

REDUCING GEAR—

Humpage New System of Reducing Gear, as applied to electric motors, described and illustrated. Electrical World and Engineer, New York, September 23, 1899.

223

The Automobile Magazine

SPEED VARYING GEAR—
Canello Dürkopp's System, described and illustrated. La Locomotion Automobile, Paris, September 7, 1899.

Henriod System, described and illustrated. The Automotor Journal, London, September 15, 1899.

STEAM AUTOMOBILES—
Brown-Whitney Steam Automobile, described and illustrated. The Autocar, London, September 9, 1899.

Description of the Stanley Steam Vehicle. The Autocar, Coventry, September 16, 1899.

STEAM-BOILER—
Tower & Company's Boiler for Steam Automobiles, described. La Locomotion Automobile, Paris, September 7, 1899.

STEAM MOTOR—
Jaquemin Patent, producing directly the circular motion in steam-engines, described and illustrated. La France Automobile, Paris, September 10, 1899.

STEEL TUBING—
Their use in automobiles, by C. G. Canfield. Automobile Magazine, New York, November, 1899.

STEERING—
The Steering of Automobiles With Divided Axles (Du Virage des Automobiles à Essieux Brisés). D. Dujon. An analysis of the action of independently adjustable steering wheels, showing the method adopted for the prevention of slippage on short turns. Revue Technique, Paris, August 10, 1899.

STEERING AXLES—
A Series of Articles on Steering Vehicles by Divided Axletree. La Locomotion Automobile, Paris, September 21, 1899.

STEERING GEAR—
Canello Dürkopp's System, described and illustrated. La Locomotion Automobile, Paris, September 7, 1899.

Priestmann & Wright's Steering Gear, described and illustrated. The Automobile Magazine, New York, October, 1899.

STEERING MECHANISM—
Henriod System, described and illustrated. The Automotor Journal, London, September 15, 1899.

SWITCH AND CONTROLLER FOR ELECTROMOTORS—
Elieson's Patent, described and illustrated. The Automotor Journal, London, September 15, 1899.

TIRES—
Description of a New Tire, which is a combination of the pneumatic, the cushion, and the sponge types. The Cycling West, San Francisco and Denver, October 5, 1899.

Julian Tire, described and illustrated. Automobile Magazine, New York, November, 1899.

TRIAL TESTS—
Project of an Automobile Race from Paris to St. Petersburg. La Locomotion Automobile, Paris, September 7, 1899.

French Trials of Electric Delivery Wagons. Electrical World and Engineer, New York, September 23, 1899.

The Paris Cab Trials. Translated from Le Génie Civil. An account of the June trials, with description of the vehicles. Automotor Journal, August, 1899.

The Agricultural Hall Exhibition Efficiency Trials. Gives complete results of these trials. Automotor Journal, August, 1899.

The Liverpool Heavy Motor Vehicle Trials. A full account of the trials, with illustrated description of the vehicles. Automotor Journal, August, 1899.

The Liverpool Motor-Wagon Trials. Brief account of the second annual series of trials of motor-vehicles for heavy traffic. Describes vehicles and tests. Transport, August 4, 1899.

VALVE GEAR—
Henriod System, described and illustrated. The Automotor Journal, London, September 15, 1899.

WAGGONETTE—
Burns "Endurance" Waggonette, described and illustrated. The Autocar, September 9, 1899.

The Automobile Magazine

VOL. I   DECEMBER 1899   No. 3

CONTENTS

AGENCY FOR FOREIGN SUBSCRIPTIONS :

INTERNATIONAL NEWS COMPANY

BREAMS BUILDINGS, CHANCERY LANE   STEPHAN STRASSE, NO. 18
          LONDON, E. C.   LEIPSIC

Price 25 Cents a Number ; $3.00 a Year
Foreign Subscription $4.00, Post-paid

"Jumbo" to the Rescue

The Automobile
MAGAZINE

VOL. I   DECEMBER 1899   NO. 3

The Horseless Fire Engine

By Captain Cordier

Technical Engineer of the Paris Sapeurs-Pompiers Regiment

NOW that the automobile is coming into such general vogue that all the traffic of our cities bids fair to be revolutionized thereby, we must look for a corresponding reform of our great municipal fire departments. Steps in this direction have already been taken in Paris, where every fire brigade is to be equipped with horseless engines and apparatus,—in Buda-Pest, where self-propelled fire engines have become a familiar sight, and in New York, where the Chief of the Fire Department can daily be seen darting to points of danger in a specially constructed automobile of his own, with which he can distance all other vehicles on the way. In New York, as well as in other American cities, such as Boston, Hartford, Pittsburg and Buffalo, powerful steam fire engines which can be run without the assistance of horses have indeed been in use so long that they are no longer considered in the light of experiments.

This is but as it should be, since fire engines run and operated by steam must in truth be regarded as the first practical forerunners of the modern automobile. Long before automobile pleasure carriages were heard of powerful steam-traction engines were employed to fight fires, and some of them were brought to such a point of perfection that even from our present *fin-de-siècle* point of view they leave but little to be desired.

Thus a daring American engineer, Captain John Ericsson, of Monitor fame, as early as 1840 constructed a steam-propelled fire engine which proved a complete success. Twenty years

The Automobile Magazine

later the more distant suburbs of the City of Boston were reached by means of a steam fire engine steered with a pole—the invention of a local fire chief, Frank Curtis. In later years the famous American Amoskeag Engine Company built self-propelled fire engines, which were used in New York for some time, until they were abandoned by reason of the heavy snows which rendered them impracticable in winter. In summer, it was claimed, they ran so fast that they were declared to be dangerous to traffic.

At the present time the City of Hartford, Connecticut, is in the proud possession of the largest and most powerful locomotive

Boston Fire Engine Responding to Call
Copyrighted by G. B. Brayton

steam fire engine in the world, named "Jumbo" (*see frontispiece*). Over 10 feet high and 17 feet long, it weighs 8½ tons, and can throw 1,350 gallons of water per minute. The boiler contains 301 copper tubes. This engine, at her first trial threw, through 50 feet of hose, 3½ inches in diameter, a horizontal stream of water a distance of 348 feet, and threw two streams, each as large as that thrown by an ordinary fire engine, a distance of over 300 feet. The road driving power of the engine is applied through two endless chains running over sprocket wheels on each of the main rear wheels, permitting these wheels to be driven at

The Horseless Fire Engine

varying speeds when turning corners. The engine may be run either forward or backward, and can be stopped inside of 50 feet when running at full speed. When in the house the boiler is connected with steam pipes from a heater in the basement, and steam is always kept up to about 95 pounds, which would run her about a quarter of a mile. The fire box is kept full of material ready for lighting, and a steel arm under the engine carries a quantity of waste saturated with kerosene oil in close proximity to a card of matches in a holder under a scratcher, the latter being attached to a cord tied to a ring in the floor. At an alarm of fire

Horseless Engine at its Station

the steam pipes are disconnected, the throttle opened, and, before the engine has moved six inches, the cord pulls the scratcher and the rod carrying the blazing waste swings around under the fire box, igniting the shavings and wood. Cannel coal is burned and steam enough can be generated in two minutes to run the engine at a speed of thirty-one miles an hour.

The horseless fire engines now operated in Boston are likewise of the Amoskeag type. From the ground to the top one of these engines is 10 feet; its length over all is 16 feet 6 inches, and the width over all is 7 feet 3 inches. The weight, equipped for service, is 17,000 pounds. The boiler is upright and tubular in style, with a submerged smoke box, and is expanded at the lower end to increase the grate surface. It is made of the best quality of steel plate, with seamless copper tubes, and is thoroughly riveted and stayed. It is jacketed with asbestos and has a lagging of wood which supports the metallic jacket. The connections with the steam cylinders are simple and have the advantage of being entirely unexposed to the air. The steam cylinders are cast in one piece. They are firmly secured to the boiler and framing and are covered with a lagging of wood, with a metallic jacket on the outside. The main shell of the pump is in one solid casting. It is a double acting and vertical pump and its valves are vertical in their action. The pump is arranged for receiving suction hose on either side and has outlets on either side for receiving the leading hose. The connection between the steam cylinders and water cylinders or pumps may be made by the old and familiar link motion and link block, or the equally familiar cross-head and connecting rod plan, both giving excellent results for ordinary steam fire engines. In the self-propelling engine, where the engine power is transmitted to the driving wheel through the main crank shaft, which is not the case when this power is transmitted to the pumps, the cross-head and connecting rod plan has many advantages, and is therefore adopted for self-propelling engines.

The manner of handling the self-propellers is very simple. The chief engineer rides on the fire box of the engine and has directly under his hand the various levers and wheels which start, stop and regulate the speed of the machine. The assistant engineer rides on the driver's seat, and by means of the large steering wheel he steers the machine in exactly the same manner as the rear wheels of the long ladder trucks are governed through a system of bevel and worm gearing. The engine can be turned around in an ordinary street with ease.

The road driving power is applied from one end of the main crank shaft to an equalizing compound, and two endless chains running over sprocket wheels on each of the main rear wheels permit these rear wheels to be driven at varying speeds when turning corners. The driving power is made reversible, so that the engine may be driven forward or back at the will of the operator. When it is not necessary to use the power of the engine for driving purposes, the driving mechanism can be disconnected by the removal of a key, so that the pumps may be worked with the engine standing still. An extra water tank is carried at the rear of these engines to supply the boiler until connections

The Horseless Fire Engine

can be made with a hydrant. The self-propeller can travel on a fair level road at a maximum rate of twelve miles an hour. It can climb any ordinary grade; in fact, any one that a team of horses can climb with a heavy load.

It appears strange, in view of the great progress of electricity in America, that so little in that line has been attempted in connection with fire engines. Only very lately, it is reported, the Fire Department of New York has entered negotiations with the La France Fire Engine Company for a complete fire engine searchlight plant, consisting of a standard New York fire engine, which, in place of the pumping machinery, will be equipped with a multipolar dynamo direct connected to a high speed Forbes engine. On a platform just behind the driver there will be mounted two 18-inch Rushmore marine projectors. Each projector will be fitted with a special diverging lens with which the beam of light may be spread out to cover as wide an area as desired, so it will be equally effective at long or short range. The projectors may be quickly removed from the engine and set upon the ground at any distance from the engine to which it is connected by a flexible cable carried upon a reel under the driver's seat.

This engine will answer alarms with the other apparatus and by instantly lighting up the scene of a fire will greatly reduce the delay and dangers of fighting fires in the congested business district.

The City of Paris, in this respect, is more advanced even than the cities of the New World. Already many self-propelled fire engines, hose wagons and hook-and-ladder trucks of various types of construction have been put in operation, and others are added as soon as available, until every important fire station shall have this modern equipment.

The first of these machines to receive its baptism of fire was Porteu's automobile fire engine, constructed by M. Cambier, of Lille.

The frame of this machine is of U-shaped bar iron and supports both the motor and the pump. The gasoline motor is placed in the rear and consists of four explosion cylinders, C (Figs. 1 and 2), in pairs, placed symmetrically with respect to the longitudinal axis of the vehicle. It is of 22 horse-power. The ignition is electric, and the carbureting is effected by the Longuemare apparatus.

The rods of the four pistons are assembled in two pairs upon two cranks and communicate motion to the shaft, A, which is provided with flywheels, V V, at its extremities. This shaft,

The Automobile Magazine

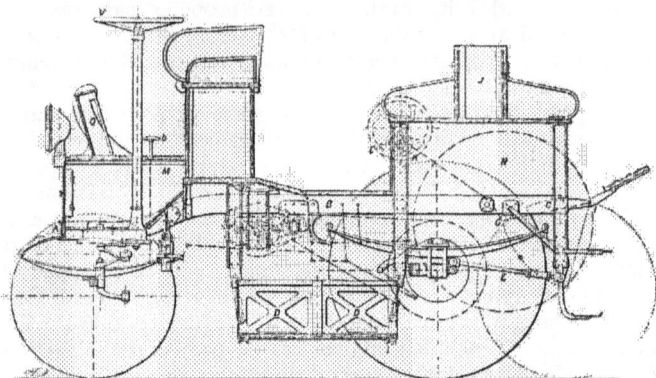

Section of French Gasoline Fire Engine

through a train of gear wheels, transmits motion to an intermediate transverse shaft, *A'*, which might be called the distributer of motion, since it actuates either the propelling apparatus or the pump, as may be desired. It communicates motion to the vehicle through the intermedium of drums, *M N O*, the different diameters of which furnish two running speeds that correspond to about nine and five miles an hour. These drums actuate a series of fixed pulleys, *b d*, and idle pulleys, *b'd'*, keyed upon the shaft that carries the chain sprockets, *p p*.

Transmission from the drums to the pulleys is effected by cross belts, the shifting of which from the fast to the idle pulleys is produced through the action of the rods, *h h*, which are maneuvered from the front of the vehicle by a lever, *L*, mounted upon the same axis as the steering wheel, *V'*.

The intermediate pulleys, *N* and *c*, belong to the mechanism that produces a backward motion, and are connected by a belt, upon which acts a stretcher, *T*, maneuvered through a handwheel, *V''*, placed under the front seat. All the belts are surrounded by an iron plate jacket that protects them from the action of the water coming from the pump.

The pump is actuated by the shaft, *A'*, which communicates motion to the driving wheel, *M*, through a pinion keyed upon the shaft.

The throwing of the drums into gear and of the pump out of gear is effected instantaneously and automatically. A fire may thus be attacked as soon as the engine arrives upon the spot, with-

The Horseless Fire Engine

out its being necessary to modify the running of the motor in
any way whatever.

The pump, which is of the Thirion system, is capable of
throwing 3,000 gallons of water a minute. Power is transmitted
to it wholly by gearings, so that the inconveniences that would
result from the use of belts, through the stretching of the leather
under the action of the water coming from the pump, are done
away with.

The steering of the engine while running is done by means
of the handwheel, V', which acts upon the pivots of the front
wheels, $R'R'$, through the intermedium of rods, t, and levers, r.
Finally, the vehicle is provided with a brake, K, which is actuated
by a pedal, B, placed under the foot of the driver.

As such an apparatus must carry a number of firemen with it,
the fore-carriage is provided with two seats, s and s', each capable
of accommodating three persons, and with a platform, J, in the
rear, upon which several men can stand.

This new engine has been submitted to numerous experiments
in the presence of the Chief of the Paris Fire Department and
his staff, and to many other persons competent to judge in such

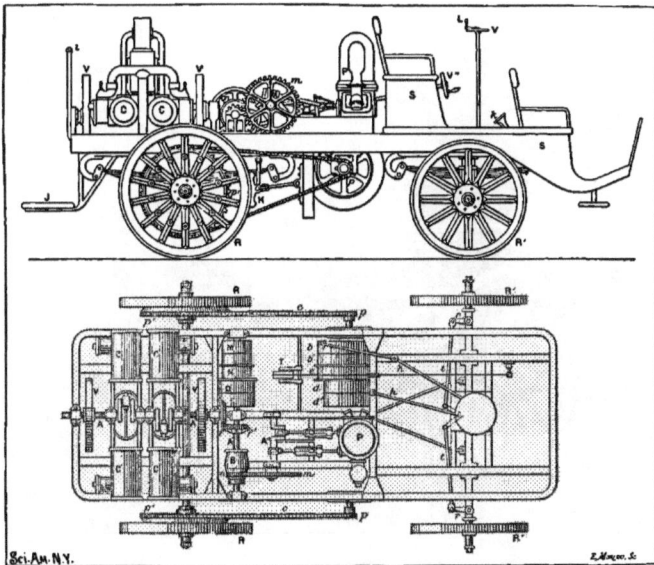

Section and Plan of Porteu's Automobile Fire Engine

matters, and has been found to operate in the most satisfactory manner.

Paris firemen have also lately been out making successful experiments with the automotor constructed especially for the service of the pompiers, after the plans of Colonel Krebs. The vehicle, which is painted red, was filled with hose and general material, and, with six men on board, went for several miles along the quay in the direction of Notre Dame Cathedral. It accomplished the journey backward and forward in a very expeditious manner and without a hitch. The motor may be seen in the fire brigade station near the Law Courts on the Boulevard du Palais. The officers of the establishment in question believe that the new cars which are in course of construction will enable firemen to answer calls in half the time required by horses.

Before this an electric hose carriage was already in use by the Paris Fire Department. The machine weighs over a ton, or with equipment, nearly three tons. The equipment consists of six men, apparatus to operate three lines of hose, a scaling ladder,

French Electric Fire Department Wagon

The Horseless Fire Engine

French Gasoline Fire Engine En Route

apparatus for wire in cellars, and one for life-saving purposes. It can travel fifteen miles an hour on good roads.

To be more exact, it should be stated that this car is of a gross weight of 1,800 kilogrammes, and of a total weight of 2,550 kilogs. (including six men), the material necessary for operating 3 hoses, 24 metres of piping, a ladder, cellar fire appliances and rescuing apparatus.

It can proceed 60 kilometres at a speed of 15 kilometres per hour, without being recharged and with an expenditure of 40 to 45 amères. On a good road it proceeds 22 kilometres with an outlay of 55 ampères. The start is well performed—instantaneous.

The motor is of 4,000 volts; the battery of accumulators weighs 520 kilogs., and has a serviceable capacity of 180 ampères.

The body of the car measures 3.25 metres in length and 1.95 metres in width.

The body and all the motor parts are raised upon a steel frame in *U* bent. The front rests upon a short axle by a triplicate spring support. The rear is supported upon an ordinary axle by a simple spring support. The traction of the motor is brought to bear upon the frame by the connecting rod, which also serves as coupler.

The conductor takes his place upon the carriage box, having on his right the speed-handle (4 turns forward 2 to the rear) and in front of him the direction-handle.

In front are placed the measuring appartus: the voltimetre, ampèremetre, circuit stop and a bracket for 2 arc lamps for illumination on dark nights.

In the body of the car there is a winder, an independent machine. Between the two axles, at the lower part is suspended by springs the cylinder of accumulators.

Hartford Horseless Fire Engine

This car is the first in which electricity has been applied in the traction of fire engines.

While French and American cities are instituting these reforms in their fire services little or nothing has been done so far in Germany or England. The state of public opinion on this subject in England may be judged from the following extract taken from a British journal:

"A very interesting review of the London County Council Fire Brigade was held at Peckham, when 149 officers and men, with five obsolete horse-drawn steam fire engines, four obsolete horse-drawn escapes, and four obsolete horse-drawn vans, all under Commander Wells. R. N., were " inspected," with evident

The Horseless Fire Engine

satisfaction, by the Fire Brigade Committee and their friends. The usual stock manœuvres were gone through, and a " display," more theatrical and meretricious than practical, was given of the means employed for rescuing persons.

From a technical point of view the show was disappointing; the appliances exhibited were for the most part obsolete, or, at any rate, not in accordance with the latest ideas of scientific fire prevention. Not a single mechanically-propelled fire vehicle was present, for the very simple reason that the Fire Brigade does not possess one. The reason that the Fire Brigade Committee of the County Council does not adopt automobile fire-engines is that that body lacks the necessary technical advice and knowledge of modern fire prevention and extinction."

Hartford Automobile Fire Truck

THE PROBLEM OF WHEEL-BASE

Length of wheel-base—*i. e.*, the distance between the centres of the front and back wheels—is an automobile problem worthy of study. A long wheel-base gives ease in steering. A short wheel-base causes great difficulty to the driver, and is dangerous.

The width between the front wheels should be the same as the width between the back wheels. A double track is a great fault.

The size of wheels is also important. Front wheels should be large, not small, so as to ride easily over irregularities in the road surface.

If the wheels of a carriage do not fit the wheel-ruts of country roads, driving is very difficult and unpleasant.

237

The Napier Motor

THE Napier motor is of the two-cylinder, vertical type, with electric ignition. The admission and exhaust valves are so arranged as to be readily accessible for cleaning purposes: they are easily removed merely by unscrewing two nuts and taking out a bolt.

The cylinders are cooled by a water-jacket supplied by a small centrifugal pump.

In starting the motor, which operation is performed by means of a hand-crank placed at one side, a special mechanism is employed in order to open the compression-valve. The moving parts are all well lubricated, and the cranks turn in oil.

The motor is provided with a ball-governor which automatically closes the admission-valve of one of the cylinders when the speed becomes excessive.

In some motors of this type Longuemare burners have been substituted for the electric ignition devices, since they are preferred by many *chauffeurs*.

The motor normally makes 780 revolutions per minute, at which speed 7-horse-powers are developed. The number of revolutions can be increased or diminished by regulating the ignition.

Santa Claus Gets a New Plaything

By Sylvester Baxter

DEAR Old Santa Claus this year will keep in touch with the spirit of the times, and on the eve of the last Christmas but one in this memorable century will make the greater portion of his rounds by automobile, in place of his famous team of reindeer drawing a sleigh. So, young friends, keep your ears wide open for the signal either of a gong's clattering staccato or the sounding of a pneumatic horn from his motor-carriage, as well as for the musical jingling of his sleighbells!

The genial old Saint has been giving a deal of thought to this matter, and has concluded that he can expedite things very materially by making use of the new system of locomotion. He has found his reindeer altogether too slow for his rapidly increasing business. They get tired too soon and he has had to have an enormous number of relays in order to make his rounds in time. All this proved expensive and inconvenient, particularly now that the dealers in reindeer-moss have caught on to the financial racket and organized a trust to put up the price on him, while lots of time has been wasted in hitching up fresh animals.

Of course he could not use reindeer everywhere, since the celebration of Christmas after his fashion has spread so extensively through the world and included so many snowless regions. So he has had to get used to horses and carriages for those parts of the world, and these have made him a deal of trouble in various ways, since somehow the reindeer was the only animal he could get used to handling without difficulty. One thing that has long caused him much vexation is that when he has planned very carefully to use reindeer in the locations where he had the most work to do, there would either come a thaw and melt all the snow, or the expected snow would not fall at all, and the ground would remain bare and frozen. Then at the greatest inconvenience he would have to change his plans at the last moment, give up his reindeer and leave them idly eating their heads off on expensive moss imported at exorbitant freight charges, and arrange for horses. He wanted to use the automobile this year altogether, he was so much pleased with it, but since the automobile sleigh has not yet been perfected he has decided to stick to the reindeer for those parts where the drifts are too deep for wheeled vehicles and make his rounds by automobile everywhere else,

even where there is considerable snow. For he has found that the broad pneumatic tires of the latter enable it to get through pretty heavy snow.

Saint Nicholas authorizes the statement that he has organized his work for this Christmas very thoroughly, and expects to get through it with unprecedented ease and celerity. He has taken a course of lessons at the All-Saints Automobile Driving School, and has become a thorough expert. He has converted many of his brother Saints to an enthusiastic devotion to the new form of locomotion, though for a long time he could not induce Saint Anthony the Abbot, the special friend of animals, to trust himself to a spin with him.

" The horse is good enough for me," said Saint Anthony, " and I will not go back on him."

Santa Claus Gets a New Plaything

But when Saint Nicholas demonstrated to him how much happier the horse's lot was becoming by reason of the lightening of his labors through the new forms of mechanical traction, the kindly Abbot saw a new light and allowed that the inventors, manufacturers and users of the automobile were men after his own heart, and he blessed their work as that of true friends of the horse.

Saint Anthony had some time before consented to be the special patron of good roads, because of their invaluable service in making easier the work of horses and all other draught animals, and he was particularly pleased when he learned that it was a Pope who originated the good-roads movement in the Great Empire of the West. So when Saint Nicholas told him that the same Pope had gone in with equal enthusiasm for carrying the automobile into universal use, after already having emancipated millions of horses from hard work by inducing men and women to carry themselves about on a curious machine called a bicycle—doing all the work themselves by pushing the wheels around with their own feet—Saint Anthony said: "That settles it! If a Pope takes such things into his own hands there is good reason why an Abbot should not be backward in the good work."

Therefore it came about that though Saint Anthony the Abbot was the last of the halo-wearers to attend the Automobile School, his belated enthusiasm made up for his tardiness, and he was much disappointed when he learned that the manufacturers were so far behind with their orders that there would be a long delay in filling his, it had come in so late. Saint Nicholas, however, promised to do what he could for him after the Christmas rush was over. Whereupon Saint Anthony the Abbot offered to take the automobile under his special protection, stretching a point or two and regarding it as an animal, although it was a machine. So he took out the catalogue of his pet animals and turned over the leaves to find a place to write it down.

"I will substitute it for one of the extinct species," he said. "Ah! here is just the place for it, the place filled by the Great Auk; 'Auk, the Great' is the way it was listed. The poor old Great Auk! That was a good bird. He used to come a long ways to listen to my sermons and always seemed so interested! It is the way of all flesh. But may it be a long time before the horse becomes extinct."

"That brings the Automobile well to the front on the list." remarked Saint Nicholas, looking over his friend's shoulder as the latter wrote the name down.

"Yes; I keep my system strictly alphabetical." said Saint

Anthony the Abbot. "It brings the Automobile just after the Ass."

"But where is the Austrich?" queried Saint Nicholas. "I should think that ought to stand just before."

"You will find the Ostrich among the O's," was the reply. "I fear your spelling is getting a little rusty, Brother Nicholas."

"I must have caught it from so much children's spelling on Christmas bundles. Besides, I never was very strong in English orthography." And the jovial Saint laughed good-naturedly.

It was on the practice-grounds of the school that this occurred. Just then Saint Patrick came along with an automobile built specially for him on the lines of an Irish jaunting-car. "Well, Tony, me boy," he cried, as he drew up, "so we've got you here at last. But what are you doing here with that beast-book of yours?"

They told him, and Saint Patrick agreed that it was a most admirable thing to do. "If 't isn't an animal it ought to be," he said; "it knows so much; but while you are revising your list

of craytures you had better strike the snakes out of it; just call your book Ireland and I'll drive 'em out meself."

Saint Anthony the Abbot explained that no noxious animals were on the list of his friends; he had included only the harmless snakes, like the garden ones—such as the striped kind, for instance—that were friendly to man and ate up vermin that preyed on his crops.

"But a snake's a snake," replied Saint Patrick, "except when he's in a man's boots, and then he's the jimjams. I'll have none of 'em on my visiting list."

Now and then a member of the Saintly hierarchy would come

Santa Claus Gets a New Plaything

speeding along on his automobile, hugely enjoying the sport. It was interesting to note the preferences of various individuals as to make and type of machine. "Here comes Brother Elijah, making a Holy terror of himself, with his bald head and long white beard, streaking it along on a clumsy, old-fashioned thing that coughs fire with every exhaust, and making a noise like a tugboat," said Saint Patrick. "He was stuck on the thing at first sight and would have no other. He said it reminded him of the Chariot of Flame that brought him up here, and he enjoyed every minute of it. But by the looks and the sound of it the blazing old rattlety-bang seems more befitting the Boss of the place down below—especially when old Baldy goes tearing along the country roads at dusk trailing fire like a comet.

"And there's Brother John the Baptist! He swears by steam just because the machine carries a water-tank!"

To return to the Christmas plans of old Santa Claus. As we have learned, he has organized his work on an automobile basis very thoroughly. He will use relays of automobiles just as he has had to with reindeer and horses. In this case, however, it is a matter of choice and not of necessity. He could easily fix it to make all his rounds on one automobile, the only cause of delay being the replenishment of motive-power material, and that could easily be arranged for without difficulty, except in the case of electricity. But there are so many good makes that he did not want to show any partiality, and so he concluded to give each approved make its turn. Moreover, he did not want to forego the pleasure of trying the various styles of electric carriage, all of which had been placed at his disposal by the Electric Vehicle Company and other makers.

Another consideration that determined him in favor of the relay plan was the fact that a great many automobiles were to be given as Christmas presents. So all he has to do is to dash up to the door on the motor-carriage in question, leave it there to be run into the stable and jump into another one that stands ready for him, to be left for somebody a little farther on. His work will be enormously accelerated by his ability to travel light, for the vehicles he goes in himself will never be freighted. But he will be followed by a marvellously long train of automobile delivery-wagons, motor-trucks, etc.—many of them lent for the occasion by the greatest department stores, the German and French postal authorities, etc., and all heavily laden with gifts, and coming in such order as to preclude all confusion in their distribution.

There was one thing about the automobile service, as he

arranged it, that gave particular satisfaction to the good Saint. He had always had to bundle up a great deal about his feet to keep them warm in the biting cold weather that prevailed through so large a part of his circuit, and this made it very clumsy and awkward for him with all the climbing around he had to do, getting down chimneys, etc., to reach the rooms where the children slept. And he also lost a deal of time in changing to the pajamas that constituted his costume for Australia, New Zealand, South Africa and other parts of the Southern Hemisphere, where Christmas comes in the summer—as well as in the tropics, where he was wont to conform to the customs of low latitudes and luxuriate in bathing trunks only.

Now, with all the automobiles, where heat was used to produce the motive-power, he had the exhaust steam or the hot products of combustion carried through coils at his feet. And for the electric vehicles he had a little electric foot-warmer attached. Under ordinary circumstances this comfort would hardly be practicable in the case of electricity, for the consumption of the power for heating would shorten the radius of the storage battery altogether too much. But since the relays were to be so frequent it made no difference in his case. Consequently one of the greatest hardships of his task was overcome and he looked forward to having not only his feet as warm as toast, but his whole body as well. Therefore, since this year he would not have to wear his weight in heavy clothing to keep out the cold, he is going to get about with unprecedented agility, and, with all the facility of a weasel, get through numerous crevices that he never ventured to tackle before. All this pleases him so that he promises to leave automobiles in as many stockings as he can—not only metaphorically, in the case of big hose for grown-up people, but literally, in the shape of a lot of beautiful automobile toys for children.

Our First Club Run

By Edgar S. Hyatt

L AST month New York had a foretaste of the century to come. For the first time since the early Dutch settlers of New Amsterdam astonished the natives with their wheeled carts and wagons, the people of Manhattan beheld some of their prominent fellow-citizens gliding through the town in a noiseless procession of swift self-propelled vehicles. It was the first public run of the new Automobile Club of America, and proved as brave a sight as New York has witnessed in many a day.

Coming as it did immediately after a futile attempt to bar horseless pleasure vehicles from the parks of New York, this demonstration of the club was a complete success. With the exception of one punctured tire there was no accident to mar the day, nor did a single horse kick over the traces when this parade of all conceivable makes and patterns wound its way in and out of the mazes of Fifth avenue and the Riverside Drive. What is more, it was a demonstration that was appreciated, for, thanks to the perfect weather of early November, thousands of people thronged the entire route of the parade, amongst whom any runaway would have been bound to create serious havoc.

The greatest crowd assembled at the Waldorf-Astoria Hotel, the present headquarters of the club, and by the time the start was to be made more than a thousand curious people were massed in

President Andrews and Jefferson Seligman in the Lead

front of the carriage entrance. From north, east, west and south
the autos came squeezing through the throng and came to a halt
near the hotel, while the club members partook of luncheon within.

All styles of machines and vehicles put in an appearance.
There were victorias and phaetons, hansoms, wagonettes and
runabouts. Every type of carriage seen in the Park daily was
there, but only the horses were lacking. Uniformed and liveried
grooms were perched behind, and fair women, in gowns of latest
automobile cut, sat beside the drivers, who, like gentlemen whips,
were perched on the boxes of their own rigs.

Luncheon over, the automobilists emerged from the hotel and
the bicycle policemen serving as outriders strove to clear a way.
There was much confusion at the start, owing to the jam in front
of the hotel, and the automobilists aroused the admiration of the
crowds on the sidewalks by their clever manœuvres in and out
between the mettlesome horses of many private carriages and hansoms waiting at the hotel entrance until Fifth avenue was reached.

Our First Club Run

In the first carriage, which was an electric victoria, was Avery D. Andrews, President of the club and Adjutant General of the State. He was accompanied by Jefferson Seligman, who handled the levers. In the second vehicle, a gasoline road wagon, were George F. Chamberlin, Vice-President of the club, and S. H. Valentine. Third place in the line was given to Whitney Lyon, founder of the club. He had his electric dos-a-dos trap, and was accompanied by Mrs. Lyon, General George Moore Smith and Miss Smith. In the next carriage, a locomobile, was Amzi L. Barber, with S. T. Davis, Jr.

Albert C. Bostwick, who had the next place in line, had started from his stable in an electric victoria, but having met with a mishap, went back for a locomobile runabout, in which he attracted much attention in the parade. Next to Mr. Bostwick was Winslow E. Buzby, who is regarded by his fellow club members as a martyr for having submitted to arrest in order to obtain the admission of automobiles to Central Park. Mr. Buzby had an electric phaeton and was accompanied by Mrs. Buzby. Mr. J. Post, Jr., followed, and Mr. Le D. L. Barber, with the Secretary

Messrs. Whitney Lyon and Winslow E. Buzby

247

The Automobile Magazine

of the Club, in a steam motor, came next. A. L. Riker, in an electric trap, and having W. G. McAdoo and K. B. Conger as his guests, followed.

Edward H. Lyon, in his steam automobile, was next, and he was followed by an electric surrey guided by Frank Pusey, who had as his guests General G. M. Dodge, J. Horace Harding, W. L. Brown and George B. Hopkins. J. A. Blair had an electric brake on which were Mr. and Mrs. Edwin Curtis and Miss Blair. Dr. John D. Zabriskie was next, in a gasoline carriage. Next was A. O. Krieger, with a De Dion motor. F. A. La Roche, in a gasoline road wagon, had W. J. Sergment as his guest. W. H. Hall also had a gasoline motor. George Isham Scott followed on a De Dion tricycle. John H. Hallock had as his guests in an electric surrey Mrs. John Wood, Mrs. C. J. McDermott, Mrs. Johnson and J. H. Wood. Thomas H. Thomas, in a locomobile, with W. H. Thomas as his guest, followed. Mr. J. C. McCoy, in his gasoline victoria, had E. B. Bronson as his guest. A. L. Whiting, in an electric, and William E. Burroughs and A. E. Bardwell, in the same type of carriage, followed.

Alexander Fischer and Mrs. Otto, in an electric vehicle, followed by Curtis P. Brady and E. B. Brady, in an electric run-

Marshall P. Wilder and Col. J. Marceau

Our First Club Run

The Parade Passing the Waldorf-Astoria

about, were next. Alan R. Hawley had as guests Miss Hawley and William Hawley. Colonel Theo. C. Marceau, in his fine electric carriage, was accompanied by Marshall P. Wilder. C. J. Field, in his gasoline carriage, had W. B. Atkins as his guest. Dr. E. C. Chamberlain was in his electric carriage, with Miss Louise Faulkner as his guest.

Others in the parade were Clarence W. Wood, accompanied by L. T. Gibbs; A. E. Chandler, Eleazer Kempshall and Mrs. Kempshall, in their electric hansom, Alphonso Pelham and Mrs. Pelham, Albert R. Shattuck, Charles T. Yerkes, G. L. Richard, F. M. Krugler, and the editor of this magazine.

The start of the parade was watched with interest by Admiral Dewey's bride-elect, who stood for some minutes at a window in her apartment in the Cambridge Hotel. When the procession of automobiles had fairly started on its way, many of the onlookers followed its route in carriages or on bicycles, just to see how it progressed.

The first objective point was the new Naval Arch glistening white at the intersection of Fifth avenue and Broadway. Without entering the Court of Honor the parade here turned northward. Going up Fifth avenue the automobiles received applause from

The Automobile Magazine

the spectators who fringed the sidewalks. Each carriage rolled along its way at an easy pace of four miles an hour. Owing to the crowded condition of the thoroughfare it was found to be impossible, however, to preserve the order of march.

Above Fifty-ninth street General Andrews increased the speed of his carriage, as there was an abundance of room and smooth asphalt for blocks. Then the run took on life and animation. Members left the line and spurted ahead to get a look at the other carriages, and at times the parade would be in a bunch instead of a single line.

Having passed all the forbidden entrances of Central Park and Mount Morris Park the column wheeled to the left around the Park and followed Fifth avenue to One Hundred and Twenty-sixth street, where it turned west to Lenox avenue, the popular drive leading to the Harlem Speedway.

There was soft mud here, which retarded the speed of some of the carriages, so that quite a gap was opened up when the half dozen leading vehicles turned west and presently swung into the Riverside Drive. Moving up the Drive the automobiles closed

Messrs Curtis and Blair with their Ladies

Our First Club Run

Gen. Andrews Reviewing

up and on passing Grant's Tomb they were cheered by several hundred persons.

At the Claremont Hotel, at the head of the Drive, there was a halt of an hour for refreshments. Members of the club gathered in groups and congratulated each other on the success of the first run under the auspices of the new organization.

General Andrews reviewed the parade from his carriage opposite Grant's Tomb at the beginning of the return run, late in the afternoon. Excellent time was made down Riverside Drive. Some who were chafing at the retarded pace increased their speed, and in groups of three or four abreast the motors dashed down the Drive. The bicycle policemen found themselves unable to follow and could not have coped with the racers had these not slowed up of their own accord after their little brush.

After the return to the Waldorf-Astoria the success of this first run was celebrated by a series of informal dinner parties among the club members. Before the paraders returned home the leading spirits among them had already arranged for another more ambitious run, this time to Ardsley, forty miles up the Hudson River.

The Oakman Automobile

THE Oakman Automobile, illustrated herewith, is a vehicle of novel construction, a gasoline phaeton, weighing about 500 pounds. The front or steering wheels are similar in type and attachment to the bicycle. The motor is placed under the rear of the body; it consists of a pair of cylinders, cast in one piece, the connecting rods taking the same crank wrist.

The interior mechanism is made accessible for examination by the entire hinged metal back of the carriage being raised. Directly under the cylinders is the muffler for the exhaust, having

a small elbow turned downward at one end. It deadens the sound of the exhaust most effectively. To the right of the cylinder is the small dynamo for sparking, the armature of which is rotated by frictional contact with the main shaft fly wheel; located on the extreme left is the spark coil, and under the seat is a storage battery. The current for sparking is taken from the storage battery, the latter being kept charged by the dynamo when the carriage is in motion. Above the engine cylinders and under the seat are two tanks separated by a small space; the left is for the

The Oakman Automobile

storage of gasoline, the right for water. The rear axle is of peculiar construction, in the shape of the letter U; the single springs that support the body at the rear are suspended from stirrups depending from the wheel axles. The driving wheels have an interior annular driving rim, against which the grooved driving pulleys of the main driving shaft impinge and impart the power of the engine to the wheels by friction. This shaft is manipulated forward or backward by the single lever rising upward in the centre of the carriage and is one of the features which make the vehicle distinctive. By means of a latch lever attached to the driving lever the operator starts the engine from his seat by engaging the latch lever in a ratchet wheel under the seat attached or geared with the main shaft, so arranged that when the driving lever is drawn suddenly back it will cause the ratchet wheel to rotate the engine enough to allow the sparking, and thus cause the needed explosions. After it is started the latch lever is released and the driving lever pushed forward, which brings the driving grooved pulleys into contact with the driving wheel rims. The speed may be regulated by this frictional contact or by rotating the top of the handle of the driving lever with the hand, which admits or cuts off the air supply to the engine. A backward motion of the driving lever applies the brake.

New Fifth Avenue Electric Stage

The Coaches' Overthrow;

Or, A Jovial Exaltation of Divers Tradesmen, and Others, for the Suppression of Troublesome Hackney Coaches

An Elizabethan Ballad

As I passed bye, this other day,
  Where sack and claret spring,
I heard a mad crewe by the way,
  That loud did laugh and sing:
Heigh downe, dery, dery downe,
With hackney-coaches downe!

The very slugs did pipe for joy,
  That coachmen hence should hye;
And that the coaches must away—
  A mellowing up to lye:
Heigh downe, dery, dery downe,
With hackney-coaches downe!

The world no more shall run on wheeles,
  With horses, as't has done;
But they must take them to their heeles,
  And try how they can run:
Heigh downe, dery, dery downe,
With hackney-coaches downe!

The elder brother shall take place—
  The youngest brother rise;
The middle brother's out of grace—
  And every tradesman cryes:
'Twould save much hurt, spare dust and dirt,'
Were they clean out of towne!

Coach-makers may use many trades,
  And yet enough of meanes;
And coachmen may turne off their jades
  And helpe to drain the fens:
Heigh downe, dery, dery downe,
With hackney-coaches downe!

'Tis an undoing unto none,
  That a profession use;
'Tis good for all—not hurt to one—
  Considering the abuse:
Then heigh downe, dery, dery downe,
With hackney-coaches downe!

The World on Wheels

By Edwin Emerson, Jr.

"This is ye rattling, rowling, rumbling age, and ye world runneth on wheels."—Taylor's "*Oddes Twixt Coaches and Carts*," 1623.

EVERY orthodox wheelwright knows that the earliest vehicles on record were those which King Pharaoh misguidedly drove into the Red Sea. Even John Taylor, the first and most formidable foe to all wheeled contrivances, in his bitter arraignment of the early hackney coaches of Queen Elizabeth's time was constrained to make this puritanical concession to Bible lore: " 'Tis to be supposed how Pharaoh's charriots that were drowned in ye Red Sea, were no other things in shape & fashion, then our Coaches be in this time, and what sore pitty is it, that ye makers and memories of them had not been obliviously swallowed in that Egiptian downfall?"

For a wonder the most learned archæologists, as well as the devoutest of theologians, are inclined to agree with the simple wheelwrights that the Egyptians were indeed the first men versed in the art of carriage making. Ezra Stratton, the author of the most exhaustive English book on the subject, thinks with Aristotle that the ancient Egyptians evolved their chariots from sledges, similar to the horse-drawn tepee poles of our North American Indians, by dragging them over solid rollers.

If this was the case it must be admitted that the gulf from the common flat roller to the wheel, as we know it, was very completely bridged, for the earliest example of an Egyptian wheeled vehicle, brought forward by Mr. Stratton in support of his hypothesis, shows an elaborate affair with a wagon body, tongue and four eight-spoked wheels. Notwithstanding the weight of this writer's acknowledged authority, it seems more than likely that the enthusiasm of his convictions on this subject seduced him into accepting this obviously late representation of

The Automobile Magazine

Egyptian Hunting Chariot in Florentine Museum

a wagon for an earlier type. It is highly improbable that a wagon on four wheels, with eight spokes, should antedate the well-known primitive chariot of the Egyptian kings which the early sculptures at Karnac reveal to us, charging into the thick of the fray, borne by two slender wheels with but six spokes apiece. By the same evidences it is made plain that six spokes were only used as an extra allowance for fighting chariots, whereas all the common carts had wheels with only four spokes. As it happens, a perfect specimen of an Egyptian hunting chariot is preserved in the Florentine Museum, which proves that these early sculptured pictures are singularly correct in detail. A still more valuable piece of proof is the fragment of an early Egyptian war chariot found in a mummy pit at Dashour, which is now preserved among the most precious relics of the New York Historical Society.

The Oldest Wheel
(Preserved by the New York Historical Society)

It is a wheel two feet eleven inches high without, and three feet three inches with its *wooden* tire. The hub, which is fourteen and a half inches long, five inches through the middle, and four and a half inches at the ends, has not the least appearance of ever having been burdened with an iron box, and the jagged pin ends look as though they had undergone many hard rubs from the linchpin (probably a wooden pin) while revolving around the axle tree. On the face of it it is quite clear that this particular ancient wheel rotated around its

Egyptian Plaustrum (B. C. 1500)

The World on Wheels

Rimming a Wheel

Bending Timber for a Chariot

axle and not with it, as some learned archæologists would have us believe.

The spokes, six in number, have a peculiar ornamental finish of their own. That which strikes the eye of the practical wheelwright most forcibly is that they have square " tangs " where they enter the felloes, a peculiarity now only found in old-fashioned heavy wheels. The tire-shoeing is unique, no evidence of the use of iron in any form being used thereon. This proves the wheel to be of the most ancient type, since the Egyptians are known to have used iron early in their history. In the British Museum is an Egyptian iron anvil, like those now in use in our smithies, which was unearthed among other remains known to be more than three thousand years old.

The manner in which ancient Egyptian wheelwrights plied their trade has fortunately been transmitted to us in the bas reliefs on the walls of an edifice erected by Thothmes III., at Thebes. Among the tools represented in the hands of the sculptured figures the most notable are a small wood saw and a semi-circular knife for cutting leather, both of which implements are still used by modern mechanics in the selfsame form as was here pictured more than three thousand years ago.

The chariot of the Assyrians was heavier and lower than that of the Egyptians, and the wheels were set farther back, on the theory, evidently, that this would make the body of the chariot more stable. What the vehicle lacked in grace of build it made up in more elaborate ornamentation. The wheels were higher and appear to have had iron spokes. In the sculptured representations of Assyrian chariots, at all events, the spokes appear

Making the Tongue and Yoke

Finishing the Sides and Trimmings

The Automobile Magazine

very light and frail, whereas the rims look as thick as modern pneumatic tires. This surmise is corroborated by the fact that

Assyrian War Chariot

the fragment of a car wheel has been found at Ninevah on the concave side of which still remain the iron roots of spokes.

The best of these chariots were inlaid with gold and silver, as recorded on a statistical tablet at Karnak, which boasts of "thirty chariots worked in gold and silver and painted poles," brought as trophies from that country. From this passage in Zachariah it would seem that ancient spans of horses were paired according to their color, just as nowadays:

"And behold, there came four chariots out from between two mountains; and the mountains were mountains of brass. In the first chariot were red horses, and in the second chariot black horses, and in the third chariot white horses, and in the fourth chariot grizzled and bay horses."

While the Greek and Etruscan wheelwrights in their productions of vehicles clung to the essentials of the chariot type evolved by their predecessors, the ancient Persians branched out and produced new types of their own. Not only were they the first to employ comfortable four-wheeled vehicles with carriage tops and cushions inside—to wit, the harmamaxa, denounced as effeminate by the Greeks—but they also startled the world by the most

Ancient Persian Harmamaxa

The World on Wheels

Modern Persian Cart
(From Ezra M. Stratton's Work)

ingenious and formidable war engines yet contrived on wheels. These were the famous scythe-and-sickle wagons which they precipitated down upon their invaders from mountainous heights. Just how they were constructed has been the despair of such learned writers as Quintus Curtius, Plutarch, Roger Bacon, Drakenborch, Ginzrot, Scheffer and Stratton. While these scholars have puzzled over the probable arrangement of the murderous cutlery upon the wheels and body of the wagons, it has been reserved to M. Carteret, the author of " The Genesis of the Auto-

Native Cart in Madras

Hindoo Zebu Wagon

mobile," to point out that these curious vehicles were in fact the earliest instance of self-propelled vehicles.*

The harmamaxa of the Persians was ultimately adopted by the Romans and gave rise to various four-wheeled types, such as the curriculus, rheda, benna, clabularius, arcera, carucca and the four-wheeled plaustrum. After the overthrow of Rome, when the Barbarians swept over the ancient world, these various well

Hindoo Pony Cart

* See October number of the "Automobile Magazine" for M. Carteret's treatise and for an illustration of one of these fighting machines.

The World on Wheels

Japanese Jin-Riksha

developed types of pleasure vehicles and workaday wagons degenerated once more into crude carts, such as still survive in their most primitive form in the Roman campagna and the Persia of to-day. The best of these, so all travelers are agreed, is the old fashioned *volante* of Cuba, a two-wheeled vehicle hung in leather springs after the fashion of our forefathers' gigs.

Sicilian Cart

The Automobile Magazine

Similar rudimentary bullock carts and pony gigs are to be found among various semi-races and peoples of more or less arrested development throughout all over the world. A peculiar

Ox Sledge of Madeira

exception are our North American Indians, who, adaptable horsemen as they have proved themselves, have never taken to the use of anything that runs on wheels.

This was likewise the attitude of the knights and nobles of the Middle Ages. Since "chivalry," as the original sense of the word indicates, resolved itself into horsemanship, your mounted cavalier had a fine contempt for any other conveyance. The furthest he would condescend was to the horse-litter, a mode of conveyance which is still used by the Catholic clergy in Spain and Portugal.

In England the Norman horse-litter displaced the ancient chariots of the Britains and the chariot-like carts and four-wheeled hammock wagons of the Saxons. After a couple of centuries, when the three races had become thoroughly blended, the Saxon carts, now called chares, had their innings again, and horse-litters presently fell into disrepute. The latest mention of them is in the diary of Evelyn in 1640.

The first mention of the new conveyance is found in the Chaucerian ballad of the Squyr of Low Degree, who is preferred by the gentle Princess of Hungary to the promise of a ride in a royal chare:

> " The chare shall be covered with velvet red,
> With a fringed canopy overhead,
> And curtains of damask, white and blue,
> Figured with lilies and silver dew.
>
> We still have the Soldan's harness, sweet
> The housings hung to the horses' feet,
> The saddle-cloth is sown with moons,
> And the bridle-bells jingle the blithest tunes.
>
> ' But I would rather have,' saith she,
> ' My loving sqyr of low degree.
> Not a gaudy chare nor days of chase
> Reward me for his absent face.' "

Another kind of carriage in use at this time was called a whirlicote. In 1380, it is related that Richard II., " being threat-

A Cuban Volante

ened by the rebels of Kent, rode from the Tower of London to Mile-end, in a whirlicote of *olde time.*"

Plainly the whirlicote was out of fashion, being superseded by coaches, newly imported from France. Their introduction in England is credited to Queen Eleanor, who died in 1291. Her death was thus amiably discussed in a ballad of the time, enti-

Medieval Long Wagon

tled " A Warning Piece to England Against Pride and Wickednesse, being the fall of Queen Eleanor, wife to Edward, King of England, who, for her pride, by God's Judgements, sunk into the ground at Charing Cross, and rose at Queenshithe:

> She was the first that did invent,
> In coaches brave to ride;
> She was the first that brought this land
> To deadly sin and pride."

Later testimony, no less direct, credits the first introduction of coaches to Sir Thomas Chamberlayne, English Ambassador to the Court of Charles V. of Spain; and again to the Earl of Rutland, for whom Walter Rippon, Queen Elizabeth's subsequent coachmaker, is recorded to have built the " first coach that ever was seen in England, Anno Domini 1555."

However this may be, a great stir was certainly made in England by the magnificence of the royal coaches which Queen Bess received as presents from her admirers abroad. Her admirers at home followed suit. Thus Sir Philip Sidney risked his knightly reputation by appearing in a coach, instead of on horseback, riding into Shrewsbury in his wagon, " with his trumpeter blowynge, very joyful to behold." Good Bishop Hall, in his " Satires," published shortly afterward, took a less joyful view:

> " Is't not a shame to see a groome
> Sit perched in an idle chariot roome.
> That were not meet some panel to bestride,
> Sursingled to a galled hackney's hide. "

Queen Elizabeth's Coaches
(From Hufnagel's Print)

The World on Wheels

In spite of such criticisms the use of coaches grew apace, so that young Shakespeare could earn a livelihood by holding the coach horses of the ladies that drove to the Globe Theatre. In 1601 a bill was actually brought in Parliament " to restrayn the excessive use of Coaches within this realm of England." In 1610 a royal patent was granted for the first English stage coach. Licenses for hackney coaches followed soon after, and became so numerous that the

Eighteenth Century Brouette

Thames watermen in London rose in riot against them. By 1630 there were more than a hundred. Acts of Parliament followed in rapid succession, limiting the number of hackney horses first to 200, then 400, then 600, and 800, until the thousand mark was passed before the end of the century.

As the coaches increased in number they became better in quality. Samuel Pepys, in his gossipy diary, records such innovations as the first trial of carriage springs and sliding glass

Twentieth Century Victoria

windows, and dilates with pride upon the fine upholstery he was able to put into his own private carriage, the possession of which brought him under suspicion as a bribe taker.

From that time until the present day English speaking men have excelled in the art of carriage building, and the highest peers of the realm even vie with one another in this field. Thus it has come about that such high names as those of Lord Brougham, Hansom and the Earl of Stanhope have become almost more familiar to the world as the immortal names of carriages, than of English noblemen. Even the name of Queen Victoria, we dare say, is kept more green by the familiar style of carriage known as the Victoria.

Now, however, a new vehicle has arisen, different from all that has gone before, with a future opening before it so bright and resplendent that the triumphs of past ages must pale and sink into insignificance.

A Modern Tally-ho

Columbia Electric Prize Racer

Gallery of American Automobiles

Haynes Gasoline Buggy

A Stanley Buggy

Gallery of American Automobiles

Winton Gasoline Phaeton

Winton Gasoline Racer

Gallery of American Automobiles

Overman Tea Cart

Electric Phaeton

Mechanical Propulsion and Traction

By Prof. G. Forestier

I.—THE CONDITIONS OF ANIMAL TRACTION RECALLED.—
There are numerous vehicles of different natures drawn upon
roads by animals, that satisfy various exigencies.

In the first place, there is the transportation of merchandise,
which is effected through two-wheeled carts or four-wheeled
trucks, to which are harnessed horses of slow gait, whose speed
varies from 2 to 2½ miles an hour. This service, which was
formerly of a most important character, was assured in France
in two different manners:

(1) By *ordinary conveyance*, in which the same horses made
the entire trip in walking for about eight hours every day and
resting the remainder of the time in order to start again the next
morning—the daily travel being nearly 19 miles.

(2) By *quick conveyance*, in which the speed was not sensibly
greater, but in which the horses succeeded each other by relays—
the wagon running day and night. The distance made in twenty-
four hours was about from 55 to 60 miles.

Concurrently with the conveyance of merchandise there was
at one time the carriage of sea-fish, say, between Dieppe and
Paris, for example. This was effected by two-wheeled vehicles
to which were harnessed five horses moving on a trot and succeeding each other by relays 7 miles apart. The speed of the animals
was about 5 miles an hour, and the distance covered amounted
to 114 miles in twenty-four hours.

Afterwards came the carriage of passengers along with
merchandise, a service which was formerly assured upon all the
national highways by stage coaches, to which were harnessed
five horses moving on a trot at an average speed of 5 miles an
hour. The mean distance apart of the relays was 7 miles. The
distance covered was here again 114 miles in twenty-four hours.

The carriage of despatches was effected more expeditiously,
say at a speed of about 7 miles an hour. Later on, when the
service of making connections with the railway had to be assured,
the despatches, and even the passengers, were carried at a speed

of 9½ miles an hour. Such a speed, which could be obtained only by putting the horses to a gallop for a part of the time, wore the animals out very quickly.

In addition to these public services there is the conveyance of individuals by hired coaches or private carriages. The speed of the hired coach varies between 5 and 7 miles. The distance traveled per day is very variable, say from 25 to 40 miles, according to the load and the value of the team. As a type of such carriages we have the hacks, which formerly had an average speed of 5 miles, but which, under the pressure of the exigencies of the public, now cover as many as 9 miles an hour, when they are hired by the distance. When they run by the hour, however, they make scarcely more than 6 miles.

As for private carriages, their speed never descends below 7 miles, and very frequently reaches 10. There are some, even, which attain 12 miles an hour; but, at such a gait, their team cannot, in a regular manner, make more than 15 miles every day.

Such speeds appeared for a long time to be amply sufficient; but, at present, railway trains are attaining a commercial speed of from 45 to 48 miles, while the public is demanding 60, and even hoping for 75 miles an hour. The passenger who lands from an express train in order to take a transfer coach finds the 5½-mile speed of the vehicle that is taking him to his destination by far too slow.

On another hand, the speed of the pedestrian is from 2 to 3¼ miles at the most; but, with the bicycle, it is now a common thing to obtain a speed three times greater, say 10 miles an hour—a performance better than that of the private horse.

The need of speed has therefore entered into our mode of life.

Now, all that we have just recalled shows that it is not possible for animal traction to satisfy this requirement economically. The horse, in fact, is incapable of making, day by day, at a fast gait, the somewhat lengthy journey of 10 miles an hour without standing a chance of becoming worn out in a very short time.

The need of going faster and faster, in covering longer and longer distances, can be satisfied by mechanical traction alone. Aside, too, from the question of speed, the carriage of heavy loads in bulk is, under certain circumstances, if not impossible, at least too onerous with animal traction. In fact, the average load that can be imposed upon a horse decreases very rapidly with the number of the animals forming the team.

In former transportations a total load was reckoned as:

3,168 lb. for 1 horse of 790 lb., say, on an average, 3,168 lb. per horse.

Mechanical Propulsion and Traction

6,327 lb. for 2 horses of 790 lb., say, on an average, 3,163 lb. per horse.

8,652 lb. for 3 horses of 790 lb., say, on an average, 2,884 lb. per horse.

11,220 lb. for 4 horses of 790 lb., say, on an average, 2,805 lb. per horse.

11,935 lb. for 5 horses of 790 lb., say, on an average, 2,387 lb. per horse.

11,972 lb. for 6 horses of 790 lb., say, on an average, 1,995 lb. per horse.

12,058 lb. for 7 horses of 790 lb., say, on an average, 1,723 lb. per horse.

12,056 lb. for 8 horses of 790 lb., say, on an average, 1,507 lb. per horse.

Such diminution of the mean load, while the number of horses composing the team decreases, is due to two causes : In the first place, to the fact that the more horses that the driver has to manage, the less he is capable of keeping them on the move and of making them act well in unison; and, in the second place, to the fact that the mean load imposed upon each horse depends upon the effective stress that he will be able to exert at certain parts of the journey; at the curves met with upon the road, for example.

In fact, a team composed of several horses may be considered as forming a regular polygon. The pull exerted by the first horse is not parallel with the direction of the second, and therefore gives a component at right angles therewith. In like manner the sum of the traction of the second horse and of the component of that of the first, parallel with that of the second, will be oblique to the direction of the third horse, and so on.

It is easy to see that the different components at right angles and parallel with the positions occupied by the successive horses are given by the relations :

$$T_n = T \sin \alpha \, (1 + \cos \alpha + \cos^2 \alpha + \ldots \cos^n \alpha)$$
$$T_p = T \, (\cos \alpha + \cos^2 \alpha + \ldots \cos^{n+1})$$

Where n is the order number of the horse counted as beginning from the leader, for which it is 1; and α is the angle at the centre corresponding to the side of the regular polygon formed by the team as a whole.

It results therefrom that while the component at right angles with the direction of each horse increases very rapidly (thus interfering with its useful action), the sum of all the useful components is notably less than the number of the animals of the team.

The relations:

$$\text{Sin } \alpha = \frac{l}{R}\sqrt{1 - \frac{l^2}{4R^2}}$$

$$\text{Cos } \alpha = \sqrt{1 - \frac{l^2}{4R^2} + \frac{l^4}{16R^4}}$$

—where l is the distance from collar to collar, and R is the radius of the curve, show that this disadvantageous influence of the curves increases with l and decreases, on the contrary, when R increases (usually $l = 8.2$ feet and R $= 130$ feet upon the national roads of France).

If, instead of presenting a curve of a wide enough radius to allow the team to spread itself without inconvenience, the road makes an abrupt turn (as happens in the streets of a city), the shaft-horses alone will be capable of pulling efficaciously; and if the load is somewhat heavy, the team, as numerous as the horses that compose it may be, will not be utilized, unless the precaution has been previously taken to impart sufficient speed to the vehicle to allow the live force acquired to permit of making the difficult passage.

As a usual thing, the number of horses given to a teamster to drive in tandem is but five. The total maximum load would therefore be about 11,935 pounds, if the weight of the horses should remain at an average of 790 pounds.

The average maximum load that five strong Boulonnais horses can be made to draw is now notably greater, especially when the journey is not too long. Aside from the fact that the trucking business has much heavier, and, consequently stronger, teams at its disposal, the roads offer it more satisfactory conditions of viability. Thus, at Paris, there exist heavy trucking enterprises, especially for the carriage of dressed stone, in which five or six horses at the most haul a total load of 26,400 pounds. The daily travel is, on an average, 36 miles, in four trips—two with a load and two without. The truck weighs 7,300 pounds. The five or six horses are driven by one teamster. The weight of the shaft horse is 1,617 pounds and that of the tandem horses 1,450. The average load varies between 4,400 and 5,280 pounds.

A fine example of regular service that may be cited in this connection is the carriage of bags of sugar for the Say Refinery. One hundred bags of 220 pounds each are carried at a load. This effective load of 22,000 pounds is placed upon a truck weighing 9,900 pounds. The total weight is therefore 31,900 pounds. In

Mechanical Propulsion and Traction

order to draw this vehicle, it takes but five horses, of which the average load amounts to 6,380 pounds. The weight of the animals is:

Tandem horses .... 1,470 pounds.
Shaft horse .... 1,600 "

On another hand, the team works, under a load, for from two and a half to two and three-quarter hours at the most, rests during the unloading of the truck, and then returns without a load in order to begin again a second similar trip. The co-efficient of traction in the routes followed in Paris may be estimated at more than 55 pounds per ton. Each horse therefore exerts upon a level a maximum stress equal to from 10 to 11 per cent. of its weight. For a trip of about 5¼ miles, the speed is, on an average, about 2⅓ miles an hour.

As an example of the carriage of an exceptionally heavy indivisible mass, no better example can be selected than that of the transportation of the bell called the "Savoyarde" from the station of La Chapelle to Montmartre Hill at Paris, in 1895. The bell weighed 36,300 pounds, and the truck upon which it was carried 13,200 pounds. In order to haul this total weight of 49,500 pounds over a road with quite pronounced declivities, a team of 28 horses in tandem had to be employed. This corresponded to a mean load of about 1,760 pounds. It was possible to obtain such a stress only by giving each horse a driver. In order to carry the bell to the inclined plane over which it was to be moved to its place, not only was a selection made of a route that presented the fewest sinuosities, but upon a portion of the way the team had to be urged to extreme speed.

In case the carriage of such loads were frequent, there would be a manifest economic interest in discarding animal traction and substituting mechanical traction for it.

Rapidity in the conveyance of passengers and despatches, and economy in the moving of heavy indivisible masses; such are the reasons that justify the substitution of mechanical for animal traction upon roads.

II.—COMPARATIVE CONDITIONS OF OPERATION OF MECHANICAL TRACTION UPON RAILS AND HIGHWAYS.—Some persons might think that in order to succeed in substituting mechanical for animal traction upon roads, it would suffice to copy more or less intelligently the engines used for traction upon rails. Many experiments were for a long time made in this direction, but all of them proved dismal failures. The conditions, in fact, are very different; and we shall pass them rapidly in review. This study will permit us at the same time to understand why mechanical traction upon roads, the first in date, has remained so long in limbo, while mechanical traction upon rails has so rapidly become flourishing.

Mechanical traction upon rails was at once provided with a trackway admirably adapted to its requirements. Struck by its economic power, capitalists recoiled before no sacrifice in order to make it a success. It was therefore protected against the jarring and violent shocks resulting from obstacles that have to be surmounted upon public roads. The rails, laid at the outset almost always on a level and in a straight line, or according to curves having a radius of over three thousand feet, assured it an almost constant stress, and, at the same time, permitted it to use a relatively heavy motor, separate from the cars to be hauled. Such separation allowed of the use of stiffer springs upon the motor than upon the passenger cars. This, along with the absence of abrupt vibrations, allowed of a rigid connection of the cylinder (fixed to the frame) with the driving axles—the elasticity of the steam permitting of sufficient play to prevent possible slight variations in distance.

As the rails assured a rectilinear direction, the engineer had merely to occupy himself with the management of his locomotive and furnace. The track, carefully isolated by continuous fences, protected the engine and its mechanism against all accidental collisions. The essential parts of the motor could therefore be placed upon the exterior. At the same time, the absence of dust permitted of leaving them in the open air and easy of inspection. In like manner, as the wheels had to move in a straight line, it was possible to key them upon the axles, since both had to revolve with the same velocity. Moreover, protected against lateral shocks, they could be plane. The absence of all violent jarring obviated the necessity of giving them elasticity and permitted of the adoption of metal for their manufacture.

Finally, although the necessity of placing the platform of the cars or the floor of the passenger coaches at a slight height above the ground, in order to facilitate the ingress and egress of passengers or the handling of freight, rendered imperative the use of wheels of small diameter for both cars and coaches, the stability of the latter permitted of the adoption of easily filled oil-boxes that greatly diminished the resistance to friction, despite the relatively large size of the journals necessitated by the heavy load carried by the axles.

Hence the gigantic strides that mechanical traction upon rails has been able rapidly to make, thanks to the pecuniary resources

Mechanical Propulsion and Traction

of all sorts that were justified by its influence upon the economic government of nations.

On the other hand, let us see what obstacles have had to be surmounted by mechanical road-traction, which, more modest in its results, has been able to make an advance only through the enthusiasm (sometimes injudicious) of inventors.

The resistant work to be overcome by the motor does not consist solely of the obstacles to traction that are offered by the road and by the sliding friction of the journal in the axle-box, due to the dead weight and effective load, but comprises, also, as with locomotives, sliding and rolling frictions of the various parts of the motor itself and of the transmission of motion from the latter to the driving wheels.

In traction upon common roads, such causes of loss of motive power are notably aggravated by the very peculiar difficulties of transmitting the power of the motor to the driving wheels. Here, in fact, the frame to which the motor is fixed rests upon the axles through the intermedium of springs, which, in order to overcome shocks due to jolting, must be elastic enough to protect the mechanism, as well as the passengers, against troublesome jarring. The elasticity of the springs increases the amplitude of the relative displacements of the extreme points of the transmission. Hence the latter must comprise distortable parts that render the co-efficient of utilization of the initial power of the motor more feeble.

On another hand, although the straight line is the normal direction of the vehicle upon rails (requiring the keying of the wheels at the extremities of a rigid axle), it may be said that the sinuous path is the rule for the vehicle upon roads. In fact, the two wheels of the same axle never meet with the same conditions at the same time, and, moreover, cannot be actuated with the same velocity. Hence the necessity of transmitting to each wheel, at every instant, a variable proportion of the motive power. Hence, too, a special difficulty of transmission—a new cause of loss. From the moment at which the power of the motor is inevitably so poorly utilized, it is important that the initial resistances applied to the wheels shall be reduced to a minimum.

The success of mechanical traction upon roads does not, therefore, depend so much upon the motor, properly so called, as upon obtaining the following desiderata: (1) A reduction to a minimum of the external stresses that oppose themselves to the motion of the vehicle; and (2) a possibility of transmitting simultaneously to each wheel that portion of the power that is necessary to it, according to the resistance that is momentarily opposed to it.

In order that, without detriment to the safety of those that it carries and of those that it is liable to meet, the traction vehicle may acquire all the speed that is compatible with its stability, it is well, besides, that it shall be provided with strong as well as efficient brakes, and with a steering apparatus that can be maneuvered with certainty and rapidity.

All such questions solved, there will be reason for seeking a motor (1) that shall be as light as possible; (2) that shall utilize an easily supplied source of energy; (3) that shall be capable of developing a power variable with the necessities of the traffic; and (4) that shall be as well balanced as possible.

Before passing successively in review the solutions proposed for these different problems, we shall give a succinct history, not of all the vehicles that have been experimented with one after another, but only of such trials made by different inventors as seem to us to have most influenced the conditions that at present confront the construction of a power carriage.

III.—Succinct History of the Successive Progresses Made by Mechanical Propulsion upon Roads.—The first person to construct a mechanically propelled vehicle was Cugnot, a French military engineer, whose experiments date back to 1769. His mechanical vehicle, with a load consisting of four persons, was capable of running upon a level at a speed of from 2 to 2½ miles an hour.

Unfortunately, its boiler did not possess an adequate vaporizing power, and, at the end of a quarter of an hour, it was necessary to stop for almost the same length of time in order to allow the pressure to rise. Moreover, as the furnace door was placed in the rear, the vehicle had to be stopped when fuel was to be put into the furnace. Nevertheless, the tentative was judged so interesting that Minister de Choiseul gave an order for the construction of a vehicle capable of attaining a speed of 2½ miles an hour with a load of from 8,000 to 10,000 pounds. This vehicle, which was finished in 1770, cost about $5,500. It is preserved at Paris in the collection of the Conservatoire des Arts et Metiers.

We omit everything that concerns the boiler and engine, notwithstanding the great interest presented by the latter, and shall occupy ourselves with the vehicle only. As may be seen in Fig. 1, this is a tricycle of which the single front wheel is both a driving and steering one. The apparatus consists of two distinct parts:

(1). The vehicle, properly so called, formed of a strong frame provided at the back with an axle and two large wheels, and in

Mechanical Propulsion and Traction

front with a circular plate bearing upon two rollers fixed to the following piece;

(2). A metallic frame carrying the boiler and engine, and resting like the beam of a balance upon the extremities of the axle of the front wheel through two pieces of bronze analogous to the axle guards of our present cars.

The boiler, placed projectingly in front of this frame, would have had a tendency to cause the front part to tilt had there not been fixed to the latter two lateral metallic pieces carrying rollers upon which the vehicle, properly so called, rested through the intermedium of its rolling plate.

In order to steer the vehicle, the driver had at his disposal a horizontal bar provided with two grips and mounted upon a vertical shaft, which, at the level of the floor of the vehicle, carried a pinion that geared with a toothed wheel mounted upon another vertical shaft passing through the frame of the vehicle, beneath which it carried a pinion that geared with a toothed sector fixed to the metallic pieces to which were adjusted the axles of the rollers.

This very intricate arrangement did not permit of a rapid enough action. At the same time, it is probable that, under the influence of jolting, the gearings would not have continued to mesh with each other. Thus is too easily explained the legend according to which, on the day of the experiments, the vehicle, upon coming from the shop in which it had been constructed, took on a speed such that its driver, powerless to steer it, could not prevent it from running against a wall that happened to stand in its way. This accident put an end to the only trial that was ever made of it.

Another point in the construction of this vehicle is worthy of attracting our attention. In order to keep in easy relation the motor and the wheel that he desired to be both a driving and steering one, Cugnot found no other means than that of placing the steam cylinders above the axle of this wheel, so that the whole displaced itself at the same time. Being unacquainted with flexible joints he had to proceed in the same way with the boiler, and hence a considerable overhang of the latter in front, which could not have been without an influence upon the difficulty attending the maneuvers of steering.

After Cugnot, numerous other inventors made researches upon the mechanically propelled carriage. Among the experiments that took place up to 1828 there were some that were of the most curious character, from the view point of the history of

The Automobile Magazine

the steam engine, but that brought to light no fact affecting the vehicle itself.

On the contrary, we have to point out several very interesting arrangements, in connection with the subject that occupies us, in the steam carriage designed for running upon ordinary roads, and that was patented at Paris in 1828 by Onesiphore Pecqueur, superintendent of the shops of the Conservatoire des Arts et Metiers (Figs. 2 and 3).

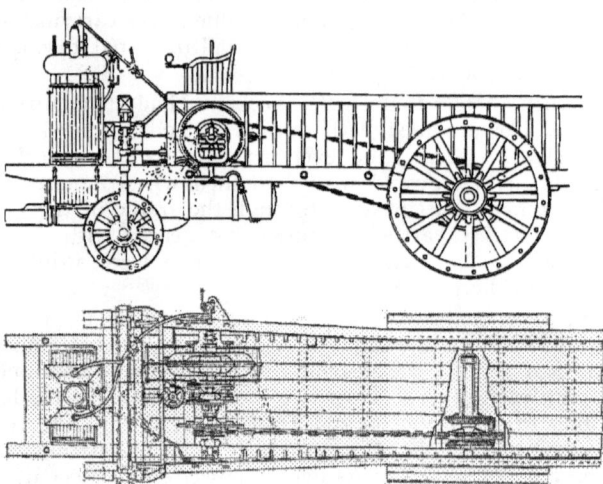

Figs. 2 and 3. Elevation and Plan of the Pecqueur Steam Wagon.

The driving wheels, which were two in number, were keyed upon the hind axle, which consisted of two parts connected by a satellite gearing, the original of the differential gearing now in use (Figs. 4 and 5). The following is the principle of it, according to the arrangement described in the patent:

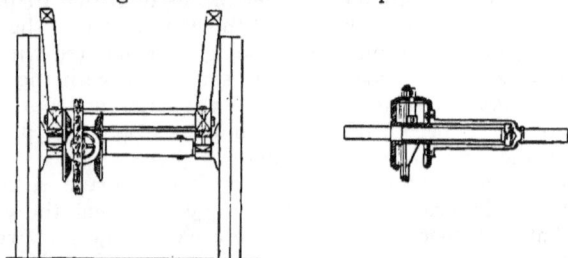

Figs. 4 and 5. Pecqueur's Satellite Gearing.

Mechanical Propulsion and Traction

Upon one of the two parts of the axle was keyed a bevel wheel. The other part carried a sleeve to which was keyed another bevel wheel symmetrical with the first. Between the two revolved a grooved pulley, around one of the spokes of which was capable of turning a bevel pinion that geared with the two symmetrical bevel wheels. The motion of the motor was transmitted to the pulley through a chain that passed through the groove. This pulley, therefore, revolved, and the pinion mounted upon one of its spokes actuated the two bevel wheels fixed to the two segments of the axle.

When either of the wheels experienced a resistance different from that of the other, the speed of the motion transmitted was the same for both, since the teeth of the pinion carried both of them along. If, on the contrary, one of them, in consequence of a curved trajectory, could not revolve freely, the teeth of the pinion displaced themselves (since the pinion was movable around a spoke of the pulley) to a sufficient degree to allow the relative velocities of the two wheels to become what was required by the trajectory followed.

The journals of the front steering axle were not in the prolongation of the latter, but were mounted upon vertical pivots movable in forks placed at its extremities. Unfortunately, these journals were connected in such a way that the wheels always remained parallel instead of having conjugate motions such that their prolonged axes should concur at the same point of the hind axle.

The steering fore-carriage was movable around a king-bolt through the aid of a toothed sector that geared with the lower pinion of the vertical shaft of the steering bar, as in the Cugnot vehicle. In like manner the fore-carriage carried the boiler and the rotary engine. It alone was provided with springs.

It will be seen that the Pecqueur carriage possessed (in an embryo state, at least) all the parts adapted for making a perfect automobile of it. Nevertheless, for want of a knowledge of this precedent, many inventors are going to wear themselves out in vain in useless experiments.

In 1835 an English carriage was introduced into France by Asda, after passing through Belgium. On the 10th of February, starting from Rue du Mont-Blanc, it made the trip from Paris to Versailles and return in four hours and a half, with a stay of forty-two minutes at Versailles. The outward trip lasted one hour and twenty-seven minutes, and the homeward one hour and twenty-one minutes.

The pressure of the steam in the boiler reached 11 atmos-

pheres. The engine was of 14 horse-power. The weight of the
carriage and its load was 9,900 pounds. On the 15th of March,
1835, the same carriage made a trip from Paris to Saint-Germain and return in four hours and twenty-nine minutes, inclusive
of stoppages at Nanterre going and coming, for the renewal of
water, as well as a stop at Saint-Germain, say in all sixty-two
minutes.

By reason of the fact that the boiler was replaced by a system
of water tubes, the breakage of which would not have been
attended with any danger, the *Constitutionnel*, in giving an
account of these experiments, asserts that " the least shadow of
a danger to passengers has been made to disappear in this English
carriage, which, in the ingenious arrangement of its mechanism,
presents all the improvements that twelve years of experiment
have successively indicated to English engineers."

" Not the slightest accident disturbed the voyage," says the
Constitutionnel, which adds: " What is certain is that, in a
short time from now, one or more regular services will be established from Paris to Saint-Germain and Versailles."

In France, the automobile vehicle seems to have been abandoned for the study of hauling-vehicles. In this order of ideas
we may mention, among other inventors, Dietz, who, in 1835,
took out a patent for a steam carriage styled a " traction-engine
running upon ordinary roads." The first type was a tricycle,
but, contrary to what was found in the Cugnot vehicle, the two
hind wheels were drivers, and the single front wheel was a
steering one. The carriages afterward constructed by Dietz had
as many as three pairs of intermediate wheels mounted in guiding
plates to permit of following the sinuosities of the road. The
general arrangement was evidently inspired by what is practiced
on railways. However, Dietz merits special mention, since he
seems to have been the first to have a presentiment of the utility
of elastic tires. It is useless to say that, having at his disposal
no practical means of realizing such desiderata, he had to try
various makeshifts, such as interposing, in the first place, a layer
of tarred felt, then cork, and then rubber, between the felly and
the tire, and holding the same in place by lateral cheeks bolted
to the felly. Dietz conceived the idea also of uniting the wooden
spokes of the wheels in a metallic box forming a hub.

(To be continued in our next issue.

NOTE.—Specially translated for the Automobile Magazine from *Le Génie Civil*

The Automobile and Public Health

By James J. Walsh, Ph. D., M. D.

IT is evident that the present movement in automobilism will
soon bring us to a practically horseless era in our cities. It
is interesting, therefore, to anticipate some of the effects on
the public health of large centres of population that the absence
of the horse is sure to have. The sanitary benefits that will
accrue from his removal, though entirely unlooked for by most
people, are rather easy to foresee. Of themselves these prospective sanitary advantages are enough to make the coming of
the horseless era a boon, and it is surprising that the advocates of
automobiles have not made more of this most telling point in
their favor. For one thing, the absence of the horse will probably entail the absolute eradication of tetanus—lockjaw, as it is
commonly called—from our cities, at least. The disease, though
fortunately not common, is by no means rare. Some 80 cases
were reported from the neighborhood of New York City alone
last summer. The very high mortality of the disease, from 60
to 90% of those attacked, makes it a dreaded visitor in our hospitals, a most unwelcome claimant on the attention of physicians.
Medical and surgical intervention so far has not been able to
check its effects. The discovery and preparation of a specific
anti-toxine for the disease some years ago it was hoped would
lessen its fatality, but the hope has proven vain.

It is the usual story every year to have a series of cases of the
disease under treatment at the hospitals of all our large cities just
after the Fourth of July. The history of these patients is practically the same in all the cases. There was a wound—often a
very slight one, but nearly always incurred from a burn while the
patient was on the street. This became contaminated by street
dirt. It healed more or less kindly, and sometimes was almost
forgotten, when about a week or ten days after the accident some
stiffness of the neck and jaw muscles began to develop. It was
the premonitory symptom of dreaded lockjaw.

Usually, in spite of every medical effort, the spasm spreads to
other muscles until all of the muscular system is involved. Consciousness remains, but as the result of the heat developed by the
constant muscular contraction, high temperature sets in and
exhaustion supervenes, if the patient does not succumb before
this to a spasm of the muscles of respiration.

The cycle of existence of the tetanus germ outside the human body is extremely interesting. The bacillus of tetanus, while itself discovered by the Japanese bacteriologist Kitasato, is one of a class of micro-organisms whose isolation and demonstration by Pasteur many years ago destroyed certain *a priori* assumptions with regard to the conditions indispensable for life. The bacillus is anærobic, that is, lives without air. Not only can it live and multiply in the absence of air, but the presence of that substance absolutely prevents its growth. Its favorite habitat is garden soil. Here, at a certain depth below the surface, it finds in summer time the warmth and moisture and nutritive materials necessary for its luxurious multiplication. Whenever on the stems of grasses, or the like, it is carried from the soil into the air, it enters upon a special stage of its existence and forms spores, seed like bodies which are especially resistent to extraneous unfavorable influences. In this form it is able to retain its vitality despite the presence of air.

Incidentally, the liking of the bacillus for the farm makes wounds that are incurred from farming implements especially liable to be complicated by tetanus. While in this spore stage the bacillus of tetanus is carried on hay, straw and the like to our cities as fodder for horses. After being eaten the tetanus spores find in the horses' intestines an ideal breeding place, with just the warmth, moisture and absence of free oxygen that are so favorable to them. At once they begin active reduplication. They do not affect the horse himself, for after all they are not inside the animal. They are not absorbed. They are only for the moment within the hollow cylinder that every animal, in its simplest expression, really is. Largely increased in numbers they pass out with the excrement, to dry on our city streets and be blown hither and thither by the winds.

Knowing all this, one might be surprised that tetanus is not more common than it is, or might wonder why nearly every wound incurred on city streets is not followed by tetanus. Owing to its absolute anærobic life, however, a certain combination of circumstances that, fortunately, is not very frequent, must conspire to give the bacillus a foothold in human or animal tissues. Even on human blood serum the tetanus bacillus will not grow in the presence of air. For successful inoculation the wound of entrance must be of such a character that the bacillus finds its way beneath the surface into the tissues and away from the air. Such wounds are characteristically those made by penetrating farm implements, as garden rakes, or pitch-forks, or the well-known rusty nail, or the lacerated wounds produced by toy or

The Automobile and Public Health

other pistols. In the depths of such wounds the tetanus bacillus finds a safe refuge and a favorable nidus for breeding, out of reach of the disturbing oxygen of the air.

In our cities the absence of the horse would practically do away with all danger from the bacillus. It would render unnecessary the importation of farm products, like the grasses on which the tetanus germ flourishes by preference; and it would do away with the breeding places of the bacilli in our cities, namely, the warm, moist droppings of the horse, in which they find an abundance of organic material for their nutrition and the necessary absence of air.

Besides tetanus there are other diseases which can be traced directly to the horse, but which are only communicated by actual contact with the animal. Actinomycosis, for instance, and certain acute coryzas, as epizootic, are common to men and horses. As they effect only those, however, who are associated a great deal and very closely with the horse, they may be properly passed over in an article on public health.

We have come to think in recent years, however, that there are other forms of bacteria besides the tetanus bacillus that find in the animal gastro-intestinal tract an ideal incubator ready to hand, besides an abundance of suitable nutrient media. Prof. Nothnagel, of Vienna, probably the most distinguished living specialist on diseases of the digestive tract, states that more than three-fourths of the fæces of the human being is made up of bacteria. In the herbivorous animals there is much more residue from the food, more indigestible products as cellulose to be found in the fæces, but still it is a perfectly safe approximation to say that considerably more than one-half of the intestinal excrementitious material of even the plant eating animals consists of microorganisms. It is not difficult then to realize the immense number of microbes that find their way into the air from equine excrement. Not all of these are harmful; that is, disease producing or pathogenic, as it is called. On the contrary, only a very small proportion of them are liable to produce any disturbance in the human system, except under special circumstances. It is well known, however, that a certain number of the intestinal bacteria that cause disorders of disgestion in man find a favorable breeding place in the gastro-intestinal tract of the horse. A form of the colon bacillus have been demonstrated there, and, while this is a most variable bacteriological family, at times possessing a good deal of virulence, at others being almost harmless in action, and apparently a normal inhabitant of the human intestine, there is at least an ever present danger from their plenteous distribution in the air through the almost constant presence of drying horse droppings on our streets.

Especially are the intestinal bacteria thus fostered liable to affect the weaklings of our population—irresistive invalids, chronic sufferers from intestinal troubles and infants. The unwelcome addition from this source in the summer time to the already abundantly luxuriant flora of the child's gastro-intestinal tract is surely a fruitful cause of digestive disturbance. Milk, water, the hands, bottles, nipples and the like become contaminated with these bacteria, unless most scrupulous and not always perfectly possible precautions be taken. Something of the prevalence of cholera morbus and dysentery among adults in the summer time is due to the same cause. The horse is practically the only animal that has the freedom of our city streets now, certainly the only one whose intestinal bacteria are scattered plentifully enough in crowded centres of population to constitute a source of danger.

Finally there remains an indirect way in which the horse has an influence on public health. Some one said not long ago that if the horse were done away with, it is probable that we would escape entirely, or at least in great measure, the plague of flies that afflicts us every summer. Flies find their favorite breeding places in stables. They find on the animals themselves a great source of nourishment. If all the stables were removed from our cities it is more than probable that the total eradication of the little pests would be a question of but a short time and very little trouble.

We have learned to regard flies in recent years as much more important factors in the spread of disease than we used to think. It is not so long since all infectious material was supposed to find its way through the atmosphere of itself, by a process of diffusion as it were, or by actual transportation in currents of air. Of late, however, we have come to consider this spread of disease without the mediation of some agency other than the air itself as extremely rare. Epidemics travel no faster than our means of communications between cities. Diseases are contagious not through the air, or only very rarely so, but because of actual conveyance of infecting material from the sick to the well.

The most active agents for such convoy are flies. Long ago it was known that they carried the germs of anthrax. Malignant pustule and other diseases have gradually been added to the list. During our late war with Spain it was demonstrated beyond all doubt that they can serve as the carriers of typhoid fever, dysentery and other intestinal diseases, and probably also of yellow

The Automobile and Public Health

fever. Their habit of lighting on all sorts of material and then carrying off portions of it clinging to their feet and bodies makes it easily understood how they may be carriers of infectious material. Gelatine plates over which flies have been allowed to walk, after having been permitted access to infectious material, swarm with colonies of virulent bacteria.

When the biological significance of small amounts of a living contagious material was not realized, it was difficult to understand how such minute quantities of even intensely toxic substances as flies might carry, could have serious consequences. Now that we know that literally the seeds of contagion are thus carried and that these immediately proceed to multiply under favorable conditions with a rapidity scarcely to be imagined, it is easy to comprehend how flies may carry with them to infect milk, water, butter and other articles of food and drink, the germs of practically any contagion.

Their eradication from our cities would be a distinct gain for public health in more ways than our present limited knowledge of bacteriology and sanitation can make clear to us. There would come, as a direct result, at least a marked reduction in the number of cases of typhoid fever. Suppuration in wounds and the suppurative infections would become less common, and the number and severity of intestinal diseases, especially those of the choleraic type so common in our cities in the summer time, would be greatly diminished.

The automobile, then, should meet with a hearty welcome from the professional sanitarian and from all those who are sincerely interested in municipal health. Whatever can be done to advance the day that will usher in the horseless era for our city streets, will be just so much done in a great humanitarian cause. It will lead to a distinct lessening of human suffering, as well be it said parenthetically, to a most welcome diminution of animal suffering, and will prove another link in the chain of sanitary improvements that in our day is lengthening the average of human life so notably and making it ever more and more liveable, because more healthy.

Women and Automobilism

By Miss N. G. Bacon*

I
T is incumbent upon the intelligent woman to be interested in
life, all its phases and developments. Automobiles and
automobilism are not only fascinating subjects for study
from the point of view of pleasure, but they offer a marvellous
opportunity for the practical utilization of any mechanical talent
or ingenuity. Thoughtful women have come to the front during
the past few years to study all branches of life's work, and to
endeavor, so far as possible, to educate themselves to fill positions
of responsibility and trust. Indeed, the march of advancing
womanhood towards all points of central energy is one of the
most interesting features of this century. Doors that have been
closed since the world has been, are to-day open. Professions
that were in the years of our foremothers considered above or
beneath the capacity of a woman are now recognized as compatible with a woman's dignity and power.

But in speaking of automobilism, we enter the arena of outdoor pastimes and occupations. All of us here can remember the
advent of the bicycle, and its reception by that estimable old
gentleman, Mr. Grundy.

The old fogeys of Rome could not have been more shocked
at Virginia's appearance in the Coliseum than were our " fine old
English gentlemen " at the sight of their womankind—self-propelled. This horror, as you will all remember, was real, and it
resulted in so strong an opposition to bicycling that it was by no
means an uncommon thing for a girl not to dare to ride near her
own house, lest the vials of paternal wrath should be poured on
her devoted head. But we fought against masculine prejudices
and the allied nuisances. Having wheeled for some seventeen
years I could speak at length upon this subject if time permitted,
but suffice it to say that women owe to the bicycle a freedom and
a power never before enjoyed.

The pastime of cycling is all very well, but the motor vehicle
gives a foretaste of something better to come. Automobilism
offers an advance in the future as inconceivable to the novice of
to-day as cycling afforded the uninitiated wheeler of the past.
The question naturally presents itself to the mind, what will the

* Paper read at Lady Harberton's house in London.

Women and Automobilism

automobile do for our womanhood? This is a large question, and cannot be answered in an off-hand or slip-shod manner. In any reply that can be made, automobilism must first be divided into departments of pleasure and profit. The automobile is, and may only be, the rich woman's toy; *i. e.*, it may be useful from the point of view of pleasure only, but it can also be considered the professional woman's friend, if viewed from its profit-earning side. As wealth holds a prominent place in this world, it seems desirable to deal first with motoring as a pastime. The efficiency of the motor vehicles of to-day leaves much to be desired, for it offers little scope, if any, for the lady automobiliste who seeks enjoyment of an unique kind. Driving a car in company with a mechanic seated in close proximity to oneself is scarcely agreeable, nor is it yet found to be satisfactory to have the man, no matter where he may be seated, in livery, to act as mechanic at one moment and as footman the next, for the motor-vehicle, by its construction and its peculiar mechanism, requires occasionally special care and attention *en route,* which only a skilled engineer can give. Hence, it is most desirable that women should study the design of horseless vehicles, for comfort is a very essential item and one that should not be despised. I have seen a considerable number of vehicles, but not one as yet that appears to be likely to yield much comfort and ease for long and short distances.

A car that is liable to continuous breakdown is unsuited to the requirements of women. A really efficient automobile, one that performs all that its manufacturer claims, although it may be full of limitations and shortcomings as to speed, vibration, noise, smell, etc., is a more desirable vehicle than one that falls lamentably short of the manufacturer's guarantee.

After design and efficiency come mechanism and propulsive power. Learning to drive a car is a comparatively simple matter, but to understand its working parts sufficiently to have them in full control, and, in case of disaster or breakdown, to regulate its apparently incomprehensible ways, and to restore, without loss of temper or patience, its running powers to a normal condition, require trained skill. There is at the present moment no place where women can be educated to handle tools, or to adjust the machinery of the car they wish to drive. A superficial knowledge may be given by enterprising manufacturers to purchasers of cars that will enable them to drive, and even to understand the general working of the machinery, but more than this is required before lady motorists can be responsible for the manipulation of their cars. It is difficult indeed for experts to detect errors of adjustment, and the cause of the imperfect working of the machinery, therefore it is imperative that those women who seek to become practical motorists should devote time and skill to the study of the mechanism of automobiles at least sufficiently to enable them to detect what is wrong in case of breakdown, and how to remedy same. It is admitted, I think, that more time is generally spent in discovering the cause of a breakdown than in removing it, for the services of a skilled mechanic can be brought into requisition directly knowledge is obtained as to the nature of the breakdown.

The propulsive energy I refer to last, although it should perhaps come first, for neither the design nor the efficiency of automobiles can be considered until it has been decided definitely whether steam, electric or petrol cars are favored. I cannot here attempt to go into details concerning the driving power of vehicles, for the subject is a deep one, and requires the most careful study.

Granted, then, that women should study the automobile before attempting to enjoy it, I pass to the nature of the pastime. Those who have enjoyed the fascination and the exhilaration of driving through the air, along our public highways, with little or no muscular effort, up hills and down dales, at a high rate of speed, can speak with eloquence in praise of its enjoyment; but even the most eloquent generally finish their eulogistic remarks by saying that no words can adequately describe the sensation. To really appreciate what an automobile is, you must run one. There are no half measures. It is " To be, or not to be." There are, of course, various phases of enjoyment. The most ecstatic I should imagine to be that of whizzing through the air at a breakneck pace, regardless of all else but speed. But, it is whispered, with hand uplifted, that cannot be. By the laws of the land any speed exceeding twelve miles an hour is prohibited. The uninitiated say, " Surely that is enough "; but those who have tasted the delights of motoring, solemnly, and somewhat regretfully, shake their heads and protest, in as mild language as is possible for their feelings, against unnecessary restrictions. Apart from the speed craze, the pleasure of pottering along sweet lanes, surrounded by landscape beauty, must not be despised. It is impossible to touch even the fringe of the subject here, and, therefore, I leave it alone, and simply appeal to the imagination of my hearers in the hope that they may catch the tiniest glimpse of forthcoming pleasures. For my own part my appetite for the automobile has only been sharpened by what little experience I have enjoyed of motor-vehicles. A few years hence I may speak with more knowledge, perhaps with even greater enthusiasm,

Women and Automobilism

and, at any rate, I hope with less opposition, for to me it seems strange that any one should consider it unprofitable for women to study the automobile and automobilism.

I come now to the professional women. Many of us are deeply interested in all agricultural and horticultural pursuits. Various colleges exist for the instruction of women in the arts connected with the cultivation of the land and its produce. Gardening, fruit growing, bee-keeping, dairy produce, and poultry-keeping, are occupations now considered to be adaptable to women's labor, and I think statistics will prove that motor-vehicles are less costly for haulage of heavy traffic. The question of transit of the produce of the land from the door of the producer to the markets is one of special interest to women, for until the nationalisation of our railways is arranged, the problems in connection with rural life are very perplexing.

Sceptics may smile, and render the world unpicturesque by means of their unseemly jokes, jests and caricatures of women driving such vehicles. Women should study the whole question dispassionately and with intelligence, in order to test for themselves whether the motor-vehicle will or will not be useful to them in their various agricultural and horticultural callings, for those who laugh last generally laugh best.

I think I have now covered the whole ground of my campaign. Those of us who seek to form a Ladies' Automobile Club have very unpretentious claims. Indeed, we are modest, for our knowledge is so meagre that all we seek is an opportunity for studying the whole matter, and to do this some centre should be formed around which women interested can gather. The Automobile Club of Great Britain claims the distinction of being " a centre of information and advice on matters pertaining to motor-vehicles, for those who are not owners as well as for owners of motor-vehicles," and yet it closes its doors to more than half the adult human race. Professor Vernon Boys wrote to me lately: " In our membership one touch of motor makes the whole world kind," but I fail to appreciate the logic of such a remark, for how can " the whole world be kind " when, out of a population of some forty millions of people only 500 odd men are members of this Club?

Who could deny even to 1,000 men the privilege of having a *social* club for any purpose whatsoever? But they cannot logically expect to form the centre of information and advice if they exclude women from their membership. One consolation, indeed, women are offered by the administration of the Automobile Club. That is, they are classified with minors—not infants—

therefore a woman now can surely claim equality with youths who have lived twenty years and eleven months, and even a score more days.

Women may be very weak and silly creatures, but they represent at least half, if not more, of the human race. True, the gentlemen members are most kind and considerate in taking women for drives like children in their motor vehicles, but it is an odd mixture indeed to have, on the one hand a club, which has been founded to be a centre of information, and, on the other, a rigid rule for the exclusion of adults in consequence of the sex disability. To make the situation perfectly logical, the trade should refuse to sell vehicles to " ladies and minors," as being only fit for the use of gentlemen. Yet I have read, continually, advertisements—indeed, it was precisely a press notice that led me first to the study of motor-vehicles—declaring that certain cars are so simple that any lady can drive them. When first the suggestion of the formation of a Ladies' Automobile Club was mooted, women were accused of desiring to intrude upon the privacy of men enjoying the comforts of their own club, but seeing that this was false, and that we only desired to form a very unpretentious centre for the study of motor-vehicles. it has been asserted that not enough women interested in automobilism exists. I put the challenge here! Are we, or are we not interested?

If we are, let us start from a centre, and study all that pertains to automobiles and automobilism and see where we stand. The sooner a start is made the better it will be for all concerned, for this is a question that affects the interests of the entire human race.

STORAGE BATTERIES FOR CANAL-BOATS

At a meeting of the Executive Committee of the Erie Canal Electric Traction Company, held recently, it was decided to adopt for use on canals, including the Erie Canal, the storage-battery motor, subject to the approval of the New York State Superintendent of Public Works. These motors are put out by the Electric Vehicle Company, of Hartford, Conn., and the Electric Storage Battery Company, of Philadelphia. These organizations are controlled, it is understood, by the Widener-Elkins-Dolan-Whitney Syndicate, which control all patents, both foreign and domestic, covering storage battery and motors and devices. It is stated that contracts for canal-boat storage batteries were let involving more than $1,000,000.

The Houpied Igniter

By Paul Sarrey

THE electrical apparatus employed in most hydrocarbon motors for discharging the explosive mixture of gas and air comprises a primary or accumulator battery, an induction coil and a vibrating armature whereby sparks of sufficient heat are produced to insure regular explosions within the cylinder.

Fig. 1

But the care required by primary cells and accumulator batteries has caused many a *chauffeur* to turn to the magneto-electric machine. Among the electric ignition devices which employ such magneto-electric machines may be mentioned the Houpied igniter, which forms the subject of the accompanying illustrations.

All the parts of the Houpied apparatus are inclosed in a wooden box Z of such small size that it can be readily carried from place to place. For greater convenience the induction coil A and the condenser C can be arranged in a second box on the automobile. Between the poles M^1 M^2 of the magnet p turn the poles S S of a Siemens armature mounted on the shaft X provided at its end with a pulley which receives power from the motor by a strap or belt. The current produced by the rotation of the armature is rendered constant in the usual manner by means of brushes B^1 B^2 connected with a split-ring collector D.

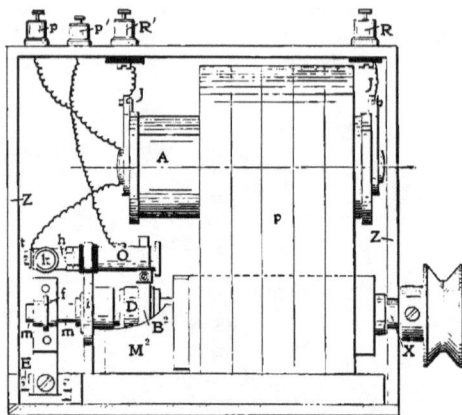

Fig. 2

Fig. 4 is an end elevation, showing the interrupter in its circuit-breaking position; in which position a current is induced in the fine wire of the induction coil A. Two vertical, parallel springs e e^1 mounted on a strip of copper, are held on the cam m and are joined by a crosspiece. The contact at the end of the spring e of the interrupter is composed of a small piece g of platinum. The spring e^1 is provided with a wheel f, turned on its axis by frictional engagement with the cam m. Connected with the brush support i is a binding post, through which passes a screw K, the

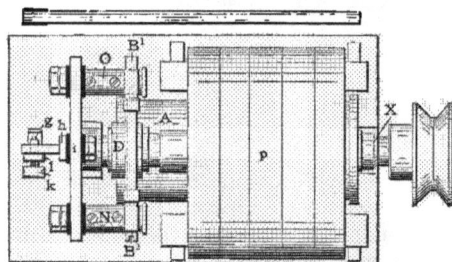

Fig. 3

point of which is adapted to engage the platinum contact g of the spring e. When the platinum g and the screw point g^1 are in contact, the circuit is completed; when the platinum and screw point are out of contact, as shown in Fig. 4, the cur-

The Houpied Igniter

rent is interrupted. The spring e carrying the contact g is thus caused to make and break the circuit by the rotation of the cam m extended from the shaft X.

One end of the secondary winding of the coil A is secured to the binding post h by a screw t; and the other end is connected with the binding post B on the exterior of the box. The binding post B, furthermore, constitutes one of the terminals of the primary coil, the other terminal being connected with the binding post p^1 and with the brush support N. The interrupter forms part of this circuit. Since the speed of most motors varies considerably, some means must be provided for regulating the length of the spark and the frequency with which it passes between the terminals. For this reason a rheostat is generally employed.

When the pulley on the end of the shaft X is turned, the armature will induce a current as it rotates between the poles M^1 and M^2. This current is collected by the brushes and conducted by the connecting wires. Whenever the cam forces the contact g into engagement with the point g^1 of the screw K, the circuit is completed; whenever the points g and g^1 are out of contact the current is interrupted. The circuit is broken twice in a single revolution of the shaft X; hence there are produced a number of induced currents which cause as many sparks to pass between the terminals of the wires running from the binding posts R and R^1.

It will be observed that in the figures the induction coil A is placed within the magnet p of the magneto-electric machine, in order to reduce the size of the apparatus. When thus arranged the coil is necessarily incased within an insulating envelope and mounted on an insulating support.

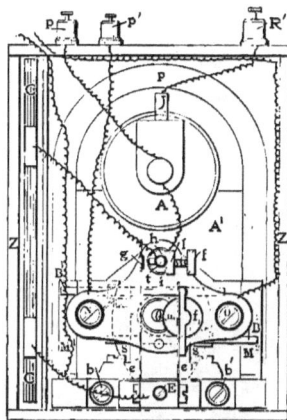

The Automobile Magazine

Fig. 4

The New Sport Abroad

(By Our Own Correspondent)

PREPARATIONS for an adequate representation of automobile interests at our great Exposition are well in hand,
and the programme thus far developed gives assurance
that this department will constitute one of the most absorbingly
attractive features of the occasion. The Automobile Club of
France has very appropriately been
given practical charge of the matter by
the committee on the Automobile Section. One of the important attractions
will be the long-distance races to be held
throughout the Exposition. They will
be so arranged that, instead of taking
place between Paris and Bordeaux or
Paris and Hâvre, as has always been the
case hitherto. the long course will cover
a route which, while it will be several
hundred kilometres in length, will have
a trefoil shape, and therefore will be
kept within easy reach of Vincennes,
where the Automobile Department will
have its headquarters, so far as practical
demonstrations are concerned. This
innovation for the long-distance contests will be advantageous both to the
racers and the public in various ways.
Those desiring to witness a race in its
progress and under any of the various
conditions that prevail according to the
character of this or that portion of the
route will not have to go far from Paris to reach any part thereof.
And many will naturally give themselves the pleasure of going to
the scene by automobile conveyance. The nearness of all parts of
the route to headquarters will also give the advantage of keeping
the public constantly informed as to the progress of a race, which
will be recorded at Vincennes upon little electric tables representing the route in miniature. with the automobiles of the contestants
shown in small models. In this way the race can be accurately

A Parisian Chauffeuse

The New Sport Abroad

Two Up-to-Date Automobilistes

followed throughout in a manner similar to that in which baseball games are represented in America.

A special track and grand stands will be constructed at Vincennes at a cost of 100,000 francs, and special prizes will be offered in addition to the regular Exposition awards. It is promised that some of the contests will be of a unique character. For electric automobiles a charging station will be established close to the track. Particular attention will be given to providing opportunities for testing automobiles of various kinds, and intending purchasers will be furnished facilities for making attractive little excursions, and thereby gain an idea of the qualities of the make which they desire to try. We understand that American manufacturers will make a particularly attractive showing in this department. For the contests the competing vehicles are to be divided into four classes: Heavy vehicles of various kinds, cabs, victorias and voiturettes—the latter including motocycles, tricycles and bicycles.

Automobile fêtes will be a great feature at Vincennes. The charm of this form of entertainment was made evident in the foretaste we had in the great fête held in the Bois de Boulogne very recently—an affair full of interest—all the vehicles participating decorated with flowers, and many of them splendidly decked out at great cost; the whole affair full of life and movement, fun, gaiety and excitement from beginning to end. A practice ground was established at the Longchamps track, covered with all sorts of stuffed lay figures and other obstacles, through which the auto-mobilists had to thread their way. There was also a contrivance so arranged that in passing they were expected to carry away at the point of a lance little wreaths of flowers hung there. Above some of these wreaths buckets were balanced, the merest touch upsetting them and deluging the passer with confetti. Some of the automobilists were boys and girls, and the skill which they showed was often marvelous. A little fellow of nine years wound his way successfully at the highest speed by the most intricate course past every obstacle. Most of this young-folks' work was done on motocycles and the like, but there was one young lady of sixteen, Mlle. Richards, who ran a large automobile with extra-ordinary facility. There was a quadrille, beautifully wheeled by eight victorias, and at the end there was a graceful farandole, with all the vehicles present—something like seventy in number—twisting their way in single file around the track.

One of the features of these fêtes, most interesting to ladies, were the latest automobile fashions displayed by some of the fair contestants. Most of the gowns then worn showed the brilliant maroon or crimson known among women's tailors and modistes as " automobile red."

Ring Sticking Contest

Now that the cool weather of autumn and early winter has set in, a new-fangled Pelisse, worn exclusively by automobil-istes, has come into fashion. This cloak, which resembles the overcoats of fashionable coach-men, is trimmed with fur, pref-erably of brown or gray, when worn with a waterfall cape. The cloaks that have no cape are long and close fitting, and are covered with fur from the top to the bottom, giving a de-cidedly arctic appearance to their wearers.

The New Sport Abroad

We have all been laughing at a comical occurrence in which the well-known Bordeaux automobilist, M. Lanneluc-Sanson, took leading part. He might well have fancied himself Don Quixote and his automobile a *fin de siècle* Rosinante, gallantly coming to the rescue of a fair dame fallen into great danger. On his way back from a service of twenty-eight days as *chauffeur* to the commanding general of the Eighteenth Army Corps, he reached the village of Mèrignac, in Gironde, at nightfall. There he found the whole place in commotion. People were running about. Inquiring as to the cause of the excitement he learned that a cow had fallen into a well. Repairing

Confetti Dropping

to the spot he found a crowd of peasants at their wits' ends, vainly trying to hoist the cow out with a rope. But the cow was heavy and the poor creature bellowed and groaned in vain. A brilliant inspiration seized M. Lanneluc-Sanson. He attached her halter rope to his auto, mounted his machine and started ahead. The rope grew taut and under the weight at the other end the automobile almost came to a stop. But gradually it went ahead, coughing mightily with the exertion. A great shout of joy arose from the assembled peasants. The cow's head came peering above the rim of the well and she was landed securely at the surface.

In Brittany there has been an interesting conversion. A good priest, the curé of a hamlet, had conceived a most profound hatred for the new invention, and when automobiles went by he would mutter maledictions upon them, cursing them as instruments of the devil, imaginings of hell, devices of heretics! When he encountered one on the road he would cross himself, and in the odor of petroleum wafted therefrom he would fancy he detected the sulphurous vapors of the infernal regions—and to tell the truth it may be doubted if the latter smell any worse than certain " petrolettes " I have in mind.

Time went on, and one evening the good father was called upon to administer the last sacrament to a dying man two leagues away in the country. It was cold and blustering; the raw wind carried driving rain. The curé shuddered as he thought of the long walk before him, the wet blast in his face.

Just then an automobile drew up at his door. " Pardon, Monsieur le curé," called the occupant; " the route to Saint-Gothbert, I pray you."

Saint-Gothbert! The very hamlet where the dying man was awaiting his ministrations! The priest hesitated a moment, but his desire proved stronger than his repugnance, and he replied: " Will you allow me, monsieur, to show you the road? Is there room for me beside you?"

Automobile Pelisse

Twenty minutes later the good curé was anointing with the holy oil the dying one, who soon sank into the dreamless sleep.

Now, when the worthy father is asked what he thinks of the automobile he will make answer: " Ah, monsieur, an invention blessed of the Lord!" And he will tell the story of that stormy evening.

Of course the experiment of the Post-office in the collection of letters by automobile is successful. The results show an average gain of forty-five minutes over the time taken in collections in the old way, on foot. A remarkable application of the automobile to postal work has been arranged for the French Soudan, whither a

Body of Partinium Automobile

9 horse-power Dietrich vehicle has just been forwarded for the purpose. The automobile, very naturally, is particularly adapted to the requirements of a level desert country, where the dryness makes good roads an easy possibility. A new colony, like a new house, has the advantage of the latest modern improvements in its first organization. In the heart of Africa I am told that it is no uncommon spectacle to see stark naked negroes dashing around on bicycles, and it would seem that in the Soudan our colored brother is to obtain one of his first lessons in civilization through experience with the automobile.

Among the most curious innovations in the manufacture of automobiles are two newly invented vehicles of French make. One derives its driving power from acetylene, a substance hitherto not used for automobile motors. The other vehicle is intended for a delivery wagon, of unusually light construction, since most of its metal parts and the body of the wagon as well are fashioned of partinium, a newly discovered metallic alloy of aluminum and tungsten.

There is beginning to be some uneasiness over the advancing price of gasoline. It has recently gone up 5 centimes a litre. The price in Paris is now 65 centimes a litre—15 centimes more than we have to pay outside the gates, all on account of the octroi.

French Postal Motor

The Automobile Club has advanced its entrance fee to 200 francs with the first of the new year. The membership of the club is now but little short of 2,000. The club has recently honored M. Forestier with a bronze medal in token of his services in organizing the club competitions, and it has also awarded to M. Georges Prade of the Vélo a medal and a diploma as a souvenir of the race between Paris and Ostend.

The work of record-breaking continues. Béconnais, riding a Phébus tricycle, recently surpassed existing records for the hour by the following figures: 10 kilometres in 8m. 41s.; 20 kilometres in 17m. 19⅘s.; 30 kilometres in 26m. 3⅕s.; 40 kilometres in

German Prize Racer

34m. 55s.; 50 kilometres in 43m. 48⅗s.; 60 kilometres in 52m. 49⅘s. And in an hour he ran 67.901 kilometres.

In Germany the following awards were made for the long-distance race of 185 kilometres from Berlin to Leipsic in connection with the recent International Automobile Exhibition in the former city: Motor carriage group—First prize of gold medal, with prize of honor in shape of a portrait of the German Emperor, to Benz & Co., of Mannheim; second prize of gold medal, with prize of honor in shape of a portrait of the King of Saxony, to Dietrich & Co., of Niederbronn; third prize of silver medal, with prize of honor in shape of a portrait of the Secretary of State

The New Sport Abroad

von Podbielski, Honorary President of the Exhibition, to Baron von Liebig, of Reichenberg, in Bohemia.

The new Deutscher Automobile Club of Berlin has for its patroness Her Imperial Highness the Grand Duchess Anastasia of Mecklenburg-Schwerin. The Grand Duchess is an enthusiastic automobiliste and has several motor-vehicles of various descriptions at her hunting castle of Gelbensand near Rostock. The Grand Duchess, with the Grand Duke, was one of the earliest visitors to the automobile exhibition, and the distinguished pair took a keen interest in the affair.

From Australia come some particulars about the largest automobile in the world, which the owner, a wealthy Queensland miner, has appropriately named the "Goliath." It is run by a petroleum motor of 75 horse-power. Its weight is 14 tons and it has a carrying capacity of about 50 tons, going at the rate of 13 kilometres an hour when loaded. The owner is the proprietor of a gold mine that lies about 600 kilometres in the interior, and the huge affair was designed to carry freight to and from the mine. Its cost was about 40,000 francs.

Grand Duchess Anastasia of Meklenburg-Schwerin

Chief Patroness of the German Automobile Club

Syner's Elastic Clutch

THE elastic clutch, invented by the Belgian engineer Snyers, consists of two disks, one of which is formed with radial grooves, and the other of which is provided with brushes composed of tempered steel wires. Our central illustration shows the two disks mounted face to face. When the two disks are thrown into engagement, the brushes enter the grooves, press against their walls, and bend slightly; the force exerted is equal to the total flexive effort of the tempered steel wires on the corresponding grooves. In order to engage or disengage the two disks, but a slight effort is required.

Whatever may be the speed of rotation of the shaft the clutch can be thrown into gear gradually and without any shock. When the disks are in engagement with each other, there is no

Elastic Clutches

danger of the parts' slipping, so long as the power transmitted is not greater than that which the clutch is capable of transmitting. The disks are instantly disengaged when the transmission-gear sustains a shock or when the power to be transmitted becomes excessive.

These results are not obtained by friction, as in most similar devices. The wear is reduced to a minimum. The clutch is capable of standing hard usage, and will operate effectively after continued service without the need of repeated inspection or repairs.

The clutch has been applied with considerable success to several Panhard-Levassor carriages.

William Rogers, C. E.

The Electric Automobile

By Prof. Félicien Michotte

(Continued from The Automobile Magazine for November)

THE DISCHARGING OF A BATTERY

ACCUMULATORS may be considered discharged when the voltage is less than 1.9 per element. *i.c.*, when it is less than the number of elements multiplied by 1.9. For 40 elements we should therefore have

$$40 \times 1.9 = 76 \text{ volts.}$$

The discharge can be forced beyond this voltage, but at the risk of deterioration to the battery. Beyond 1.8 volts the discharge should never be pushed, for then deterioration will certainly set in.

A battery of accumulators, even when not in use, will gradually be discharged. It should therefore cause no astonishment if, at the end of three days, the cells should yield no current.

Electric carriages can be divided into two classes: (1) Vehicles with fore-carriages (*avant-train moteurs*), and (2) rear-driven vehicles.

Vehicles with Fore-carriages.—When electric vehicles were first introduced, it seemed as if the fore-carriage system were the only one applicable to electric automobiles. The carriage was pulled instead of being pushed; and the small size of the electric motor apparently simplified the solution of the problem. But if the motor be easily disposed of, there still remains the difficulty of adequately transmitting the power to the wheels. Jeantaud, who first employed the system, has since discarded it. In order to overcome the difficulty, Messrs. Krieger and Doré have devised two methods.

Krieger mounts a motor at each wheel; and each motor transmits its power directly to the wheel by means of a small bronze pinion. This system has the advantage of employing two motors. If one motor give out, the carriage can perhaps still be driven by the other. On the other hand, the two motors and the very large and complex switch required, increase the cost of the carriage considerably.

In the Doré carriage the motor is mounted in front of the driver's box; and power is transmitted to the front axle by a

The Automobile Magazine

vertical shaft provided with pinions which drive a differential gear on the front axle.

Rear-driven Carriages.—Rear-driven carriages are most numerous; and their transmission-gear is similar to that of petroleum automobiles. The motor drives a shaft carrying the differential gear and having at each end a sprocket-wheel connected by a chain with another sprocket on the carriage-wheel. The use of the double chain-gear has the disadvantage of consuming considerable current owing to the unequal tension of the two chains.

ELECTRIC BRAKING—RECUPERATION

When the carriage is running on a down-grade, the current can be shut off and the carriage run by its own momentum. The motor, however, continues to turn, owing to its connection with the wheels. But, instead of transmitting power, it receives power from the wheels; and the armature, turning between the electromagnets in the contrary direction, generates a current. The generation of this current offers resistance to the rotation of the armature, which resistance is transmitted to the wheels by the transmission-gear and opposes their movement. The current produced can be conducted to the accumulators, and hence a certain quantity of electricity is stored up. What has been lost is therefore partially recuperated. The electric brake is useful and efficient, but the current produced is so feeble that there is but little to be gained by conducting it to the battery.

CHARGING-STATIONS

Carriage batteries can be charged:
1. By dynamo.
2. On lighting-circuits.

By Dynamo.—The dynamo can be driven by any motor whatever—hydraulic, steam, gas, or petroleum. The power of the dynamo should be proportional to the number of carriages to be charged. Direct-current dynamos are well adapted for the purposes of a charging-station. The voltage of the dynamo must be greater than that of the accumulators to be charged; the ampèrage should be equal or lower than that of the battery; if it be higher, it can be reduced by means of resistance-coils.

On Lighting-Circuits.—In cities, where there is a general circuit, part of the current can be shunted to the battery, if its tension be not greater than 100 to 110 volts. If the tension be higher, the usual resistance-coils can be resorted to.

The Electric Automobile

COST OF A STATION

A charging-station is far from being costly. An oxid battery can be charged by a dynamo of fifteen ampères, and a Planté battery by one of thirty to forty ampères. A switch-board, with the necessary instruments, costs but little more than $40 (200 fr.) in France.

The cost of charging by using electric-light circuits varies with the length of the feed-wire. Some companies charge exorbitant prices.

THE CARE OF AN AUTOMOBILE

The care to be given to an electric carriage is confined chiefly to the motor, accumulators, and co-acting mechanism.

Motor.—In running a motor or dynamo, the wear of the brushes should be noted, the collector and connections kept in order, and sparking prevented.

If a wire of the collector be defective, it should be immediately repaired; otherwise the motor will surely be injured. The bearings should be well lubricated.

Accumulators. — The accumulators should be frequently tested for short circuits and dead cells.

If it be observed that the voltage of a charged battery be below the normal, each cell-couple should be separately tested. If the voltage of one be lower than the normal, then in this particular group defective cells will be found.

The elements of the cells should be tested by a low-reading voltmeter (capacity, 3 volts). Cut out all inactive cells; examine the connections frequently and keep them in order; see to it that the insulation of the retaining-case is perfect. Keep the electrolyte up to the standard strength, either by the addition of water or acid.

In repairing defective connections, care should be taken to arrange the wires exactly as they were originally. It is advisable to prepare a diagram of the wiring for purposes of verification.

If the plates become coated with sulphate, the battery should be sent to the manufacturer to be restored to its former active condition.

Mechanism.—The care of the mechanism of an electric automobile is exactly similar to that of a petroleum-motor carriage.

The Automobile Magazine

TOOLS TO BE CARRIED

A special wrench for the wheels.
A monkey-wrench.
One pair of pincers.
One pair of cutting-nippers.
Wire; insulating material for defective connections.
Lubricating-oil for the motor.
Emery-cloth.
Extra brushes and coil-springs.
Rags.
At the charging-station, acid (20°), pure water, a Beaumé aërometer, and one or two plates should be kept.

DISTANCE COVERED BY AN ELECTROMOBILE

The distance which a carriage can cover is limited by the capacity of its accumulators and the efficiency of its motor. The former is the more important; for the capacity varies with the system and with the weight of each element.

Experience has shown that a weight of 10 to 12 kilos per element, or a total weight of 690 to 780 kilos, gives the best results.

With a Planté battery weighing 780 kilos, a carriage can cover 60 to 65 kilometres with a good driver, and 50 to 55 kilometres with an ordinary driver. The former knows how to utilize his current better than the other. Hence the difference.

Distances of 100, 150, and 180 kilometres are said to have been covered; soon we shall hear of runs of 200 kilometres. These are but harmless pleasantries, and their authors have probably never ridden in an electric carriage.

HOW TO RUN A CARRIAGE

An electric carriage can be more readily controlled than any other vehicle. The driver merely manipulates a number of levers and need take no especial precautions in starting or stopping.

In the petroleum or steam carriage, the question of fuel does not disturb the *chauffeur;* for he can readily replenish his supply whenever it may be necessary. But the driver of an electric automobile is hampered by the limited capacity of his battery.

The Electric Automobile

When his power gives out he must seek a source of electricity and lose considerable time in recharging. For this reason he must learn how to utilize his current to the utmost profit—a matter which is simple enough, but which requires a little study. He should keep an eye on the road before him, increasing his speed on an up-grade, shutting off the current on a down-grade, stopping his carriage by allowing the motor to run down, and not by means of the brake, and avoiding all unnecessary, excessive discharges. By handling his current and motor thus judiciously he can add ten kilometres to the distance which his carriage can normally cover.

These rules—increasing the speed on an up-grade, shutting off the current on a down-grade, and allowing the carriage to travel by its own momentum—are simple enough, and are easily learned and applied. In coming to a stop, shut off the current at the proper time, and allow the motor to run down of its own accord: use the brake only when the momentum acquired will carry the vehicle too far.

Hill Tests

The most important point on which the purchaser of an automobile should be satisfied is the hill-climbing power of the motor vehicle submitted to him. Only not is it necessary that the carriage should take its full load up a steep hill, but it is essential for satisfactory touring that a steep hill should be taken at a good speed.

Many automobiles are so under-powered (that is, the weight of the carriage body and load is too great for the power of the motor) that on a hill of any steepness they cannot pull their load at a speed of more than four miles an hour.

If one of these carriages mount a hill of steep grade at four miles an hour and descends it even at the high speed of thirty miles an hour, its average for the two miles would be, in spite of the illegal and break-neck rush down-hill, only a shade over seven miles per hour.

If a buyer finds that the motor he is inspecting has not been submitted to the Automobile Club test, he should insist on the seller's carrying out a hill-climbing test in his presence.

A purchaser should, after the hill test, take the time over a mile on the flat, to see that the car as geared for hill-climbing will also make good time on the level.

Steam Carriages of the Société Européenne d'Automobiles

EVERYONE who visited the *Salon du Cycle et de l'Automobile* remarked the steam tricycle-cart exhibited by the *Société Européenne d'Automobiles*. This little automobile attracted attention not only because of its light appearance, but also because of its simple motor—a steam-engine of novel construction patented by Messrs. Tatin and Tanière.

Steam Tricycle Cart

In our sectional and plan views of the carriage, *r* is a cylindrical petroleum-reservoir located beneath the seat. By means of an air-pump *n*, the pressure within the reservoir is so regulated that, upon opening the proper cocks, the hydrocarbon will flow through the tube *b* to the Bunsen burner B, mounted beneath the coil S, which constitutes the steam-generating portion of the boiler C. The boiler, in addition to its small size, possesses the merit of generating steam in an exceedingly short time. The water contained in the reservoir L is fed to the boiler by a pump, the quantity supplied regulating the speed of the motor M, driven by the steam generated in the coil S and conducted by the tube or pipe *t*. The products of combustion escape beneath the carriage through the passage *c*.

The motor M is a single-cylinder, double-acting engine, the

piston-rod of which drives the crank v and hence the shaft A, carrying the fly-wheel V. At both ends, the motor-shaft is provided with gear-wheels Ec, each of which is operatively con-

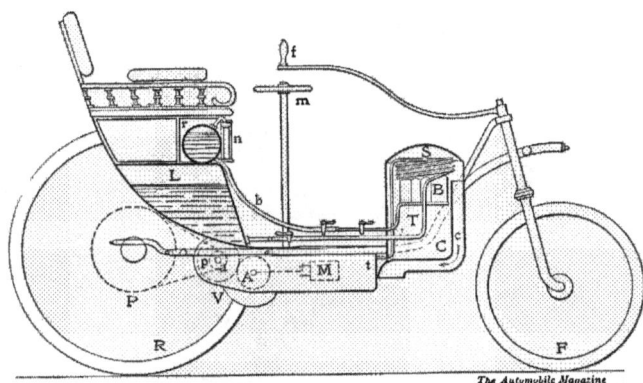

Sectional View of Steam Tricycle Cart

nected with a carriage-wheel R, through the medium of a pinion p, connected by a chain with the sprocket-wheel P.

The arrangement is evidently extremely simple, and is all the more noteworthy because the usual differential-gear has been discarded. But a single wheel is driven at a time. Hence there are two changes of speed, the one being obtained by the large

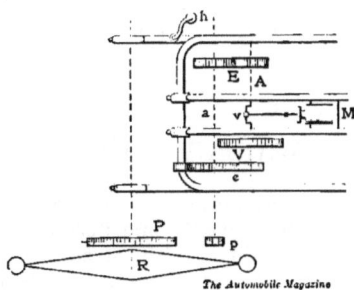

Plan of Steam Tricycle Cart

gear-wheel E, and the other by the small gear-wheel c. The motor is thrown into gear by means of the hand-wheel m; and steering is effected by means of the handle-bar f, controlling the small front wheel F. The brakes are operated by means of the pedal *h*. The motor can be readily reversed so as to drive the carriage backwards.

The carriage is in every respect equal to the petroleum-motor *voiturette*, and is more easily operated and controlled. By in-

creasing the amount of water fed to the boilers and the quantity of petroleum supplied to the burners, the speed can be increased. The carriage can easily be driven at a rate of sixteen miles per hour. The water and petroleum reservoirs have a capacity sufficient to enable the carriage to run five hours.

THE AWAKENING OF RUSSIA.

Automobiles have now obtained official recognition in St. Petersburg. So far only twenty-one permits have been issued for travelling by autocar through the streets of the capital. The rules which automobilists must observe have been drawn up and only await the sanction of the Duma, or Municipal Council. Before the permits are granted the *chauffeurs* have to pass an examination and their vehicle is carefully inspected. They must go at a speed of not over twelve versts (about eight miles) an hour, without causing any smoke or steam.

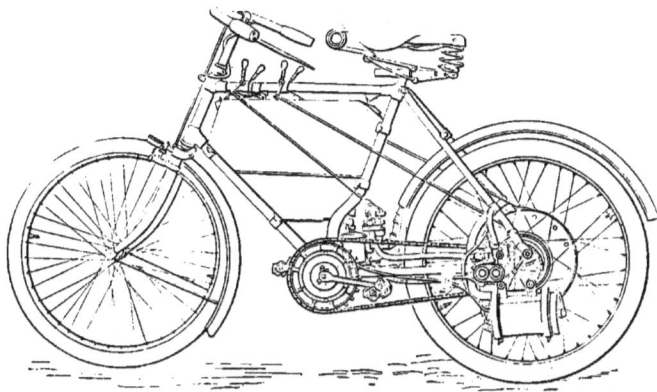

The Butikofer Motocycle

The Butikofer Motocycle

A UTOMOBILES have now been in use for several years, but so far as external appearance is concerned, little improvement has been made over the first types introduced. The public cabs which go winding in and out among the wagons and cars that crowd our business-streets, seem to have been built upon the lines of a steam-roller; their huge tires, their clumsy wheels, their low tops, have often enough offended the eyes of those accustomed to the slower, but more graceful horse-drawn hansoms. In motocycles the awkwardness of appearance is even more apparent and is largely due to the attempt to bring two radically different elements into harmony. A motor is one thing, a bicycle a totally different thing. And the attempt to apply the one to the other without some changes in design, must necessarily produce a combination which, esthetically, leaves much to be desired. A glance at one of the more recent types of motocycles, of which we present illustrations, will prove that decided changes in design must be made before the motor can be successfully applied to the bicycle. But, whatever may be the faults of existing motocycles, it cannot be denied that considerable ingenuity has been displayed in their details. As a typical motocycle we have selected a Butikofer vehicle, in which both the faults and the merits mentioned may be found.

The Automobile Magazine

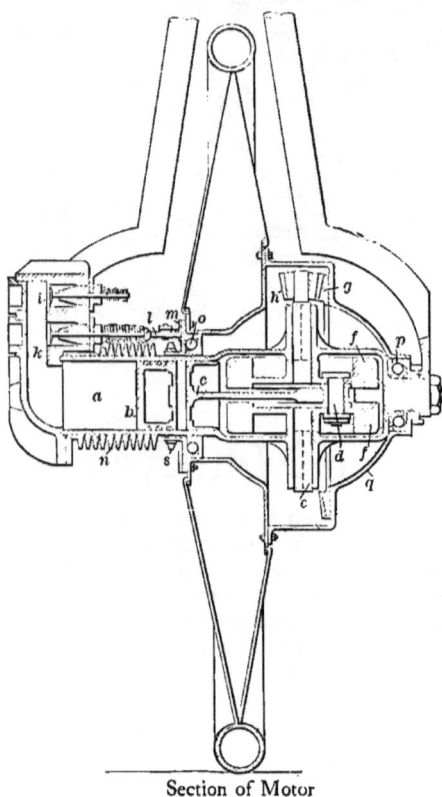

In our sectional view of the Butikofer motor, the cylinder is represented by a, the piston by b, the piston-rod by c, the crank by d, and the fly-wheel by f. At the end of the motor-shaft e is a bevel-gear h, engaging a second bevel-gear g, secured to the casing q, forming part of the bicycle-wheel hub mounted in ball-bearings o, p. The sprocket-wheel s is chain-connected in the usual manner with the pedal-sprocket, and is provided with four symmetrically-disposed cams, by which the exhaust-valve k is operated. Any form of ignition device can be used. The remaining details of construction require no explanation.

Section of Motor

STEERING-GEAR

The steering-gear is of vital importance. The point is not whether the car be guided by a tiller or a wheel, for a tiller may govern an admirable gear and a wheel may control a dangerous form of transmission of direction. A steering-gear, to be safe, must be such that the steering-wheels are not deflected by accidental causes, such as stones in the road. The worm-gear used in French racing cars has proved to be very safe and efficient.

Ernst Petroleum-Motor Carriage

ONE of the most attractive carriages exhibited at the *Exposition Internationale des Tuileries* was a *voiturette* made by Ernst et Cie., which was noteworthy both for its handsome appearance and for the novelty of its mechanism.

The *voiturette*, as our illustration shows, is a light, graceful, two-seated vehicle, which, although it weighs not quite three hundred pounds, is nevertheless remarkable for the durability of its construction.

Ernst Voiturette

The motor used develops 2¼ horse-power, and is enclosed in a casing mounted in the rear of the carriage, and so disposed that the carriage loses nothing in appearance. So perfect is the operation of this motor and its carbureter, that no odors whatever are given off. The carriage can be started gradually, and can be driven up fairly steep grades at a good speed. Two changes of speed are provided.

In the Ernst three-seated carriage, which we illustrate in perspective and plan, a four horse-power motor is used, having two vertical cylinders, CC'. The motor is mounted in the forward portion of the carriage and, through the medium of a friction-clutch, E, turns a longitudinally-extending shaft, A, which drives a friction-wheel, P.

This friction-wheel forms part of a mechanism comprising

315

Ernst Carriage

two disks, pp', rotating in a plane perpendicular to that of the friction-wheel, P, so that there will be two points of contact instead of one, thereby insuring the transmission of the movement of the wheel, P. The friction-disks, pp', are so mounted on an auxiliary shaft, a, that by means of a shifting device they

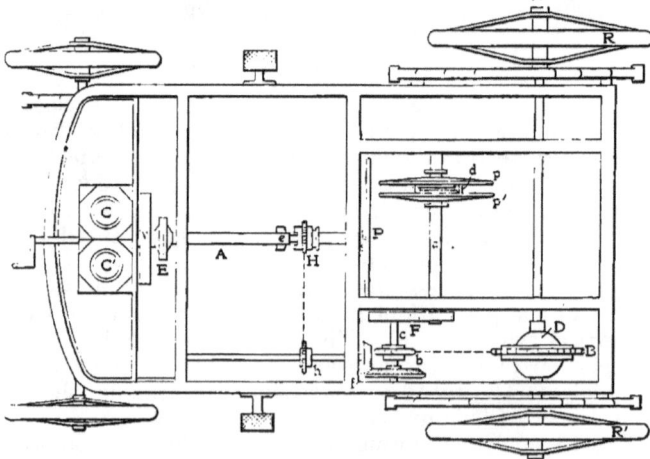

Plan of Ernst Carriage

Ernst Petroleum-Motor Carriage

can be moved in order to vary the speed. This speed will depend upon the relative positions of the disks, pp', and the friction-wheel, P. When the disks are in contact with the periphery of the wheel, the speed will be at its maximum; when the disks are in contact with the centre, the speed will be zero; and when the disks are shifted beyond the centre, the motion of the shaft, a, will be reversed. It is evident that by this ingenious arrangement of friction-disks and wheels the carriage can be driven backward without any shock or jar. Since the two disks, pp', touch the wheel, P, at two different points, it follows that they are driven at different speeds. This inequality is compensated for by a small differential gear, d, mounted between the disks.

Ernst Hydrocarbon Tricycle

The movement of the shaft, a, is imparted to the shaft, c, at a reduced speed, by means of connecting gearing, F. A small sprocket-wheel, b, on the shaft, c, is connected by a chain with the sprocket, B, of the differential gear, D, mounted on the axle of the driving-wheels, RR'. The frictional engagement between the disks, pp', and the wheel, P, may prove inadequate to drive the shaft, c. For this reason the shaft, A, has been provided with a clutch mechanism, c, whereby the shaft, c, can be directly driven through the medium of the bevel-gear, f, and chain, Hh. In ascending a steep grade, the automobilist throws the clutch, c, into operative position, so that the motor will surely drive the carriage even though the friction-disks may slip. On level roads the disks and co-acting wheel, P, are used.

The Sanciome Petroleum-Bicycle

THE Sanciome petroleum-bicycle, of which we present an illustration herewith, comprises an extra light motor of four-cycle type, which develops about 1-horse-power and weighs not quite twenty-one pounds. The carbureter and ignition devices have been simplified in construction, for the purpose of saving weight and of rendering their control easy. The explosive mixture is electrically discharged either by primary batteries or accumulators, as in all motocycles.

The motor can be used on any bicycle and does not affect the stability of the wheel. It can be automatically thrown out of gear when the bicycle is running on a down grade. The shocks and jerks due to bad carburation have been avoided.

The Automobile Magazine

The Sanciome Petroleum-Bicycle

The bicycle, including the petroleum-supply and battery, weighs from sixty-one to sixty-six pounds, depending upon the size of the frame, and can cover twenty miles on a level road, and ascend grades of seven and eight per cent. without the aid of the pedals. No difficulty is experienced in steering.

The parts have all been constructed with a view to withstand long and hard usage, and are capable of being readily repaired whenever it may be necessary.

The Automobile
MAGAZINE
An *ILLUSTRATED* Monthly

VOL. I   NO. 3   NEW YORK DECEMBER 1899   PRICE 25 CENTS

The AUTOMOBILE MAGAZINE is published monthly by the United States Industrial Publishing Company, at 31 State Street, New York. Cable Address : Induscode, New York. Subscription price, $3.00 a year, or, in foreign countries within the Postal Union, $4.00 (gold) in advance. Advertising rates may be had on application.

Editorial Comment

THE Automobile Club of America has public sentiment very generally on its side in its endeavor to have New York's Central Park opened to the class of vehicles represented by its name—forcing the issue by the test case instituted by two of its members, who made a dash into the Park on an automobile and of course got arrested straightway by a Park policeman. The fact that the party was headed by Mr. Winslow E. Buzby, the banker, augurs well for the success of the effort, for it was Mr. Buzby who secured the opening of both Riverside Drive and Prospect Park to automobiles. There appears to be no good reason why an exception should be made of Central Park. It is doubtful, however, if anything can be accomplished merely by contending against the authority of the Park Commission to determine by rules and regulations the sort of vehicles that may be admitted to the pleasure grounds in their charge, however we may feel the injustice of a given restriction. For that matter appears to have been pretty thoroughly established under the bicycle agitation of years since, and it is just as well not to cherish any illusions. Public sentiment, however, triumphed then as it will triumph now. It would seem, however, that there is really no effective regulation applying to automobiles in general in Central Park, if, as stated, the only authority for making the arrest lay in a clause of the Park regulations dating back to 1873, prohibiting the entrance of " steam engines, fire engines and heavy drays " into the Park. At the most that would hardly apply to anything more than steam automobiles. The aggressive action taken by the Automobile Club in the matter demonstrates the great value of such an organization, not only to the interests more immediately concerned, but to the community at large, through earnest and systematic promotion of one of the greatest instrumentalities of modern progress. The triumph of the Club in this contention is inevitable, as indicated by the irresistible advance of the bicycle in face of precisely similar obstacles. Therefore we may confidently look to see in the near future the parks everywhere freely opened to the automobile, subject only to such regulations as may be for the true interest of its users as well as those of the public at large.

The special automobile ordinance, however, which was simultaneously introduced in the Municipal Council of New York, is not likely to help matters materially, and in some respects it might prove a serious hindrance. Since there is no law against automobiles, as the court had already ruled, there was no more call for a special ordinance permitting their use than for one permitting pedestrians to walk on the sidewalk. Moreover the ordinance makes this permission contingent upon the conduct of a person driving an automobile, who is enjoined to come to a full stop at request of or signal from a driver of a vehicle drawn by a horse or horses, should the latter show signs of alarm. This provision is too vague. Experience with similar restrictions in England show that it puts it into the power of persons maliciously inclined or hostile to automobiles to cause most serious and unjustifiable annoyance to drivers of the latter by feigning that their horses are alarmed. In England, for instance, the police authorities of certain villages have instituted a systematic persecution of automobiles by sending out detectives with horses purposely refractory. Another objectionable feature of the proposed measure is its limitation of speed to eight miles an hour. While low for a maximum. for in the vast and all but deserted outlying parks of the suburbs it would forbid the taking advantage of conditions where the way is unimpeded and there is nothing to justify restriction to so low a rate.

Editorial

The Next Postal Reform

There is a most useful field for the automobile in the postal service, particularly in the expediting of collections and deliveries in cities and towns. In various other ways it has proven a most valuable adjunct to postal work in Germany and France. In Paris and other European cities there are late automobile collections for outgoing steamers. A similar convenience would be of much value in this country. The value of the automobile in collections has been demonstrated by experiments in Buffalo, Detroit and elsewhere. Postmaster Dorr, of Buffalo, is enthusiastic about the automobile, which, at a dinner given recently to the Postmaster General in Chicago, he made the special subject of a most interesting speech. He declared the automobile " a machine which will contribute more to the advancement of business and the pleasure of man than can now be estimated."

As to its capacity he said: " I am impressed with the idea that the automobile can do the work of six horses at the cost of keeping one. It may not always be possible, or advisable, to so arrange the work as to realize such a percentage of advantage, but if a locomotive can do the work of a thousand horses, and an electric car can do the work of fifty horses, I am willing to stand here and say that a good automobile can do the work of six horses at the same expense as the care and keeping of one good horse usually requires." His investigations as to forms of motive power led him to the opinion that for rural free delivery the gasoline is more fit, and the electric more satisfactory in city collection and transferring.

Mr. Dorr thus gave the result of his own experiments in collection:

" Last April was when first I began seriously to think of the applicability of automobiles to department work. First I consulted the superintendents of the city delivery division of the Buffalo Post-office. Well I remember the answer of Superintendent Leib after having suggested to him that boxes could be collected at the rate of one in two minutes. His answer was: ' You might collect two or three boxes at that rate if they were close together.' Well, this same Superintendent Leib surprised himself by collecting thirty boxes in thirty minutes in a run of four miles. I expect soon to be able to collect sixty boxes in sixty minutes, running a six-mile route. I hope the department will provide opportunities for automobile advancement in all branches of the service, and if expectations are fully realized, then the pneumatic tube will be face to face with a dangerous competitor."

The Automobile Magazine

An Up-to-Date Phantom

A N institution, as a rule, demands age before it can be
expected to be the subject of romance, poetry, legends,
mysteries and strange happenings. Take the aspect of
phantoms, for instance! There are numerous accounts of
phantom ships, phantom horsemen, phantom stage-coaches, etc.,
but not until pretty recently have phantom railway trains appeared
on the scene, and we have not yet heard of any phantom wheelmen, well adapted as the silently flitting bicycle is to act a part
in ghostly apparitions. But the automobile gives new testimony
to its extraordinarily rapid development, to its phenomenal pre-
cocity for a freshly fledged creation of modern invention, by
already becoming the subject of a mystery that has been puzzling
many people. Possibly this exception of the automobile from the
rule may be due to the fact that though the invention has only
just now been made practicable in its application, its beginnings
date back for more than a century, while the idea itself has so
long been in the mind of men as to represent, perhaps, a prehistoric aspiration! However this may be, there comes from
Boston—very appropriately, since that is the stronghold of the
American Society for Psychical Research—an account of what
would seem to be a veritable Automobile Flying Dutchman!

Like the most famous of phantom ships, whose legend
inspired Wagner to compose his *Fliegender Holländer,* this
mysterious motor-carriage has a habit of turning up at the
most unexpected places, and never being found when looked for.
Actually, it seems that several lawyers, together with numerous
policemen, have been endeavoring to find this particular automobile for the past four or five months, but have never yet succeeded
in getting hold of it, although it is repeatedly seen on the suburban
roads, now in this place, now in that. It is described as of a
low rakish type, driven by a gasoline motor, with seats for four
persons, back to back, and sheltered by a light canopy. It is
commonly seen just at dusk or soon after nightfall, running at a
stealthy pace over some suburban road, and occupied by two
muffled figures. It leaves a strong trail of very unpleasant sulphurous smell behind it; the odor lingers so long in the air, of a
calm evening, that some persons who have chanced to encounter it
several times have nicknamed it " The Flying Skunk." Inquiries
have been made of all known automobile manufacturers, and
they all say that they have made no vehicle of such a description.
It is supposed that it must, therefore, have been made in some

Editorial

small shop, perhaps by two mechanics, who go out evenings to experiment with it. Some there are who attribute its origin to the Evil One.

The reason why discussion is rife is that two ladies were out driving some months ago and in one of the suburbs encountered an automobile of that description, carrying two men. Their horse showed signs of fright, and they asked the men to stop and wait until they got by. Instead of stopping—according to the statement of the ladies—the men answered impertinently to the effect that women had no business to be out driving by themselves, and they kept on their way. In consequence the horse backed and the carriage went off an embankment, a smash-up ensued and the ladies were injured. The automobile kept on and its owners, or occupants, are now wanted in court, but the law has not been able to lay hands on them. Whether the non-success of the effort forms another testimonial to the traditional reputation of the average police, or to the extraordinary efficiency of the automobile as an instrumentality for eluding pursuit, can hardly be said as yet.

Unhappily, this incident indicates that the automobile may contribute a new variety to the Road-Hog species. The most numerous variety of the species was long found among drivers of horses—as many a wheelmen will testify—with the Carriage-Hog and the Wagon-Hog as sub-varieties. But of late years the Bicycle-Hog has become exceedingly numerous, and possibly is now in a large majority, making himself as obnoxious—and even a greater source of danger—to his fellow wheelmen, as to other frequenters of the road. Probably the evolution of the Automobile-Hog is to some extent inevitable, but the sentiment of automobilists in general will co-operate with the law to make him as rare an object as possible, and by wise regulation reduce his capacity for harm.

AUTOMOBILE LICENSES

In Europe every automobile and every automibilist is licensed as a measure of precaution. The vehicle is subjected to a careful examination by the proper official; and the safety of every part is ascertained before a license is granted. The automobilist must prove that he is capable of driving a motor-carriage and that he is thoroughly familiar with the working of the mechanism. Then, and not till then, does he receive permission to drive his carriage through the public thoroughfares.

The Automobile Magazine

A similar two-fold examination might be instituted in American cities. An automobile motor cannot successfully be operated by a novice in mechanics; its construction must first be thoroughly understood before an automobilist can hope to guide his carriage through the streets with safety.

A PLEA FOR THE ELECTROMOBILE

It seems to be the fashion among automobile journals to sneer at the electric carriage, to cast slurs on its supposed incapability of making long tours, and to exaggerate its faults. The well-worn arguments of small accumulator-capacity, short runs, and necessity of recharging at frequent intervals have long outlived their usefulness. To be sure the electromobile has its faults; so has every motor-carriage. But that it is far from being the thing of shreds and patches which many of our contemporaries would have us believe, is proven by a few runs recently made in France with carriages driven by electricity.

The distance between Paris and Rouen is somewhat more than 84 miles. Nevertheless with a single charge of the accumulator-battery of his carriage, Comte de Chasseloup-Laubat covered the entire distance in seven hours and fifteen minutes—a performance which should in itself be sufficient to refute the objections made against the electromobile. At the *Criterium des voitures électriques* in Paris, the first electric-carriage race ever held, a Columbia vehicle again proved what the electromobile was capable of doing both in the way of speed and economy of power. Bouquet and Garcin, makers of the well-known B. G. S. electric carriages, made the round of all the towns in lower Normandy with a despatch and facility that caused many a skeptical Parisian *chauffeur* to open his eyes in astonishment.

With accumulator-batteries of still greater capacity than those used in the runs cited, we may hope for even better results. Let us not forget that it is but a very short time since Planté devised the first practicable method of storing up electrical energy, and that great improvements can still be made in the construction of accumulators for automobile traction. That the last word has not yet been spoken for the storage battery is demonstrated by the many patents granted each year for new forms of cells. Indeed, the French automobile press is even now commenting upon the possibilities of a new battery which, with its great capacity and comparatively small weight, will probably add new laurels to those already earned for electric traction by Chasseloup-Laubat, Bouquet, and Garcin.

324

Editorial

The electric motor, regarded as an ingenious toy a quarter of a century ago, has developed into one of the most perfect mechanical contrivances of its kind. And the storage-battery, the much-abused storage-battery, possesses many features which have ever proved attractive to those who appreciate its merits, whatever may be its faults.

The strength of the electromobile lies in its simplicity. To set the motor in operation is mere child's play. By means of a simple switch the carriage can be started, made to run at any desired speed, stopped, and propelled backwards—all without any noise, without any offensive odor, without any shock.

The electromobile has shown what it can do both within the limits of a city and in long-distance traveling. The sphere of its application can be still further broadened by the adoption of a certain standard for storage-batteries. By making cells of only one size so that all electric carriages would use like batteries, many of the present objections to the electromobile would disappear. If a certain uniformity were observed it would be possible for an automobilist to exchange his exhausted cells for new ones at almost any place, without being dependent upon the electric-light circuit of a large city. In these days when there is a general cry for simplicity in all forms of mechanical devices, it is not too much to hope that the storage-battery will be thus constructed; for the step is one which can only benefit electric traction.

Metal Wagon-Bodies for Automobiles

Despite the rapid strides made by the automobile industry in France and in other countries, the largest manufacturing-firm in Europe cannot produce more than about forty vehicles in a month. Often a vehicle must be ordered a year in advance; and the fortunate possessor of a new carriage is sometimes enabled to make a neat profit by selling his automobile to some *chauffeur* too impatient to wait until the manufacturer reaches his order. In France the cause for this unnecessary delay in filling orders is to be found in the lack of proper machinery and the difficulty of obtaining skilled labor.

In the United States the best automatic machinery in the world is made and the quality of the labor employed is above reproach. The most formidable obstacle with which American manufacturers are confronted is the great difficulty of obtaining wagon-bodies. The trucks and necessary motors are all ready; but wagon-bodies are so greatly in demand, that the American manufacturer is not much better off than his French *confrère*.

The Automobile Magazine

The question naturally arises: Is it necessary to use wooden bodies? Why not build automobiles entirely of metal? That the plan is feasible has been fully demonstrated by the partinium (an aluminum-tungsten alloy) omnibus recently made by De Dion et Bouton. Aluminum and its alloys enter considerably in automobile construction in Europe; and there is no apparent reason why their use should not be extended to the making of all-metal carriages. Besides the greater strength and durability secured by the employment of aluminium, there is also a considerable reduction in weight. The De Dion-Bouton omnibus already mentioned weighs less than 500 pounds, and nevertheless has a seating-capacity of twenty-six. The use of steel tubing for the same purpose has already been advocated in these pages.

An Expert's Opinion

Among those who have followed the development of the automobile in America it has for some time been known that Prof. Elihu Thomson has been giving much attention to the practical problems connected therewith. A few weeks ago he stated some of the conclusions which he had reached, in the course of an informal talk before a local club in the Boston seaside suburb of Swampscott, where he lives. Since Prof. Thomson is not only one of the greatest of living inventors and the chief consulting expert for the General Electric Company at its great Lynn works, but is perhaps the most eminent authority on physical science in this country, his views have a particular value for everybody interested in the movement. He pronounced the present situation a peculiar one, inasmuch as it involved an enormous demand for something not yet developed. This fact of the demand for a new invention arising so remarkably in anticipation of the supply was unlike anything else he knew of in the nineteenth century. He said that higher ideals in the matter were sought in this country than in France, where the development had been so active, but where so many people were contented to ride anything at all. The possibilities of the field were enormous; the amount of traffic awaiting the automobile was almost beyond computation, for not only would the invention displace existing wheeled vehicles very extensively, but many who do not now own a horse and carriage would demand automobiles. Its utilization in business would be great, outside of pleasure-driving.

Prof. Thomson's characterization of a good automobile was to the effect that its requirements were ease of control, ease of

Editorial

hill climbing, abundant power, ease in steering, safety, cleanliness and unobjectionable character as to noise. As to speed, he said that people would be content with fifteen miles an hour. He did not approve of the French habit of high speed. In this country the automobile-scorcher would be stopped by law, as the bicycle-scorcher has been.

The Coming Show

Under the auspices of the National Cycle Exhibition Company a cycle and automobile show will soon be held at Madison Square Garden, which will offer an excellent opportunity for gathering under our roof all the different makes of American motor-carriages. The Garden will be divided into 143 spaces on the main floor and into 81 additional spaces in the first balcony. Makers of parts used in automobile construction will be well represented.

Without in any way detracting from the advantages of a combined automobile and bicycle exhibition, we cannot help remarking that a specific motor-carriage show, without any bicycles or bicycle appliances, would better serve the interests of the horseless carriage—especially if such an exhibition were organized by the Automobile Club of America.

Combined bicycle and automobile shows have attained neither in Paris nor in London the success which has characterized the specific automobile exhibition. The number of automobile makers in the United States is limited, to be sure: but there are enough of them to insure success to an independent show.

Steam-Automobiles

It is our intention to publish in our coming issues two articles on steam-carriages, which should be of considerable interest to our readers. The first article will be written by Prof. Thurston, the Director of Sibley College, at Cornell University, one of our foremost American authorities on steam-engineering. Under the guidance of the author this article will be supplied with illustrations of the best and latest types of steam-motors for the open road, as used in this country as well as in Great Britain and on the Continent.

Our succeeding article on the same subject will be in the nature of a critique by the first French authority on steam-automobiles.

UNCLE SI'S CHRISTMAS PRESENT

1. UNCLE SI: "That's an allfired queer buggy that Rube sent me. Guess he forgot to send the shafts. I'll put some on."

2. "Get ap, Dobbin! I reckon this handle is ter steer with, er maybe it's a brake."

3. "Gee whiz! Whoa, Dobbin!"

4. "Gosh! I'm runnin' over the horse!"

5. "Will this thing ever stop?"

6. "Wall, I'll be gol derned!"

(*From the New York Sunday Journal*)

Automobile Cartoons

AUTOMOBILE TALK

"He has a great faculty for putting the cart before the horse."

"Oh, I wouldn't say that. Say he has a habit of trying to make the wheels run the motor."

The Park Commissioner's Nightmare

(Morland in La Comedie Politique)

Park Commissioner Clausen of New York said that if automobiles were permitted to run free it would soon be impossible for pedestrians to go unmolested. When he consented to enter Central Park in an automobile he was surrounded by an escort of mounted policemen, two riding in front, one on either side of the carriage, and three bringing up the rear.—*Press Despatch.*

Press Notices and Book Reviews

In a recent number of the *Zeitschrift des Vereins für Motor Sport,* Herr Berdow discusses the automobile and its effect upon transportation. After having described the first attempts made in the way of mechanical traction and after having mentioned the great speeds attained of late with motor-carriages, he remarks— perhaps not without justice—that the public cares not so much for high-speed vehicles as for automobiles durable in construction and somewhat cheaper than those now in use. Safety, not speed, is the first requirement to his mind.

In the opinion of Herr Berdow, as well as of many other auto-mobilists, the hydrocarbon-motor carriage, under present conditions, is the only automobile suitable for long-distance traveling. Owing to the great weight of its accumulators and the necessity for frequent charging, the electric carriage has a very limited field. But in large cities the difficulties attending the use of heavy storage batteries are readily overcome, by reason of the many charging-stations met with almost everywhere; here, at least, the electromobile holds its own, if the many cabs and other electric public conveyances be any proof.

Judging from the many horseless delivery wagons which are daily seen on the streets of a large city, it would seem that the automobile is rapidly gaining favor as a means of transporting merchandise. For this purpose no great speeds are necessary, elegance is a minor matter—although some of our delivery wagons compare very favorably in appearance with automobiles used only for pleasure—and the greatest attention is paid chiefly to the construction. Besides being relatively less costly than the pleasure carriage, the automobile transportation wagon is partic-ularly well adapted for connecting out-of-way villages with rail-ways.

Instead of the three or four trains per day placed at the disposal of the public on local lines and small railways, the auto-mobile would enable more trains to be run, owing to the relatively small number of passengers necessary for the success of the enterprise.

Our German contemporary states that in Germany at least seventy-seven passengers must be carried by each train to cover the running expenses of a road; and even if the cost of main-taining the rolling-stock and repair-shops only be reckoned, twenty-one passengers would still be required. An automobile line could be conducted at the same rates with ten, or with twenty passengers at the most.

Messrs. E. Bernard et Cie have just published the twelfth edition of the *Notes et Formules de l'Ingenieur, du Constructeur-Mécanicien, due Métallurgiste, et de l'Électricien,* compiled by a committee of engineers under the editorship of Charles Vigreux and Charles Milandre.

The new edition has been completely revised, enlarged, and brought up to our present standard of technical knowledge. After the fundamental laws of the various subjects discussed have been tersely and clearly stated, the theoretical formulæ derived from these laws are given, together with the method of applying them practically. In each part of the book, the results which are based upon theoretical calculations are supplemented by those obtained by practical means.

This new edition is concluded with a dictionary in three languages—French, English, and German—in which technical terms most commonly used are translated.

Our French contemporary, *La France Automobile,* has published a most interesting account of the projects of Felix Dubois, the Soudanese explorer. At the *Concours des Poids Lourds* of last year it was proposed to use automobiles in the French Soudan for the transportation of supplies to posts situated between the Senegal and Niger rivers. The project is now about to be carried out. After the very conclusive experiments made by M. Dubois, it has been decided to establish a line of automobiles between Toukoto and Bamakou, separated nearly 200 miles. M. Dubois left France on November with the intention of making all preliminary arrangements. He was accompanied by Captain Ostermann, who has had some experience in similar enterprises, and by two foremen and eight engineers, besides a staff of seven assistants.

The rolling-stock for the line will consist of fifty Dietrich trucks of 9–10 horse-powers, which have been shipped to Saint Louis, together with one million litres (220,000 gallons) of oil. Of these fifty vehicles, five can be transformed into passenger-carriages.

M. Dubois and other members of the Commission arrived at Saint Louis on the 11th of November. The trucks and accessories were transported by boat to Kayes, and thence by rail to

The Automobile Magazine

Toukoto. Here the head of the line will be located. Workshops and store-houses will be erected, so that repairs can be made without any delay. M. Dubois will take ten of his trucks to Bamakou, establishing oil-stations along the route as he proceeds.

When the line is in complete operation, convoys of automobiles will transport necessary supplies to the posts, as well as the goods of French traders. Coming back the trucks will be laden with native products, such as rubber, cotton, ivory, gum, and the like.

The most remarkble feature in the whole project is the selection of Chinamen as *chauffeurs*. It seems that besides their carelessness, which would in itself be sufficient to render them incapable of acting as automobilists, negroes have an uncontrollable passion for drinking the petroleum used in the carbureters. Assuredly a more extraordinary vice has not been heard of. The automobile will therefore serve as the direct means of an invasion of the dark continent by the yellow race.

The second edition of the *Manuel pratique du conducteur d'automobiles,* by Messrs. Pierre and Yves Guédore, has just been published.

This new edition contains descriptions of all recently introduced automobiles, and is therefore to be considered as supplementary to the first edition. The work is to be recommended to all those interested in automobiles.

"Le Littoral Sportif" is an illustrated sporting guide of the Marseilles, Toulon, Cannes, Nice, and San Remo districts. It is published by "La Côte d'Azus Sportive," in Cannes, France.

AN IDEAL AMBULANCE

An automobile ambulance has been made for St. Vincent's Hospital, New York City. It is propelled by electricity, and is intended for a model of its kind. Electric power, it is claimed, is more advantageous for propelling a vehicle where it is essential to have a very steady motion. The large pneumatic tires, it is expected, will also contribute in no small degree to the comfort of the patients.

The Automobile Index

Everything of permanent value published in the technical press of the world devoted to any branch of automobile industry will be found indexed in this department. Whenever it is possible a descriptive summary indicating the character and purpose of the leading articles of current automobile literature will be given, with the titles and dates of the publications.

Acetylene Automobiles—

Description of two acetylene gas-motor vehicles, a victoria and a truck; with two illustrations. "Scientific American," New York, November 11, 1899.

Illustrated description of a new carriage built by the Auto-Acetylene Co. "Electricity," New York, November 15, 1899.

Accumulators—

A serial article on accumulators for automobiles, by E. C. Rimington. One illustration. "The Automotor Journal," London, October 16, 1899.

Aluminum—

An article on the possibilities and uses of aluminum as applied to carriage building, by Mr. Percy W. Northey. "The Automotor Journal," London, October 16, 1899.

"Amongst American Motor Men"—

A serial article on the automobile situation in the United States, by Mr. Henry Sturmey. "The Autocar," London, November 4, 1899.

Automobile Hansom—

A New Type of French Automobile Hansom. Illustrates and describes a vehicle exhibited at the recent Exposition in Paris, which is arranged to carry four people and is mounted from the front. "Electrical World and Engineer," September 16, 1899.

Automobile Regulations—

Synopsis of automobile regulations in Belgium. "Electrical World and Engineer," New York, October 28, 1899.

Balancing of Motors—

A serial article on the balancing of motors, by Mr. H. E. Wimperis. With two illustrations. "The Automotor Journal," London, October 16, 1899.

Clubs—

The Automobile Club of France. Francis P. Mann. Gives the history of the club and the work carried out. "Electrical Review," New York, September 13, 1899.

Electric Automobiles—

Description of Belknap Company's light automobile. "The Motor Age," Chicago, October 24, 1899.

Description of an electric Stanhope built by the Kensington Bicycle Co. Illustrated. "The Cycle and Auto. Trade Journal," Philadelphia, November 1, 1899.

Description of the new type of electric cab now in use in Berlin. Two illustrations. "The Motor-Car Journal," London, November 3, 1899.

Illustrated description of an electric automobile vehicle designed by Mr. W. H. Chapman. "Electrical World and Engineer," New York, November 11, 1899.

Electric Motors—

A short description of automobile electric motors built by the Siemens & Halske Co. "Electrical World and Engineer," New York, November 4, 1899.

Hydro-carbon Automobiles—

Description of Raouval's carriage, with 12 illustrations. "La Locomotion Automobile," October 5, 1899.

333

The Automobile Magazine

Description of Underberg voiturette, with two illustrations. "The Motor - Car Journal," London, October 13, 1899.

Illustrated description of the Hugot voiturette. "Motor-Car Journal," London, October 27, 1899.

New hydro-carbon carriage of Hereford, England. Described and illustrated. "The Autocar," London, November 4, 1899.

Hydro-carbon Motor—
Illustrated description of the "Abeille" motor. "La France Automobile," Paris, October 22, 1899.

Loutzky's hydro-carbon motors, 2, 3½ and 5 H. P. With nine illustrations. "Le Chauffeur," Paris, October 25, 1899.

Illustrated description of the Crest gasoline motor. "The Motor Vehicle Review," Cleveland, October 31, 1899.

Igniters—
Description of the "Spiral Igniter," with two illustrations. "La France Automobile," Paris, October 22, 1899.

International Automobile Exposition in Berlin—
Report of the Consul General of the United States, with five illustrations. "Consular Reports."

Liquid Fuel Burner—
Description of the Musker Automatic Liquid Fuel Burner. Four illustrations. "The Automotor," London, October, 1899.

Motocycles—
Illustrated description of Eadie quadricycle. "The Motor-Car Journal," London, October 20, 1899.

Motocycles and How to Manage Them—
A serial article by A. J. Wilson, concerning use of motocycles. Three illustrations. "The Auto-Car," London, November 4, 1899.

Motor Vehicles in the Stock Market—
A long editorial putting special emphasis on the over-capitalization of motor-vehicle concerns. "Engineering News," New York, November 2, 1899.

Motor Wheel—
Description of the Walters wheel for gasoline motors. Two illustrations. "The Motor Age," Chicago, October 31, 1899.

Oil Motor—
Illustrated description of McLachlan heavy oil motor. "The Motor-Car Journal," London, October 13, 1899.

Postal Automobile—
The Loutzky automobile for postal service, described and illustrated. "Le Chauffeur," Paris, October 25, 1899.

Racing Vehicles—
Description of the new Vallée racing carriage. with five illustrations. "The Autocar," London, October 28, 1899.

Illustrated description of the Winton racing vehicle, "Scientific American," New York, October 28, 1899.

Illustrated description of the Benz racing carriage. "The Motor-Car Journal," London, November 3, 1899.

Steam Automobile—
Description of Simpson & Bodman steam lurry, with fifteen illustrations. "The Automotor Journal," London, October 16, 1899.

Steam Generator—
Description of Toward's water-tube steam generator for steam motor vehicles. with four illustrations. "The Automotor Journal," London, October 16, 1899.

Steering Gear—
Description and illustration of the Iden steering gear for automobiles. "The Motor-Car Journal," London, October 27, 1899.

Trials—
Commission's Report on the Test of Heavy Motor Vehicles, Held at Versailles, in October, 1898. Gives the conclusions of the tests made. "Engineering News," September 7, 1899.

French Trials of Electric Delivery Wagons. Illustrates the two electric wagons that stood the Versailles test. "Electrical World and Engineer," Sept. 23, 1899.

The Automobile Magazine

VOL. I   JANUARY 1900   No. 4

CONTENTS

AGENCY FOR FOREIGN SUBSCRIPTIONS :

INTERNATIONAL NEWS COMPANY

BREAMS BUILDINGS, CHANCERY LANE   STEPHAN STRASSE, NO. 18
LONDON, E. C.   LEIPSIC

Price 25 Cents a Number; $3.00 a Year
Foreign Subscription $4.00, Post-paid

A Traction Train in the Transvaal

The Automobile
MAGAZINE

VOL. I   JANUARY 1900   NO. 4

The Automobile in Traction
By Robert H. Thurston
(Director of Sibley College, Cornell University)

T HE term "automobile," in the widest sense of the word, covers every form of self-moving vehicle, every application of stored energy to the useful work of propelling wagons or carriages, whether on the common road or on the rail. It has come to signify, however, only the former of these two classes, although the essential features of construction and the vital principles of successful action are the same in both. In this restricted sense, the vehicle which first became recognized as a practical success was a "steam-carriage," and its success was complete and triumphant, from the point of view of the engineer and constructor, two generations ago. The early steam automobiles of Hancock and Gurney, of Sir Charles Dance, and of Gordon and James, in the "early thirties," were most satisfactory constructions, when the state of the mechanic arts at the time is considered, and were promising a great commercial and financial success when, a few years later, adverse legislation and direct interference, in the interests of the stage-coach proprietors and related enterprises, and the rise of the railway, brought these great pioneers to an abrupt halt.

The conditions of complete success are simplicity and power in the machinery, safety and economy in operation, and good and unobstructed roads. These conditions were fully met in the work of those early constructors. Steam in "water-tube" or "safety" boilers, at pressures of 200 and 300 pounds on the square inch, was employed, with entire success in all respects.

The Automobile Magazine

Steam-engines were made so light and so strong that no difficulty arose in their employment in motor-carriages. The combination of the boiler and engine with the carriage was an admirable illustration of ideal engineering for the time.

These carriages were in operation for months at a time, and ran over roads of, often, very considerable inclination, and through the most crowded London streets, conveying crowds of people, and without danger to the passengers or to drivers of horses met on the way. They attained speeds of 20 and 25 miles an hour, and made long journeys to various distant towns and cities. They carried thousands of passengers and traveled with them thousands of miles. In 1833, about twenty of these automobiles were regularly traversing the streets of London and its suburbs.* The hostile legislation which interrupted the wonderful development, at the time, of this promising application of invention and the mechanic arts to transportation on the highway has only recently been repealed to such extent as to permit the beginning of a revival to be made. Even now, some discrimination against automobiles still exists in the legislation of Great Britain. On the Continent, less difficulty has arisen on this score, and there has consequently been, especially in France, a more rapid and extensive introduction of the later inventions in this line than elsewhere.

The experimental work of the first half of the century, however, established these facts, if we may accept the report of a Parliamentary Commission of that time, and the testimony given before it by Farey and other great mechanics and engineers of that period: These automobiles were speedy, safe and commodious; they were light of weight and powerful as to motor; they could traverse any roads on which horses could work; they were a less costly conveyance than vehicles drawn by horses at the same speeds; they improved rather than injured the roads, and they did not, in any serious degree, frighten horses or impede common traffic.

After the automobile employed in passenger conveyance was driven off the common road, the "road-locomotive" remained the only representative of this class of automobiles. It was built as an agricultural engine mainly, and employed in hauling the plough and in towing trains of wagons about farms and from point to point in the country, and gradually came to assume considerable importance. By the year 1875 there were a number of firms on both sides the Atlantic building this class of automobile; among whom that of Aveling & Porter, still well-known in this

*History of the Growth of the Steam-Engine; p. 169.

The Automobile in Traction

Gurney's Steam Coach

trade, had by that time built up a considerable business in this direction, and supplied such machines not only to buyers in Great Britain, but in all her colonies, and some to the United States. As the pioneers of the heavy automobile for heavy traffic, these engines retain great interest, as well as because of the fact that their manufacture still continues with increasing extent of distribution.

The writer, called upon to make a formal trial of one of these road-locomotives at about that time, found that an automobile "tractor" weighing about five tons was capable of drawing a load of over ten tons, on a good road, up a gradient of 533 feet to the mile at the rate of four miles an hour, and nearly thirty tons up a rise of 225 feet to the mile. The construction was substantially that of a light steam-locomotive, as commonly built for the steam-railway, with such special modifications as were compelled by the roughness of the highway, which it must be able to traverse smoothly and safely. The principal special devices observable were its system of suspension, its arrangements for steering by pivoting the front axle, and a peculiarly ingenious device for permitting the hind wheels, while both serving as drivers, to travel independently of each other in turning corners or following a devious path. This same construction is still to be seen in nearly all the road-locomotives and traction-engines of our own time.

To-day, the automobile for traction is constructed in a great variety of forms and for a variety of special purposes, and steam, air, vapor and electric drives are employed; the latter finding extensive use in the cities, where current is easily obtained for charging its batteries; the first-named motor-fluid is used for the

The Automobile Magazine

heavier work, and petroleum vapors serve well in long-distance work of a lighter character. All are in a tentative stage in the sense that no one can say yet which, if either, will ultimately prove the most generally useful; but it would now seem extremely probable that all will continue, for a long time to come, to find employment in one or another of the many branches of automobile work looming up before the mechanical engineer. At present, it can only be said that, for heavy work, such as is the subject of our discussion, and especially for long-distance transportation, steam seems likely to retain that pre-eminence which it has acquired during the two-thirds of a century in which it has been growing up to its task and evolving satisfactory forms and proportions of mechanism. The electric motor and traction-engine is proving itself capable of doing good work wherever current can be found with certainty, of the right kind and in ample quantity whenever demanded, and the later vapor-engines are coming into use for intermediate conditions, where current is not to be obtained as wanted and where rapid motion over long routes is desired. Steam road-locomotives are in use in thousands and have been numerous, wherever extensive operations are carried on, for many years. Petroleum vapor machines are now built in hundreds, and the electro-mobile traction-engine is coming to be a familiar machine in cities, at steam-railway tunnels and in mining. The compressed-air automobile finds a place

An Automobile Omnibus

The Automobile in Traction

The Layland Lorry

where air is stored for use in accessible locations and ample volume.

A parade of fifteen steam road-locomotives, towing forty trucks, as a supply train to be shipped to South Africa, recently, at Aldershot, England, gives some suggestion of the extent to which that form of automobile has come to be employed in ordinary heavy highway transportation. A line of automobiles is reported to have just been planned for Porto Rico, between Ponce and San Juan, carrying both freight and passengers. A system of automobile traction on the Erie and other canals is another of the signs of progress; while the shipment of scores of motor-trucks to England and to France by our own manufacturers indicates that the United States will soon lead in this department of mechanical engineering, as it already does in that of electric railways and their machinery, contracts for which are now coming to our builders in million-dollar bargains.

The economy of the automobile system comes out in high relief when the working of the heavy classes of machines for business purposes is studied. The costs of maintenance and of repairs and the estimates for depreciation are large as percentages of the original costs of purchase; but in comparison with horse-power, the only comparison of interest in this connection, it is found that the fuel-account of the machine and that of the animal differ so enormously in favor of the former as to decide any question of profit quite apart from the consideration of the rapid deterioration of the horse in heavy work. The fact that street-car horses have but two to four years of profitable employment and

The Automobile Magazine

meantime depreciate fifty per cent. and more in value obviously gives sufficient evidence that the machine may not be relatively, even if absolutely, short-lived; while the continued working of locomotives on the railway for a generation, with slight diminution of efficiency, may be taken as proof that good care in management and maintenance may insure long life and a comparatively small percentage of total depreciation for the automobile.

The Judges of the Liverpool Trial of the heavy automobiles exhibited by the "Self-propelled Traffic Association" of last year report some instructive deductions from their experiences. Maintenance is estimated by them at 20 per cent. and 30 per cent., according to character of automobile and its work, and depreciation at 15 per cent. on the prime cost, or a total of not less than 35 per cent. per annum. This assumption that one must be prepared to replace the capital invested, practically, every three years, seems somewhat intimidating; yet it is found to be the fact that, even so, the gain by the introduction of the heavy automobile for the performance of work formerly entirely carried on with draught-horses is a very considerable net return on the investment; it being understood that the automobile is given ten hours' work a day—it may be worked twenty-four hours a day—and on good paved roads. With a rougher road the profit would become still greater, comparatively, though absolute costs would rapidly increase with the roughness of the track; it being further understood that the traffic is carried on over ordinary and fairly maintained highways, well within the limits of the machine. The life and efficiency of the automobile increases rapidly, however, with improvement of the roadbed. It will be interesting to compare this report with that of the Parliamentary committee of the time of Hancock and Gurney and Gordon. Its substance is as follows:*

The automobiles experimented with were all capable of substitution for horse-drawn vehicles with economy; loads up to four tons weight were transported successfully over roads thirty to forty miles in length in the neighborhood of Liverpool; the costs of transportation in the cases examined were less than the charges for railway transportation in the same district, though the committee is not certain that this apparent gain can be taken as probably general, since the severity of the work of the road-vehicle may make costs of maintenance excessive after a time.

It was found that motor-vehicles were then somewhat liable to breakdowns; but it was believed that, after some little experience, the evolution of correct proportions of parts and the use of

*Ibidem; p. 171.

The Automobile in Traction

materials precisely adapted to their peculiar requirements may be expected to reduce this risk. This risk of accident was, however, less a fault of construction of the machine than of the road, in the usual case, and the construction of good roads was, in this case, as in that of the bicycle, the prime prerequisite of successful utilization, with maximum economy and profit, of the automobile of whatever form or class. Wet, muddy and heavy roads increased costs of operation, as well as of repair and maintenance and the risks of use, of these vehicles enormously, and improvement of the roadbed similarly increases the economy and the safety of operation of the machine in very large proportion. Under the ideal conditions of city work on smooth and well-cared-for pavement, the automobile becomes at once safe and very profitable. It was found to be the fact, however, that the automobile could traverse and manœuvre on roads of any sort that horses could work on with profit or were ordinarily required to haul loads over. Control, stopping, starting, steering and reversing were as readily accomplished as with horse-drawn wagons—often more so. Hills were climbed much more readily than with horses; especially where, as should always be the case, variable motor speeds were provided for fast and light and for slow and heavy traffic. Improvements may be expected in these various directions, and especially in the reduction of the care and attention required, in operating the automobile on the part of the person in charge.

With so splendid a beginning the future of this industry can hardly be now imagined.

From the earliest days of the automobile and of invention in this department of mechanical engineering, the obstructions have been largely legislative, and the laws of Great Britain to-day, as sixty-five years ago, by restrictions on weights and speeds, greatly impede progress in this direction. The attempt to forbid the use of the automobile in Central Park is a more modern illustration of the fact that apprehension and ignorant protest may do much to retard advancement of the most promising of arts and inventions.

The proportions of the later designs of heavy steam-automobiles are given in the following table, which represents the judges' figures for the trials of the last summer. It is seen that the weight usually ranged in the neighborhood of three tons, although one approximates two tons. A "trailer" weighed two-thirds of one ton. The Thorneycroft vehicles were built by a firm in which the great torpedo-boat builder was the moving spirit:

Particulars of Steam Motor-Vehicles.

Official No. of Vehicle.	Extreme Dimensions Over All. Length.	Breadth.	Height.	Dimensions of Platform. Length.	Breadth.	Available Platform Area.	Diameter of Wheels.	Width of Tires.	Height of Floor Line of Platform. Light.	Laden.	Length of Wheel Base.	Driving Wheel Gauge.	Angle of Lock.	Capacity of Boiler to Mean Glass.	Capacity of Water Tank.	Capacity of Bunkers.	Capacity of Oil Tank.	Tare without Fuel or Water.
	ft. in.	ft. in.	ft. in.	ft. in.	ft. in.	sq. ft.	ft. in.	in.	ft. in.		ft. in	ft. in.	°	gals.	gals.	cwts.	gals.	t. c. q. lb.
1. Thorneycroft..	16 0	6 6	8 5	10 0	6 6	65	F. 2 9 R. 3 3	4⅜ 5¼	3 9½	..	8 11½	5 6½	34	19.25	114	3	..	2 19 3 19
2. Thorneycroft..	18 1	6 6	9 0	12 0½	5 7½	67.7	F. 2 10 R. 3 3	4¼ 5¼	3 9	..	9 11	5 6½	35	25.0	160	5¾	..	3 3 1 9
3. Coulthard	15 7	6 7	9 4	10 5	5 5	56.4	F. 2 9½ R. 2 11½	4 5	3 8½	..	9 6	5 8	30	15.0	*56	..	22	2 4 3 1
4. Trailer	10 7	6 2	3 9	10 5½	5 6	57.5	F. 2 8 R. 2 11	3 3	3 7	..	5 6	5 3	..	..	..	..	..	0 14 2 27
5. Leyland.......	18 2	6 5	10 1	12 6	5 8	70.8	F. 3 3 R. 3 3	4 5	3 9½	..	9 11	5 3	26	22	50	..	20	2 17 0 1
6. Clarkson	17 9	6 5	10 10	11 0	6 4	69.7	F. 3 2 R. 3 5	4½ 4½	4 0	..	9 11	5 3	31	..	..	..	27	2 19 3 20
9. Bayleys	16 5	6 6	8 7	9 9½	6 2⅛	60.8	F. 2 8½ R. 2 11	4⅛ 5¼	3 9½	..	8 5½	5 8	25	19	82	3	..	2 19 1 7

* Hot well in addition.

The Automobile in Traction

This list seems a remarkably short one, when it is considered that it is reported that there are no less than 110 builders of automobiles in Great Britain; but they commonly confine themselves to the construction of the lighter classes of carriages.*

A sample English construction of this class is seen in the accompanying illustration, the first on the list, built on the Thorneycroft system. Like the majority of these heavy automobiles, it is a steam-motor. A steel frame carries a platform of 65 square feet area, for merchandise, supported by springs, as in railway constructions, and on the forward part of the frame is mounted the motor-machinery. The boiler has 75 square feet area of heating surface, encloses 19 gallons of water, and has a steam-pressure of 175 pounds, which can be safely

A Typical Traction Autocar

raised, if desired, to 200. About 114 gallons of feed-water and 350 pounds of fuel (coke) are taken on at starting. The latter is carried in bunkers enclosing and guarding the machinery at the forward end of the frame. Compound engines, 4 + 7 x 5 inches, in general dimensions, with a maximum speed of 450 revolutions per minute, are geared down to two speeds, through an intermediate shaft and differential gear, to the axle. The power is transmitted over springs directly to the tires, and not through the spokes, as is often done by other builders.

A machine built for work in South Africa is that of Messrs. Coulthard. It is comparatively light, carries a working pres-

*It is reported that there are about 600 builders in France, 110 in England, 80 in Germany and 60 in the United States. The automobile "shows" in London and in Paris have drawn together about 200 exhibitors each. Thirty of the larger companies in the United States have a capitalization averaging over $5,000,000 each.

sure on its boiler of 200 to 225, testing to 450 pounds, per square inch. It carries an air-condenser, which produces considerable economy. Oil-fuel is employed. The engine is triple-expansion, developing 15 horse-power at 500 revolutions per minute, its cylinders measuring $2\frac{3}{4} + 4\frac{1}{8} + 6 \times 5$ inches. The transmission consists of spur gearing and friction clutches. The other vehicles differ in details, but all properly fall into one class. That they have been given practicable and practical forms and are admirable constructions may be inferred from the fact that their trials were upon a macadam and cobble-stone pavement and on inclines of from 1 in 14 to 1 in 9; the mean being 1 in 12.3. The long-distance trials were over neighboring country roads, and a distance was covered of over 35 miles. On both trials, and

Columbia Electric Omnibus

whether loaded or unloaded, it was found perfectly practicable to stop, start and manœuvre readily, both on the level road and on the steep inclines; the latter being, in fact, too steep for use of horses and loaded vehicles.

In the latest trials of vehicles of this class, heavy automobiles, weighing from 5 to 8 tons, were sent over long-distance routes, with loads of from 2 to 4 tons, at speeds of 4 to 6 miles an hour, with consumptions of fuel ranging from 125 to 200 pounds of coal or coke, and of 3 to 4 gallons of petroleum, or from 6 to 10 pounds and from one-eighth to one-fifth gallon, per mile; making the 20 miles in four to five hours, usually. One heavy auto-

The Automobile in Traction

Thorneycroft's Steam Wagon

mobile, on a speed-trial, made an average of nearly 27 miles an hour for one mile. These figures seem to the ordinary reader remarkable, and yet they are not so far different from those of the great automobilists of seventy years ago. The fact that " history repeats itself " is in this case true in a perfectly literal sense.

Another year remains before the close of the XIXth century, the century having progress in mechanical engineering as its salient characteristic, the century of the birth and marvelously rapid maturing of every system of modern industry. The life of the automobile began with the century, and a short hundred years —in fact, two widely separated decades of the century, the years 1825–35 and 1890–1900—have produced practically all that we now recognize as valuable in the introduction of self-propelled vehicles on the highway; although it is, of course, obvious that advances in detail in all other directions, in the application of the motors of various classes to a variety of purposes have had much to do with the final success of this class of inventions. The conclusions of the Parliamentary Committee of sixty-five years ago have been fully confirmed, and the contemporary automobile is known to be safer, speedier, cheaper, more sanitary, in all ways better, from the commercial and business standpoint, than horse-drawn vehicles. Many millions of dollars are being invested in the manufacture and in the purchase and use of these remarkable vehicles, and they are now evidently come to stay, as practical and economically satisfactory machines. The stage of experimentation is now largely in the past.

347

The Automobile Magazine

In use, the automobile is finding its way into rivalry with the horse on every field of competition, and in none more successfully than in the transportation of merchandise and the haulage of heavy loads. The work begun at the beginning of the century by Trevithick and developed so finely by Hancock and his competitors has never entirely ceased to progress in this department. It had become a large and steadily growing permanent business long before the revival of the automobile for passenger conveyance in either its heavier or its lighter forms.

Automobiles of every sort and fine roads will work together for a common good. The introduction of the self-propelled vehicle will insure the constant pressure of commercial interests in favor of the improvement and extension of systems of well-built roads throughout the country, and the progress of the existing movement in favor of good roads will be effective in the promotion of the introduction of the automobile of every class. The effect of the general use of the bicycle has been enormous in this direction; yet that is in large degree a toy, the minister of pleasure simply, while the automobile may be expected to become a necessity in every department of business and to exert vastly more persistent and more important influence upon legislation and public improvement. Urban traffic must, ere long, become mainly dependent upon the new system of transportation, and suburban greatly modified and improved.

An Automobile Dump Wagon

A Fast American Run

By Hiram Percy Maxim

I N returning from Atlantic City, New Jersey, to Philadelphia,
on November 18 last, Mr. Justus B. Entz and the writer,
on a Columbia electric automobile, made what is probably
the fastest long-distance run yet accomplished by an electric
vehicle. What was possible with the carriage was by no means
accomplished for reasons that will appear hereinafter. The trip
back to Philadelphia concluded a four days' outing, during which
several rather elaborate technical determinations were made. No
special arrangements had been made for making an official run
of it, as had been done on some of the previous runs, so that
for other than the start the writer's notes must be taken.

The start was made from the Windsor Hotel in Atlantic
City at precisely 11 o'clock A. M. The clerk of the hotel. Mr.
Poffenberger, kindly acted as starter and wrote in my note-book
the time, cyclometer reading and his signature. The course
from the hotel was through Illinois avenue, to Pacific avenue
to Missouri avenue, to Baltic avenue, to the Meadow Turnpike,
the sole wagon road out of Atlantic City. At 11.06.10 we
stopped one minute at the toll-gate to pay our fare of 15 cents

The Automobile Magazine

Awaiting Signal to Start

and to wait for the drawbridge to close and the resulting congested traffic to straighten out. Slow running was, of course, imperative. Out of the traffic, no easy matter to accomplish at high tide, as the water at this time actually stood over the entire highway in many places, we were able to move at a little better pace, although the fearful holes and rough bridges and four badly frightened horses caused many slow-downs, not to speak of four full stops. At the turning of the 5.71 mile mark by our cyclometer, we passed the outer toll-gate, began the ascent of about 500 feet of 4% grade, and gained the main land and wider roads. The time here was 11.25.35, making the rate of speed 13.4 miles per hour. Two minutes later we passed Pleasant-ville Post-office.

On the wide and hard road good time was possible, although there were many bad spots having deep holes and rough places, resulting several times in nearly throwing us out of the carriage.

At Absecom the Pennsylvania Railroad crosses the highway, and when we arrived we found a gravel train drawn up across the road and engaged in ballasting the roadbed. Absolutely nothing could be done but to stop and wait until they were finished, even lurid language being en-

An Unwelcome Stop

A Fast American Run

tirely lost on the Italian laborers who were engaged in the work. A photograph was taken to show the trying situation. A dead loss of twelve minutes occurred here, we being permitted to cross at 11.48.45, having stopped at 11.36.40. One minute later we passed the Absecom Post-office, and stopped to secure the signature of the most important looking of the bystanders standing outside of the Post-office, together with the time at which he wrote his signature in my note-book. We then started again, and experienced no interruptions, so that the times and distances recorded became more representative of what the carriage could do.

The first reading after leaving Absecom was the 11.42 mile mark, which was turned at 11.58.55. The average speed from the start, including all the losses of time, was, therefore, at the rate of 11.6 miles per hour. As the slow running necessary across the meadow road caused an unknown delay, the actual running time, deducting the thirteen minutes accounted for, is not computed.

From this on the miles passed are as follows:

*Miles.	Time.
11.42	11.58.55
12.57	12.02.45
13.70	12.06.40
14.86	12.10.20
16.00	12.13.35
17.15	12.16.53
19.42	12.24.22
20.55	12.27.45
21.71	12.31.32
22.85	12.35
24.00	12.38.48
25.14	12.42.12
26.28	12.45.20
27.42	12.48.43
28.57	12.52
29.71	12.55.40
30.85	12.59.10
32.00	1.03
33.14	1.07.10

The fastest mile was seen to be at the rate of 21.8 miles per hour. From the 11.42 mark to the last one recorded above the

---

* The uneven mileage readings are caused by our cyclometer being a 28" and our wheel a 32", making a multiplication of ⅞ necessary to get the actual mileages as above given.

H. P. M.

351

The Automobile Magazine

Where the Fastest Time Was Made

total distance was 21.72 miles, which had been done in the time of 1 hour 8 minutes, or at an average rate for the entire 21.72 miles of 19.2 miles per hour. The last two miles of this was through the Town of Hammonton, where several right angle turns had to be made in the center of the town, where, of course, speed was necessarily slower than could have been run, and furthermore, between the 12.24.22 reading and the 12.31.32 reading a continuous stretch of repaired road was encountered, consisting of loose stones merely thrown in the worn portions of the road.

Single mile readings were discontinued at the last one mentioned above, in order to partake of a lunch which was eaten *en route,* speaking more or less favorably for the steering qualities of the vehicle, as the driver was easily able to eat while running at speeds between fifteen and twenty-five miles an hour, where traffic was light.

At 1.31, after the lunch had been completed and preparations were being made for a continuance of the mile readings. the worst delay of the entire trip occurred. It was brought about by the endeavor to pass a horse-drawn vehicle which was going in the same direction, but

Waiting for a Ferry

352

A Fast American Run

much slower than we. The bell was rung continuously, and the occupant of the buggy, after what appeared to be the most thorough deliberation, drew out to one side of the road to let us pass. Having a clear way, we turned

Crossing the Ferry

on the power to get past, when, without any warning, the driver of the vehicle pulled suddenly directly back into the road. A ditch of considerable depth occupied the other side of the road. Between it and the horse-drawn carriage we struck the latter with most appalling results. Almost a complete capsize of the buggy resulted, we striking it almost head on and receiving no injury ourselves whatever. The occupant of the buggy, however, was, it is needless to say, startled past all comparison by suddenly finding an automobile sitting on the seat beside him. Both right buggy wheels were entirely collapsed, the reach bent and the carriage generally dishevelled. The horse, strange to relate,

Slow speed

took the incident as an entire matter of course, and had it not been for our having to unharness him, would doubtless have fallen asleep. The buggy was found to be occupied by a country Doctor wending his way home. It

353

The Automobile Magazine

was necessary to find a place where his buggy could be left, and to find some one who could be hired to lead his horse home, which was distant six miles. In time this was arranged, and the Doctor persuaded to allow us to take him home on our automobile. After rather involved explanations and exchanging of cards, the three of us packed ourselves away on the little carriage and started for Berlin, the home of the Doctor. The start was effected at 1.58, making a total loss of 27 minutes, which speaks for our diplomatic powers. The Doctor's home in Berlin was reached at 2.20, and we were composed enough to again read the cyclometer, which indicated 46.3 miles. A further delay of two minutes was occasioned in Berlin leaving the physician at his house.

Thereafter it became ill-advised to run at the full speed of the machine, as no more accidents were desired, and as we approached Philadelphia the traffic, of course, became thicker. At exactly 2.40 we passed the 50-mile mark from the start at Atlantic City, and it is interesting to note that the cyclometer, which had been used on the previous day for an official 100-mile run, at this 50-mile mark accurately matched as we passed it this time. The cyclometer was a Veeder and arranged on the right forward wheel. We passed the mark at a speed of about 12 miles per hour, and by pre-arrangements, Mr. Entz decided to shout as the front hubs passed the line, and the writer to shout as the cyclometer turned the 50-mile point. The two shouts were simultaneous.

The Gloucester Pike was passed at 2.55, the cyclometer reading 54.7. At this point we took the road to Gloucester, which for two miles was deep in sand, making even fair speed entirely out of the question. It was decided to take the Gloucester Ferry to Philadelphia. As we approached the ferry the roads naturally became very bad. many of them being the most wretched cobblestones. The Gloucester Ferry was reached at 3.26.35, the cyclometer reading 60.7. The total time, therefore, was 4 hours, 26 minutes. 35 seconds, and the speed, including all stops. 13.6 miles per hour. Deducting every stop, the total time was 3 hours 41 minutes and 35 seconds, which gave an average rate of 16.4 miles per hour. Considering the slow running necessary on the bad roads and in the congested traffic of the first and the last five miles, we may say roughly that the intermediate fifty miles of the run were covered at an average speed of between nineteen and twenty miles per hour.

As it stands, it establishes both the world's record for speed on a long distance run for an electric automobile. and also the Atlantic City-Philadelphia record for any automobile.

Motor Cycle Racing

By Al Reeves

THE racing of motorcycles may still be described as one of the sports of the future. The great success of these races in America last season is proof positive that the new sport will have a large following among those people who delight in athletic competition and speed trials, especially when great danger is apparent.

The terrific speed at which these machines whirl around the banked bicycle tracks; the quick, nervous puffing of the motors and the skill and nerve exhibited by the riders who steer and operate them, all seems very fascinating and inspiring to the spectator. Capable of railroad speed when pitted against one another, motorcycles are sure to find favor with the excitement-loving public of this country, particularly those who are interested in motor vehicles. The machines, whether tandems or otherwise, are all of the automobile order and naturally interest the people who proclaim any style of self-propelled machine wheels the vehicle of the future.

Popular as the sport has been from its introduction in this country by Henri Fournier, of Paris, the thrilling pastime of racing on these diminutive locomotives grows even more interesting as faster times are accomplished. Up to the close of 1899 the fastest mile covered by a motorcycle in America has been one

Zimmerman following Fournier's "Infernal Machine" at Asbury Park

minute and nineteen seconds. The feat was accomplished at the Garfield bicycle track in Chicago by the steam tandem which is the property of Major Taylor, who followed it on a bicycle at the same rate of speed. In view of the present early stage of construction, it is evident that, on a properly constructed track, with weather conditions favorable, a mile a minute is the inevitable record of the future.

So far motorcycles propelled by electricity have been shown to be scarcely practical. When such machines do work the speed

Steam Tandem Pacing McDuffee at 1.19 per mile

Motor Cycle Racing

Caboillet and Marks Pacing Linton
Copyrighted by E. Chickering

is wonderful, but the uncertainty of the power tends to preclude their use for competition. In fact, some of the European tracks have decided to bar electric tandems from races, since contests have been rendered uninteresting by their failure to operate. Of

Champion Elkes Paced by Petroleum Motor Tandem

course, on larger vehicles where lightness is not so essential the electric power acts very differently.

In America, as in Europe, it has been found that gasoline and petroleum motors are best suited for light vehicles. This country has always been well in the van where the use of steam was concerned, and the few steam motorcycles used here have met with fairly good success. One of them has covered the mile in one minute and nineteen seconds, the fastest ever made on a track, although a petroleum machine of American manufacture did the same distance over a straightway course near Boston in one minute and sixteen seconds.

The credit of introducing motorcycle racing in this country belongs to Henri Fournier, at one time the champion bicycle rider of France, and a motor cyclist of great daring. The manner in which he circled the little Madison Square Garden track last winter made him a great favorite with the public, and all during the racing season his work, on what he termed the "infernal machine," was greatly admired. At Asbury Park early in the spring he astonished the followers of this sport by operating his machine at the rate of one minute and a half per mile.

Although a number of impromptu motorcycle races occurred before this, the first real contest of machines in America took place at the Baltimore Coliseum in August. There were three

Petroleum Motor Quad. Steered by Caboillet, with Dudley Marks, Operator

Motor Cycle Racing

Motorcycle Paced Race at Manhattan Beach
Copyrighted by George Hare

crews on the motorcycles, and Fournier, with his mate, C. S. Henshaw, scored a victory after speeding around the saucer-shaped six-lap oval for a little more than twenty-four minutes, or an average of 1.37 for each of the fifteen miles. The race caused six sturdy athletes to face disaster and possible death over every inch of the journey. It was admitted, by men of uncommon *sang froid,* to have been the greatest cycling spectacle ever seen in America, and one fraught with a thrill of excitement that caused the four thousand spectators to follow every movement of the machines on their fearful journey. Wheel to wheel the racing tandems fairly flew around the oval, the seemingly nerve-less men gripping the handle-bars while the motors sputtered at a wild rate—sounding not unlike firecrackers on the Fourth of July. The slightest turn of the front wheel, a pebble on the track, or the puncturing of a tire by a splinter from the board surface meant death or terrible injury. At the finish but a few feet separated the racers. It is such exciting incidents that fasci-nate a large class of Americans, and for that reason the sport sprang into favor at once.

Then came motorcycle contests on all the tracks of the East as well as contests between the middle distance champion bicy-

The Automobile Magazine

clists aided by those speedy pacing machines. At Manhattan Beach, New York, in the 100 mile race, Elkes and Pierce had a hard and even struggle up to 50 miles, after which the former could not stand the strain and Pierce came out a winner. The arrival of Caboiller and Marks, the former a famous French steersman and the latter the manager of Tom Linton, tended to increase the interest in the sport, especially when the pair paced Linton to victory over Harry Elkes, the American champion.

All in all the season ended with greatly increased interest in this style of competition. The final race at Manhattan Beach resulted in a victory for J. W. Judge, of New York, and Charles Miller, the six-day champion. The latter returned from abroad in the spring with a petroleum tandem of French manufacture and great power, which proved to be faster than the American product. In consequence the pair had no trouble in winning almost all the motorcycle races of the season. A different story may be told next season, however, when American manufacturers will have had more experience in building these speedy affairs.

How motorcycle racing of the future will be conducted can only be surmised. The high-banked tracks built for cycling cannot be used with safety at faster than a 1.25 pace, and the miles which have been covered by motorcycles in faster time were at the imminent risk of the lives of those manning the machines. It is not improbable that the future will see tracks banked as high as forty degrees around which the machines can rush at railroad speed—one mile in sixty seconds being considered nothing extraordinary—so that the two-wheel vehicles will appear to stand out almost horizontally from the speed course. Special tires that must be treated with ice-water to prevent their getting too hot, and fastened to the rims by various lugs and other appliances, will doubtless be a part of the future racing machines. Riders themselves, who would be unable to breathe while traveling at such a pace without protection, may be fitted with masks and other appliances, while wind shields to cut through space will be rigged in front of these virtual flying machines. Men who steer or operate them will require nerves of steel, while possible spectators who are subject to heart trouble may be refused admittance to the tracks where such exciting contests are conducted. There is every prospect of some thrilling sport with these miniature automobiles—if we might so term them—and contests may be witnessed in which human and artificial powers will be matched.

(To be followed by a Special Article on Automobile Races in our next number)

The Gordon Bennett Automobile Cup

THE conditions regulating competitions for this cup have now been issued by the Automobile Club of France, the present custodian of the trophy. The following is an official translation of these rules:

I.—All foreign automobile clubs recognized by the Automobile Club of France are entitled to challenge and compete against the club who holds the cup.

II.—The recognized clubs are: The Automobile Club of America, the Automobile Club of Austria, the Automobile Club of Belgium, the Automobile Club of Germany, the Automobile Club of Great Britain, the Automobile Club of Switzerland, the Automobile Club of Turin. Any club not appearing on this list and desirous of challenging will have to be unanimously accepted by the above clubs. Its name would then be added to the list to judge the validity of the subsequent challenges. It is nevertheless understood that on the proposition of one of the clubs to erase the name of another club this could be done if all the other clubs were unanimous.

III.—Any qualified club wishing to challenge the club holding the cup should give notice before the 1st of January of each year by registered letter addressed to the President, giving the number of automobiles which will compete for the cup. The amount of 3,000 francs should be deposited with the club holding the cup, this amount to be returned if one of the automobiles engaged is present at the start. The President of the Automobile Club of France should also be informed by a registered letter even if this club were not to be one of the competitors.

IV.—The cup can be competed for every year between May 15 and August 15. If two or more clubs of the same country were accepted in the list of challengers, it is understood that this country could only be represented by three automobiles at the most, and in such case the clubs of the said country would have to choose these automobiles. In case they could not agree, the automobiles would be chosen in order of entry. The exact date of the race to be fixed of a common accord between the clubs interested.

V.—Each club can be represented by one, two, or three automobiles at will, but the fact that only one or two automobiles of a club should start cannot deter the other clubs from their right to have three cars to compete.

VI.—In case the club holding the cup should receive in time the challenge of several clubs, it could choose to hold one race only, in which each club would be represented by three automobiles at the most.

VII.—The automobiles qualified for these races are those coming under the description given in the Racing Rules of the Automobile Club of France, published in 1899, as follows:

The automobile must weigh more than 400 kilos. and carry at least two passengers, seated side by side, of an average weight of 70 kilos. each, it being understood that in case the average weight should not reach 76 kilos. the deficiency would be made up by ballast.

The 400 kilos. are counted for the weight of the automobile when empty, this being without passengers or supplies (such as coal, petroleum, water, accumulators), or tools, spare fittings, luggage, clothing, or provisions (Chap. I., Rule 9).

VIII.—The automobiles must be constructed entirely and in every part in the respective countries of the competitors.

IX.—The automobiles must be constructed by members of the competing clubs and the two seats occupied during the whole time of the race.

X.—Each automobile competing to send one delegate to form the Committee, Mr. Gordon Bennett to be member of this Committee and also the Automobile Club of France, even when not taking part in the race, to be represented by one delegate. The delegate to name a President, chosen outside of their own body, who, in case of the votes being equally divided, would have the casting vote. If they could not agree on the choice of a President, he would be named by Mr. Gordon Bennett, or, in his absence, by the President of the Automobile Club of France. The Committee to name, not necessarily from among themselves, a starter, a judge, and the timekeepers. The Committee to see that the rules are strictly adhered to and judge all the incidents which might take place.

XI.—The race, which would be a " go-as-you-please " race, without any stipulated stoppage, to take place on the road, and the distance to be from 550 kiloms. minimum to 650 kiloms. maximum. This distance to be chosen between two towns or apportioned into several outward and homeward journeys, but in the latter case the minimum distance of each part not to be less than 150 kiloms.

The Gordon Bennett Automobile Cup

XII.—The race to take place in the country of the club holding the cup, with option for this club to hold it in France.

XIII.—The start to be given at the same time to all the automobiles, which would be placed in the following order: In front one of the automobiles belonging to the club holding the cup, then one automobile of each of the clubs taking part in the race, beginning by the club whose challenge was first received; after these automobiles, the second automobile of the holders, followed again in the same order by the second automobiles of the other clubs, and again after these the third automobiles in the same order.

XIV.—The automobile passing the post first to be declared the winner, entitling its club to the cup, even if it were the only automobile of this club to cover the full distance.

XV.—In case of a dead-heat between the club holder of the cup and one of the challengers the former would keep the cup.

XVI.—In case of a dead-heat between two of the challengers, these two clubs would have to race again together under the present rules before the expiration of two months from the date of the first race, it being understood that in such case the race could be held after the period fixed under Rule III.   In case the two clubs could not agree on the choice of a road for the race, a draw would decide.   If one of the clubs should refuse to take part in the second race, the other club would, *ipso facto,* take the cup.

XVII.—The cup to be left with the previous holder until the two clubs who have made the dead-heat meet again.

XVIII.—In the event of one of the clubs who had challenged being alone represented at the start, one of his automobiles would have to cover the distance stipulated in the maximum time of 24 hours, failing which the cup would be kept by the club challenged.

XIX.—It is understood that no club can ever become owner of the cup, but only hold it on the conditions fixed by the present rules.

XX.—Should the club holding the cup become extinct, the cup to be returned to Mr. Gordon Bennett, or in default to the Automobile Club of France.

XXI.—All races for the cup, either taking place in France or in any other country, to be run under the racing rules of the Automobile Club of France, published in 1899.

XXII.—Any club becoming holder of the cup, or challenging for it, undertakes to abide absolutely by the present rules and also by the rules of the Automobile Club of France on road racing, published in 1899.

Gallery of American Automobiles

Riker Electric Cabriolet

Riker Hotel Bus
364

Gallery of American Automobiles

Graut Bros.' Hydrocarbon Trap

Electric Vehicle Company Coupé

A Simple Steering Device

A PATENT on a new steering connection has just been issued to R. W. Jamieson, of Rochester. The object of this device is to cause the individual steering wheels to turn about their vertical axes in the centre of the wheels, and to overcome the jiggering action of the steering lever, which is a result with other forms of steering connections.

When the wheel encounters an obstruction, head-on, the tendency of the short stud axle is to turn about its connection, but this action in this device is entirely overcome, as the arc-shaped bearing members act as a locking device, and entirely prevents the wheels from turning out of their natural course, and overcomes the whipping of the lever from the hand, when rough obstructions are encountered.

This device has been thoroughly tested on a steam automobile, a photograph of which is herewith reproduced.

The Automobile Section at the Paris Exposition

M. JEANTAUD recently laid before the proper committee of the Paris Exhibition of 1900 his elaborated programme for a series of competitions, races and *fêtes* to be organized this year in connection with the automobile section of the Exhibition. This section will be held at Vincennes, as the space available at the Champ de Mars would be quite inadequate for the requirements of the new industry. The programme, including a budget of estimated expenditure and receipts, was unanimously adopted by the committee, whose official report it thus became. It was then read to the members of the Committee of the Automobile Club, who decided to accept the organization of the programme as provided for in the report, but only on condition that the directorate of the Exhibition do not modify in any way the figures set out in the budget, and that the credit of £5,000, as required by M. Jeantaud, is duly made. The programme, as will be seen, comprises competitions for all classes of automobiles, and should prove one of the most attractive features of next year's exhibition. Briefly summarized, it provides for the following events: In the month of May—A competition for touring cars, consisting of daily trials of 150 kilomètres during five days. The speed will not be permitted to exceed twenty kilomètres per hour in towns and villages and thirty kilomètres per hour in the open country. The points governing the trials will be: (1) Consumption of fuel; (2) action of motor; (3) comfort of vehicle; (4) facility of steering. The competitors will be divided into four categories: (1) Cars of two seats, weighing more than 400 kilos; (2) cars of four seats; (3) cars of six seats; (4) cars of more than six seats.

In the month of June—A competition for cabs and delivery vehicles carrying loads up to 1,200 kilos. Five daily trials during one week, the distance per day being 60 kilomètres in Paris.

In the month of July—A race for all these vehicles, divided into three categories, which are defined in the Automobile Club's racing rules of 1899. During one week, Thursday excepted, five races of from 300 to 400 kilomètres each will be run. Starting from the enclosure at Vincennes the vehicles will promenade to Joinville, from which place they will commence to race. On the return they will be timed at Joinville, and will then proceed to Vincennes and promenade twice round the Daumesnil track. Particulars of their movements during the races will be exhibited at the track by means of semaphores, etc.

The Automobile Magazine

In the month of August—A competition for voiturettes not weighing more than 400 kilos, and carrying two persons either side by side or tandem. The same programme as that for the touring cars.

In the month of September—A competition for light delivery vehicles carrying a maximum load of 100 kilos. The same programme as that for the cabs.

In the month of October—A competition for heavy vehicles coming under the following categories: (1) Vehicles for the transport of passengers; (2) vehicles carrying merchandise weighing more than one ton; (3) vehicles carrying a minimum load of 1,250 kilos. The trials will consist of five daily runs of 50 kilomètres each, starting from and returning to Vincennes.

In addition to these competitions there will take place on the track surrounding the lake Daumesnil, in the Bois de Vincennes, a series of sixteen automobile *fêtes*, comprising races and gymkhana events. Competitions for decorated cars will also be held, and toward the end of October all the competitors in the various events will process from the Exhibition to the Automobile Club's villa in the Bois de Boulogne, returning to Vincennes the same day. This should prove to be a most interesting sight, as every type of touring and racing cars, cabs, voiturettes, cycles and transport automobiles will be represented.

Every provision will be made to enable exhibitors to demonstrate their vehicles, and a special track will be constructed for the running of those cars unprovided with pneumatic or rubber tires.

The estimate of expenditure as provided for in Monsieur Jeantaud's budget is about $40,000, while the receipts are calculated at about $15,000, showing a deficit in round figures of $25,000. These figures, of course, only relate to the programme of *fêtes* and races, and do not refer to the Exhibition itself.

The opinion generally expressed is that the report is a truly excellent one, and should tend to make the automobile section of the Exhibition one of the most interesting features of next year's show. The attractiveness of the programme has gone far to dispel any fear which may have been entertained by some automobile manufacturers that the distance of Vincennes from the site of the Exhibition proper would be detrimental to the attendance of visitors. The opportunity afforded of almost daily witnessing the speed of the fleetest vehicles in their respective classes will assuredly be welcomed by next year's visitors to Paris, and once at Vincennes they will give plenty of attention to the manufacturers' latest productions.

The Automobile Club of Great Britain

By a Founder Member

UR classical scholars were righteously indignant when it was decided to call it the " Automobile " Club. " Why," said they, " should we adopt this wretched mongrel word, a horrible mixture of French and Latin? Is it because the French have been guilty of burdening the industry with such an abomination that we should perpetuate it?" But there were

The Automobile Magazine

others, and those others were in the majority, who said: " To the winds with derivations! Let us bear the same name as our big brother ' the Automobile Club de France! ' "— and we do.

At the end of 1897 a Mr. Frederick Simms suggested the formation of the Club and guaranteed its preliminary expenses. He is now one of the Vice-Chairmen.

The Hon. Evelyn Ellis, who was the first man to introduce a motor carriage into England, became the other Vice-Chairman, and Mr. Roger W. Wallace, Queen's Counsel, the Chairman. The Presidency remains unfilled, for the Club has a hope that some day the first gentleman in Europe will be pleased to give it his patronage. A Committee was formed of noblemen and gentlemen who were interested in the movement, and of representa-

The Club Office

tives of the chief automobile manufacturers. Rules were framed, the guiding principle of which was that the Club should not promote the interests of or be connected with any particular firm, company or commercial undertaking, but that its purpose should be the general advancement of the motor cause.

It was difficult at first to persuade people that the Club was pure, for there was in existence a so-called " Motor Car Club," which was, as the " Automotor " neatly put it, " a commercial adjunct " to a group of company promoters. Gradually, however, the government authorities began to recognize that the Club was not connected with finance. There was in existence previous to the formation of the Club an excellent institution bearing the unfortunate name of the Self Propelled Traffic

The Automobile Club of Great Britain

Association, presided over by Sir David Salomans, Bart., a gentleman who had been largely instrumental in the passing of the Light Locomotives Act, which gave motorists their rights. They did much useful work by holding trials of the heavier motor vehicles intended for commercial transport. In July, 1898, this association amalgamated with the new club.

The history of the first year of the Club's existence may be of interest to American automobilists, since their Club is only recently formed.

The year 1898 was marked by continuous activity. In the winter season there were discussions and lectures, including the following subjects:

The best means of promoting the utility of the Automobile Club.

The Billiard Room

The bearing of past invention on future motor car design.

Motor car machinery.

Motor touring at home and abroad.

The practical working of the Benz Motor.

In the summer there were club runs and tours:

At Easter there was a tour by Guilford, Winchester, Chichester, Littlehampton, Worthing, Brighton, Tunbridge Wells and Sevenoaks, a distance of 221 miles.

The Whitsuntide Tour was by Hertford, Cambridge, Newmarket, Norwich, Cromer, Ipswich, Colchester, Chelmsford. The tour occupied five days and comprised 330 miles. And again in the autumn there was a third tour by Maidstone, Rye, Hastings, Eastbourne, Tunbridge Wells and London.

The Automobile Magazine

There were frequent one-day runs. The Club had special enclosures at Henley Regatta, at Sandown Park Race Course, and elsewhere. There were garden and houseboat parties given in its honor. These Club runs and tours were very delightful. The main roads of England are crowded with historical and romantic associations and were explored by many members for the first time by means of the tours. Members brought their wives, and very festive, indeed, were the Club dinners at country hotels where sixty or more sat down to dinner of an evening and recounted the strange experiences of these delightful days. Episodes occurred then, when all were learners, which can never occur again in such number and variety, and they are the cause of many a hearty laugh now. How well the afternoon is remem-

Lounging Room

bered when a popular member of the Club was outside the Club door for two hours trying to start his Benz, and after pulling his machine almost to pieces, it was found that the cause of his difficulty was that his *oil-can was touching an electric wire*, hence, he had no spark in his cylinder!

At the beginning of 1898 there were 163 members, at the end there were 380 (an increase of 217) ; in November, 1899, there are 541, an increase of 161 since the 1st of January last.

In order to meet the extraordinary expenditure connected with the organization of the Club a Guarantee Fund was formed, to which members generously subscribed.

In the winter house dinners are held monthly. Recently the Second Annual Dinner has been held. Nearly two hundred

The Automobile Club of Great Britain

guests sat down, and among the speakers were the Hon. G. Shaw Lefevre, M. P.; the Hon. John Scott Montagu, M. P.; Lord Ampthill; Major Holden, and others.

This year, 1899, the Club has undertaken much important work. In addition to its tours it held, in June, official trials of vehicles on a course of 50 miles and on a steep up-gradient. The result of these trials has recently been published in an interesting volume.

In June the Club held an exhibition of motor vehicles, which lasted a week, and undoubtedly gave great stimulus to the industry. It is now holding a series of 100-mile trials, and is organizing another exhibition and a 1,000-mile trial from London to Edinburgh and back, with one-day exhibitions at 9 big provincial cities during the course of the trial.

There is a Club journal issued to members free once a week.

A glance at a programme recently issued by the Committee of the work it has in hand, will perhaps give the best idea of what the British Club is about.

The Committee have now under consideration the following matters. Special Committees will meet and press forward with these matters.

1. Recommendations to the Local Government Board as to amendments in the Regulations of the Board affecting Light Locomotives on Highways.
2. The formation of a Society for protection against vexatious prosecutions and vexatious actions at law.
3. The organization of a race for touring and racing carriages in France next spring.
4. The organization of the Second Automobile Club Show, to be held from or about 24th March to 7th April next.
5. The framing of Rules affecting Automobile Racing.
   NOTE.—At the last meeting between the Racing Rules Committee of the Club and the representatives of the National Cyclists' Union, highly satisfactory advance was made toward a settlement of this matter.
6. The representation of British manufacturers at the *Paris Exhibition of 1900, and at other Exhibitions.

---

* The Automobile Club of Great Britain has, at the request of the Royal Commission for the British Section of the Paris Exhibition of 1900, agreed to co-operate and act as the Advisory Body in respect to the British Automobile Section.

The Automobile Magazine

7. Arrangements for the accommodation of Members of the Club who may be visiting the Paris Exhibition of 1900. It is proposed to rent a small villa near Paris.
8. The compilation of an accurate list of Motor-spirit Stores, Repairing Firms, etc.
9. The compilation of a small Pocket Book giving useful information of a non-technical nature.
10. Arrangements whereby manufacturers or Members may, at any time, submit motor vehicles to trial, and obtain a Club certificate as to their capabilities.
11. The formation of Branches of the Club in Scotland, in Ireland, and the provinces.

The comfortable premises occupied by the Club are at Whitehall Court, London, but members are looking forward to having in London, at no distant date, a house nearly if not quite as magnificent as the palatial abode of the Automobile Club de France.

The qualification for membership is not so very strict, as far as social status is concerned. Speaking broadly, cycle agents and professional cyclists would not be admitted as members, but automobile manufacturers and the better class of authorized agents of automobile manufacturers can become members of the Club, On the other hand, there is a sporting element consisting of many well-known members of the Royal Yacht Squadron, Masters of Fox-hounds and other sporting men, well known shots, etc., who have become enamored of automobilism. The Duke of Newcastle, the Marquis of Anglesey, the Earl of Carnarvon, the Earl of Shrewsbury and Talbot, Earl Russell, Lord Robert Cecil (son of the Marquis of Salisbury), General Sir Arthur Ellis (for years an equerry to H. R. H. the Prince of Wales), Lord Suffield (Master of the Horse to the Prince of Wales), and other noblemen, are members of the Club. The Club is recognized by and has reciprocal relations with the Automobile Clubs of France, Belgium, Switzerland and Austria, and hopes soon to have similar relations with the Automobile Clubs of Germany, Italy, and last, but not least, the Automobile Club of America.

Messrs. Hugo de Bathe and Moffat in Racing Rig

The New Sport Abroad

By Our Own Correspondent

THE Automobile Club of France is giving much attention to the arrangements for the contest for the International Challenge Cup to be held next year, some time between May 15 and August 15. This is to be an annual event, and limited to recognized clubs. At present there are eight: those of France, Great Britain, the United States, Germany, Austria, Belgium, Switzerland and Turin. The Belgian club is the first to send notice of its intention to enter the contest. The course will have a length, according to the final draft of the rules recently made, of between 544 and 640 kilometers by road, or 340 to 400 miles. The first contest will naturally be held in this country, but after that the defending club will have the option of deciding whether it shall be held in its own country or in France, which, with its universally good roads, offers the most favorable conditions for such a race. Should the cup go to Germany, Great Britain, or Italy, a good course could easily be found in those countries. I fancy, however, that it would be difficult at present to find in the United States 340 consecutive miles of good roads adapted to automobile purposes, although I should say that it will not be long before such a route could easily be mapped out in

The Automobile Magazine

either of your two States of Massachusetts or New Jersey, where great progress has been made in road-building, I am informed. So, should the cup go to your country at the start, as it will be quite likely to, if we may judge from the American record in yachting, bicycling and other matters, I suppose the contest would be held in France the next year—unless, with the supreme audacity of your countrymen in carrying out huge projects, it should be determined to construct a system of perfect highways expressly for the purpose! Should the Belgian club be the winner it could also hold the race in its own country, for little Belgium is full of good roads, though the course would have to be pretty labyrinthian to keep it within the national confines! But fancy Switzerland the winner and holding the contest among the Alps! The roads are undeniably good in Switzerland, but with the curves and gradients there obtaining it would be a contest of skill rather than speed, and the hazard would not be slight.

According to the rules each club may enter three vehicles. The net weight must not be more than 880 pounds—that is, without reckoning the weight of accumulators, fuel, water, petroleum, or luggage of any kind. Each vehicle must be driven by a member of the club that enters it and must also carry another person. The start will be given simultaneously to all the contestants. The cup will go to the club whose colors are carried by the winning vehicle, but will never become the permanent property of the winner. Challenges must be sent to the defending club before January in each year, and a deposit of $700 made, to be refunded at the start. All participating clubs will share equally the expenses of each contest.

A sign of the times is the action of the swell bicycle club of Paris, the Rallye-Velo, in taking up the automobile and building an automobile stable in connection with its club-house in the Rue de Chatres. This example has been followed by the Bicycle Club of Vienna. The enormous growth of automobile interests here in Paris is manifest in the fact that the Automobile Club proposes to celebrate the admission of its two thousandth member with a monster festival. As the membership is now 1,967 the affair may be looked for in the near future.

The growth has, indeed, become so great, that at last there is serious talk of restrictive measures. It is notable that, while restriction is attempted at the outset in your American cities, here the agitation has not arisen until there is something to restrict. Reckless speed, noise, and the undue crowding of the parks are all subjects of complaint, and the municipal council has taken the matter in hand. A committee of the council, appointed to con-

The New Sport Abroad

sider the subject, finds that the main purpose of the parks—that of providing the public with opportunities to seek quiet and escape from the confusion of the city amidst pleasant rural scenes, is interfered with by the manner in which some automobiles go tearing through the Bois de Boulogne and other pleasure-grounds, has reported in favor of restricting this use in various ways. It is therefore proposed to confine automobiles to the regular carriage roads in the Bois and in Vincennes, excluding them rigorously from the Pré Catelan, which is a popular resort for children, and from all the cross allées and wood paths, while on the Allée de Longchamps they would not be allowed between one and seven o'clock in the afternoon. Moreover, speed in any part of the city must not exceed 15 kilometers an hour, and in crowded or narrow thoroughfares it must not exceed the walking-rate. There has, indeed, been much recklessness of late, and accidents have grown too numerous. The best automobilists admit this, and agree that there is a necessity for judicious restriction. The Automobile Club proposes to lend its influence toward bringing to an end the abuses that have sprung up, though it contends that the restriction of the speed limit to 15 kilometers is too low for the outlying parks.

The noise of the exhaust made by many motor-cycles is another source of complaint. This is something that can easily be remedied, and would not exist to-day except for carelessness.

The Automobile Club has honored M. Meyan with its grand medal of bronze for organizing the Tour de France. M. René de Knyff has sold the voiture with which he won the Tour de France for the enormous sum of 66,000 francs. This indicates that a famous winning automobile has a value equivalent to that of a famous winning horse!

Imperial Russia was in evidence for the automobile on a recent morning in the Bois de Boulogne, when a landau containing the Grand Duke Alexis, Prince Sergius Galitzine, and MM. Zouroff, Potemkin and Heilmann made its appearance in tow of a Heilmann motor. The entire imperial family, with the exception of the Czar, is said to be devoted to the automobile. No wonder President Loubet prefers Count Pototsky's automobile to the erstwhile gala coach of the French Chief Magistrate. Speaking of Russia, it is of interest to state that an automobile omnibus route is to be established in the Caucasus, between Choucha and Khan Bagin, M. Charron having just sent to Tiflis a 12 horsepower Panhard-Levassor vehicle for the purpose. The Sultan of Morocco has recently ordered an automobile of the same make, with a motor of 7½ horse-power.

The Automobile Magazine

The number of automobiles in Paris on November 1, according to the official registration, was 1,795. So rapid is the increase that the beginning of the new year will probably see the number increased to at least 2,000.

A novel event that took place here late in October was the first automobile wedding—that of M. Richard Dupont and Mlle. Jeanne Constant at the church of Saint-Germain l'Auxerroix. The entire party appeared at the church portals in 20 brand-new automobiles, the bridal carriage elaborately decorated with chrysanthemums and ferns.

A notable automobile tour is that recently made by the Countess Corsini Sforza, from Paris to Rome and Naples. The route lay through Morvan, Lyons, Nimes and Marseilles; thence over the Corniche to Genoa, and through Florence and Perugia. The vehicle was an 8 horse-power Peugeot.

I hear from Nice that your compatriots, Mr. and Mrs. William K. Vanderbilt, Jr., have arrived there in their automobile and are enjoying delightful trips along the Riviera. Probably by the time these lines appear they will be in Egypt enjoying the novelty of automobiling about the Pyramids.

You will soon have a visit from our eminent inventor M. Serpollet, who has made the greatest advances in bringing the steam automobile to the present high degree of practicability. The Gardner-Serpollet is the name of the system which includes M. Serpollet's inventions. Your countryman, Mr. Frank Gardner, is the senior member of the firm. While not so much has been said about steam as about other forms of motive-power in automobile work, it has really made great strides here and is particularly in demand for omnibus purposes, wagons, and other forms of the heavier traction vehicles. The extent of the demand for such vehicles may be inferred from the fact that the Gardner-Serpollet house has under consideration two contracts for 500 vehicles each—one from the post-office, for mail-wagons to be used here in Paris, and one from Bordeaux. Good methods of filling large orders do not appear to be understood in this country, and our mechanics find it difficult to adapt themselves to it. Mr. Gardner says that his factory can turn out only 150 vehicles, but by having the parts manufactured in the United States the same number of men can put out 500 in the same time. A very large business will naturally be done between France and the United States in this way, and your automobile export trade will inevitably enormously promote American industry. Your light, graceful, and beautifully made American work is already in high favor here. The absence of demand in

The New Sport Abroad

your country, when your first concerns to undertake automobile manufacturing were engaged in their pioneer work, led the makers to seek a market here, where even then the demand far exceeded the supply. That made us acquainted with the quality of American work and it has come to be well appreciated. So, although your manufacturers are now overwhelmed with home

Winter Auto Costume

orders, they would do well to cultivate the foreign market that is eager for their products.

You are doubtless aware that the first prize, the gold medal, at the Berlin Automobile Exhibition, was awarded to the house of de Dietrich & Co., for its remarkably rich exhibit, which included a wide range of types, such as the " Spider," " Petit Duc," phaeton, omnibus, hunting, touring and racing vehicles.

The Automobile Magazine

A Berlin Texameter Droschke

Since the firm belongs in Niederbronn, Alsace, Frenchmen feel a sort of satisfaction in the result. The head of the firm, Baron de Dietrich, has won many contests in his own automobile. Two of the Dietrich vehicles, a phaeton of six and a break of nine horse-power, made the trip from Paris to St. Petersburg, by way of Berlin, without accident, and three ladies, Mme. Thevin, Mme. Houry, and Mlle. Houry, were of the party. In another Dietrich vehicle Dr. von Stern and his wife, of Vienna, made the journey from that capital to Paris and back to Salzburg, and a second journey from St. Gilgen, near Salzburg, to Berlin, to visit the automobile exhibition, and return. Last summer, on August 6, Dr. von Stern and his wife went in the same vehicle over the highest Alpine pass, the Stilfserjoch, 2,700 metres above the sea.

In Berlin, Herr Thien, who has long been prominent in local transportation interests, has begun a new motor-cab service. The pleasing German name of these vehicles is "automobiletexame-terdroschken." The pictures that I have seen show them to be remarkably handsome and graceful carriages, despite their cumbersome name. This name, preposterous as it may sound, does not come up to the Flemish word for automobile, which reads, "Snelpaardelooszoonderspoorwegpetroolrijluig." In summer they have a half-top that is thrown back in the usual fashion, while in winter the carriage is entirely closed in as a coupé. It is driven by a "Phœnix" Daimler motor. It is adjusted to four speeds. In the city the rate is limited by police-regulation to 14 kilometers an hour, or about 8¾ miles. Even at the highest speed it can instantly be brought to a stop. The

The New Sport Abroad

designation of the first of these cabs to begin service, "No. 8941," indicates that in Berlin but little short of 9,000 cabs drawn by horses will be replaced by motor-vehicles in the course of a few years. And, as motor-cabs will be more extensively patronized than those with animal-traction, in a degree commensurate with the popularity of electric tram-cars as compared with horse-cars in your country, by the time the substitution is completed there will doubtless be at least 15,000 motor-cabs, or droschkes, in the great German capital, beside the many motor-omnibuses. And when we consider the similar demands in Paris, London, Vienna, Rome, Madrid, and all the other cities of Europe, great and small, it may be seen that there is no likelihood of any dull times in the automobile industry during the present generation.

A curious use of the automobile in Berlin is for advertising purposes—motor-cycles with voitures running about the streets with placards and also with advertising legends on the backs of the driver and his companions.

The Vienna Bicycle Club, the oldest and largest in the city, has established an automobile section.

In Graz, the capital of the Austrian province of Steirmark, the Steirmärkischer Automobil-Club has been organized. Graz may be called the Austrian Dresden, having a large number of

German Advertising Motor

381

well-to-do persons of leisure, from all parts of Europe, among its residents, and the automobile naturally finds a great popularity there.

The following from a letter from Lady Jeune to the London *Telegraph,* after an automobile trip to Weston-super-Mare with her husband, Sir Francis Jeune, gives an English woman's views of the charms of motor-touring, as they call it in her country:

Lady Jeune, who, with Sir Francis Jeune, paid Weston-super-Mare a visit in a motor-car about three weeks ago, contributed an interesting article on Motor Touring to last Saturday's *Daily Telegraph.* Her ladyship says:

" There is no more delightful mode of seeing England than driving through it on a motor-car. The speed at which it goes enables one to visit a wider range of country than would be possible in a carriage. There is no road too hilly or too bad for it to undertake, and the speed can be regulated to whatever pace suits. Eighty miles a day is not too hard a task to set one's motor, and, unlike a horse, it goes better as the distance increases and the day wears apace, and it comes in almost discontented at being stopped at night-fall, for it works best in the cool evening air. There is no limit to the willingness of a motor. It never stops, it never flags, and its noise is a sort of song of anxiety, uncertainty and emulation. To the driver and those who appreciate it, it has almost the intelligence of a creature, while no higher tribute can be paid to its virtues and qualifications than the fact that by those who know it best it is always spoken of as ' she.'

" Automobilism is doing a great deal in the direction of increased health and enjoyment for us, but a hundred years hence, how shall we get about? We who are watching the birth and infancy of improved locomotion shall all then be sleeping after our long journeys here are over. But while we may, let us develop and popularize as much as we can any new power which enables the tired and hard-worked ones of the earth to get easily and quickly away into communion with Nature and the peace she bestows, which will always be the greatest and most merciful of all her gifts." Certainly Lady Jeune's enthusiasm is great.

On November 15 the London Automobile Club held its fourth annual meet with a run to Brighton. About 50 vehicles participated. The start was from the Hotel Metropole and the route by way of Reigate, where luncheon was enjoyed at the White Hart Hotel. The procession through the Brighton streets was an imposing one. A banquet in the evening closed the ceremonies, the return the next day being informal.

An Artist's Appeal

By Frances S. Carlin

EVERY man and woman (particularly woman) who has ever sketched out of doors knows what weary work it is just to transport oneself and one's paraphernalia to the "point of view." The distance is often great and the very simplest kit heavy, and it really seems as if it were spelling Art with too big an A to attempt anything except from one's veranda or upper windows. But the prettiest little streams, the most picturesque old mills or inviting stretches of beach are always " Waal! about two mile further on," any other measure of distance being incomprehensible to the bucolic mind. And then the difficulties and discomforts so often incidental to sketching. The dusty road or wet grass to sit on. The havoc that a high wind will make of easel and umbrella; to say nothing of a wet canvas. The utter impossibility, sometimes, of finding even a crevice for the umbrella stick, amongst the rocks. The " no thoroughfare " that a willful cow can make of a shady lane. All these things are most trying to a would-be landscape painter.

The writer will never forget the two weary hours that devotion to art and a particularly pretty " bit " compelled her to spend, balanced on one foot on the bottom rail of an old fence. Paint-box and sketching block in one hand, pencil and brush in the other; her water pail hung on a convenient but rusty fragment of nail. And mosquitoes, with the unerring instinct that seizes an opportunity, holding high carnival. Or another time, when parasol, artist and camp stool had to be strapped together so that the masterpiece might be finished before the incoming tide swamped everything.

Some artists, Monet, for instance, more luxurious than their fellows, have a boy and a push-cart. We, in modest imitation, raked out of the garret of our French *pension* an old perambulator that served us many a good turn. Last summer we essayed a wheelbarrow, a most unruly steed, that balked up a hill and bolted down. Some kind friends not in the profession suggest a horse and carriage. Well! in the first place, artists, as a rule, do not have horses and carriages at their command. And if they do, horses have to be unhitched and fed, a great waste of time with the light changing every minute. So, then, why not have an automobile, especially constructed for an artist's needs. Low and open in front like an invalid's chair, but with the seat wide enough for two for an occasional friendly spin. A parasol suspended from a hook, like a baby carriage. An easel adjustable to the brake. Places under the seat for color box and canvases.

And the cost low enough to bring the trap within the means of the most impecunious artist. We do have so few pennies left after paying our landlords and frame makers.

As speed would be a secondary matter a low rate of horse-power would be sufficient, and the carriage could be very light and simple in construction, so that it could be trundled home in case of a breakdown. As a rule, artists are not practical mechanics, as are many of the present owners of automobiles. The motor of this artist's delight could be gasoline or kerosene, the smell of which is only offensive to the people behind in the road, and not any worse than some other team's dust. As the floor of the carriage would be only a foot or so from the ground an explosion or a runaway need not necessarily be fatal, but simply a little hastening of the exit, which brings me to the end of my description.

And now, good people of the automobile companies and Trusts, will you not make for us some such carriage as this? Think what the future of American Art would be, unhampered by conditions. Think of the undying gratitude of that vast body of men and women, the artists of this country, of whom there are three thousand in New York City alone, and make for us a practical and reasonably economical Artists' Automobile.

A DISCERNING ANIMAL

That some horses should take affright at the uncanny appearance of their most formidable rival seems but natural, considering the highly excitable temperament of our four-footed friend. A new phenomenon, however, is a French horse which makes it its business to attack unsuspecting *chauffeurs*. This amiable beast lives at Cosne, and has already destroyed three automobiles. Its last victims were MM. Demay and Bossu, who were surprised by a sudden violent assault of the infuriated animal, which jumped into the carriage and proceeded to kick the whole machine into smithereens. When the two *chauffeurs* disentangled themselves from the ruins of their automobile. one had his leg broken. while the other had been kicked sprawling over the dashboard. After this slight diversion the horse gave an exhibition of speed on its own account.

Liquid Air

By Harrington Emerson

LIQUID air cannot have any commercial value for the production of either power or cold. A few elementary considerations will make this plain. One pound of coal can evaporate ten pounds of water. As it is converted into steam the water expands to about 1,700 times its volume as water. Ten pounds of liquid air expand to only 800 times the volume as liquid (under an increase of temperature up to sixty-five degrees Fahr.). It would therefore take 20 pounds of liquid air to yield the same power as can be obtained from one pound of coal. When 20 pounds of liquid air are as cheap as one pound of coal, then, and not before, liquid air can compete in cost. But even then it cannot rival coal. A piece of coal will keep almost indefinitely. A piece of anthracite coal could be carried from tropics to pole for a hundred years and without deteriorating. It would ever be ready to yield its stored energy. Who could keep a pound of liquid air even for twenty-four hours?

Weight for weight coal is twenty times as powerful, and a thousand times as durable. It is as yet a hundred times as cheap and a thousand times more easily taken care of.

Liquid air has little value as a commercial refrigerant. It shows intensity of cold but no volume. We do not attempt to heat our house with the arc light, yet its focus is the highest temperature known. It also has intensity, but no volume. Throw a quart of liquid air at a girl twenty feet away, and it is all dissipated before it reaches her, so that she scarcely feels coolness. A pound of ice in a refrigerator would cool it more than a pound of liquid air. Ice costs three dollars a ton, liquid air more nearly three dollars a pound, but were its cost but a cent a pound it would still be seven times dearer than ice. Even a small piece of ice, if wrapped in a blanket, will keep all night. A similar amount of liquid air, put in a can and that wrapped in the blanket, would disappear in a few minutes.

If under these elementary conditions liquid air cannot hold its own with either coal or ice, what chance would it have when violently used in a complicated machine?

Mechanical Propulsion and Traction

By Prof. G. Forestier

(Second Paper)

P ROFESSOR OLIVIER, of the Ecole Centrale des Arts et Manufactures, in a recent report read at a session of the Société d'Encouragement, declared in 1840 that the problem of steam locomotion upon roads should, from the standpoint of mechanics, be considered as solved. This too optimistic opinion, as well as that of the *Constitutionnel* upon the carriage imported by Asda, should render us slightly skeptical as to the excellent results said to have been obtained by Goldsworthy, Gurney and Walter Hancock, who organized regular passenger services assured by steam carriages, of which the malignity of the railway companies alone caused the disappearance in 1836.*

* This opinion, greatly amplified in the second volume of *Vehicles Automobiles*, by M. Louis Lockert, is also that of Count de Chasseloup-Laubat in the *Poids lourds* of the official catalogue of the Exposition of Automobiles at Paris in 1898.

Yet, in a pamplet entitled "Voitures à Vapeur sur routes ordinaires," published by H. Fournie, Paris, 1835, the authors, Gaby-Cazalat and C. Muyaud, inventors and manufacturers of another vehicle, after remarking that the carriage imported in 1835 by Asda was, as regards mechanism, like that of Gurney, make the following declarations :

(1). The apparatus that produces, distributes and utilizes steam could not be applied with advantage to the service of ordinary streets until it has received fundamental improvements that the English have been unable to make.

(2). The disarrangements and disconnections of which the carriage imported by Asda gave examples, and which resulted from the vicious conception and vicious arrangement of the mechanism, prevent all continuous locomotion, even in a transportation of slight extent.

(3). Finally, the expenses occasioned on the one hand by the disarrangements and disconnections, and on the other, by the great loss of coal and power, due to the vicious construction of such engines, robs them of all chance of success, since, even though no frequent interruptions in running were to be feared, they would not fulfill the first of the conditions, without which there is no success possible —economy.

Coming from a competitor, such criticisms might seem to be exaggerated ; but the fact must not be lost sight of that the object of the pamphlet cited was to answer the criticisms of the *National* and *Les Debats* formulated against the Gaby-Cazalat and C. Muyaud carriages in taking as a basis the accidents that happened to the Asda or Gurney carriage.

On page 23 of an English work, entitled "Motor Cars or Power-carriages on Common Roads," by A. J. Wallis-Taylor, C. E., we find a reflection of the opinion of MM. Lockert and Chasseloup Laubat in these words : "Steam carriages constructed by these inventors (Gurney and Hancock) were run for some time in different parts of England and Scotland, *with considerable success as far as the mechanism was concerned*, but failed to prove profitable as commercial undertakings, chiefly, no doubt, owing to the *heavy tolls* which were levied upon them, and to the obstructions that were everywhere thrown in their way, their very success as practical mechanical contrivances having raised up a horde of enemies against them." On page 27, on the contrary, we read : "*Internal discussions and disputes* occurring among the members of the company that had been formed to work these steam engines resulted, however, in the stoppage of the undertaking."

Moreover, if we refer to the conclusions of the jury on tests of the Liverpool competition of 1898, we shall see that even in the present state of the somewhat too light construction resulting from the legal prescriptions in England, the material is still exposed to frequent interruptions of service that are as onerous as they are unfortunate.

The reflections of the departmental engineer-delegates to the Competition of Heavy Weights at Versailles in 1898, only too fully justify the heavy tolls imposed upon power carriages by the corporations whose business it is to keep the roads in repair.

The Automobile Magazine

However this may be, the following are a few data as to the conditions under which the English carriages were constructed in 1832.

At the outset Gurney employed a steam traction carriage of which the principal arrangements are shown in Figs. 6 and 7.

The driving wheels were in the rear and their cranked axle was connected directly with the rods of the pistons placed in front of the carriage. Arms keyed upon the extremities of this axle

Fig. 6. Vertical Section of Gurney's Steam Carriage.

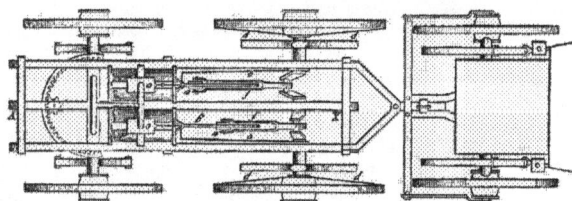

Fig. 7. Plan of Gurney's Steam Carriage

acted upon the fellies of the wheels. In turning about it was possible to act upon the springs that held these arms and to cause the action of the arms that carried along the interior wheel to cease.

In this first carriage the steering was done through a pinion that acted upon the toothed sector of the fore-carriage, which was movable around a king-bolt. In his subsequent experiments Gurney discarded the traction-carriage and replaced the method

Mechanical Propulsion and Traction

of steering by a three-wheeled bogie fore-carriage. The proper motions of this fore-carriage were obtained by the driver through the intermedium of a lever that acted upon the front wheel, which was placed at the end of a pole fixed at right angles with the axle of the fore-carriage properly so-called. This somewhat complicated arrangement must have given, and, in fact, often did give, rise to disarrangements.

Hancock's first carriage was a tricycle, of which the front wheel was both a driving and steering one. In the following ones (Fig. 8) the hind axle was the driver and was connected through a chain with a wheel actuated by the connecting rod of a piston.

The steering fore-carriage was movable around a king-bolt. The driver acted upon this fore-carriage through the intermedium of a steering bar mounted upon a vertical rod, which, beneath the floor, carried a pinion connected through a chain with a toothed sector mounted upon the fore-carriage. The pictures

Fig. 8. Hancock's Steam Carriage

of this carriage give us the impression that it was a heavy and pretty badly arranged vehicle.

The automobile carriage for passengers and the automobile truck have been neglected by inventors in France, where particular attention seems to have been paid to the road locomotive.

As long ago as 1856 the Lotz establishment, of Nantes, acquired a well deserved reputation for its agricultural road locomotive, designed to perform the double role of a motor for actuating threshing machines, plows, etc., and a traction engine for hauling such machines from the farm-house to the field and from one field to another, etc. In the first type the steering fore-carriage, which was two-wheeled, was maneuvered through a chain set in motion by a hand wheel actuated by the engineer standing in the rear, as in the present compressing rollers.

In 1865 M. Lotz, with a view to rendering his road locomotives particularly adapted for hauling upon canals, patented a method of steering through a single wheel placed in front and in the centre of a horizontal face-wheel. This latter was actuated by a pinion mounted at the bottom of a vertical rod, at the top of which was fixed a pinion with helicoidal toothing that geared with the endless screw of the shaft of a vertical rudder wheel. The results obtained with these road locomotives were such that many thought that herein was to be found a solution of the question of cheap transportation upon roads. The requests for permission to run road trains reached the Administration in so large numbers that the latter thought it necessary, on the 30th of April, 1866, to address to the Prefects a ministerial decree designed to regulate matters concerning the new method of transportation.

Although more particularly devoted to the manufacture of traction engines, M. Lotz has nevertheless produced a genuine passenger steam carriage, provided with a single steering wheel in front, and in which is found the general arrangement that has finally become prevalent. In this vehicle the engine and vertical boiler were placed in front in a sort of semicircular fore-body supported by the steering wheel. The engineer, standing between the motive gear and the passenger compartment, had within reach of his hand all that was necessary for the management of the engine and furnace as well as for the steering of the carriage. Motion was transmitted to the rear driving wheels through a chain.

While recognizing the fact that road locomotives have rendered and are capable of still rendering services in a few very special cases, we shall not dwell upon the different types of engines of this kind that have successively been put in service. Nevertheless we shall mention the Thomson road locomotive (constructed at Edinburgh in 1869), because it is the first vehicle of the kind in which, to our knowledge, vulcanized rubber tires were used for the driving wheels, as well as for those of the steering fore-carriage. The good effect of these was such that Thomson was able to *dispense with the interposition of springs between the axles and the carriage frame.*

We give, too, figures of the English road locomotive of Aveling and Porter, because they show the application of the Pecqueur satellite gearing to the transmission of the motion of the motor to the driving wheels (Figs. 9, 10 and 11).

Mechanical Propulsion and Traction

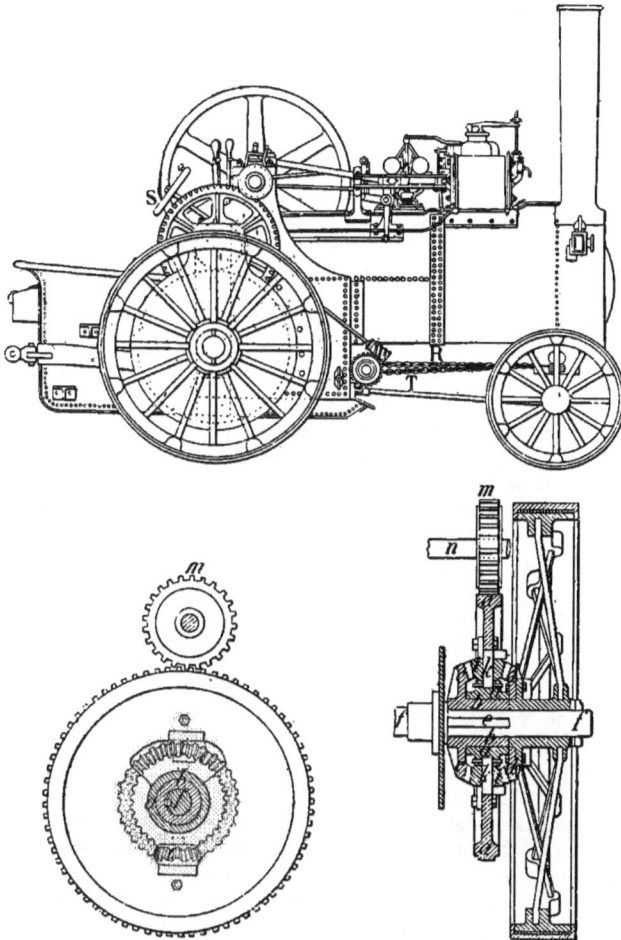

Figs. 9, 10 and 11. The Aveling and Porter Road Locomotive

R, chain for setting the steering axle in motion ; S, crank for setting in motion the endless screw that actuates the pulley placed under the boiler, and around which wind the extremities of the chain R ; T, iron rod for strengthening the fore-carriage ; *m*, a pinion keyed upon the driving shaft *n* ; *a*, toothed wheel loose upon the axle *f* ; *ii*, gearing mounted upon the spokes *k* of the wheel *a* and transmitting the motion of the latter to the bevel wheels *dd*, which are fixed to one of the driving wheels loose upon the axle *fcc*. At the other extremity of the axle *f* is keyed the other driving wheel.

In France Baron Seguier, in a study of mechanical vehicles, in 1866 put forth the idea that it would be well to render the driving wheels independent and actuate each of them by a sepa-

391

rate motor. He judged that it would be easy to co-ordinate the motions of these two motors in such a way as to give each of the wheels the motion adapted for obtaining the direction that might be desired. This idea was carried out by Michaux in 1870.

In France the automobile carriage for passengers was taken up again about 1873 by Bollée, of Mans. The experiments of this manufacturer appear to us to be worthy of mention, because he claims the honor of having devised one of the steering arrangements by pivoted wheels having properly conjugated motions. The starting point of his invention, according to his patent of April 28, 1873, was as follows: Having remarked with what facility the Lotz three-wheeled road locomotive performed its evolutions, he was desirous that each of the two wheels of his fore-carriage should be capable of revolving around a vertical axis like the Lotz single front wheel. This arrangement became the starting point of one of the most widely used steering apparatus. We shall return to it in detail further along.

The rapid vaporizing boiler, which was of the Field system, was placed in the rear. The two-cylinder horizontal engine, placed in front and incased so as to protect it against dust, actuated a differential gearing mounted upon an intermediate shaft of which the extremities carried sprockets connected through chains with sprockets keyed to the hubs of the driving wheels.

We here bring to a close a historic resumé, of which the object has not been to study the different motors that have been successively utilized for mechanical propulsion, but solely to inquire into the conditions of construction proposed for the mechanically propelled road vehicle. We have, in fact, come to an automobile carriage which, to a still further degree than the Lotz steam carriage, is provided with all the parts necessary. In the following chapters we shall evidently have to mention many more improved arrangements of detail and to describe general arrangements that are preferable; but, upon the whole, we are in possession of a vehicle that realizes our desiderata: the possibility of transmitting simultaneously to each driving wheel that part of the power that is necessary to it. according to the resistance that is momentarily opposed to it: a steering apparatus of a sufficiently certain and rapid maneuver: a relatively light motor, and a boiler with rapid vaporization. The vehicle with which we have just closed our brief history may be applied either to the carriage of passengers or to the transportation of any merchandise in bulk that can be loaded upon any sort of truck.

Specially translated for the Automobile Magazine from *Le Génie Civil.*

(To be continued in our next issue.)

The Truth about Motor Horse-Powers

By Georgia Knap

DESPITE the honesty and the good faith of most manufacturers, the author of the present article knows beforehand that the following explanation of the horse-power of automobile motors will cause many of them to raise a general hue and cry of reprobation against him. To such he replies that the brake tests upon which he has based his exposition were made in the presence of many spectators; that they were recorded in a report signed by the experts in charge, and that the tests in question have been repeated several times under the best possible conditions of carburation, compression, etc.

"But," you will say, "what good is it to us if you have shown that most automobile motors do not develop the energy in horse-power claimed for them, provided they will drive a carriage or tricycle at the rate of twenty miles an hour on a level road and at a good speed on heavy grades? What difference does it make whether they be called four horse-power or eight horse-power motors?" Your argument is good, it must be admitted; but, on the other hand, this incorrect denomination, if you take it literally, will lead you to consequences which you hardly expect. Your motor, for a given piston area and stroke, should develop a certain horse-power; but when it is subjected to the brake test you will be astonished to find that it develops but half the supposed energy. Because a certain motor has the same diameter of cylinder, the same stroke as yours, and because it develops 4½ horse-power, therefore you conclude that your own motor should yield at least 4 horse-powers. And you will rack your brain because your motor develops only 2½ horse-powers; and you will work three months, six months, seeking the two missing horse-powers, changing every part—ignition tubes, carbureters. etc., but the result in the end will always be the same.

There is still another matter to be taken into account—the possibility of an action with your customer: for he may sue you because his motor does not yield the energy claimed for it at the time it was purchased. Recently a *chauffeur* conceived the idea of using his carriage motor for driving a dynamo to light his house. The dynamo required 2½ horse-powers under good con-

The Automobile Magazine

ditions; the motor—a water-jacketed engine—was rated at 3¾ horse-powers. When the two had been set up and properly connected the motor absolutely refused to start the dynamo. After a thousand vain attemps to generate a current, he finally decided to have his motor tested by some competent engineer, who found that it was capable of developing only one horse-power and a half. Perplexed, our *chauffeur* went to one of his friends and told him his woes.

" Your motor is out of order," the latter replied; " I have the same carriage and the same motor and mine almost runs away. But are you certain that your carbureter is good? Do you know that with good carburation you can double the horse-power of your motor?"

Unconvinced by these brilliant arguments, our *chauffeur* proposed to subject his friend's motor to the brake test. The result proved that, as in the first instance, the second motor likewise yielded only 1½ horse-powers. The two friends, now both enraged, promised each other to "make it warm" for the manufacturer. And nevertheless, up to that time neither had uttered a single complaint. The engine worked comparatively well; grades were ascended at a small speed, it is true, but still they were overcome. Nevertheless, a simple brake test demoralized the two *chauffeurs*.

If it be considered how large is the amount of work represented by a horse-power it should cause no astonishment to find that a horse-power and a half are sufficient to drive a 400 kilog. (880 pounds) carriage at the rate of 18 to 20 kilometres (11.18 to 12.42 miles) per hour on a level road, and at the rate of 8 to 10 kilometres (4.968 to 6.21 miles) on up-grades. In factories, two and three horse-power gas engines are often used to drive machine-tools on which eight and ten men are employed.

Tables given at the end of this article will show exactly what power is required to drive vehicles at speeds varying from 15 to 40 kilometres (9.3 to 24.84 miles) per hour. By referring to the data there collected and tabulated the *chauffeur* and the manufacturer can determine with reasonable accuracy the power of a motor, provided the weight and speed of the carriage be known. For a 150 kilog. (330 pounds) carriage, the weight of the passengers being included, a 1½ horse-power motor will be required to attain a speed of 25 kilometres (15.52 miles) per hour on a level road. For a *voiturette* weighing 300 kilog. (660 pounds), including two passengers, a motor of 1½-horse-powers will be necessary to cover 25 kilometres on a level road, and a 1.9 horse-power motor for a speed of 30 kilometres (18.63 miles).

The Truth About Motor Horse-Powers

For a two-seated carriage weighing 400 kilog. (880 pounds), passengers included, a 1.3 horse-power motor will be required for a speed of 20 kilometres (12.42 miles) per hour on a level road.

But since all these vehicles have changes of speed, it follows that the power need not be increased to ascend grades. Hence we shall find that if our 300 kilog. *voiturette* encounter a grade of 5c. per metre, our horse-power and a half will drive it up the ascent at a speed of 15 kilometres (9.315 miles) per hour. The 400 kilog. carriage will climb the same grade at about 10 kilometres (6.21 miles); for our table shows that to ascend a grade of 5c. per metre a 400 kilog. carriage with a speed of 15 kilometres will require a 1.9 horse-power motor; and since we have but 1.3 horse-powers at our disposal, a very small down-gearing will be necessary. Thus, with the speed-changing gear, grades can be climbed with a motor which would absolutely refuse to run if the speed of the vehicle were not reduced and to make the number of revolutions for which it was built. The tables referred to will be indispensable in calculating the dimensions of the parts of speed-changing gears so that these parts will correspond with the power of the motor. They will, furthermore, indicate what speeds can be attained for a vehicle of a given weight and for a grade of a given number of centimetres per metre. It will be a simple matter to calculate the required down-gearing from the data furnished by these tables. Thus, for example we find that a carriage weighing 650 kilog. (1,430 pounds), including the passengers, and driven by a 5 horse-power motor, will cover 35 kilometres (21.735 miles) on a level road. But if an up grade of 4 centimetres per metre be encountered, then the speed will fall to 24 or 25 kilometres (14.904 or 15.525 miles) per hour; for our table shows that in order to cover 25 kilometres in an hour on an up grade of 4 cent. per metre a 650 kilog. carriage will need a 5.2 horse-power motor. It is, therefore, required to ascertain the down-gearing desirable for a speed of 24 kilometres per hour without changing the number of revolutions of our motor. If the grade be still steeper, 7% for example, which is unusual, then our table shows that a 650 kilog. carriage, in order to ascend a 0.07 grade at a speed of 15 kilometres per hour, will require a motor of 3.9 or 4 horse-powers. There still remains a single horse-power at our disposal, but since it is possible that a heavier grade may be encountered, a grade of 9 or 10% for example (which is not rare in mountainous regions), it would be wise to retain our last speed of 15 kilometres per hour, unless we add a fourth speed of 12 kilometres per hour and change our 15 to 18. The calculation will therefore be reduced to a very simple expression and will enable us to arrange changes of speed with certainty and with every chance of success in our favor.

From the foregoing explanation it is evident that a considerably smaller power is required for automobile traction than is generally supposed. It will also be seen that most 4, 5, and 6 horse-power motors are overrated; and that if these horse-powers be really developed there will be a corresponding increase in speed.

For various loads on level roads a 3 horse-power motor should drive a carriage weighing 300 kilog., including the passengers, at a speed of 40 kilometres per hour, and force it up grades of 5 cent. per metre at the rate of 26 kilometres per hour. If it be considered how small is the passive resistance of the transmission gears in actual use on carriages the speed will evidently be inappreciably reduced.

Double-cylinder motors are usually rated at nearly their true horse-power; we say usually, for there are also exceptions to the rule. We have known double-cylinder motors rated at 6 horse-powers which actually developed the full power; but we have also known motors ostensibly of 10 horse-powers which hardly yielded 6. If the fever for overrating the energy of motors continue, manufacturers will soon sell us motors of 25 and 50 horse-powers.

THE BEST BRAKE TEST FOR AUTOMOBILE MOTORS.

The brake which we shall presently describe is undeniably the simplest and cheapest of its kind.

The Prony brake, well known to all motor makers, can readily be used for measuring the power of steam-engines, but when it is applied to a high-speed motor working with the irregularity and violence of an explosion-engine its use is accompanied by great difficulties; and good results are obtained only by the exercise of considerable care. The brake, during the test, must be constantly lubricated; soap suds are splashed in all directions, often covering the experimenters and rendering the manipulation of the apparatus extremely difficult. If the motor be abruptly stopped, the load is hurled forward, to the danger of those present. These are the disadvantages which have led to the invention of a rope-brake of such simplicity that it can be made by any one. The brake can be used alike for slow steam-engines, making 150 to 200 revolutions, and for electric motors, making 2,000 and 3,000 revolutions per minute.

The operation of the brake is as simple as its construction :

The Truth About Motor Horse-Powers

Around the fly-wheel A of the motor a rope is passed, coated with graphite, to prevent overheating and to facilitate the operation. The rope is attached to a steelyard or preferably to a dynamometer. The dynamometer D is secured to a fixed point B, and to the other end of the rope a weight P is fastened.

When the motor is in operation, the weight P, by reason of the pull which it exerts, will give rise to a rubbing friction which will cause the rope to cling to the fly-wheel. The rope will, furthermore, tend to raise the weight. In order to measure the work it is therefore necessary to multiply the weight P by the circumferential velocity of the fly-wheel.

Let R be the radius of the fly-wheel and r the radius of the rope. The power of the motor will then be expressed by the formula:

$$T = \frac{\pi\ 2\ (R + r)\ n\ (P - p)}{60 \times 75}$$

Example—Given a motor developing $1\frac{1}{2}$ horse-powers at a speed of 1,200 revolutions, the diameter of the fly-wheel being 30 centimetres, let us suppose that the value of P is 8 kilogs., that of p, 1,200 kilogs., and the diameter of the rope 6 millimetres. The pull on the steelyard or dynamometer D will then be

$$T = \frac{2 \times 3.14 \times 0.153 \times 1200}{60 \times 75} \times 6.8 = 1.74 \text{ H. P. } (\textit{chevaux-vapeur}).$$

It is important that p be not neglected; for the value of P alone will give a power far greater than that actually possessed by the motor. In the foregoing example we should thus obtain 2.05 horse-powers instead of 1.74.*

When the experiment is at an end the motor should not be stopped by means of the brake still applied to the fly-wheel; for an abrupt stop would injure the steelyard or dynamometer. It is merely necessary to detach the dynamometer and the weight, allowing the rope to slide from the fly-wheel. Too many precautions cannot be taken; serious accidents may happen if the rope becomes coiled around the motor-shaft and the hook caught in the clothes of the experimenter.

In order to secure the yoke, I, to the rope, copper rivets are used, plaited in the strands.

The weight P should be gradually increased until the motor evidently works with an appreciable effort; but the load should never be so great that the motor either runs by starts or not at all. The weight P can be as heavy or as light as it may seem desirable.

* The *cheval vapeur* is equal to 0.9863 H. P., or almost equal to the English and American horse-power.—EDITOR.

the only limitations to be observed being those previously mentioned. If the motor have but a small weight to lift, it will turn more quickly than if the weight be heavier. In either case the value expressed in horse-powers will be the same.

It is important for the success of the test that a good speed indicator and an accurate chronometer be used, and that the experiment last at least fifteen minutes.

The tables which follow, as well as the formulæ upon which they are based, have been prepared by Messrs. Boramé and Julien,* and give the force exerted tangentially to the driving-wheel of an automobile and the work in horse-powers (*chevaux-vapeur*) developed, the calculation being based on the load (weight of the vehicle included), the grade and the speed. It will be evident from a cursory examination of these tables that the data there collected will considerably simplify the calculations of automobile-manufacturers. The table of forces will enable the dimensions of intermediate parts for the transmission of power to be readily figured; and the remaining tables will indicate the motive power required for a given vehicle.

WORK IN HORSE-POWERS DEVELOPED DURING THE RUN OF AN AUTOMOBILE

Calculations based on the load (weight of the carriage included), the grade, and the speed, independently of the passive resistance offered by the system of transmission adopted.

SPEED OF 15 KILOMETRES PER HOUR.

H. P. = French *cheval-vapeur*, 0.9863 H. P.

GRADES.	LOADS PULLED, WEIGHT OF VEHICLE INCLUDED.											
	300	400	500	650	800	1000	1250	1500	1750	2000	2500	3000
	H. P.	H. P.	H. P.	H. P.	H. P.	H. P.	H. P.	H. P.	H. P.	H. P.	H. P.	H. P.
om.00....	0.6	0.8	1.1	1.3	1.6	2.0	2.5	3.0	3.5	4.0	5.0	6.1
0.01................	0.8	1.1	1.3	1.7	2.0	2.6	3.2	3.9	4.5	5.1	6.4	7.8
0.02	1.0	1.3	1.6	2.1	2.5	3.1	3.9	4.7	5.5	6.3	7.8	9.4
0.03	1.2	1.5	1.9	2.4	2.9	3.7	4.6	5.6	6.5	7.4	9.2	11.1
0.04................	1.3	1.7	2.2	2.8	3.4	4.3	5.3	6.4	7.5	8.5	10.6	12.8
0.05................	1.5	1.9	2.5	3.1	3.8	4.9	6.0	7.2	8.4	9.6	12.0	14.5
0.06................	1.7	2.2	2.7	3.5	4.3	5.4	· 6.7	8.0	9.4	10.7	13.4	16.1
0.07................	1.8	2.4	3.0	3.9	4.7	6.0	7.3	8.9	10.4	11.9	14.8	17.8

*The Boramé-Julien method of calculating horse-powers was described in the *Automobile Magazine* for October.—EDITOR.

The Truth About Motor Horse-Powers

H. P. = French *cheval-vapeur*, 0.9863 H. P.

GRADES.	LOADS PULLED, WEIGHT OF VEHICLE INCLUDED.											
	300	400	500	650	800	1000	1250	1500	1750	2000	2500	3000
	H. P.	H. P.	H. P.	H. P.	H. P.	H. P.	H. P.	H. P.	H. P.	H. P.	H. P.	H. P.
0m.00	1.0	1.3	1.6	2.0	2.5	3.0	3.8	4.5	5.3	6.0	7.5	9.0
0.01	1.2	1.6	1.9	2.5	3.0	3.8	4.7	5.6	6.6	7.5	9.4	11.2
0.02	1.4	1.9	2.3	3.0	3.6	4.5	5.6	6.7	7.9	9.0	11.2	13.4
0.03	1.6	2.2	2.7	3.4	4.2	5.3	6 5	7.9	9.2	10.5	13.1	15.6
0.04	1.9	2.5	3.0	3.9	4.8	6.0	7.5	9.0	10.5	11.9	14.9	17.9
0.05	2.1	2.8	3.4	4.4	5.4	6.7	8.4	10.1	11.8	13.4	16.8	20.1
0.06	2.3	3.1	3.8	4.9	6.0	7.5	9.3	11.2	13.1	14.9	18.6	22.3
0.07	2.5	3.4	4.2	5.3	6.7	8.2	10.2	12.3	14.4	16.4	20.5	24.5

H. P.=French *cheval-vapeur*, 0.9863 H. P.

GRADES.	LOAD PULLED, WEIGHT OF VEHICLE INCLUDED.											
	300	400	500	650	800	1000	1250	1500	1750	2000	2500	3000
	H P.	H. P.	H. P.	H. P.	H. P.	H. P.	H. P.	H. P.	H. P.	H. P.	H. P.	H. P.
0m.00	1.4	1.8	2.2	2.8	3.4	4.2	5.3	6.3	7.3	8.3	10.4	12.4
0.01	1.7	2.2	2.7	3.4	4.2	5.2	6.4	9.7	8.9	10.2	12.7	15.2
0.02	2.0	2.5	3.1	4.0	4.9	6.1	7.6	9.1	10.6	12.0	15.0	18.0
0.03	2.2	2.9	3.6	4.6	5.6	7.0	8.7	10.4	12.2	13.9	17.3	20.7
0.04	2.5	3.3	4.0	5.2	6.4	8.0	9.9	11.9	13.8	15.7	19.7	23.5
0.05	2.8	3.6	4.5	5.8	7.1	8.9	11.0	13.2	15.5	17.6	22.0	26.3
0.06	3.1	4.0	4.9	6.4	7.9	9.8	12.2	14.6	17.1	19.4	24.3	29.0
0.07	3.4	4.4	5.4	7.0	8.6	10.8	13.3	16.0	18.8	21.3	26.6	31.8

SPEED OF 30 KILOMETRES PER HOUR.

H. P. = French *cheval-vapeur*, 0.9863 H. P.

GRADES.	LOAD PULLED, WEIGHT OF VEHICLE INCLUDED.											
	300	400	500	650	800	1000	1250	1500	1750	2000	2500	3000
	H. P.	H. P.	H. P.	H. P.	H. P.	H. P.	H. P.	H. P.	H. P.	H. P.	H. P.	H. P.
0m.00............	1.9	2.4	3.0	3.7	4.6	5.7	7.0	8.3	9.7	11.0	13.7	16.4
0.01............	2.2	2.8	3.5	4.5	5.5	6.8	8.4	10.0	11.6	13.2	16.5	19.7
0.02............	2.6	3.3	4.1	5.2	6.3	7.9	9.8	11.6	13.5	15.5	19.3	23.0
0.03............	2.9	3.7	4.6	5.9	7.2	9.0	11.1	13.3	15.5	17.7	22.1	26.4
0.04............	3.2	4.1	5.2	6.7	8.1	10.1	12.5	15.0	17.4	19.9	24.8	29.7
0.05............	3.5	4.6	5.7	7.4	9.0	11.2	13.9	16.6	19.4	22.1	27.6	33.0
0.06............	3.9	5.0	6.3	8.1	9.9	12.3	15.3	18.3	21.3	24.3	30.4	36.4
0.07	4.2	5.5	6.8	8.9	10.7	13.4	16.7	20.0	23.2	26.6	33.2	39.7

SPEED OF 35 KILOMETRES PER HOUR.

H. P. = French *cheval-vapeur*, 0.9863 H. P.

GRADES.	LOAD PULLED, WEIGHT OF VEHICLE INCLUDED.											
	300	400	500	650	800	1000	1250	1500	1750	2000	2500	3000
	H. P.	H. P	H. P.	H. P.	H. P.	H. P.	H. P.	H. P.	H. P.	H. P.	H. P.	H. P.
0m.00............	2.5	3.2	3.8	4.9	5.9	7.2	9.0	10.7	12.4	14.1	17.5	20.8
0.01............	2.9	3.7	4.5	5.7	7.0	8.5	10.6	12.6	14.6	16.7	20.7	24.7
0.02............	3.3	4.2	5.1	6.5	8.0	9.8	12.2	14.5	16.9	19.2	23.9	28.6
0.03............	3.6	4.7	5.8	7.4	9.1	11.1	13.8	16.5	19.2	21.8	27.1	32.4
0.04............	4.0	5.2	6.4	8.2	10.1	12.4	15.4	18.4	21.4	24.4	30.3	36.3
0.05............	4.4	5.7	7.0	9.1	11.1	13.7	17.0	20.4	23.7	27.0	33.6	40.2
0.06............	4.8	6.2	7.7	9.9	12.1	15.0	18.6	22.3	25.9	29.6	36.8	44.0
0.07............	5.2	6.7	8.3	10.7	13.2	16.3	20.2	24.2	28.2	32.1	40.0	47.9

The Truth About Motor Horse-Powers

SPEED OF 40 KILOMETRES PER HOUR.

H. P. = French *cheval-vapeur*, 0.9863 H. P.

GRADES.	LOAD PULLED, WEIGHT OF VEHICLE INCLUDED.											
	300	400	500	650	800	1000	1250	1500	1750	2000	2500	3000
	H. P.	H. P.	H. P.	H. P.	H. P.	H. P.	H. P.	H. P.	H. P.	H. P.	H. P.	H. P.
0m.00	3.2	4.1	4.9	6.2	7.5	9.1	11.2	13.3	15.5	17.6	21.8	26.0
0.01	3.7	4.7	5.6	7.2	8.7	10.6	13.1	15.6	18.1	20.6	25.5	30.5
0.02	4.1	5.3	6.3	8.1	9.8	12.1	15.0	17.8	20.7	23.5	29.2	34.9
0 03	4.5	5.9	7.0	9.1	11.0	13.6	16.8	20.0	23.2	26.5	32.9	39.4
0.04	5.0	6.4	7.7	10.1	12.2	15.0	18.6	22.2	25.8	29.5	36.6	43.8
0.05	5.4	7.0	8.4	11.0	13.4	16.5	20.5	24.4	28.4	32.4	40.3	48.3
0.06	5.9	7.6	9.1	12.0	14.6	18.0	22.3	26.7	31.1	35.4	44.0	52.7
0.07	6.3	8.2	9.8	12.9	15.7	19.5	24.2	28.9	33.6	38.3	47.7	57.1

Formula : $Nch = \dfrac{f + e}{75}$

Nch = expression of the work in horse-powers; f = force in kilogramme-exerted tangentially to the driving-wheel; e = distance in meters covered in a second by the carriage.

TABLE OF FORCES.

Exerted tangentially to the driving-wheel of an automobile, calculations being based on the load (weight of vehicle included), the grade, and the speed.

SPEED OF 15 KILOMETERS PER HOUR.

GRADES.	LOAD PULLED, WEIGHT OF VEHICLE INCLUDED.											
	300	400	500	650	800	1000	1250	1500	1750	2000	2500	3000
	K.	K.	K.	K.	K.	K.	K.	K.	K.	K.	K.	K.
0m.00	11.5	15.0	19.0	24.0	29.0	36.0	45.0	54.0	63.5	72.0	90.0	109.0
0.01	14.5	19.0	24.0	30.5	37.0	46.0	57.5	69.0	81.0	92.0	115.0	139.0
0.02	17.5	23.0	29.0	37.0	45.0	56.0	70.0	84.0	98.5	112.0	140.0	169.0
0.03	20.5	27.0	34.0	43.5	53.0	66.0	82.5	99.0	116.0	132.0	165.0	199.0
0.04	23.5	31.0	39.0	50.0	61.0	76.0	95.0	114.0	133.5	152.0	190.0	229.0
0.05	26.5	35.0	44.0	56.5	69.0	86.0	107.5	129.0	151.0	172.0	215.0	259.0
0.06	29.5	39.0	49.0	63.0	77.0	96.0	120.0	144.0	168.5	192.0	240.0	289.0
0.07	32.5	43.0	54.0	69.5	85.0	106.0	132.5	159.0	186.0	212.0	265.0	319.0
	0.75	0.80	0.85	0.90	1.00	1.10	1.20	1.35	1.50	1.65	1.90	2.20

Approximate areas, in square metres, on which the resistance of the air acts.

SPEED OF 20 KILOMETRES PER HOUR.

GRADES.	LOAD PULLED, WEIGHT OF VEHICLE INCLUDED.											
	300	400	500	650	800	1000	1250	1500	1750	2000	2500	3000
	K.	K.	K.	K.	K.	K.	K.	K.	K.	K.	K.	K.
0m.00	13.2	17.0	21.0	26.7	33.0	41.0	51.0	61.0	71.2	81.2	101.5	121.3
0.01	16.2	21.0	26.0	33.2	41.0	51.0	63.5	76.0	88.7	101.2	126.5	151.3
0.02	19.2	25.0	31.0	39.7	49.0	61.0	76.0	91.0	106.2	121.2	151.5	181.3
0.03	22.2	29.0	36.0	46.2	57.0	71.0	88.5	106.0	123.7	141.2	176.5	211.3
0.04	25.2	33.0	41.0	52.7	65.0	81.0	101.0	121.0	141.2	161.2	201.5	241.8
0.05	28.2	37.0	46.0	59.2	73.0	91.0	113.5	136.0	158.7	181.2	226.5	271.3
0.06	31.2	41.0	51.0	65.7	81.0	101.0	126.0	151.0	176.2	201.2	251.5	301.3
0.07	34.2	45.0	56.0	72.2	89.0	111.0	138.5	166.0	193.7	221.2	276.5	331.3
	0.75	0.80	0.85	0.90	1.00	1.10	1.20	1.35	1.50	1.65	1.90	2.20

Approximate areas, in square metres, on which the resistance of the air acts.

SPEED OF 25 KILOMETRES PER HOUR.

GRADES.	LOAD PULLED, WEIGHT OF VEHICLE INCLUDED.											
	300	400	500	650	800	1000	1250	1500	1750	2000	2500	3000
	K.	K.	K.	K.	K.	K.	K.	K.	K.	K.	K.	K.
0m.00	15.0	19.2	23.6	30.0	37.0	45.7	57.0	68.0	79.0	90.0	112.2	134.2
0.01	18.0	23.2	28.6	36.5	45.0	55.7	69.5	83.0	96.5	110.0	137.2	164.2
0.02	21.0	27.2	33.6	43.0	53.0	65.7	82.0	98.0	114.0	130.0	162.2	194.2
0.03	24.0	31.2	38.6	49.5	61.0	75.5	94.5	113.0	131.5	150.0	187.2	224.2
0.04	27.0	35.2	43.6	56.0	69.0	85.5	107.0	128.0	149.0	170.0	212.2	254.2
0.05	30.0	39.2	48.6	62.5	77.0	95.5	119.5	143.0	166.5	190.0	237.2	284.2
0.06	33.0	43.2	53.6	69.0	85.0	105.5	132.0	158.0	184.0	210.0	262.2	314.2
0.07	36.0	47.2	58.6	75.5	93.0	115.5	144.5	173.0	201.5	230.0	287.2	344.2
	0.75	0.80	0.85	0.90	1.00	1.10	1.20	1.35	1.50	1.65	1.90	2.20

Approximate areas, in square metres, on which the resistance of the air acts.

The Truth About Motor Horse-Powers

SPEED OF 30 KILOMETRES PER HOUR.

GRADES.	LOAD PULLED, WEIGHT OF VEHICLE INCLUDED.											
	300	400	500	650	800	1000	1250	1500	1750	2000	2500	3000
	K.	K.	K.	K.	K.	K.	K.	K.	K.	K.	K.	K.
cm.00	17.0	21.5	26.5	33.7	41.1	51.0	63.0	75.0	87.0	99.2	123.5	147.5
0.01	20.0	25.5	31.5	40.2	49.1	61.0	75.5	90.0	104.5	119.2	148.5	177.5
0.02	23.0	29.5	36.5	46.7	57.1	71.0	88.0	105.0	122.0	139.2	153.5	207.5
0.03	26.0	33.5	41.5	53.2	65.1	81.0	100.5	120.0	139.5	159.2	198.5	237.5
0.04	29.0	37.5	46.5	59.7	73.1	91.0	113.0	135.0	157.0	179.2	203.5	267.5
0.05	32.0	41.5	51.5	66.2	81.1	101.0	125.5	150.0	174.5	199.2	248.5	297.5
0.06	35.0	45.5	56.5	72.7	89.1	111.0	138.0	165.0	192.0	219.2	253.5	327.5
0.07	38.0	49.5	61.5	79.2	97.1	121.0	150.5	180.0	209.5	239.2	298.5	357.5
	0.75	0.80	0.85	0.90	1.00	1.10	1.20	1.35	1.50	1.65	1.90	2.20

Approximate areas, in square metres, on which the resistance of the air acts.

SPEED OF 35 KILOMETRES PER HOUR.

GRADES.	LOAD PULLED, WEIGHT OF VEHICLE INCLUDED.											
	300	400	500	650	800	1000	1250	1500	1750	2000	2500	3000
	K.	K.	K.	K.	K.	K.	K.	K.	K.	K.	K.	K.
cm.00	19.4	24.5	29.7	37.7	46.0	56.1	69.5	82.7	96.0	109.0	135.3	161.5
0.01	22.4	28.5	34.7	44.2	54.0	66.1	82.0	97.7	113.5	129.0	160.3	191.5
0.02	25.4	32.5	39.7	50.7	62.0	76.1	94.5	112.7	131.0	149.0	185.3	221.5
0.03	28.4	36.5	44.7	57.2	70.0	86.1	107.0	127.7	148.5	169.0	210.3	251.5
0.04	31.4	40.5	49.7	63.7	78.0	96.1	119.5	142.7	166.0	189.0	235.3	281.5
0.05	34.4	44.5	54.7	70.2	86.0	106.1	132.0	157.7	183.5	209.0	260.3	311.5
0.06	37.4	48.5	59.7	76.7	94.0	116.1	144.5	172.7	201.0	229.0	285.3	341.5
0.07	40.4	52.5	64.7	83.2	102.0	126.1	157.0	187.7	218.5	249.0	310.3	371.5
	0.75	0.80	0.85	0.90	1.00	1.10	1.20	1.35	1.50	1.65	1.90	2.20

Approximate areas, in square metres, on which the resistance of the air acts.

SPEED OF 40 KILOMETRES PER HOUR.

GRADES.	LOADS PULLED, WEIGHT OF VEHICLE INCLUDED.											
	300	400	500	650	800	1000	1250	1500	1750	2000	2500	3000
	K.	K.	K.	K.	K.	K.	K.	K.	K.	K.	K.	K.
0m.00	21.7	27.6	33.1	42.0	50.5	61.5	76.0	90.1	104.5	119.0	147.0	176.0
0.01	24.7	31.6	38.5	48.5	58.5	71.5	88.5	105.1	122.0	139.0	172.0	206.0
0.02	27.7	35.6	43.5	55.0	66.5	81.5	101.0	120.1	139.5	159.0	197.0	236.0
0.03	30.7	39.6	48.5	61.5	74.5	91.5	113.5	135.1	157.0	179.0	222.0	266.0
0.04	33.7	43.6	53.5	68.0	82.5	101.5	126.0	150.1	174.5	199.0	247.0	296.0
0.05	36.7	47.6	58.5	74.5	90.5	111.5	138.5	165.1	192.0	219.0	272.0	326.0
0.06	39.7	51.6	63.5	81.0	98.5	121.5	151.0	180.1	209.5	239.0	297.0	356.0
0.07	42.7	55.6	68.5	87.5	106.5	131.5	163.5	195.1	227.0	259.0	322.0	386.0
	0.75	0.80	0.85	0.90	1.00	1.10	1.20	1.35	1.50	1.65	1.90	2.20

Approximate areas, in square metres, on which the resistance of the air acts.

Formula, $F = P (0.025 + 0.0007 v + p) + s v^2 + 0.0048.$

F = force tangential to the driving-wheel; P = weight in kilogrammes of the entire load; 0.025 = co-efficient of the resistance offered by an ordinary road to wheels having an average diameter of 0.80 metres; v = speed in kilometres; $0.0007 + v$ = term proportional to the speed and relative to the resistance due to jars occasioned by inequalities in the road; p = grade per metre; s = area of surface in sq. metres, upon which the resistance of the air acts; $s v^2 + 0.0048$ = expression of the resistance of the air.

WHY THESE RACES?

The French and American motor races are practically trial trips for speed, strength, and reliability, and are giving the French and Americans their superiority. The recent races in France have been attended by scores of German and American engineers. The last two contests were, it is pleasing to note, witnessed by English engineers also. I noticed several of them examining the winning carriages, camera and note-book in hand, at Ostend the other day, and many were at Boulogne. The foreigner has borrowed so much from us that a little reciprocity is quite in order. No one wants to travel forty miles an hour, but everyone desires an absolutely reliable engine. These races, over all kinds of roads, have revealed the weak points in construction, and have proved the worth of new inventions.—*London Chronicle.*

An Improved Power-Transmitting Mechanism

By E. E. Schwarzkopf

I N order to transmit the power imparted to the piston of an engine with a greater degree of efficiency than has been attainable by means of the directly-connected crank, inventors have devised many curiously-constructed gears which are designed to overcome the difficulties ordinarily encountered in power-

Stationary Engine with New Power-Transmitter

transmission. Among these devices the rolling, cogged crosshead mounted between a fixed rack and a moving rack driven by the piston-rod, has figured prominently. There has also been used, with more or less success, a toothed, fulcrum-wheel rolling on a stationary rack, with the piston-rod coupled to the wheel-journals and the crank connected with the wheel at a point less distant from the center of the latter than the length of the crank.

A mechanism in which both of these systems have been combined, has been invented by Mr. W. A. Pitt, of Stamford, Conn., and has been employed with noteworthy success in engines and foot-power appliances. The improvement consists essentially in mounting the crank-shaft with its axis parallel with the fulcrum-wheel, but in a different plane. Not only is the old problem in engine-building of overcoming the dead centers effectually solved, but an evenness of turning moment is obtained which adds much to the efficiency of the engine.

The new power-transmitting mechanism, as indicated in Fig. 1, a side elevation representing the device applied to a double-cylinder engine, comprises essentially two similar cranks C and C^x set oppositely at the ends of a driving-shaft A. Coupled to the cranks are connecting-rods F and F^x pivoted at their lower ends to lever-arms G and G^x. These lever-arms are driven by gear-wheels H and H^x, meshing respectively with

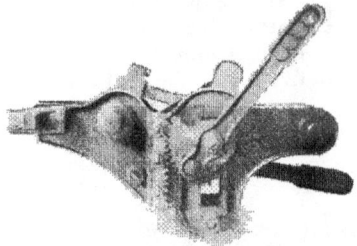

Food Power Appliance

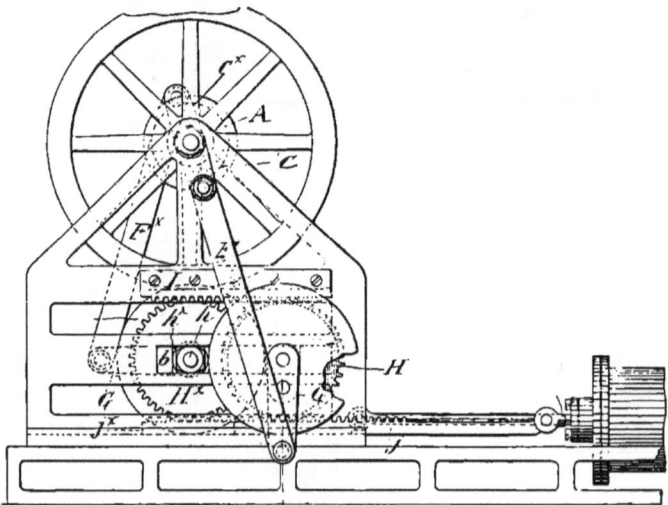

Fig. 1 Side Elevation

Improved Power-Transmitting Mechanism

fixed racks I and I^x and with sliding racks j and j^x, connected with the piston-rods.

The connecting-rods are five times as long as the cranks C and C^x. The lever-arms G and G^x are twice as long as the cranks and therefore make one-quarter of a revolution for two-thirds of a revolution of the cranks. The size of the fulcrum or gear wheels H and H^x is such that in rolling on the fixed racks through a quarter of a revolution, the centers of the wheels will travel in the slots b a distance equal to the length of the lever-arms G and G^x. The stroke of each piston will be equal to the distance traveled by the center of the corresponding gear-wheel, plus one-fourth the circumference of the wheel measured on the pitch-circle.

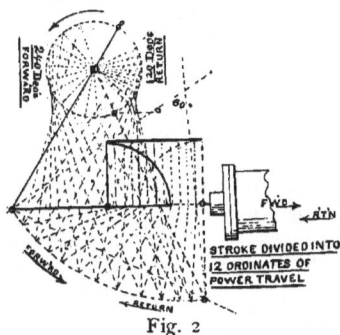

Fig. 2

As the piston-rods reciprocate the sliding racks, the wheels H and H^x roll along the upper fixed racks and swing the lever-arms G and G^x through one-quarter of a revolution. When one of the cranks, C or C^x, is on the upper dead center of its path, the corresponding lever-arm will be in a horizontal position, and its extremity will be two crank-lengths beyond the center of crank motion in a plane parallel to a horizontal line drawn through that center. The piston-rod will then have reached the end of its outstroke. As the piston-rod returns, it moves with a constantly-accelerated movement, drawing with it the sliding rack and causing the corresponding gear-wheel to roll upon its fixed rack, the lever-arm to swing through 90° to a vertical position, and simultaneously to be drawn through a distance equal to the diameter of the crank-pin-path circle, in a plane parallel to a horizontal plane passing through the center of the crank-motion. Hence the extremity of the lever-arm describes an epicycloid (Fig. 2) measuring an arc equal to one-sixth of a circle, the radius of which circle is equal to the chord connecting the extremities of the epicycloid.

For one piston-stroke, each crank will be rotated 240°. And since on the return stroke of the piston the crank has to move only through 120°, the piston returns double the velocity of the forward stroke. In a two-cylinder engine, the driving-shaft is alternately acted upon; while one piston controls or sustains the load, the other imparts motion. In a single-cylinder engine the crank also rotates 240° for every forward piston-stroke, with the dead centers 240° apart on the one side and 120° on the other.

Let us compare the operation of an engine fitted with this new power-transmitting mechanism with that of an ordinary engine having the usual directly-connected piston and crank. In the ordinary engine the power is first exerted on a minimum length of crank leverage, reaches a maximum at approximately the middle of the stroke, and finally falls to a minimum at the end of the stroke. In the new contrivance a minimum power is exerted at the beginning of the stroke, and the crank is pulled upon with a maximum momentum; at the end of the stroke a maximum power is exerted, the momentum being a minimum.

Other means than a sliding rack may be employed to roll and reciprocate the gear or fulcrum wheels along the fixed racks. In the engine illustrated a segment-gear is used in place of the full gear-wheel and connected with the piston-rod by means of an ordinary sliding cross-head. The principle of operation is obviously the same as that first described.

When a two-cylinder engine is used, the cranks are never on dead centers at the same time; and when one crank is on the dead center there will always be sufficient leverage on the other to pull it off. When only a single cylinder is employed, the motor is made to act through considerably more than one-half the crank-travel, so that the flywheel is required to carry the shaft through less than half a revolution. Besides overcoming the dead centers, the mechanism possesses the merit of applying its power tangentially to the crank, thereby obtaining the greatest possible efficiency for a given expenditure of power.

A BOON FOR PHYSICIANS

Motor-cars, according to the London *Daily Mail,* are becoming daily more popular with the medical profession. In fine weather a motor-tricycle serves the purpose of a country medical man excellently, and enables him to cover his rounds with great saving of time. For wet weather a hooded phaeton, driven by a motor, with plenty of power, enables a doctor to see his patients without the anxiety attendant on leaving valuable horses standing in the rain.

The motor is always ready to start at a moment's notice to attend to a night call.

The Ravel "Intensif" Motor

By Gustave Chauveau

I N his new motor, M. Ravel, whose name is well known to specialists, has endeavored to obtain a higher power within a definite bulk and weight; and hence the name " Intensif " applied to his last production.

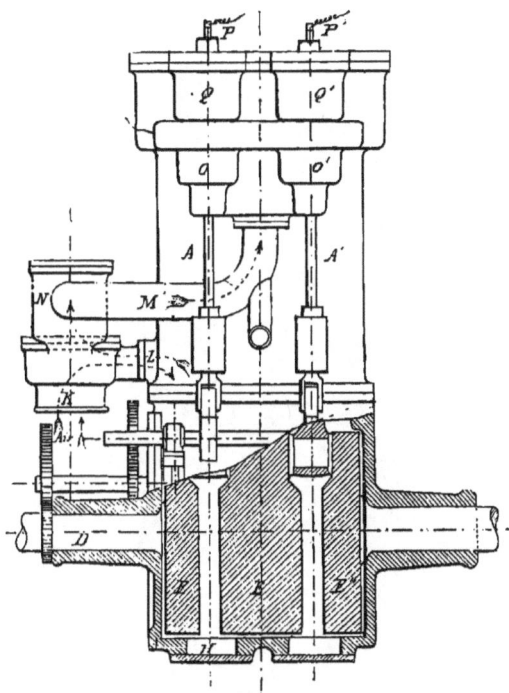

Fig. 1.   Front View

Before proceeding to a description of this motor, we shall enter into a few considerations that we regard as indispensable.

The gasoline motors of what is called the four-cycle type are. as well known, almost exclusively adopted at present, especially for use upon automobiles. Such motors comprise, in principle, a cylinder in which moves a piston that actuates the driving-shaft, and above the upper dead-point of which is situated a chamber that contains the distributing parts and the igniting device. The operation of such a motor is as follows: The piston being supposed to be at the end of its up-stroke, at the time of the admission-period, the chamber above is full of the products of the preceding combustion. When the flywheel causes the piston to

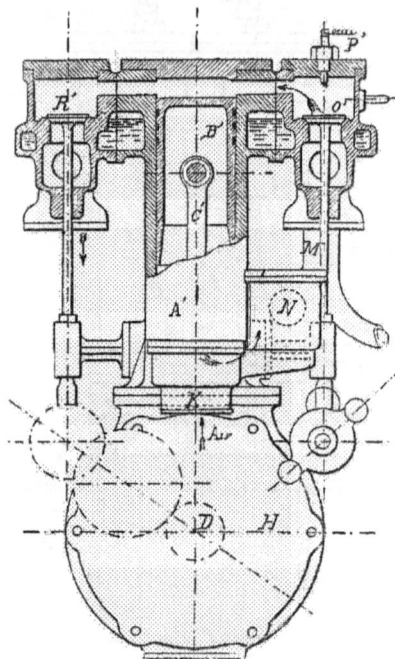

Fig. 2. Side View

descend, a partial vacuum is formed in the rear of the latter, the admission-valve opens mechanically (or automatically, rather) under the action of the vacuum, and the explosive mixture enters the cylinder and drives before it the products of combustion, which follow the piston. At the end of the down-stroke, the piston again rising under the impulsion of the flywheel, the admission-valve closes, and the explosive mixture admitted is forced into the chamber above.

The Ravel "Intensif" Motor

As the volume of the latter is less than that of the space in the cylinder unoccupied by the piston, there results a compression of the mixture varying in degree with the ratio between the two volumes just mentioned.   In automobile motors, the compression reaches, practically, from about 35 to 45 pounds to the square inch.

When the mixture is ignited in an appropriate manner in the vicinity of the upper dead-point, there ensues an explosion which develops a pressure of from 175 to 220 pounds to the square inch. The piston, upon being driven downward through such pressure,

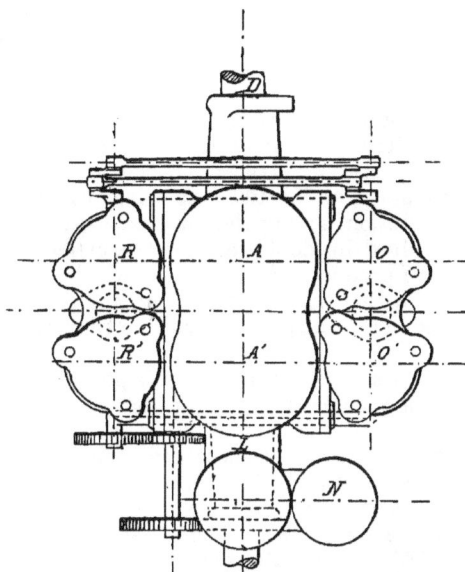

Fig. 3.   Plan

gives the flywheel a sufficiently strong impulsion to permit it to effect the other strokes, while the products of combustion expand ; that is to say, lose their temperature and pressure.   Just before the end of the down-stroke, a communication with the exterior being established, while the pressure is still in the vicinity of 45 pounds to the square inch, the exhaust begins, and the piston, in rising, forces to the exterior, as far as to the end of its up-stroke, the products of the combustion that has just taken place.   A new admission then occurs, and so on.

We thus see that if we have a free cylinder space of a definite volume, and force the mixture into a chamber of another definite volume, we shall obtain a compression of, say, from 35 to 45 pounds that will produce an explosive pressure of from 175 to 220 pounds, the expansion bringing the pressure to the neighborhood of 45 pounds and the exhaust to that of the atmosphere. The mean pressure during the motive stroke will be practically from 65 to 75 pounds per square inch of piston surface, corresponding to a definite power for a definite velocity. Upon the whole, under definite conditions of velocity and piston surface, of richness of the explosive mixture, and of ratio between the volume of the cylinder space and that of the compression chamber, we shall have for a given cylinder charge expanded in one cylinder a definite mean pressure corresponding to a likewise definite power.

If, then, through some arrangement or other, we succeed in having two cylinder charges instead of one, of the same explosive mixture, and compress both in one double chamber, we shall finally have a double mass of an identical explosive mixture, at the same degree of compression, giving the same explosive power; but if, instead of expanding this double mass in two cylinders, we expand it in but one, the products of combustion will begin to escape at a pressure much higher than the preceding, and the mean pressure will be increased in definite proportions, and be, practically, nearly doubled.

We shall thus succeed in obtaining a power that is sensibly double, within a volume and weight that remain sensibly the same.

This is what M. Ravel has succeeded in doing in his " Intensif " motor (Figs. 1, 2 and 3), which consists of two vertical cylinders, $A A'$, placed side by side, and the pistons, $B B'$, of which move at the same time. These pistons, through the connecting-rods $C C'$, actuate the driving-shaft D, which carries crank-plates, $E E' E''$, that act as flywheels. The whole is placed in a casing, H, the free space in which is reduced as much as possible by the bulk of the pieces which it contains.

This casing is capable of communicating with the exterior through a butterfly-valve placed at K, and a conduit, L; and, with the compression chamber of the two cylinders, through the same conduit, L, and another conduit, M, between which is interposed a carbureter, N.

The conduit M debouches beneath the admission-valves, $O O'$, above which are located the igniters, $P P'$. The exhaust-valves are on the other side, at $R R'$.

The Ravel "Intensif" Motor

A circulation of water assures the cooling of the valves, and a governor keeps the velocity within definite limits. The admission and exhaust-valves, as well as the igniters and the governor, are controlled by a series of gearings, shafts, cams and appropriate contacts, of which the arrangement may be easily imagined. The intermediate shafts, of course, revolve at half the velocity of the driving-shaft.

Let us now see how the motor operates: One of the pistons, say B', for example, being at the end of its up-stroke during the working period, the other one, B, will be likewise at the upper dead-point; but, at the time of the admission-period, the piston B', forced downward, will carry B along with it, and both will drive before them the air that fills the lower part of the two cylinders. At this moment, the admission-valve O being open, this air, passing through L and M, will enter the cylinder A, after becoming carbureted by traversing N in part. At the lower dead-point, the exhaust will be continued in A', while the admission will cease in A. The pistons, then returning in an opposite direction, will form a vacuum; and pure air, entering through the valve K, will flow into the casing H and fill the lower part of the cylinders, while the exhaust that follows an admission will finish in A', and the compression that follows an explosion will take place in A.

We thus see the four cycles completed in each of the cylinders alternately, the bottom charging the top.

But what should be well remarked is that the two cylinders that suck the air admit at the same time two charges of explosive mixture, which are forced into a single compression chamber of double the volume of an ordinary one. The principle enunciated above is therefore well worked out.

Since the governor closes the admission-valve when the velocity becomes too great, there is less of an admission, and, consequently, a feebler compression, explosion and mean pressure, which re-establish the equilibrium.

A reduction of the weight and bulk of automobile motors is so important a matter and the means employed by M. Ravel to reach the result are so ingenious that we have not dwelt at any too great a length upon the "Intensif." There is, nevertheless, reason for remarking that the feeble expansion of this motor will be attended with a poor rendering, and consequently a high consumption, and that the exhaust mechanism will be submitted to very high temperatures.

Let us wait, then, until practice, our great mistress, shall furnish us with a decisive opinion.

The Buchet Head for Working-Chambers

THE Buchet cylinder-head which is represented in the annexed engraving is applicable to all motors of the De Dion and Aster type. The increase in power gained by means of the attachment is said to be as large as thirty per cent.

The admission-valve, the exhaust-valve, the igniter, and the relief-valve are of the usual kind; but instead of placing the first three in a recess formed in the wall of the cylinder and communicating with the latter by means of a passage, the inventor has

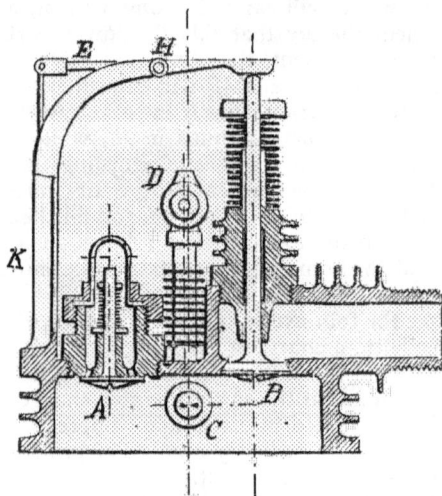

mounted them in a chamber constituting a prolongation of the cylinder, with the same diameter. The valves are directly placed on the cylinder head, with the relief-valve at D. The stem of the valve B is upwardly prolonged, so that it can engage the free end of a lever E fulcrumed at H and supported by a casting K on the cylinder.

At first blush it would seem that the arrangement of parts is open to criticism, because it might be difficult to obtain a pure mixture of air and gas about the igniter. But the tests to which the invention has been subjected contradict the supposition.

An Improved Explosion-Engine

By Baudry de Saunier

A FRENCH engineer, M. Marmonier, of Lyon, has invented a hydrocarbon-motor which presents many advantages over the explosion-engines at present in use on automobiles. Chief among these advantages may be mentioned an expenditure of power proportional to the speed of the vehicle and a simplicity of construction which is so often conspicuously lacking in automobile-engines.

The motor is composed of two cylinders *A A,* mounted parallel to each other and open at both ends. Each cylinder incloses, besides, a piston *B,* connected with the motor-shaft by a rod *b* and a crank *m,* a counter-piston *P* connected with the correspond-

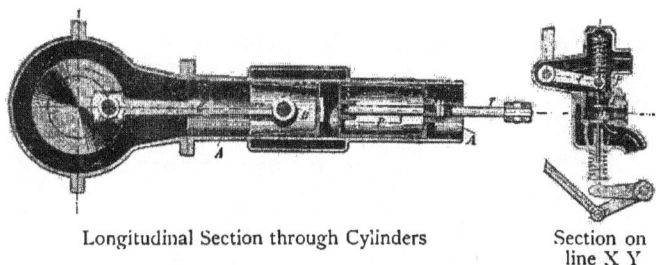

Longitudinal Section through Cylinders   Section on
                                                     line X Y

ing piston of the other cylinder by means of a rod *T* and a cross-head *G.* The crank-shaft carries a pinion meshing with a gear-wheel, the parts being so proportioned that the gear-wheel rotates with half the speed of the pinion. This gear-wheel, lettered *D* in Fig. 2, is provided on each of its lateral faces with a cam *d* engaging a friction-roller *E* connected by means of a rod with the link *F,* oscillating on a pivot *Z.* The shaft *L* of the governing-wheel is formed at one end with a worm *M,* controlling a slide adapted to move vertically in lateral flanges of the slide. This movable slide is connected with a rod *H,* pivoted at its outer extremity to the cross-head *G.*

Let us suppose that the engine is about to begin the first cycle. The piston, as it moves, opens the admission-valve; and the cam *d,* through the medium of the roller *E,* shifts the rod *K* and causes the link F to swing from right to left. If the slide be in such a position that the rod H is in line with the axis of the cylinder, the counter-piston, P, will remain stationary, in which case the quantity of mixture admitted will be equal to the volume displaced by the piston. But if the slide occupy a lower position on the link, the rod H will force the counter-piston into the cylinder simultaneously with the intake stroke of the piston B; the quantity of gas admitted is then diminished. The lower the position of the slide on the link, the smaller will be the quantity of gas admitted; and when the slide is in its lowermost position the gas will be entirely cut off.

If the slide move above the axis of the cylinder, the link being swung in the opposite direction, the counter-piston will move

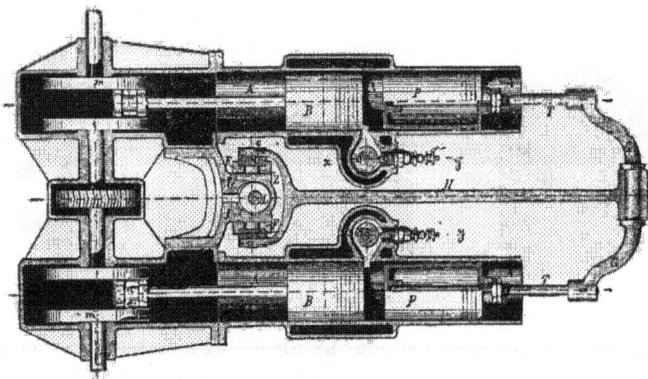

Plan and Horizontal Section

away from the piston B as the slide moves up, so that the counter-piston will suck in an additional quantity of fresh air.

The second cycle—that of compression—now begins. The rollers E continue to turn on the cam d. The link swings from left to right; and the counter-piston returns to its initial position.

During the third and fourth cycles, the rollers turn on the concentric part of the cam d; and the link F, and, hence, the counter-piston P remain stationary. The stroke made by the piston B during this period is a working-stroke, followed by the expulsion of the burnt gases.

Only during the first two cycles is the link oscillated and the counter-piston displaced. During the two remaining cycles both link and counter-piston are stationary; and the slide, and, hence,

An Improved Explosion-Engine

the counter-piston, remain in the positions assumed at the end of the first two cycles.

If the distance between the piston and counter-piston be one-half less than the normal, it follows that, (1) the quantity of burnt gases left in the cylinder at the end of the exhaust-stroke is also one-half less, that (2) the quantity of gas admitted is one-half less, and that (3) the compression will remain the same,

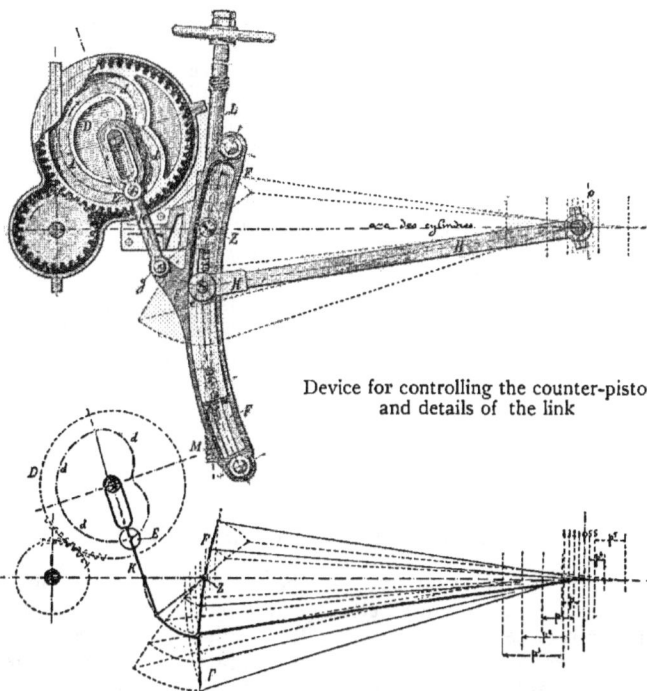

Device for controlling the counter-piston and details of the link

Diagram of the operation of the link

since there will be a volume of gas one-half less, within a space one-half less than the normal. The proportions of the constituents of the explosive mixture remain the same in all cases.

Conversely, when the slide passes above the pivot Z and the counter-piston moves away from the piston, the quantity of gas admitted will be greater, and the compression will be effected within a larger space.

417

The Automobile Magazine

When the slide is in its lowermost position, at which time there is no admission of gas, the counter-piston P approaches the piston B and follows it during its inoperative, intake stroke.

From the foregoing explanation it follows that the link in this engine of Marmonier's is similar to that of a steam-engine, and that it is the principal distributing device.

It has been stated that as the slide nears the end of its travel, the admission-valve is gradually closed. Hence, it follows that the pressure of the gases at the time of their cut-off might be less than that of the atmosphere. In this case the greater pressure of the atmosphere would cause the intake valve to open and admit a new supply of explosive mixture to the cylinder. The method of overcoming the difficulty is shown in the detail section forming part of Fig. 1. During each expansion of the gases in the cylinder, the spring of the intake valve is mechanically subjected to a greater tension than that of the escape-valve. Hence it follows, owing to the greater resistance offered by the intake-valve spring, that the escape-valve will be opened; and only pure air will be admitted to the cylinder.

The valves are thus operated by means of the cam previously mentioned, which cam, during the third cycle, besides controlling the escape-valve, also actuates the lever l, whereby the collar p is caused to tighten the spring.

AUTOMOBILE STREET SWEEPER

An automobile street sweeper, the invention of MM. Thomas and Lerocher, has recently been approved by the head of the street cleaning service of Paris, and will doubtless be in general use before the end of the present summer. The propelling machinery of the new automobile is placed in front, while underneath and behind are a tank and the sprinkling devices; for the wagon sprinkles the streets as well as sweeps them. The brushes come in the rear, and can be raised clear of the street while it is being sprinkled. The design is to run the car over the street to be cleaned, first sprinkling it, and then passing over it again to sweep the dust into small piles. It is the intention of the Paris authorities to also have electric wagons to remove the dust after it has been gathered into the heaps.

The Old and New Bollée Voiturettes

ALTHOUGH Léon Bollée's tricycle-voiturette has met with much favor in France by reason of its strong construction, speed, and simplicity, it cannot be denied that the vehicle is open to criticism. The tandem position of the passengers, the lack of comfort, the low construction and consequent exposure to dust and mud, and the frequent slipping of the single rear wheel offset the merits of the vehicle.

With the purpose of overcoming these defects the Société Anonyme des Voiturettes Automobiles, by whom Bollée's *voitu-*

Bollée's Four Wheeler

rette patents are controlled, have designed a new, four-wheeled model, in which the seats are placed side by side and the frame sufficiently raised to overcome the objections to which we have referred. The simplicity which characterized the construction of the old model has been preserved; a single lever starts, changes the speed, and brakes the carriage.

Of our illustrations, Fig. 1 is a perspective view of the new *voiturette*, reproduced from a photograph. Figs. 2 and 3 are two plan views in which the operating mechanism of a three and four wheeled *voiturette* are diagramatically compared, the dotted lines representing the carriage-bodies.

Those familiar with the tricycle-voiturette will recall that its frame consists of drawn-steel tubes, the two longitudinal members of which, $A\,A$, are connected by three lateral braces $B,\,B',\,B''$, the last of which, B'', is provided at its ends with two bearings $C\,C'$ and an arm D. Two pivots are mounted to turn in the bearings $C\,C'$; and by these pivots the shafts or axles of the front, steering wheels $D'\,D''$ are carried, which wheels are in turn connected by the divided axle $E\,E'\,E''$ and controlled by the wheel G through the medium of a pinion, a rack G', and a lever H. The driving-wheel I can be shifted forward or backward.

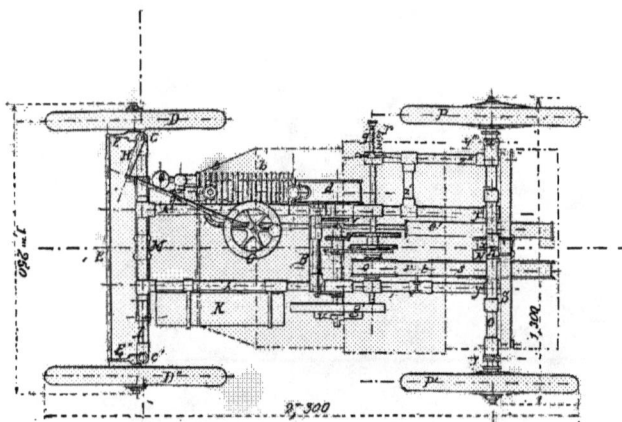

Bottom Plan of Four Wheeler

The motor consists of a flange-cooled cylinder b, a piston, and a head c inclosing the explosion-chamber, the gas-distributing and igniting devices. The connecting-rod is inclosed in a casing d; the carbureter is mounted at e; and the exhaust-muffle at e'.

The motor-shaft carries a fly-wheel g and three gear-wheels, and is journaled in bearings on the longitudinal members $A\,A'$. Parallel with the motor-shaft is an intermediate shaft likewise journaled in the members $A\,A'$ and movable laterally. On the intermediate shaft are mounted three gear-wheels which can be thrown into engagement with the pulley o. By shifting the parts laterally in this manner, the several changes of speed are obtained, a gear-wheel on the intermediate shaft being made to engage the

The Old and New Bollée Voiturettes

corresponding gear-wheel on the motor-shaft. The shifting is effected by a lever connected with the intermediate shaft by means of the pinion p and the rack q.

The lever is also capable of being swung around the shaft in order to move the rear driving-wheel I backward or forward and thereby to tighten or loosen the belt r, running over the pulleys o and s. In this manner the driving-wheel is gradually thrown into and out of gear with the motive parts. The same operation of throwing the driving-wheel out of gear causes the pulley s to be forced into contact with a fixed brake-shoe t, whereby the vehicle is stopped. It is evident that only a single lever is necessary for the various operations of starting the carriage, changing the speed, throwing the driving-wheel out of gear with the inter-

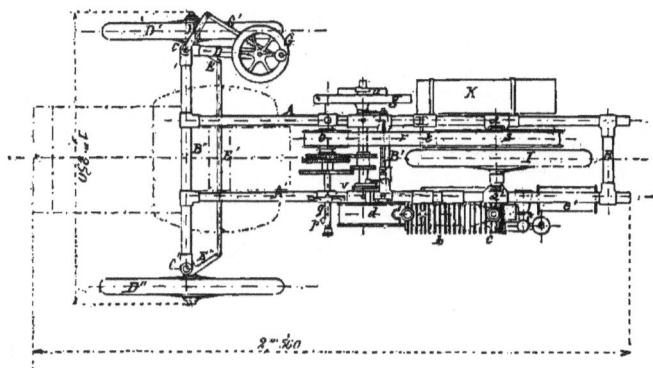

Bottom Plan of Two Wheeler

mediate shaft, and forcing the pulley s into frictional contact with the brake-shoe t.

The motor is controlled by a ball-governor connected with the exhaust-valve and inclosed in a casing u secured to the flywheel g. The exhaust-valve is operated by a cam on the gear-wheel v. The hydrocarbon to be vaporized is contained in a reservoir K. The oil supply for the burner is located behind the rear seat.

The construction of the new quadricycle is essentially the same as that described. The frame has been turned about end for end, as a comparison of the two plan views will prove. The divided axle connecting the steering-wheels has been placed without the frame instead of within; and the steering-wheels have been further connected by an axle L provided with a spring supporting the front portion M of the frame. Like the tricycle-voiturette the new carriage is steered by a wheel G, a rack G', and a lever H; but the rack and lever have been so connected with a rod that they can follow the movements of the front axle. The two rear driving-wheels $P P'$ are joined by an axle O, carrying at its middle a differential gear N, which here takes the place of the rear tricycle axle. The differential gear carries the driving-pulley and a gear for a Lemoine brake, x. As in the three-wheeled carriage the rear axle and its wheels can be shifted back and forth. A single lever is again employed for starting the vehicle, changing the speed, and braking, for which purpose the rear axle is supported by four bearings, at the points y, y', y'', y'''.

In order to hold the bearings in position a cross-piece R is employed which extends above the driving-pulley. Upon this part the carriage body is supported by means of a spring S.

The intermediate shaft has been lengthened to bring the operating-lever further inward; and the braces $z z' z''$ have been arranged accordingly. The carriage is steered with the left hand; and the lever is operated with the right hand, to both of which departures from the usual method, the driver becomes readily accustomed.

At the front of the carriage the oil supply for the burner, the lubricator, and the baggage-carrier are located.

The principal dimensions of the old and new vehicles are given in the following table:

	Tricycle.	Quadricycle.
Total length	8.40 ft.	7.54 ft.
Total width	4.10 "	4.26 "
Total height	3.44 "	4.59 "
Approximate weight (empty)....	440 lbs.	660 lbs.
Speed in miles per hour..........	6, 12, 18	6, 12, 18

FREDERICK GOUGY.

THE KILOMETRE RECORD

M. Jenatzy, who at present holds the Automobile Kilometre Record (about 65 miles an hour), has heard that machines are being prepared to wrest this record from him. He states that if his record is lowered he will build an electric car to cover the kilometre at the rate of 125 miles an hour. These machines are playthings and fulfill no useful purpose.

The Locomobile

OF the three sources of energy (steam, gasoline, and electricity) which are generally used in mechanical traction, steam is preferred by many a *chauffeur* by reason of certain indisputable advantages which it possesses over its rivals. It is true that its employment entails the instalment of a boiler, and a rather complex engine. But the many improvements which, of late, have been made in the construction of steam-carriages have so thoroughly removed these obstacles that steam-driven vehicles are among the most popular at present in use. It can hardly be denied that the interests of steam traction have been very largely benefited by the introduction, both here and abroad, of an American steam-carriage which has become widely known under the name, " Locomobile."

The underlying principles which govern the construction of the Locomobile are not essentially different from those of other steam-carriages; but the vehicle has been so ingeniously designed,

that a description of the arrangement and construction of its driving mechanism will surely prove of interest to our readers.

Steam is generated in a small, tubular boiler and is conducted to a vertical, double-cylinder motor to drive two pistons, the reciprocating motion of which is converted into rotary motion by means of the usual connecting-rods and cranks. This rotary motion is imparted to the driving wheels through the medium of two sprockets and a chain connected with the differential. The entire carriage weighs about 425 pounds empty and 600 pounds loaded.

The Carriage Frame—Our illustrations show a light carriage-body supported on four wheels by a transverse spring in front and two longitudinal springs in the rear. The frame is composed of strong steel tubes 1¼ in. in diameter. The wheels are 28 in., small in size, are made entirely of metal, are provided with ball-bearings, and fitted with pneumatic tires.

The steering mechanism is the simplest imaginable and is composed of a handle bar or wheel controlling a central connecting rod which, by means of two links, operate the front wheels of the carriage.

In Fig. 4. representing a plan view of the " Locomobile," D is the steering bar or lever and T the oil supply tank for the burner

The Locomobile

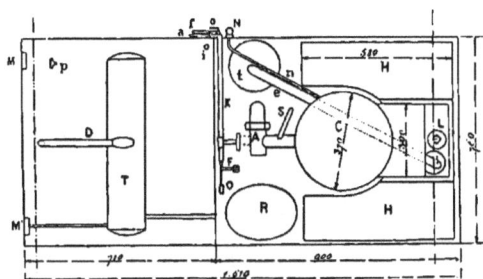

Fig. 4.

beneath the boiler, the tank being placed at the front of vehicle under the feet of the occupants. The boiler C, mounted in the rear, is flanked by the water-reservoir HH. The steam-admission valve is represented by A, the safety-valve pipe by S, and the exhaust outlet by e. The water reservoir is filled by removing the plug L and by inserting a hose. Two gauges M M^1 are provided, one of which indicates the pressure within the boiler and the other the air-pressure, whereby the oil is forced from the tank T to the burner, the pressure being maintained at the desired point by means of a hand-pump.

At the side of the driver's seat are three small levers, a, f, and o. The first lever (a) controls the steam-admission valve A and likewise serves as a means for regulating the speed, since there is no speed-changing gear as in petroleum carriages; the speed is regulated by limiting the quantity of steam admitted to the cylinders. The second lever f enables the driver to reverse the engine and drive the carriage backwards, and is connected with the steam-distributing devices F of the cylinders. The third lever o is connected with the feed-cock $O;$ by its means the water is either pumped to the boiler or returned to the reservoir H.

The Boiler—The boiler is of the tubular type and is remarkable for its very rapid generation of steam. From a technical standpoint it is probably the most interesting part of the entire carriage. It is composed of a cylindrical, sheet copper body, reinforced by two windings of steel wire. At each end of this cylinder is a steel plate pierced with holes which receive 300 copper tubes (a^1, a^2, a^3, etc.), providing flues for the hot gases arising from the burners beneath the boiler.

The body of the burner, like that of the boiler, is cylindrical and is made of sheet metal, and mounted on square supports. A second, interior cylindrical casing receives the hydrocarbon,

425

already vaporized by its passage through a feed-tube, extending through the boiler and connecting with the burners. The second cylindrical casing contains 114 short copper tubes (c^1, c^2, c^3, etc., fig. 5), forming as many air-flues about which are located 20 capillary orifices through which the flames of the burning oil escape to heat the 300 vertical tubes of the boiler.

It will be understood that the oil flowing from the tank T is heated and vaporized as it passes vertically through the boiler, then escapes through the capillary orifices and ignites as it comes into contact with the air flowing through the small air-flues. But when the boiler is first started, some auxiliary means are evidently necessary to volatilize the oil. The means consist of an auxiliary heating tube, which connects the burner with the oil-feed tube.

The steam is conducted through a pipe in the upper part of the boiler, to the admission-valve and thence to the cylinders. The products of combustion escape in the rear of the carriage.

In order to control the oil-feed of the burners, an ingenious, automatically-operating apparatus is provided, which is represented in Fig. 6. This oil-feed regulator is essentially composed of a metallic diaphragm held between the two members E. The right side of the diaphragm is subjected to the boiler-pressure; and the left side is connected with a needle-valve controlling the flow of oil, the stem of the valve being encased in the body D of the regulator. The oil flows through the tube A to the burner B, first passing through the opening commanded by the valve. The auxiliary heating-tube referred to above is mounted at C. The tube F connects the diaphragm with the boiler A so that the steam-pressure can act upon the diaphragm. When the pressure becomes excessive the needle-valve partially closes the feed opening and reduces the quantity of oil supplied to the burner. When the maximum pressure is reached the opening is almost closed.

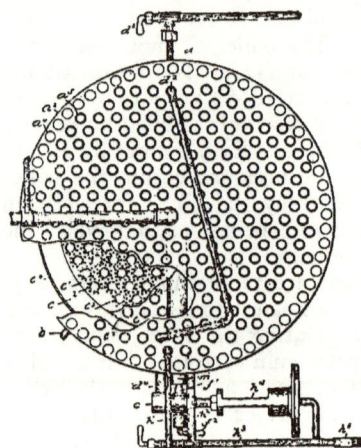

Fig. 5.

The Engine—The motor (Figs. 7 and 8) consists of two

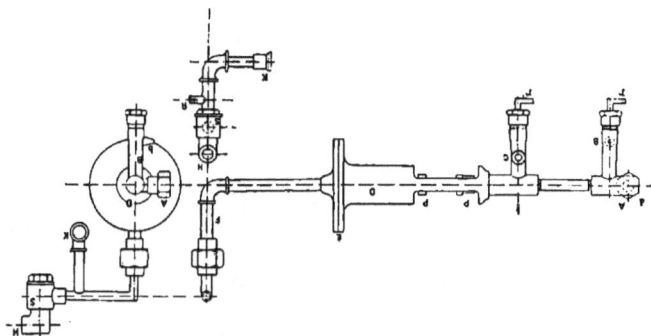

Fig. 6.

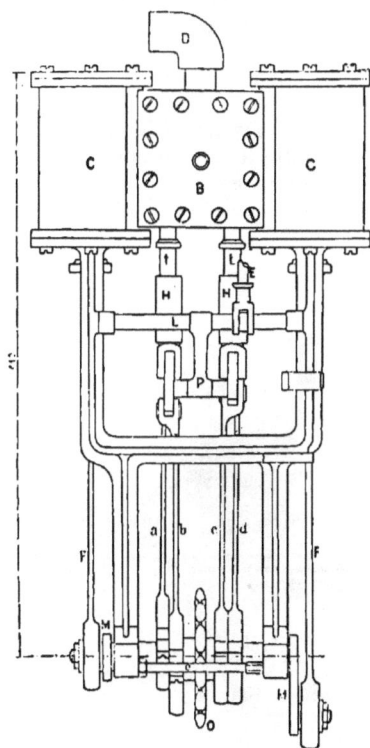

Fig. 7.

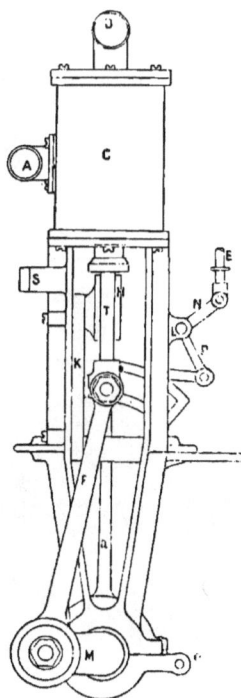

Fig. 8.

vertical cylinders *C C* and constitutes essentially a miniature
marine engine. Steam is conducted to the steam chest *B* and to
the cylinders by the pipe *D*. The slide-valve stems are represented by *t t;* their guides by *H H*. Two slides are provided,
operated by the levers *E, N, P,* which actuate the rods of the
eccentrics *a, b, c,* and *d*. *F F* are the connecting-rods; *M M* the
cranks; *o* a 12-tooth sprocket-wheel, chain-connected with a 24-
tooth sprocket in the differential gear, mounted in the center of
the axle of the rear driving-wheels. The steam is exhausted
through the pipe *A*, with no noise whatever. The cranks
and eccentrics have ball-bearings, thereby reducing friction and
noise. Two brakes are provided, one connected with the differential and operated by a pedal, the other connected with friction
bands on the driving-wheels and operated by a lever.

The boiler is subjected to a water pressure of 600 pounds to
the square inch, and the construction is such that it is impossible
to explode it.

The Locomobile is noteworthy for its easy running and total
lack of noise.

This was the verdict of Fire Chief Croker of New York, who
used this machine for several months while responding to fire
alarms. At his suggestion Fire Commissioner Scannell likewise
took with him a Locomobile when he sailed to Cuba to reorganize
the Fire Department of Havana.

The Automobile
MAGAZINE
An *ILLUSTRATED* Monthly

VOL. I NO. 4   NEW YORK JANUARY 1900   PRICE 25 CENTS

The AUTOMOBILE MAGAZINE is published monthly by the United States Industrial Publishing Company, at 31 State Street, New York. Cable Address : Induscode, New York. Subscription price, $3.00 a year, or, in foreign countries within the Postal Union, $4.00 (gold) in advance. Advertising rates may be had on application.

Editorial Comment

I F the new year witnesses the same degree of progress in automobile affairs that has distinguished 1899, then the visible effects of the work that has been done will be something prodigious. There is, indeed, no occasion for the use of the qualifying " if." It is safe to make the positive assertion that such will be the case, and that, moreover, the progress made will be even greater than that represented in the phenomenal record of 1899. The past year has been emphatically one of organization. The extent to which tremendous undertakings have been set on foot has constituted one of the most dramatic passages in the history of modern industrial development. Previous to the year 1899 everything done was in the way of preliminary work: experimentation, preparing the ground, and laying the foundations; all proceeding so quietly as to attract little general attention. Only a few far-sighted observers realized that one of the greatest of the advances in this century of achievements was at hand—an advance that men had been looking for ever since the invention of the steam engine made mechanical traction a potentiality.

When capital suddenly came to a realization of the magnificent scope of this new opportunity, it found the preliminary work thereby carried to a stage so advanced that little remained to be done in order to proceed with production on the enormous scale required by the demand awaiting to be supplied. So it was natural that, when the work of organization was taken in hand on a scale so tremendous early in the past year, the same makers should have been called to a leadership in the work which will make the bicycle industry, enormous as it has grown to be, insignificant in comparison with the automobile, and which will spread the demand for good roads into every section of the United States and Canada. It is therefore safe to predict that long before the Twentieth Century has half run its course—and in all probability before its second quarter has begun—it will be possible to start from any city of the United States and proceed by automobile in perfect comfort over perfect highways through to the City of Mexico. And that will be a trip well worth the making.

AUTOMOBILE CLUBS AND THEIR FUNCTIONS

The millions of capital that have been invested in automobile undertakings the past year—such as the establishment of enormous manufacturing plants and the formation of powerful companies in all the principal centres for the systematic conduct of the business—will, in the course of this year, begin to make a great showing for the expenditure, and it will not be long before the automobile will be as familiar a sight as the bicycle, not only in the cities, but in the populous country districts as well. Every approved form of motor-vehicle—whether electricity, steam, or petroleum, furnish the motive-power—appears now to be developing along the right lines, and each will have its due place in the great work now under way.

Another direction in which admirable progress has been made lies in the social organization of the movement. The Automobile Club of America, recently formed here in New York, is the pioneer of its kind in this country, and it almost immediately proved the immense value of associated effort by straightway acting with energy to secure the opening of Central Park to the automobile on a par with other pleasure-vehicles. Similar organizations are talked of for Chicago, Philadelphia, Boston and other great centres, and will doubtless be effected early in the course of the year.

Editorial

Among the activities in which such organizations may be expected to take the lead is the organization of meets, races, long-distance and short-distance tours, exhibitions, etc., for their respective localities, and these activities will contribute powerfully to the cultivation of a general feeling of community of interests among all concerned in the movement. An invaluable function, of course, as we have seen in the case of the New York organization, lies in securing to the automobile its proper rights in the public highways and pleasure-grounds, as well as in the defense of automobilists against the various aggressions and persecutions certain to arise here and there. Another important function lies in the securing and diffusion among members of all sorts of information concerning roads, routes, wayside taverns, etc., together with the establishing of stations for replenishment of motive-power, repairs, etc. For instance may be cited the following inquiry recently made of a technical journal:

" Do you know if such a thing is to be had as profile maps of the various roads radiating from New York, showing at a glance distances, gradients and length of same, together with descriptions of the roadbed? I have been looking for such maps, and I have so far been unable to find them, although they are in use in England, and are of great assistance, especially to cyclists and automobilists. Of course, if any one takes the trouble, he can take the ordinary contour maps and make for himself a profile plan, but the average man does not care to go to this trouble, but would rather trust to chance as to what he will find. Such books would be extremely useful to people desiring to go as far as possible on one charge."

The example of the League of American Wheelmen in furnishing to its members systematic information similar to that desired in the foregoing is one that may well be followed. For instance, the Massachusetts Road Book, supplied to its members by the Massachusetts division of the League, is a model of its kind, containing not only excellent maps of the roads throughout the State, but lists of routes, and much compact information of the kind desired as to character of roads, gradients, etc. Since automobilists are interested in many of the same things that wheelmen are it has been suggested that mutual interests would be promoted if the former would join the L. A. W. It would, indeed, be a most advisable thing to do. The return for the small entrance-fee and annual charge is very large.

Good Roads and Automobiles

An automobile street-sprinkler would be something very useful, and it would effect a great economy in the work. One form of automobile street-sprinkler is that in use on many electric street-railways, and it has proven itself very efficient. Its advantages, however, are confined to those streets where there are car-tracks. An automobile street-watering cart free to run anywhere would extend the benefits of the system to all parts of a town. It would not only do the work more evenly than with animal traction, but would accomplish it so much more quickly that one automobile-sprinkler could cover a much larger territory in the same time, and therefore do the work probably of at least two carts drawn by horses. Another great economy lies in the fact that where street sprinkling is done by contract the arrangement has to be for a season of so many days, regardless of the fact that wet weather may make sprinkling unnecessary for many of these days. With automobile watering-carts the necessity for such a waste in expenditure would be overcome. Running-expense would be confined to the time of actual operation, and not extended over the periods when horses stand idle in the stable. An official in one of the largest park departments of this country, where the cost of road-sprinkling is something over ten thousand dollars a year, says that the saving by the use of automobile-sprinklers would amount to fully half that sum. An additional advantage of the automobile-sprinkler would be the fact that, therewith, road-sprinkling would be practicable in rural communities where there is no system of public water-supply. The advantage of dustless roads would therefore be greatly extended, to the great comfort and convenience of all concerned, and it would save macadamized State highways, constructed at great expense, from the destructive influences of seasons of drought, when for the lack of binding moisture the road-bed disintegrates and is ground into fine powder that blows away. For it would easily be possible so to construct an automobile road-sprinkler that, by means of its own motive-power, it could quickly fill itself by pumping from a pond, stream, or well, so that it would not be necessary to pump water into an elevated tank for the purpose. It would thus be possible to keep the roads in large public parks well sprinkled without going to the great expense of either building such tanks at frequent intervals or of extending the mains from the public supply solely for such a use. Yet another advantage of the automobile-sprinkler would be its additional function

Editorial

as a road-roller. For it would naturally be provided with heavy wheels having a very wide tread. And the weight of the water, carried in quantity much larger than it would be practicable for horses to pull, would make a most admirable road-roller of the machine, constantly improving the surface while sprinkling it.

THE EXAGGERATION OF MOTOR HORSE-POWERS

Apart from durability and economy, efficiency is one of the most vital points to be considered in the purchase of a motor for automobile use. If our automobilists were to subject their motors to the brake-test they would find them credited with horse-powers far in excess of that which they are actually capable of developing. The evil of exaggerating the commercial rating of engines has been admirably discussed in a trenchant article from the pen of a prominent engineer, which we have published in this issue, and which will probably offer food for thought to many of our readers. The simple brake-test described in the article in question, although well known to every engineer, will probably be new to many automobilists.

American manufacturers are capable of building the finest motors in the world—motors which have earned for themselves well-merited praise wherever they have been used. It seems on this account all the more deplorable that American makers should resort to means so utterly unworthy of them in advertising their product.

WARNING-SIGNALS

What should constitute the warning-signal for automobiles? This is a very important question. In Europe the pneumatic horn is commonly used, making a noise similar to that of a steam-whistle. In this country the tendency is to employ a gong for the same purpose. This, however, is open to the objection that the equipment is similar to that of the electric street-car, so that accidents are liable to arise through mistaking the approach of an automobile for that of the latter, the passer therefore deeming himself out of danger when clear of the car-tracks. With the use of the horn, however, there would be no danger of such a mistake. If the horn were adopted for the automobile in this country, its use would then be practically universal for the automobile, and uniformity the world over in such a matter is something desirable. Perhaps legislation to that end may be found necessary.

The Automobile Magazine

MOTOR RACING

The record-breaking run of an American electric automobile over a course of 100 miles, which is so entertainingly described by Hiram Percy Maxim in this issue, serves to draw attention to the subject of motor racing. This is a sport which is bound to come into great vogue, stimulated as it has been by the recent gift of James Gordon Bennett's International Challenge Cup.

Only last July Mr. Bennett's newspaper, the New York *Herald,* published and double-leaded a cable despatch from Paris announcing that the world's record for automobiles had been broken by Count Chasseloup-Laubat, who had succeeded in making a journey of 81.6 miles on a single charge of an electric storage battery weighing 2,000 lbs., while the vehicle itself weighed 2,400 lbs. more. The *Herald* then declared that this "opened up possibilities of a far reaching character."

Mr. Maxim's run of a trifle more than 100 miles was made in 7 hours and 45 minutes. The distance was carefully measured off by three bicyclists who went in advance of the motor carriage setting up marks at every fifth mile. The time of both start and finish were recorded by disinterested spectators who were called on as witnesses. The storage battery of the vehicle which accomplished this run on a single charge weighed 980 lb., while the vehicle without the batteries weighed 1,200 lb., bringing up the total weight to 2,500 lb., with the passengers. The voltage of the battery was 98 at the start, fell to 97 during the second and third hours, 93 during the fourth hour, 92 in the fifth, 89 during the sixth, finally falling to 81.7 toward the close of the seventh hour. A good part of the distance was covered at a three-minute gait, and the average speed for the whole distance was 16.8 miles per hour. The battery charge for the 100 mile journey was 190 ampere hours, of which the cost was about $1.25. The actual cost of this amount of energy at the dynamo is 41 cents.

Mr. Maxim is to be congratulated on this brilliant demonstration of the *Herald's* prediction of American possibilities.

A WARNING TO THE PUBLIC

A timely note of warning has been raised by a writer in *La Locomotion Automobile.* This gentleman, who is a lawyer and practising member before the French Court of Appeals, warns

Editorial

intending exhibitors at the Paris Exposition of 1900 to be on their guard. Not only does he apprehend great danger from unscrupulous imitators of new devices and inventions not covered by French patent rights, but he is furthermore of the opinion that valuable trade names will be similarly appropriated.

This is a timely warning and one that is bound to be heeded by the exhibitors of automobiles and automobile appliances. In the past, as will be recalled by many American manufacturers, there have been so many complaints on that score in France that intending exhibitors should take every precaution to safe-guard their wares. We are in hearty accord with the suggestion of *La Locomotion Automobile* that a judiciary tribunal should be appointed during the Exposition year in Paris with powers to make a quick disposition of all cases of infringement of patent rights and trade-marks arising between French manufacturers and foreign exhibitors.

THE LATEST POSTAL REFORM

Since the publication of the latest annual report of the Postmaster General, new efforts have been made to carry out his suggestion for the adoption of automobiles in the postal service. By a circular request of the Assistant Postmaster, inventors and manufacturers have been invited to submit their designs for postal automobiles to the authorities at Washington. Besides this the postmasters of likely cities, such as Brooklyn, Buffalo, Boston, Washington and Chicago, have been authorized to conduct a series of tests with such automobiles as may be submitted to them by local manufacturers.

The results of these tests and of this new departure in our postal service will be submitted to the readers of our next issue in an able article by First Assistant Postmaster General Perry S. Heath, amply illustrated by photographs and drawings of postal automobiles already in use. Most of these are taken from illustrations of postal cars and delivery wagons adopted by the governments of Germany, France and England, but a few, it will be noted, are now already in active use in some of our American cities.

The Speeding of the New Year
(VICTOR GILLAM in *Judge*)

His Occupation Going
(CHARLES NELAN *in the New York Herald*,

Automobile Humors

THE MERRY CHASE

1700

1900

(E. Nicolson in *Charivari*)

One Thing the Machines Can't Do

FARMER GREEN—" If they keep on usin' bicycles an' horseless kirrages,
I'd like to know what the horses are goin' to do?"
OLD HORSE—" Don't worry; I can kick."—*Puck.*

The Automobile Magazine

EVERY DOG HAS HIS DAY.

Mirza guards the automobile while Gaston and Yvette seek a secluded nook.

A brave Dane profits by the occasion to flirt with little Mirza.

He wags his tail so expressively that he turns the lever and starts the machinery. Buzz! Buzz! goes the motor, and they are under way.

Behold Mirza and her gallant escort making the rounds of the Bois de Boulogne at the rate of 40 kilometers per hour despite all ordinances.

(B. Rabier *in the Journal Amusant.*)

Automobile Humors

Off for a New Crusade
(HORACE TAYLOR in *The Verdict*)

LOVE AND SCIENCE

We hear of horseless carriages,
  Propelled by unseen force ;
Also of loveless marriages,
  Which generate divorce.
We hear of wireless telegrams,
  A wonder of our day ;
But 'twixt them armless courtships
  Will never come to stay.

The Automobile Index

Everything of permanent value published in the technical press of the world devoted to any branch of automobile industry will be found indexed in this department. Whenever it is possible a descriptive summary indicating the character and purpose of the leading articles of current automobile literature will be given, with the titles and dates of the publications.

Accumulator Charging Plant—

A brief notice referring to the Berliner Maschinenfabrik's portable combination for the charging of electric accumulators. One illustration. "The Motor-Car Journal," London, November 10, 1899.

Accumulators—

A serial article, by E. C. Rimington, on the construction of accumulators for automobiles. "The Automotor Journal," London, November, 1899.

Automobile Fore-carriage—

Five illustrations showing the Heilmann automotor fore-carriage as applied to various vehicles. "La France Automobile," Paris, November 12, 1899.

Automobile Management—

A serial article by "Philauto," giving practical hints to the users of automobiles. With four illustrations. "The Motor-Car World," London, November, 1899.

Automobile Street Sweeper and Sprinkler—

From *La Nature*. Gives illustrations and particulars of an apparatus which has given very satisfactory results in Paris. 500 w. "Scientific American" (Supl.), October 7, 1899.

Automobiles and Public Health—

By James J. Walsh, M. D. "The Automobile Magazine," December, 1899.

Axles—

Illustrated description of the Stourmel axle for the front motor and steering gear of delivery wagons. "L'Industrie Automobile," Paris, October, 1899.

Balancing of Motors—

A mechanical study by H. E. Wimperis, Wh. Sc., with illustrations. "The Automotor Journal," London, November, 1899.

Brakes—

Daniel Dujon's new "cable brake," described and illustrated. "La Locomotion Automobile," Paris, October 19, 1899.

Carbureters—

An improved form of carbureter devised by Mr. M. H. Lepape, of Paris. Described and illustrated. "The Motor-car Journal," London, November 10, 1899; also "The Automotor Journal," November, 1899.

Compensating Crank Shaft—

A full description of Whitney's compensating crank shaft. Three illustrations. "The Motor Age," Chicago, November 23, 1899.

Compensating Gear—

Description of a compensating gear invented by Messrs. W. E. Wentzel, of Lynn, and G. E. Whitney, of Boston, Mass. Four illustrations. "The Motor Vehicle Review," Cleveland, O., November 14, 1899.

The Automobile Index

Elastic Clutch—

Description and illustration of the Syner's elastic clutch, by William Rogers, C. E. "The Automobile Magazine," December, 1899.

Electric Automobiles—

Electrically Driven Automobiles. (Ueber Elektrisch Betriebene Selbst fahrer). C. P. Feldmann. A review of the electrically driven vehicles shown at the recent Paris competitions. "Elektrotechnische Zeitschrift," October 5, 1899.

Electric Cabs in Chicago. Illustrates and describes the vehicles of the Illinois Electric Vehicle Transportation Company, giving related information. "Western Electrician," October 7, 1899.

Pullen Company's electric carriage, described and illustrated. "The Automobile Review," Chicago, November, 1899.

Description of Wood's various vehicles, with five illustrations. "The Motocycle-Automobile," Chicago, November, 1899.

The new Hurtu carriage described by Mr. A. Delasalle. Five illustrations. "La Locomotion Automobile," Paris, November 9, 1899.

Illustrated description of the "Columbia" electric omnibus. "Elec. World and Engineer," New York, November 18, 1899.

The Keating delivery wagon described and illustrated. "The Motor Vehicle Review," Cleveland, O., November 21, 1899.

Illustrated description of an electric carriage built by the United States Automobile Co. "The Motor Age," Chicago, November 30, 1899; also "Motor Vehicle Review," Cleveland, O., December 5, 1899.

Description of a new carriage built by the Hub Motor Co. One illustration. "The Motor Age," Chicago, November 23, 1899.

The latest Oppermann electrical dog-cart, fitted with its new chainless driving gear. An illustrated description. "The Motor-Car Journal," London, November 24, 1899.

The latest Columbia vehicles, described and illustrated. "The Motor Vehicle Review," Cleveland, O., December 5, 1899.

"The Electric Automobile," a serial article by Prof. Felicien Michotte (continued from "The Automobile Magazine" for November). "The Automobile Magazine," December, 1899.

Electric Motors—

Illustrated description of a new motor built by the Lincoln Electric Co. "The Motor Vehicle Review," Cleveland, O., December 5, 1899.

Electric Touring—

An article on the use of electrical automobiles by tourists in France. "La Locomotion Automobile," Paris.

Exposition—

The International Motor Carriage Exposition at Berlin. An Account of the opening of this exposition, with illustrated description of some of the vehicles. U. S. Consular Reports, No. 561, October 24, 1899.

First American Automobile Club Run—

By Edgar S. Hyatt. "The Automobile Magazine," December, 1899.

Hints on Avoiding Accidents—

Advice to drivers of automobiles on necessary conditions for safety. "Cycle et Automobile Industriel," Paris, September, 1899.

Hydro-carbon Automobiles—

The new "Star" automobile, as exhibited in the London National Show. An illustrated description. "The Autocar," Coventry, November 25, 1899.

The Beeston Light two-seated automobile, described and illustrated. "The Autocar," Coventry, November 25, 1899.

Illustrated description of a new automobile built by the Alldays & Onions P. E. Co. Its motor is of the double horizontal cylinder type. "The Autocar," Coventry, November 25, 1899.

A new voiturette invented by André Py, of Paris; described and illustrated. "L'Industrie Automobile," Paris, October, 1899.

Description of the new Rossel gasoline carriage, with four illustrations. "L'Industrie Automobile," Paris, October, 1899.

Illustrated description of the "Rochet" carriage. "Le Chauffeur," Paris, October 11, 1899.

Illustrated description of several vehicles propelled by acetylene motors. "The Automobile Review," Chicago, November, 1899.

Illustrated description of the Haynes-Apperson phaeton and of its run of 1,050 miles in twenty days. "The Motocycle-Automobile," Chicago, November, 1899.

Description of the Société Bourguignone's three-seated voiturette. Three illustrations. "The Motor-Car Journal," London, November 10, 1899.

Description of the "Esculape" voiturette, by Mr. Georges Cruchet. With two illustrations. "Le Chauffeur," Paris, November 11, 1899.

A new style of Bollée four-wheeled voiturette, described by Mr. Frederik Gougy. Three illustrations. "Le Chauffeur," Paris, November 11, 1899.

A description of the Renault voiturette. With four illustrations. "The Autocar," Coventry, November 11, 1899; also "La France Automobile," November 12, 1899.

The Humber "Sociable Quadricycle," the Daimler Co.'s "New Parisian Car," the Roots and Venables' heavy oil carriage, the Dennis light doctor carriage, and the "Swiss Mountaineer" automobile. A descriptive article, with illustrations, of these vehicles as exhibited in the National Show (London). "The Motor-Car Journal," London, November 24, 1899.

Illustrated description of the Ernst petroleum motor carriage. "The Automobile Magazine," December, 1899.

Hydro-carbon Motocycles—

The Sanciome petroleum-bicycle, described and illustrated. "The Automobile Magazine," December, 1899.

The Butikofer Motocycle. A description and illustration of same. "The Automobile Magazine," December, 1899.

The Renaux gasoline tricycle described and illustrated. "L'Industrie Automobile," Paris, October, 1899.

Illustrated description of the "Rochet" motocycle. "Le Chauffeur," Paris, October 11, 1899.

Description and illustration of Prof. Herring's motocycle called the "Mobike Tandem." "The Autobain," November, 1899.

Description of Werner's new motor-bicycle. One illustration. "The Motor-car Journal," London, November 17, 1899.

An illustration showing the new motor bath chair made by the Rover Cycle Co. of Coventry, Eng. "The Autocar," Coventry, November 18, 1899.

Illustrated description of the Enfield quadricycle. "The Autocar," Coventry, November 25, 1899.

Description and illustration of the Edmond tricycle. "Cycle and Auto. Trade Journal," Philadelphia, December 1, 1899.

Description and illustration of the Boyer bicycle. "Scientific American" (Supl.), New York, December 9, 1899.

The Girardot gasoline bicycle, illustrated. "Scientific American" (Supl.), New York, December 9, 1899.

Illustrated description of the Lamandìre et Labre bicycle. "Scientific American" (Supl.), New York, December 9, 1899.

Mechanism of the Girardot bicycle, illustrated. "Scientific American" (Supl.), New York, December 9, 1899.

The Automobile Index

The Richard-Choubersky automobile tandem, described and illustrated. "Scientific American" (Supl.), New York, December 9, 1899.

Hydro-carbon Motors—

A full description of the Henriod motor, with four illustrations. "L'Industrie Automobile," Paris, October, 1899.

Illustrated description of a multiple-cylinder motor patented by the Gesellschaft fur Automobilewagenbau, of Germany. With four illustrations. "La France Automobile," Paris, November 5, 1899.

A full description and illustration of the Dawson self-starting, reversing and power-increasing motor. "The Autocar," Coventry, November 11, 1899.

Description of a new gasoline motor designed by Mr. M. T. Minogue, of Springfield, O. Three illustrations. "The Motor Vehicle Review," Cleveland, O., November 14, 1899.

A new motor built by "La Minerve" Company. An illustrated description. "La Locomotion Automobile," Paris, November 16, 1890.

The Kuhlstein-Vollmer detachable motor, described and illustrated. "Cycle and Auto. Trade Journal," Philadelphia, December 1, 1899.

Igniters—

An illustrated description of the Houpied igniter. By Paul Sarrey, "The Automobile Magazine," December, 1899.

Illustrated description of the Gianoli & Lacoste devices for ignition in motors. "Le Chauffeur," Paris, October 11, 1899.

Simms' patent ignition gear, described and illustrated. "The Autocar," Coventry, November 18, 1899.

Mechanical Propulsion and Traction—

By Prof. G. Forestier. "The Automobile Magazine," December, 1899.

Motocycle Management—

Serial articles, by Mr. A. J. Wilson, under the title of "Motor Cycles and how to manage them." Three illustrations. "The Autocar," Coventry, November 11 and 18, 1899.

Motor Vehicles—

Horseless carriages. By George N. Crouse. Brief review of progress in France, England and the United States. "Yale Sci. Monthly," October, 1899.

Reformation of Horseless Vehicles—

An article by Miss Clara Fazan on the necessity of reforming the construction of automobiles. "The Motor-car Journal," London, October 27, 1899.

Santa Claus Gets a New Plaything—

By Sylvester Baxter. "The Automobile Magazine," December, 1899.

Speed Varying Gear—

Description of Gaillardet's speed varying gear, with three illustrations. "La France Automobile," Paris, November 5, 1899.

The Newton speed varying gear, described and illustrated. "The Autocar," Coventry, November 18, 1899.

Steam Automobiles—

The Simpson and Bodman Steam Lorry. An illustrated detailed description of vehicles of new design, the generator, and tests made. "Automotor Journal," October, 1899.

Description of the "Stanley" voiturette, with eight illustrations, by Mr. Paul Sarrey. "La Locomotion Automobile," Paris, November 2, 1899.

The Baldwin steam carriage, described and illustrated. "The Motor Age," Chicago, November 23, 1899.

Illustrated description of the Leach steam motor carriage. "The Motor Age," Chicago, November 23, 1899; also "The Automobile Review," Chicago, November, 1899.

The Automobile Magazine

Description of some steam vehicles built by E. F. Brown in 1882-83. Two illustrations. "The Motor Age," Chicago, November 30, 1899. Steam carriages of the Société Européenne d'Automobiles. An illustrated description. "The Automobile Magazine," December, 1899.

Steam Dray—

The Coulthard new three-ton steam dray, described and illustrated. "The Motor-Car Journal," London, November 10, 1899; also "The Autocar," Coventry, November 18, 1899.

Steam Motors—

Illustrated description of the Chandler device to control the distribution of steam in automobile motors. "La Locomotion Automobile," Paris, November 2, 1899.

Steam Omnibus—

A brief description of the Coulthard colonial steam omnibus now in use in West Australia. One illustration. "The Autocar," Coventry, November 18, 1899.

Steering—

Serial articles, fully illustrated, on steering by means of the divided axle. Dr. C. Bourlet. "La Locomotion Automobile," Paris, October 19 and 26; also November 2, 9 and 16, 1899.
"The Steering of the Automobile." a study on this subject by "Indus." With eight illustrations. "The Motocycle-Automobile," Chicago, November, 1899.
Robert W. Jamieson's steering device, described and illustrated. "The Horseless Age," New York, November 9, 1899.

Steering Mechanism—

Illustrated description of a new steering mechanism invented and patented by Mr. A. V. Kendall, of Hartford, Conn. "The Motor Vehicle Review," Cleveland, O., November 21, 1899; also "The Motor Age," Chicago, November 23, 1899.

The Horseless Fire Engine—

By Captain Cordier of the Paris Sapeurs-Pompiers Regiment. "The Automobile Magazine," December, 1899.

The Influence of Automobilism on Transportation—

A work by Herr Berdow, of Leipsic, Germany.

Trials—

Competitive trials of electric accumulators in Paris. Results of tests made during the months of August and September, 1899. "La Locomotion Automobile," Paris, October 19, 1899.
Competitive trials of hydro-carbon motors in Paris. Described and illustrated. "La Locomotion Automobile," Paris, October 19, November 2, 9 and 16, 1899.

Ventilation—

The Removal of Odors from Electric Battery Vehicles (Geruchbeseicigung in Akkumulatorwagen). R. Ulbricht. Describing methods of ventilating the batteries of electrical vehicles so as to prevent unpleasant odors. "Elektrotechnische Zeitschrift," September 28, 1899.

Wheels—

An article on the necessity of a type of wheel that will withstand different varying stresses. The principles of the compression and suspension types are explained in detail. "The Motor Age," Chicago, November 14, 1899.

The World on Wheels—

By Edwin Emerson, Jr. Historical review of the evolution of vehicles. Illustrated. "The Automobile Magazine," December, 1899.

Women and Automobilism—

By Miss N. G. Bacon. Paper read at Lady Harberton's house in London. "The Automobile Magazine," December, 1899.

Electric Vehicle Company

COLUMBIA AUTOMOBILES

ELECTRIC AND GASOLINE

Runabouts

Victorias

Delivery Wagons

Omnibuses

Dos-a-Dos

Landaus

Phaetons

Hansom Cabs

Broughams

Opera"Bus"

OPERATING AND SELLING COMPANIES

State of New York NEW YORK ELECTRIC VEHICLE TRANSPORTATION COMPANY, - 100 Broadway, New York City.
State of Pennsylvania PENNSYLVANIA ELECTRIC VEHICLE COMPANY, - - 815 Drexel Building, Philadelphia.
State of Illinois ILLINOIS ELECTRIC VEHICLE TRANSPORTATION COMPANY, - 1215 Monadnock Building, Chicago, Ills.
New England States NEW ENGLAND ELECTRIC VEHICLE TRANSPORTATION Co., - 527 Exchange Bldg., Boston, Mass.
District of Columbia, WASHINGTON ELECTRIC VEHICLE TRANSPORTATION Co.,
Panorama Bldg., 15th St. and Ohio Ave., Washington, D. C.
Manager for Europe, HART O. BERG, - - - - - - - - 54 Avenue Montaigne, Paris, France.

For territory not represented by local companies, all communications should be addressed to the

ELECTRIC VEHICLE CO. 100 Broadway, New York City

The Automobile Magazine

VoL. I   FEBRUARY 1900   No. 5

CONTENTS

AGENCY FOR FOREIGN SUBSCRIPTIONS :
INTERNATIONAL NEWS COMPANY
BREAMS BUILDINGS, CHANCERY LANE   STEPHAN STRASSE, No. 18
LONDON, E. C.   LEIPSIC

Price 25 Cents a Number; $3.00 a Year
Foreign Subscription $4.00, Post-paid

Modern Mail Collection in America

The Automobile
MAGAZINE

VOL. I   FEBRUARY 1900   NO. 5

Automobile Postal Service
By Perry S. Heath

First Assistant Postmaster General

A S the main object of the Post-office Department is to provide for the safe and prompt transmission and delivery
of mail matter, and as the officials who are entrusted
with the administration of this important branch of the public
service, as well as the legislators who provide for its maintenance,
desire that it shall be conducted on strictly business principles,
the different bureaus of the Department are constantly considering new inventions and devices which may prove of value, either
by increasing the efficiency or diminishing the cost of the service.

The appearance on the streets of our cities of motor driven
vehicles, and their employment by business establishments as
delivery wagons, has suggested their use for the collection of
mail from street letter boxes. The subject has been brought to
the attention of the Department, not only by manufacturers and
agents, who desire to further their own interests, but also by postmasters, who believe that this new departure in locomotion will
aid them in solving one of the most difficult problems which the
free delivery service in large cities presents—that is, the improvement of the collection service by providing for more frequent
collections of mail from street letter boxes, and for more prompt
and rapid transmission to post-offices and stations. This service
is now performed by letter-carriers, and is costing the Government over four hundred thousand dollars ($400,000) annually
for horse hire and contract wagon service, in addition to the
salaries of the carriers.

The Automobile Magazine

Practical tests made in some of the larger free delivery cities have proved the adaptability of the automobiles for this work, but these tests were made in Buffalo, Washington, Detroit, and other cities where the streets are smooth, the boxes well placed, and all the conditions most favorable. The substitution of the automobile for the horse and wagon, and the introduction of a new collection system, would necessarily involve many radical changes in a service which is now thoroughly organized and giving general satisfaction, and the Department, while aiming to keep abreast of the most progressive business interests, would not be justified in making such changes until experience has proven the utility of the new style of vehicle under all existing conditions. Several manufacturers are now working on plans for automobiles, to be used exclusively in the postal service. Every facility will be afforded them for testing these vehicles when completed, and it it believed that it will not be long before the inventive genius which has given us the horseless carriage will have overcome all difficulties. The Post-office Department will then be able to supply vehicles built to travel over all kinds of roads, in all kinds of weather, and vastly superior, as regards safety and speed, to the

Dr. Martin, of Buffalo, and his " Pioneer."
The First Test of Automobile Mail Collection in America.

446

Automobile Postal Service.

Postal Automobile in Berlin.

wagons and carts drawn by horses, now used in the free delivery service.

The employment of the automobile in the collection and distribution of mail matter was one of the first uses which suggested itself, when the possibilities of the motor vehicle for public service began to be considered. Certainly in no other single field does it hold out greater probabilities of benefit to the whole people. Traveling at a considerably greater speed than the street cars, and enabled to reach points by far more direct routes and without numerous delays, the advantages which the vehicles possess for city postal service are readily apparent. On the other hand, with rural free delivery likely to ere long become an established fact in all the more densely populated sections of the country it may be seen also that by the utilization of the new motive facilities the residents of our farms will in many cases be enabled to enjoy almost as many opportunities for quick communication as their city cousins.

The use of the automobile for postal service in Germany and France has long since passed the experimental stage. The post-office authorities of Berlin some time ago put into practical use six of the Loutzki automobiles, and so successful was the experiment that a large additional number of the vehicles were soon after ordered from the manufacturers. The experiment in Berlin so completely demonstrated the efficiency of the motor vehicle for the work that a number of the other large cities in Germany

The Automobile Magazine

immediately began preparations to introduce the system. The French government is not only regularly making use of automobiles for the transportation of the mails in Paris and other large cities, but has recently ordered fifty heavy wagons, each equipped with nine horse-power gasoline engines, for the purpose of carrying mail in the Soudan.

The initial introduction of the automobile in postal service in this country was made some months ago in Buffalo, N. Y. The vehicle used was an electric phaeton of about one ton weight, manufactured by the Pope Manufacturing Company. As a speed trial a four mile run from the main office to a sub-station was made in nineteen minutes, and the return trip consumed but eighteen minutes. During the collection trial the route covered was the same and mail was collected from twenty-two regulation boxes and eight package boxes—a total of 150 pounds—in thirty-three minutes.

This practical test of the adaptability of the automobile for the work of collecting mail from street letter boxes was made at Buffalo, under the auspices of the department, and the result, so far as related to that city and its superb streets, was entirely satisfactory. The experiment leads to the conclusion that valuable improvements for the collection branch of this service are in store through this departure in locomotion, limited at present by requisite conditions, which seems to demand asphalt or other smooth pavements.

The first automobile to be manufactured especially for mail collection service was one constructed by the Winton Motor Vehicle Company, of Cleveland, and is herewith illustrated. The test of this vehicle recently made in Cleveland was entirely successful, and was all the more remarkable from the fact that it took place during a fierce snow storm and under about the most unfavorable conditions imaginable. The test was made over a twenty-two mile route, and mail was collected from 120 boxes. Under ordinary conditions a collector with horse and

French Letter Carrier.

Automobile Postal Service

Collecting Mail in Cleveland

wagon can cover this route in exactly six hours, but even under the unfavorable conditions noted the automobile performed the work in two hours and twenty-seven minutes.

In speaking of the test Postmaster Dewston, of Cleveland, said: "The test was a very severe one, on account of the snowstorm, but the result was most satisfactory. Comparing the time taken in covering the route with the time required by horse and wagon the test speaks highly in favor of the use of automobiles in the collection of mail. It takes ten horses and wagons to collect the mail in Cleveland now, and the test shows that the work could be done with five automobiles. The post-office department is much interested in the automobile, and I think that it is only a question of time when it will be adopted."

In a recent letter to Waldon Fawcett, of Cleveland, Postmaster Dickerson of Detroit, who is one of the most progressive officials in the service in the country, says: "About sixty days ago a three-wheeled affair appeared in front of our office, and out of curiosity we tested it. We have one carrier's route that takes two hours and thirteen minutes to cover. We put the carrier on this three-wheeled vehicle, and all the boxes within that route were picked, the entire route covered and all mail collected in one hour and eight minutes, and to my mind this article came the nearest to what postmasters want for collection of mail from boxes of anything I have yet seen."

The Automobile Magazine

A similar vehicle for postal purposes is R. H. Plass' automobile, the designs for which have been favorably considered by Postmaster Wilson of Brooklyn, who was authorized to test postal automobiles.

In the drawing, A represents the body of the vehicle. Attached to the frame A' of the vehicle are the three supporting-wheels A^2 and A^3 A^3. The front wheel A^2 is suitably mounted on a standard a, mounted in a suitable socket a', provided at its upper end with handles a^2 for turning the wheel to guide the vehicle. The supporting-wheels A^3 are permanently attached to

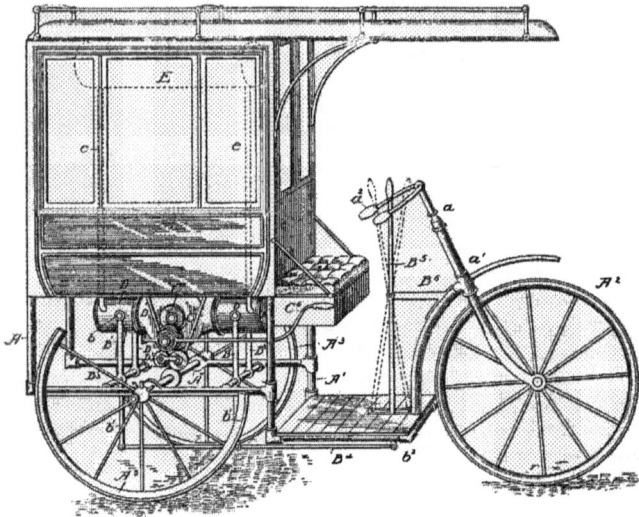

The Plass Postal Vehicle

a shaft A^4, mounted in the frame A' of the vehicle and provided at its center with a friction-wheel A^5, which is designed to receive through suitable intermediate mechanism motion from an engine B.

The engine B is mounted on upright rods B' B' B^2 B^2, which are respectively pivoted on horizontal rods B^3, attached to the frame A' of the vehicle.

The engine may be of any suitable construction, preferably for naphtha or gasoline.

Another vehicle which was thoroughly tested by Postmaster Wilson of Brooklyn was the Locomobile runabout pictured in

Automobile Postal Service

our frontispiece. After subjecting this vehicle to a series of searching tests, Postmaster Wilson declared himself amply satisfied with its practical qualities.

While still engaged in these tests Postmaster Wilson expressed himself as follows:

" The more I go into this matter the more convinced I become that the use of automobiles is practicable. We now pay 12 cents per mile for all mail carried. Our service in Brooklyn costs about $60 a day. I understand that the automobiles can be operated

German Accumulator Mail Wagon

at about three-quarters of a cent per mile. Thus you can figure out the saving. Here is a man who says we can buy gasoline for $1 a barrel of fifty gallons. He guarantees one horse-power for ten hours for each gallon. But figuring it out at the highest figures yet received, I can see a clear gain of about $25 a day, from which the interest on the investment can be paid and furnish a fund for repairs and leave a large surplus. It now costs the Brooklyn office $72,000 a year to handle the mails. We pay $16,000 a year for the electric cars, $4,000 for the carrying of the closed pouches on the ordinary cars, $10,000 for car tickets,

$28,000 for the wagon collections and $14,000 for the railway mail service. As I figure it out we can make a saving of fully 25 per cent. of this amount and also establish our own plant. We would require about eight wagons for station work, six heavier wagons for the railway service and about thirty light wagons for the collections. These wagons in the hands of our own men, who can soon become competent in the service, will add probably 25 per cent. to the efficiency of the service and make the office independent of accidents.

" The railroad people have had a good thing out of this service, even aside from the money they receive, for they have the assurance in times of strife of having their lines kept open by the United States mails. Why, I have heard railroad people say that it would be a good thing to have the mails go over their roads without charge to have the assurance that their lines would not be interfered with. I believe that Brooklyn will be the first office to adopt this system. In any event, the whole question will be inquired into very carefully."

The postmaster of Boston said: " I believe that the automobile would be of great benefit and could be utilized to much advantage in the mounted carrier service." Postmaster Samuel

Oakman Mail Cart

Automobile Postal Service

Vollmer's Postal Wagon

G. Dorr, of Buffalo, states that he expects ere long to have automobiles in constant use for mail collection in that city.

Already, in Buffalo, mail is being carried in an electric trap from the post-office to Station D, a distance of four miles, in eighteen minutes. Thirty mail boxes in that distance are tapped, and the trip is made in thirty-three minutes. In this trip 150 pounds of mail are collected.

At Baltimore a test was made with a steam carriage, the inventor of which achieved a speed of thirty miles an hour. In Washington the electric wagon was used, and tests were made with various other motors.

In Chicago several tests have been made in the collection districts, and the showing was so creditable that the Second Assistant Postmaster-General has called for bids for a permanent automobile service for the conveyance of mails and supplies between the general post-office and certain downtown stations. No contracts, however, have as yet been awarded.

An objection was originally raised to the employment of automobiles for postal service between main offices and sub-stations which was based on the claim that the employment of the automobile would make it difficult, if not impossible, to sort mail *en route*. Automobile manufacturers have, however, demonstrated to the satisfaction of post-office officials that they can construct auto cars in which this work can be done quite as easily and expeditiously as on the mail cars now in service on the street railways in many of the larger cities.

Mention has been made above of the automobile postal cars employed by the French government in the Soudan. These cars, let it be understood, are post coaches intended for passenger service rather than mail delivery wagons.

The British Colonial government, on the other hand, has taken steps to introduce automobile mail wagons in the true sense in some of the most distant crown colonies.

The steam wagon shown in the illustration below is one built by the Lancastershire Steam Motor Co. of Leyland, England, for use by the British government in the postal service in Ceylon. It is constructed to carry one ton of mail matter and will average ten miles an hour on fair roads. This autocar was put through a long and severe series of tests by the British postal authorities before it was allowed to be despatched to its destination.

Thus it may be seen that the rule of the white man over other less developed races does indeed confer upon them some of the latest and most admirable products of civilization.

Civilization in Ceylon

Race Track of Vienna

Motor Racing
By Edwin Emerson, Jr.

IF horse racing is indeed the Sport of Kings, then motor races must be the sport of millionaires. Surely it takes millionaires to furnish sweepstakes reaching the hundred thousand mark, as have been demanded of late for some of the more momentous motor races, not to mention the racing carriages themselves which have been known to fetch as much as 60,000 francs for one single vehicle.

Now that an international challenge cup for automobile races has been established by an American gentleman, not unmindful, it is fair to presume, of the fate of that other international trophy —the America's cup—this country has entered into the spirit of these races with a zest which promises the best of sport for all concerned.

On this side of the Atlantic ocean the sport of motor racing is still in its infancy. Altogether not more than a score of such contests have been waged.

Leaving out of account several unofficial brushes between the lucky owners of some of the earliest automobiles that were seen in America, the first contest that can truly be called a race was that held under the auspices of the *Times-Herald*, in Chicago, on November 28, 1895. This was followed in 1896 by the *Cosmopolitan Magazine* race up the Hudson river, from New York to Irvington. Both of these races were won by the Duryea type

Jenatzy's Racing Projectile

of gasoline-propelled carriages. After this the vast multiplication of types gave rise to new rivalries, and the motor bicycle and tricycle for track racing and pacing purposes appeared. For these machines, too, the season's sport opened as usual in France. First the Perigord challenge road race was run from Paris to Rouen and back, a total distance of 132½ miles. Girardot, one of the four starters, was the first to reach the turning point, which he did in 2h. 18m., finishing in the lead in 4h. 26m. for the entire distance, an average of nearly thirty miles per hour, not allowing for twenty minutes' necessary stops. The winner's machine was a Panhard-Levassor petroleum vehicle, weighing 1,600 pounds, driven by an eight horse-power four-cylinder motor. The same machine finished well up in the list in the Paris-Amsterdam road race.

The sport in the United States did not begin until late summer. As has been told in Mr. Reeves' excellent article on Motorcycle Racing which appeared in our last issue, a five-cornered contest for 25 miles between teams of motor cycle riders was held at Manhattan Beach cycle track on September 4, the competing teams being Fournier and D'Outrelon, Waller and Steenson, Stinson and Stafford, Ragan and Caldwell, and Judge and Miller. This last pair were mounted upon a Jaillu machine, fitted with a De Dion motor, the rest upon American Orient machines. Miller and Judge finished first in 39.58, Stinson and Stafford second in 41.17⅖, and Caldwell and Ragan third in

42.30⅗. The following are the times made in this contest, being American records for motor cycles from two miles to the finish:

Miles.	*Time.*	*Miles.*	*Time.*
1	1.36⅖	13	20.21⅕
2	3.07⅗	14	22.00⅗
3	4.40⅕	15	23.37
4	5.14⅘	16	25.13⅘
5	7.45⅘	17	26.52⅖
6	9.19⅗	18	28.28⅗
7	10.53	19	30.06⅖
8	12.27	20	31.43⅕
9	13.59⅘	21	33.20⅕
10	15.33⅘	22	34.56⅘
11	17.06	23	36.36
12	18.43	24	38.17⅗
25	39.58		

During this same time but few genuine automobile races were held in this country. The attempt to hold an international long distance race over American roads fizzled out before the project had got well under way, and another attempt at a long distance record across the American Continent resulted in ignominious

Baron Turckheim in His Dietrich Racer

failure. The event that came nearest to a *bona fide* contest was last year's race for a $2,000 sweepstake at Galesburg, Ill.,* between E. B. Snow, of Wyoming, and Dr. Morris, of Galesburg. Though the challenge had been for fifty miles only fifteen miles in all were made by the winner, owing to the breakdown of the challenger at this point. The fifteen miles were done in forty-three minutes and fifty-four seconds.

The popular idea upon which the speed of an automobile depends is, like most popular ideas, entirely wrong. By some it is held that a certain automobile is faster than another merely because its motor develops a greater number of horse-powers. Others maintain that the speed of the carriage is governed by the size of the wheels, and support their statement by comparing the automobile with the high-speed locomotive, which, with its huge drivers, is speedier than the small-wheeled freight engine. Undoubtedly there is a grain of truth in the assertion that a relationship exists between speed and horse-power; but if the subject be critically studied it will be found that the factor of speed in motor-carriages depends not upon one condition alone, but upon five—(1) the horse-power of the motor; (2) the number of revolutions made by that motor; (3) the weight of the vehicle; (4) the gearing; (5) construction of the moving parts as well

Dr. Lehwess in His Racing Vallée

* See article "An American Auto Race" in the November issue of this Magazine.

Motor Racing

M. Mors in His Own Racer

as of the entire carriage, to reduce friction as much as possible.

If the other four conditions be the same, it cannot be denied that, of two carriages, the faster will be the one having the more powerful motor.

But the speed of a motor has also an effect upon the speed of the carriage; for the greater the number of revolutions made per minute by the fly-wheel, the more swiftly will the driving-wheels of the carriage be turned by the intermediate gearing, and the greater will be the distance covered in a given time. The number of revolutions made by automobile motors varies between 600 and 1,200 per minute; the average motor makes between 800 and 900 revolutions. In all modern automobiles the number of revolutions can be increased by means of an "accelerator." If the motor were constantly run at maximum speed it would very evidently soon deteriorate, for which reason the careful automobilist will push his carriage to the utmost only when he is ascending exceedingly steep grades or when it is necessary for him to cover a given distance in the shortest possible time. As a general rule, high-speed motors are used only on pleasure vehicles; heavy trucks, in which tractive force is the main consideration, are usually driven by engines which make comparatively few revolutions per minute.

Weight is also an important condition upon which the speed of the carriage depends. Often enough it has happened that in ascending a grade one of the occupants of a vehicle has been compelled to alight in order that the motor, already running at its highest speed, might drive the carriage to the summit. Indeed, the motor is sometimes capable of driving only the vehicle up a hill, and the driver himself must perforce walk beside his

Baroness Zuylen de Nyevelt Getting Ready to Start

carriage. It is plain enough that a 12 horse-power carriage weighing only 1,500 pounds will make better time than if it weighed 2,500 pounds, and that a light, two-seated cart will be speedier than a heavier, four-seated vehicle. As an example, heavy autotrucks may be cited, which, although provided with powerful motors, run at very low speeds but develop considerable tractive force. These trucks can transport loads varying from 5 to 10 tons, depending upon the horse-power of the motor. For this reason, French manufacturers are beginning to build wagon bodies of partinium, an aluminium-tungsten alloy of very nearly the same specific gravity as pure aluminium, but of far greater strength.

The speed of automobile vehicles, whether racing machines or otherwise, has been steadily increasing. This is due largely to the public contests that have been held for the last few years in France and other places on the continent. Indeed it can be truly said that the present flourishing condition of the automobile industry in France has been largely brought about by the generally favorable attitude of the French press, aided by the energetic enthusiasm of special publications, such as " Le Vélo," " La

Count Bozon de Perigord Waiting for Starting Signal

France Automobile," " La Locomotion Automobile," and " Le Chauffeur."

Almost all the men that have figured in these events will once more come to the front during the great races that are to be run off during the time of the Paris Exposition.

This list has been recently published, and embodies many interesting features. The first event of the season will be decided on Sunday, April 15, and will be the fifth annual race from Paris to Roubaix, in which a category for motor cycles will be reserved. The distance to be covered is 288 kilometres; the entry fee is 5 francs, and the entries will be received up to mid-day, April 9. Prizes of 500, 250, 150, 100, and 50 francs are offered.

On Thursday, April 26, a competition for electrically-propelled vehicles will take place over a course from Paris to Dijon. Entries, accompanied by a fee of 100 francs, will be received up to mid-day, April 23. This competition is distinctly original, as all the vehicles entered, whether light or heavy, big or little, will compete on the same footing. Despatched from Paris they will be required to travel along the route to Dijon until they can proceed no further. The automobile which last ceases to move, provided it has averaged sixteen kilometres per hour, will be

Heavy Weight Race at Nice

declared the winner. Thus, this competition is really a test of capacity, and although many objections as to the entire fairness of the scheme will doubtless be raised, still the idea is too good to be dropped, and it is to be hoped that the promoters will receive sufficient support to enable the event to be decided.

The next automobile event will be held on Thursday, May 3, and will consist of the fourth annual motor-cycle competition between Etampes and Chartres. The distance is one hundred kilometres, the entry fee 20 francs, and engagements will be received up to Tuesday, May 1, at mid-day. Prizes ranging from 200 francs to 1,000 francs are offered.

Motor Racing

This will be followed by an event over the same course reserved for voiturettes weighing not more than 400 kilos. Thursday, May 10, is the date fixed, and entries will close on the previous Tuesday, the fee being 20 francs, and the prizes of the same value as those for the motor-cycles.

Again, on Thursday, May 17, the motor-bicycles will be afforded an opportunity to display their prowess over the same route.   This event will only be open to motor-bicycles not exceed-

René de Knyff, the " King of Chauffeurs "

ing 40 kilos. in weight.   Entries will close May 15, at mid-day; fee, 20 francs.   Wednesday, May 23, will witness the Derby for automobiles.   Paris to Bordeaux in a single stage of 568 kilometres is no light undertaking even for a French racing automobilist, but each year witnesses increased entries for the historic race.   The categories will be : Cars, entrance fee, 200 francs; small cars (less than 400 kilos.), 150 francs; motor cycles, 100 francs.   Names of intending competitors will be received up to May 19.   The final competition, which *Le Vélo* will promote next year, will be another race up the hill of Gaillon, and this

The Automobile Magazine

Count Chasseloup–Loubat

will be decided on 11th November. The categories will be—
(1) cars weighing more than 400 kilos.; (2) cars weighing from
250 to 400 kilos.; (3) cars weighing less than 250 kilos.; (4)
motor cycles weighing less than 150 kilos.; (5) motor cycles
of two seats (occupied); (6) bicycles; (7) chainless. The entry
fee is 10 francs, and names will be received up to the previous
day at 12 o'clock.

A unique contest will be the amateur road race devised by
M. Albert Lemaitre, which is to be held under the auspices of
" La France Automobile " on June 8, in the week after Witsun-
day. The distance will be from Paris to Rheims, and the prizes
are to consist solely of baskets of champagne.

The competitors must be members of the Automobile Club de
France, and must not be accompanied by a professional or amateur automobile driver or mechanic.

Further, each competitor must have a lady by his side, the
idea apparently being that under these conditions he will not take
racing risks.

Another unique contest is that which is to be held on a wager
of the Baron de Caters and Jenatzy, the well-known builder of
the projectalite racing machine which we illustrate in these pages.
These gentlemen have laid a wager that they would build and

drive an electric automobile capable of covering one hundred kilometres within an hour. If their bet is accepted the race will be run over that remarkably level stretch of road of nearly 100 kilometres, from Evereux to Lisieux, which has but one hill during its whole stretch, and that only a slight grade between Rivière and Thibonville.

The most important contest of all, needless to state, will be the great international series of races for the challenge cup given by Mr. James Gordon Bennett, of America. This contest will be held on Thursday, June 14, la Fête-Dieu of the French. Altogether five clubs, through their selected teams, will compete for this coveted trophy. They are the automobile clubs of France, Belgium, Germany, Italy, and America.

In France the first result of the offer of the international cup was that everybody aspired to the glory of defending the trophy. There was certainly good material to chose from. In the first line there are MM. Renê de Knyff, Charron, Girardot, Chasseloup-Loubat, Gilles Hourgières, Albert Lemaitre, and Levegh. These gentlemen are all veritable kings of the road, as a glance at a very few of their achievements will prove.

M. de Knyff has figured prominently in all of the many races in which he has competed, and his two principal victories have been those scored in the Paris-Bordeaux race, 1898, and Le Tour de France, 1899.

Charron in his Winning Panhard

465

The Automobile Magazine

M. Charron has had a wonderfully successful career, and among many performances his wins in the Marseilles-Nice, 1898, Paris-Amsterdam, 1898, and Paris-Bordeaux, 1899, are all particularly noteworthy. M. Girardot has always been one of the most consistent drivers in the racing world, and enjoys the distinction of having been second in big races more frequently than any other *chauffeur*. A sort of fatality seemed to follow his steps, and " seconds " in Paris-Amsterdam, 1898, Nice-Castellane, 1899, and Le Tour de France, 1899, may be instanced as his near approaches to victory. More recently he has succeeded in passing the winning-post first, as witness Paris-Ostend and Paris-Boulogne. In the former he " dead-heated " with M. Levegh.

Start of Paris-Amsterdam Race

The Count of Chasseloup-Loubat will always be remembered as the adversary of M. Jenatzy in the series of electromobile speed tests which took place in the early spring of this year. Previous to this he had won the Marseilles-Nice race in 1897, and this summer he took third prize in Le Tour de France.

M. Gilles Hourgières has experienced a lengthy career, as in 1897 he won both the Paris-Dieppe and the Paris-Trouville races. In 1898 he finished second in the Marseilles-Nice event, and gave a phenomenal performance in the Paris-Amsterdam *course*, after experiencing a breakdown in the early stages of the race. This year he has only raced once, viz., in the Paris-Bordeaux, when he was placed fifth.

M. Albert Lemaitre has scored victories in Marseilles-Nice,

1897, and in Nice-Castellane, the Nice mile race, and the de la Turbie and Pau-Bayonne races, 1899.

M. Levegh gained two brilliant victories this year in the Paris-Ostend and the Bordeaux-Biarritz events, his average time in the latter race being altogether exceptional.

In 1898, the famous French racer, " Rigal," covered a kilometre in 1.20⅔ over a level stretch. Villemain covered the same distance in 1.16⅓ on an inclined plain, averaging eight per cent. Different machines figured in these two tests. Even in the case of machines of identical construction, however, a corresponding

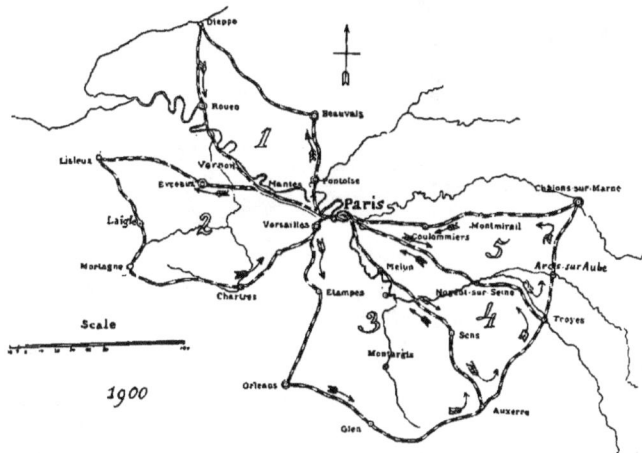

Map of French Five Day Race for 1900

1st day—Paris, Pontoise, Beauvais, Dieppe, Rouen, Vernon, Mantes, Paris.
2d day—Paris, Mantes, Evreux, Lisieux, Laigle, Mortagne, Chartres, Rambouillet, Paris.
3d day—Paris, Versailles, Etampes, Orléans, Gien, Auxerre, Sens, Montereau, Melun, Paris.
4th day—Paris, Melun, Montereau, Sens, Auxerre, Troyes, Nogent-sur-Seine, Paris.
5th day—Paris, Nogent-sur-Seine, Troyes, Arcis-sur-Aube, Châlon-sur-Marne, Mont-mirail, Coulommiers, Paris.

increase of speed had been noted. In 1898 Marot covered one kilometre in 1.42⅔ over a level stretch. At the end of last year Bardin took a motor of the same makers over a kilometre, with an inclined plain of eight per cent., at the rate of 1.36⅘.

The obvious course for the French Club would have been to avoid complications by leaving the selection of the races to merit and chance—after the manner of our trial yacht races—that is, by holding trial races and making those *chauffeurs* who would come out as first, second and third winners the defenders of the French Club. Instead of adopting this simple plan, the Com-

De Paiva in his Dietrich

mission Sportive of the Automobile Club de France preferred to be guided by its own judgment, and accordingly rendered an off-hand decision, which threw the burden of defence upon those lucky *chauffeurs* MM. Knyff, Charron and Girardot.

This decision so offended some of the *chauffeurs* who thought that they should have had an equal chance with the others that they are reported to have gone over to the enemies of France. Thus MM. Lemaitre and Velghe, who race under the *noms de guerre* of Anthony and Le Veghe, have officially informed Baron Zuylen, the President of the French Club, of their intention to fight for the cup under Belgian colors. Another member of the French Club is understood to have been won over by the German Automobile Club and will probably enter the race with a machine made by a French firm in Alsace. The alleged turncoat is M. Loysel, the winner of the first race from Bordeaux to Biarritz, and likewise the first to establish a record for the kilometre during the trials in the Park of Achères.

The American challengers for the cup are Messrs. Alexander Winton and Andrew L. Riker. Mr. Winton will race in a machine similar to that in which he made the run from Cleveland to New York last year. Mr. Riker will enter with a new machine specially devised for this race. The third challenger is believed to be the Locomobile Company of America.

Motor Racing

Unlike our American contestants the French challengers were selected from a field of nearly half a hundred candidates. Their official averages, as computed by *Le Vélo,* are as follows:

Girardot	3	4	2	0	1	38
R. de Knyff	3	4	1	0	0	34
Levegh	3	3	0	0	2	29
Lemaitre	4	0	1	0	0	23
Charron	3	0	1	1	0	20
Antony	2	1	0	0	0	14
Chasseloup-Laubat	1	1	0	1	1	12
Pinson	0	0	2	2	2	12
Giraud	0	2	0	1	0	10
Koechlin	0	0	2	1	0	8
Heath	0	0	1	2	0	7
Loysel	0	0	0	3	0	6
Broc	0	0	1	0	1	4
Jamin	0	0	1	0	0	3
Petit	0	0	1	0	0	3
Henon	0	0	1	0	0	3
De Dietrich	0	0	1	0	0	3
Archambaud	0	0	0	1	0	2
De Castelnau	0	0	0	1	0	2
Huillier	0	0	0	0	1	1
Jenatzy	0	0	0	0	1	1
Flash	0	0	0	0	1	1

Alexander Winton in his Racer

Locomobile Winning the Nice–Monte Carlo Race

The situation now appears thus: If the cup is won by M. Knyff, he will be a Belgian racing for the French Club; if by M. Lemaitre, by a Frenchman in behalf of Belgium; if by M. Loysel, by a Frenchman racing under German colors. Should we carry off the cup, on the other hand, Mr. Bennett's donation may be regarded as something of a "Greek gift."

No wonder the editor of *Le Vélo* calls the trophy a "cup of discord." But the more trouble, the merrier, for all these international complications will only serve to increase the gayety of the nations as they assemble at the new World's Fair on the Seine.

The End

The French Racing Rules

PREAMBLE.

THE ideas which have governed the drawing up of the present Racing Rules are the following:

1. The Automobile Club of France is the sole authority regulating races of automobile vehicles and motor cycles.

2. The general spirit of these rules is that the races are run and won by a combination of the machine and its riders, which must not be separated during the race.

I.—GENERAL RULES.

Article 1.—Every competitor entering for a race of motor vehicles or motor cycles is supposed to be acquainted with these rules, and undertakes to abide, without dispute, by the results to which such rules may lead.

GENERAL PROVISIONS.

Art. 2.—All automobile races and record trials organized in France shall be controlled by the Racing Rules of the Automobile Club de France.

Art. 3.—All races which are not controlled by these rules are forbidden, and all competitors therein will be disqualified.

PUBLICATION OF PROGRAMME.

Art. 4.—The programmes of races—
(1) Must be sent to the Sporting Committee of the Automobile Club de France; (2) and must be published in the Press at least five days before the races, if they be on the track, or fifteen days if they be on the road.

Art. 5.—The programme shall contain—
(1) The number of the prizes and the amounts for each race; (2) the distances; (3) the amount, if any, of entrance fee attached to each event; (4) the date and hour for closing of entries; (5) the amount of forfeit, if there be any; (6) the place at which entries are received; (7) complete and exact itinerary of road races. These itineraries shall not undergo any modifications, except from absolute necessity, in such cases notice shall be immediately given individually to each competitor.

Art. 6.—After publication of the programme no modification shall be made in it as regards prizes—the amount of which shall not be increased—or as regards the nature of the races originally announced. Mention shall be made on the first page of all race programmes that the meeting is held under the rules of the Automobile Club de France.

Art. 7.—A copy of the programme and rules shall be sent to each competitor on his entering for a road race.

CLASSIFICATION OF RACES.

Art. 8.—Races shall be " open " or " reserved." " Reserved " races shall be races confined to competitors fulfilling a definition stipulated by the promoters.

CATEGORIES.

Art. 9.—The categories officially recognized by the Automobile Club de France are as follows—

(1) Vehicles (motor cycles and small carriages) weighing under 250 kilogs. (5 cwt.) ; (2) vehicles weighing more than 250 kilogs. and carrying at least two passengers side by side of an average of not less than 70 kilogs. (11 stone) each, it being understood that if the average weight of the passengers does not amount to 70 kilogs. (11 stone) each, the deficiency may be made up with ballast. In track races and records, however, vehicles with two seats need only carry one passenger, but in road races two passengers are compulsory.

In addition, the promoters may subdivide the two foregoing categories into as many classes as they please.

Art. 10.—The Sporting Committee shall be the sole judge of the classification of all motor vehicles, as well as of questions which may arise therefrom.

ENTRANCE FEE AND FORFEITS.

Art. 11.—The amount of the entrance fee shall be fixed by the promoters, who will decide whether it is repayable or not to the competitors who have started.

Art. 12.—The forfeit is not a matter of right; it must be specified on the programme, as also its amount.

Art. 13.—Entrance fees which are repayable and forfeits, if they are not claimed within a month, shall become the property of the promoters.

The French Racing Rules

Art. 14.—Entries shall be made as follows—

(1) By letter; (2) by telegram, confirmed by letter of same date.

Art. 15.—Any entry which is not accompanied by the fee or which is sent in too late will be annulled *ipso facto.*

Art. 16.—Any competitor wilfully sending in a false statement may be prevented from starting, and will be liable to a fine.

RACING NAMES.

Art. 17.—Any competitor may use a racing name subject to approval by the Sporting Committee.

Art. 18.—The racing name becomes permanent and cannot be changed without the permission of the Sporting Committee, to whom a written request must be sent, accompanied by a fee of 20 francs.

STEWARDS OF THE COURSE.

Art. 19.—In every race upon the road or on the track the promoters shall choose three stewards, whose appointment must be approved of by the Sporting Committee, and whose names should be communicated at the same time as the programme.

Art. 20.—The stewards are entrusted with the carrying out of the programme, and with seeing that the rules are strictly observed, and are also to settle any protest that may arise out of the race.

Art. 21.—The stewards can either prevent a competitor from starting, or start him after the others, if his inexperience or the construction of his car would seem to present a danger to other competitors.

Art. 22.—The stewards have a right—

(1) To prevent a competitor from starting; (2) to publicly reprimand a competitor; (3) to impose fines up to a maximum of 200 francs (£8); (4) to disqualify a competitor for a maximum period of a month.

In these two latter cases the competitor has a right to appeal to the Sporting Committee.

Art. 23.—Should the stewards deem that a heavier fine ought to be imposed, they can apply to the Sporting Committee, which has full power to inflict any penalty after taking evidence from those interested.

Art. 24.—The starter is appointed by the stewards, and he alone judges the validity of a start.

The Automobile Magazine

Art. 25.—As a general rule the start is given while the vehicles are at a standstill, and they must start by their own means, but in certain cases a flying start will be allowed with the sanction of the stewards.

Art. 26.—The start shall take place in the order of entry, unless by special arrangement.

Art. 27.—In races upon the track the start shall be given to all the competitors at the same time, and this can also be done on the road; or the vehicles may be sent off at regular intervals between each competitor.

JUDGE AT WINNING POST.

Art. 28.—At the winning post there shall be one judge, and his decisions shall be final. If, however, there is a large number of competitors the judge is entitled to assistance, but the judge must be chosen by the stewards.

Art. 29.—The winning of a race is judged from the front of the front wheel for motor cycles and motor carriages alike.

Art. 30.—Should two or more competitors finish level the judge declares a dead-heat, and the two prizes shall be equally divided between the competitors finishing level.

Art. 31.—In distance races the competitor must cover the whole course in order to be entitled to a prize.

Art. 32.—In time races the competitors shall be placed according to the number of kilometers covered.

Art. 33.—When a single competitor starts a limit of time may be fixed by the stewards within which the course must be covered.

Art. 34.—Should a single competitor start in a race he shall have the right to the first prize.

OBSERVERS AT CORNERS.

Art. 35.—Observers chosen by the stewards shall be placed at the corners of the course to see that one competitor does not interfere wilfully or otherwise with another by wrongfully getting in front of him, or shutting him in, or by any other manœuvre which would be calculated to wrongfully affect the result of the race.

OBSERVERS IN ROAD RACES.

Art. 36.—In road races a certain number of observers shall be appointed and placed where it may be necessary to stop the competitors, or compel them to drive at a stipulated speed, and the observers shall see that these instructions are strictly adhered to by the competitors.

The French Racing Rules

Art. 37.—The measurement of the track shall be taken at 0.30 metre from the inside ropes. On all tracks the winning post must be indicated by a clearly indicated line.

Art. 38.—For the establishment of records on the track a certificate of measurement, with an annexed plan prepared by a qualified surveyor, shall be furnished.

GENERAL REGULATIONS RELATING TO RACES.

Art. 39.—Any competitor who in a race crosses in front of another, shuts in or obstructs another by any means so that the latter is prevented from advancing, may be stopped in the race or penalized by fine or disqualification, so long as the collision was not rendered unavoidable by a third competitor or the competitor who was obstructed was not himself in fault, but the fact that the collision was involuntary, or that it did not affect the result of the race, shall in no case be admitted as a valid excuse.

Art. 40.—No competitor shall be allowed to cross the course of another until he is at least two lengths of his machine ahead of such other competitor.

Art. 41.—No sign or advertisement shall be displayed on any vehicle while racing.

Art. 42.—No vehicle shall be pushed or assisted by any one other than its authorized occupants under pain of disqualification.

Art. 43.—Competitors shall be responsible for all civil and criminal penalties whatsoever.

SPECIAL REGULATIONS FOR TRACK RACING.

Art. 44.—A competitor wishing to pass another must do so on the outside, and so as to leave the competitor passed the following space from the rope, viz., for motor cycles 1.30 metres (4 feet), and for motor cars 3 metres (10 feet).

Art. 45.—A race containing too many entries may be run in heats, semi-final and final.

Art. 46.—The racing stewards shall arrange the heats, semi-final and final, and their decision shall be without appeal.

Art. 47.—No accident shall admit of a competitor running again, either in another heat or in the final.

Art. 48.—Any competitor leaving the track to get off his machine must start again from the point where he left the track.

The Automobile Magazine

Art. 49.—In road races the approach of a competing vehicle must be notified by a horn, trumpet, or some similar instrument.

Art. 50.—Vehicles which have to travel by night must display a white and green light in front and a red light behind.

Art. 51.—In road racing, competitors must conform to the traffic regulations of the police.

Art. 52.—Competitors must make themselves acquainted with the route, and no allowance will be made for mistakes they may make. Moreover, if any competitor takes a shorter or easier route than the one prescribed, he will be disqualified. The stewards shall be sole judges of the comparative distance or ease of the routes followed.

PROTESTS.

Art. 53.—The right of protest lays with the competitor, but the stewards can always interfere officially in case of necessity.

Art. 54.—Any competitor lodging a protest must always substantiate his grounds of protest, and the competitor protested against has the right of being heard in opposition to the protest.

Art. 55.—No protest will be considered unless it is put into writing. Protests must be considered by the stewards on the spot, and a decision shall be come to immediately, whenever this is possible.

Art. 56.—Protests shall be lodged at the times and in manner following: Protests as to classification of competitors and of machines, as to validity of entry and payment of entrance fees—before the race and verbally. Protests as to unfair running, errors of route, or any other irregularities on the route—within 24 hours after the race, and in writing. Protests as to the fraudulent starting of a competitor in a race for which he was not qualified—eight days after the race, and in writing. For protests in races on the road—eight days after the finish of the race.

PENALTIES.

Art. 57.—Penalties imposed on competitors in and organizers of races are recoverable immediately on their being notified to the parties concerned and on their publication in the journals officially notified by the Sporting Committee.

DISQUALIFICATION.

Art. 58.—If a competitor is disqualified in a race he loses all right to a prize.

The French Racing Rules

Art. 59.—A public and official reprimand is pronounced by the stewards or by the Sporting Committee of the Automobile Club de France, and involves the insertion in a public journal of an official notification by the Sporting Committee.

FINES.

Art. 60.—The moneys received in fines shall be paid into the funds of the Sporting Committee, to be distributed or devoted to sporting competitions.

II.—REGULATIONS AS TO RECORDS.

TIMEKEEPERS.

Art. 1.—The Sporting Committee shall appoint the official timekeepers and shall prepare a list of them every year.

Art. 2.—Timekeepers to be eligible for appointment must (1) possess a reliable chronometer stop-watch, certified as first-class by the Observatories of Besançon, Geneva, or Kew; (2) furnish the name of the maker of their chronometer stop-watch.

Art. 3.—The Sporting Committee may, when they see fit, require the timekeepers to renew the certificates as to their chronometers being first class. Certificates must be renewed every three years.

Art. 4.—The appointment of timekeepers is revocable at any time. Before appointment they must—

(1) Submit to an examination permitting of the chronometrical test (*a*) of 10 tests of 500 metres (500 yards) and under; (*b*) of 10 tests of from 500 metres (yards) to 2,000 metres (yards); (*c*) of two tests of 20 kiloms. (15 miles) at least, or a test of 50 kiloms. (38 miles), the stop-watch showing the time of each lap and the time of the total distance.

In the above tests the candidate for appointment as timekeeper shall write down on the forms, of which a model is deposited at the offices of the Automobile Club de France, the times recorded by him. At the same time a certified official timekeeper shall make similar entries, but independently of the candidate. The candidate shall remit these forms, duly filled up, to the certified timekeeper in a sealed envelope. At the end of the tests the certified official timekeeper shall forward these forms to the Sporting Committee of the Automobile Club de France, together with the results of his own checking, certifying that the examina-

The Automobile Magazine

tion has been properly conducted, and that there has been no collusion, comparison, or correction of results.

Art. 5.—The Sporting Committee shall decide on the appointment after examining and comparing the written results. A candidate who has been rejected may re-enter for election after a month.

Art. 6.—Timekeepers must sign the forms recording the times taken by them. Any timekeeper signing a record not made by himself will be *ipso facto* disqualified. He will also be disqualified by the simple decision of the Sporting Committee that his records have not been confirmed.

Art. 7.—The Sporting Committee takes cognizance of records on the track and road records. Each of these two categories comprises records of distance and of time, as well as the records for both categories defined by Article 9 of the Racing Regulations.

Art. 8.—The distances officially recognized for record racing are: On the track, 500 metres (500 yards); from 1 to 100 kiloms., per kilom. (1,094 yards); and for distances beyond 100 kiloms., per 50 kiloms. On the road: 500 metres; from 1 to 10 kiloms., by kiloms.; from 10 to 50 kiloms., by 10 kiloms.; from 100 kiloms., by 100 kiloms. The official distances in English miles: Distances of miles, 50 miles and 100 miles will be recognized.

Art. 9.—All races for records must be made from standstill, and vehicles must be started only with their own power.

Art. 10.—Races for records of 500 metres (541 yards) and of from 1 to 10 kiloms., inclusive, may be made by flying start.

Art. 11.—The time records of the Automobile Club de France are records by the hour without limit.

Art. 12.—The time records from town to town are also by the hour without limit (homologous).

Art. 13.—No record will be recognized as official unless it has been established over distances rigorously tested, and unless the time has been checked by several official timekeepers recognized by the Automobile Club de France.

TRACK RECORDS.

Art. 14.—Starts for track records shall take place from a tape.

Art. 15.—Attacks on the record shall be timed according to the laps round the track and by the hour up to 100 kiloms., by kilom., and by hour up to 200 kiloms., and by the 5 kiloms., and by the hour from 200 kiloms. upwards.

Whilst timing records timekeepers are expressly advised to

The French Racing Rules

take the times of the English distances at the half mile, mile, and all the military distances, especially the 10, 20, 30, 40, 50, and 100 miles, and above the last-named distance by the 100 miles.

ROAD RECORDS.

Art. 16.—Road records straight ahead are recognized from 1 to 50 kiloms., above that distance they are taken by the 50 kiloms. Road records permit of embracing the outward and return journey for all distances.

TIMEKEEPERS' FEES.

Art. 17.—Timekeepers are forbidden to accept any remuneration over and above the tariff fixed below, viz.: For a day, or part of a day, occupied in racing or in getting to and from a race, 30 francs (25s.).

Art. 18.—The traveling expenses of timekeepers are arranged by mutual consent.

Art. 19.—Every timekeeper must, at his own expense and on his own responsibility, procure such assistance he may require in working out his calculations, or for any other outside act or operation required, not strictly coming within the province of a timekeeper.

Art. 20.—Timekeepers may be temporarily suspended or have their appointment revoked for any act affecting their private or professional honor. Provided that this step cannot be taken unless by order of the Sporting Committee, after the accused timekeeper has been heard.

Art. 21.—No timekeeper shall be required to act as such for more than six hours at a stretch.

OPERATION OF THE REGULATIONS.

Special Article.—The present regulations shall come into force and be binding on all promoters of automobile races as from January 1, 1900.

ADDENDA TO RACING RULES.

The Sporting Committee of the Automobile Club de France believe they will be doing a useful work in bringing to the notice of organizers of road races, by means of its instructions and its opinions as regards road racing. The Sporting Committee being a branch of the Society for the encouragement and development of the automobile industry appreciate that the aim of holding

The Automobile Magazine

trials and competitions under uniform conditions is to demonstrate the excellence or the faults of the various competing vehicles. The region of true sport has nothing to lose but everything to gain by looking thus upon matters which must tend to bind together all organizers of races, and to so arrange their programmes as to secure as much uniformity as possible. With regard to races of one stage only there should be no difficulties. If the starting signal is given to all vehicles at one time the task of the judge at the finish will be greatly facilitated. If the start and the finish are timed by chronometer the times can be taken with exactitude. The Committee recommend that only their official timekeepers should be employed in all cases.

According to the regulations as to records the only homologous times are those certified by the official timekeepers. For races in several stages the question is more difficult. As a matter of fact, it is certain that the obligatory stoppage and storing of vehicles in an enclosure in the town at which a stage ends add to the complication and also somewhat to the expense, but there is no other means of obtaining a reliable test. Moreover, if there is no enclosure a vehicle arriving at the end of a stage in a damaged state and virtually unfit to continue its journey on the morrow, could be repaired, and, so to say, thoroughly renewed during the night by a body of workmen. In this way a vehicle constantly under repair might start every morning in as good condition as one never needing repair. Such a result would be unfair from a sporting view, whilst it would be injurious from the point of view of the progress of automobilism.

Another point of some nicety is the following: It is clear that even the best vehicle in the world, after doing a stage of 200, 300, or even 400 kiloms., at high speeds, requires, before proceeding again, to have its principal parts tested, adjusted, cleaned, and lubricated, to have its reservoirs replenished, etc. It is necessary, therefore, to give competitors, both on arrival in the enclosure and before leaving, a certain time for looking to their vehicles. Therefore, to comply with the spirit of what has been said above, the competitors during this time ought only to give their vehicles the above necessary and reasonable attentions, but not to do any repairs.

The question is, where do these necessary and reasonable attentions end and repairs begin?—a very thorny question. Whilst it may be said that any fair-minded and competent persons can avoid being deceived, it cannot be denied that it is not practicable to fix a definite limit. The only practical way to steer clear of complicated regulations which it would be impossible to enforce

The French Racing Rules

strictly, and which would lead to endless disputes, whilst at the same time preventing trickery, is to allow a very short space of time for attending to the vehicle, both on arrival and before departure. For example, an hour on arrival and 15 minutes or half an hour before departure.

To effect this object it is understood that if a vehicle does not put in an appearance before the starter at its specified time, or is not in a fit state to start away when the signal is given, then its time is reckoned from the moment the starting signal is given, whether or not it is ready to start.

The arrangement and supervision of the enclosure should, therefore, have the most careful attention of the organizers and stewards.

As in racing by several stages the arrivals must necessarily be timed by chronometer, it seems natural to make the starts also by time. The first day the vehicles can leave according to the order in which they were entered, but the following day they will leave in the order in which they arrived the night before.

In road racing, whether in one or more stages, competitors are allowed every opportunity of effecting repairs on the road, but the time occupied in so doing is counted as their running time. However, in conformity with the ideas just expressed, the Sporting Committee strongly recommend the sealing of all the essential parts of the vehicles, such as wheels, motors, frames, etc. Naturally competitors must not be allowed to replace parts which have been sealed. Steps should be taken on arrival, and also at certain stages, to see that these sealed parts are in place, and any vehicle which is found to have any of these parts wanting will be disqualified from the race. The employment of lead seals which are liable to be pulled out on the way are not recommended in the case in question.

These precautions are particularly recommended with motor cycles, the parts of which it is so easy to change in the course of a race.

Organizers of road races must not forget that they must obtain the authorization of the prefects and the consent of the mayors of the departments and communes traversed. Often they will be obliged to bind competitors down to certain speeds over certain districts of the journey (in towns, dangerous portions of route, etc.). In this case they must take care to enforce on the competitors the exact observance of the specified speeds over these portions of the route.

The Sporting Committee earnestly enjoin on organizers to indicate fully and plainly to competitors the routes to be followed

481

by means of notices, signboards, etc., to call attention to dangerous places, and to the places where speed must be slackened.

It is by the adoption of such measures intelligently carried out, and in accord with the local authorities, that accidents of all kinds may be avoided. Still it cannot be denied that these involve considerable work in long stages. These recommendations must not be taken as implying any modification of the regulations which provide that competitors take all risks and responsibility, whether arising from their errors or otherwise, and whether civil or penal, which may happen to them. The organizers, after having conscientiously done their best, must decline to accept any responsibility for errors of route which competitors may make, and leave entirely on their shoulders all responsibility, whether civil or penal, which they may incur.

Such are the broad lines which the Sporting Committee think should be adopted in the organization of road races to ensure order.

Additional Gordon Bennett Cup Rules

The Sports Committee of the Automobile Club of France have now published officially the rules attaching to any challenge for this trophy. They are practically identical with the draft as published by us in our January number. The only alterations are as follows:

Art. V. (ex-Art. IV.).—The exact date shall be fixed by common consent of the interested clubs before February 1st of each year.

Art. XVII. of the rules is replaced by the following: *Art. XVII.*—After confirmation of the results of the race, the cup must be handed to the victor within a fortnight. In case of a *dead heat*, and awaiting the settlement of any question concerning the *dead heat*, the cup will remain in possession of the club which previously held it.

In *Art. XVIII.* there is one slight modification to be noted. The suggestion was that the " walk-over " race was to be completed within 24 hours. The amendment states that it must be completed within a maximum time fixed by the Committee.

Art. XXII.—The cost of transporting the vehicles and their accessories, fuel, etc., shall be defrayed by the owners of the said vehicles, or by the clubs they represent.

Official Time-keeping Rules

A S will be seen from the translation of the racing rules drawn up by the French Automobile Club, and which we publish in our present issue, special regulations have been made to ensure that official time-keepers shall not only possess some technical qualifications, but also that their timepieces shall be of superior mechanical construction. No one at all conversant with time-keeping as practised at the various racing events both in England and France will, we think, affirm that it is as accurate as it might and should be, and when records are " timed " to the fifth part of a second—and this small interval of time will not infrequently cause the record to pass from one holder to another —it is seen at once how essential and necessary it is that the time officially declared as that occupied during an event shall be the true time interval. Those interested in records will perhaps be surprised to learn that, from the scientific point of view, very few of these records can be accepted as true time intervals, and, taking into consideration the very crude means employed at competitions, races, etc., to measure time, the best that can be said is that the time as deduced is but a rough approximation, which, owing to various errors that will be mentioned later, may be either very close to the true time or separated from it by a considerable time interval. When we consider that the ordinary official time-keeper is not required to pass any chronometric examination, neither is it required of him that he should have had any technical training or experience in the accurate measurement of time as practised in observatories or on board naval or telegraph cable steamships, and that the instrument he employs has in nearly all cases a variable rate, and, further, that the personal equation of the timekeeper and the rate of his timepiece are unknown, it will be seen that these so-called records can rarely be deemed correct measurements of time. For most purposes of sport an approximation is near enough, but for astronomical or navigating purposes the method of the official timekeeper would be wholly inadmissible. Still, there is no reason why cycling and motor vehicle contests should not be timed with at least some approach to scientific accuracy much greater than that which obtains at present.

For the accurate determination of a time interval we require a special observer and a special timepiece. Both are subject to certain errors, which may be ascertained and allowed for.

As regards the observer, no two persons will observe the same phenomenon at the same instant. The difference between the actual time of an occurrence and its observed time constitutes the " personal equation " of that observer. This may be .1 sec. + or —, or it may be .5 sec. + or —. It varies in the same individual according to the state of health. In astronomical and physical observatories, surveying and cable steamships, where the accurate determination of time is essential to the proper performance of the work, this personal equation is always carefully ascertained for each observer. In order to show the influence that this very small time interval might exert upon a record, suppose we learn from the sporting press that a distance of 10 kilos. has been covered in 10 m. 40 s. by rider A. The same event takes place with another timekeeper, and rider B completes the distance in 10 m. 39 4-5 s., or, as we prefer to put it, following scientific usage, 10 m. 39.8 s. Subsequent testing of the observational powers of the two timekeepers reveals that the timekeeper for A has a personal equation of — .2 sec., while that of the timekeeper for B is + .2 sec. The corrected times would, therefore, be: A, 10 m. 39.8 s.; B, 10 m. 40 s. This case, although supposititious, is by no means impossible or unlikely. Nor are such personal equations as we have used unusual, as every astronomer or scientific navigator will admit. If, then, racing records are to be true records and not mere approximations, clearly the personal equation of each official timekeeper must be known, so that accurate comparisons of observed times can be made.

Dealing now with the instrument employed to measure time intervals, we should explain that no instrument keeps exact time —another surprising statement, and one that the official timekeeper will, no doubt, except to. Exact time can, however, always be ascertained providing the error and rate of the timepiece are known. By the error is meant the difference in time on a given date between the watch, chronometer, or clock in question and a standard clock or regulator synchronized from Greenwich, and the rate is the daily change in the error. If the latter be uniform, or fairly so, the time can always be ascertained with accuracy. Thus the problem of finding the longitude at sea consists essentially in determining the time at ship and noting at the instant of observation the time as shown by the chronometer. the difference between the two is the " meridian distance," which expresses the longitude. With a good chronometer and an expert observer very accurate results can be obtained. But in all time problems, whether they consist in timing a cycle race or determining the position of telegraph cable buoy in the middle of the Atlantic, the absolutely essential condition is that the time-piece shall have a uniform rate. The " error " may be large or small so long as it is known; the rate may likewise be large or small, providing it is known, and that it be uniform, *i. e.*, does not vary. We perhaps labor this point, but we do so purposely. Now the marine chronometer is a very expensive instrument; a second-hand one good enough for ordinary navigation will cost not less than £20. In order to preserve a uniform rate the chronometer is carefully slung in gimbals and enclosed in an air-tight box, which is placed in another box, which is carefully cushioned. This latter box is then screwed down, and as far as possible the chronometer is kept free from all shock and vibration. The operation of winding is carefully performed at a stated time by a responsible person. No one but a qualified optician is ever allowed to touch the mechanism, not even to move the hands. Notwithstanding all these precautions an absolutely uniform rate, even in the best chronometers, cannot be ensured. The rate alters with temperature, and will not be the same at sea as in harbor. The variations are, however, usually very small, and it is quite possible to determine the time, either local or at Greenwich, with a limit of probable error of $\frac{1}{2}$ second, or even less. Now, our point is this. Seeing it is so difficult to ensure a uniform rate even in a specially constructed and expensive instrument as the chronometer, which will cost anything between £25 and £50, and considering also what care has to be taken to ensure fairly accurate results, is it reasonable to assume that a similar degree of accuracy can be obtained in a watch which may cost perhaps £5 or £6, or say £10, and which is daily subjected to a course of treatment which is diametrically opposed to that necessary for chronometers? If, as official timekeepers assert, they can obtain accurate results to the fifth part of a second with their what they so absurdly term " chronographs "—really they are nothing of the kind, because they do not furnish any " graph " at all—is it likely that such very commercial men as ship-owners would pay say £100 for a set of three chronometers for a steamship when, *pace* the official time-keeper, they could ensure equal results with a 5-guinea split-second watch?

Let us now consider the treatment a watch receives at the hands of such " experts " as official timekeepers. In the daytime it is usually worn on the person, and so exposed to the radiant heat of the body. Its temperature will usually be between 70° and 80° F., at times more. It participates in the oscillatory movement of the body, due to walking or riding. It also occupies a vertical position. At night-time it is usually placed on a table in a horizontal position, and exposed to the surrounding air, which may be anything from 30° to 80° F., or more. Obviously under such conditions a constant rate cannot be expected. As a matter of fact, the watch will lose in the day-time and gain at night-time. Even the best-made English watches would show a variable rate—often a *very* variable rate—under such conditions. Again, in the hands of the official timekeeper another cause of disturbance is introduced, and one which renders accurate timekeeping simply impossible. The more mechanism and movements the British watchmaker can crowd into a watch, the more the latter commends itself to an indiscriminating public, especially the official time-keeping portion thereof. Hence we have a wholly unnecessary split seconds movement, and the dial is divided accordingly, and the watchmaker and the official time-keeper delude themselves and the public that they can measure time to the fifth part of a second. " How wonderful! " says the unsophisticated. " How absurd! " says the scientific observer or navigator.

Now, a watch we may regard as consisting of a motor and a train of mechanism. This motor has no governor, but it is regulated so as to drive the mechanism at a more or less uniform rate; consequently, when we throw in gear the centre seconds movement we are putting more work on the motor, and so causing its rate of doing work to alter—in other words, we alter the time rate. Again, all mechanism possesses the quality of inertia; before that centre seconds mechanism acts an amount of time—it may be infinitesimal, but still an amount—has to be occupied in overcoming the inertia of the mechanism, and similarly in stopping the centre seconds movement, which action the official time-keeper *thinks* indicates time to the fifth part of a second, inertia still acts and tends to carry the movement on, and hence the angular displacement of the seconds hand is by no means, or necessarily so, a true time interval—it may have a limit of error of .2 sec. or .3 sec.

The centre seconds movement is rarely seen in watches in which accuracy of going is the desideratum. It is never seen in chronometers nor in Admiralty " deck " watches. At the same time the movement can be advantageously applied to ordinary watches and used to measure small intervals with some degree of accuracy provided the conditions necessary for accuracy are complied with. It will be seen then that a watch used as described is liable to three very serious errors before we can say that a record is correct to the fifth part of a second. We must know not only the competency of the official timekeeper as regards his ability to observe, but also the chronological efficiency of his

Official Time-keeping Rules

watch. Owing to the improvements in watch manufacture it is now possible to obtain watches that are really very accurate timekeepers. A watch may have passed the Kew test and may have the A certificate, but its subsequent good performance depends very largely upon the intelligent care bestowed upon it by its possessor. It does not follow that such a watch will be a good timekeeper *after* it has left the observatory, and hence time taken by it is not necessarily exact. Indeed, it is a wonderfully good watch that will maintain an efficiency of 50 per cent. in the Kew classification for a few months. The Kew certificate resolves itself into little more than that the watch to which it refers is of such mechanical excellence that if properly used the time will be correct within certain stated limits of error.

We have laid stress upon the importance of a uniform rate in any instrument used for the accurate determination of time intervals. We have shown how, even in marine chronometers, this rate is variable. In the best watches, even when used with every care, the rate can very rarely be guaranteed to remain constant for a few days. When used by lay and unscientific persons such as official timekeepers, a " rate " is quite out of the question. So far as the writer is aware, no watches are made having a split seconds movement which is used intermittently that can be relied upon to have a uniform rate. As the result of some considerable experience in rating chronometers and watches, he has found that even in the best watches the rate is usually very variable. It is in timing events that occupy some few hours that the influence of the rate is so important. Thus, suppose we are timing an event that lasts three or four hours, and are using a good watch that has a gaining rate of 5 secs. per day. Let the apparent time be one hour, then it is easily seen that this will be in error by .2 sec., or, in the nomenclature of the official timekeeper, ⅕ sec., and in an event lasting four hours this will be .8 sec., or nearly 1 sec. of time. That this is no supposititious case will be admitted, when we say that it is quite a very common thing in using the very excellent " Admiralty deck watches " to find an error of half a second + or — during a period of, say, three hours during which the watch has been taken ashore for the purpose of taking observations.

Bearing in mind the many sources of error, personal and instrumental, which, as we have seen, the accurate determination of a time interval is liable to, even when trained observers and the best instruments are employed, it will be admitted that the determination of the time occupied in racing events by means of technically untrained men using comparatively very cheap watches with complicated and fancy mechanical movements must be liable to extremely large errors. In fact, such records as we see published in the cycling press are, from the scientific point of view, worthless, as they are merely "apparent times," and to render them "true times" they require to be corrected for many errors unknown as to sign and amount, the limit of error being certainly not less than 1 second and possibly 1½ seconds.

In order, then, to obtain the time interval of a racing event, the competency of the timekeeper to observe should be determined by examination and his personal equation ascertained and checked on the day of the race; the watch employed should be compared immediately before and after the event with a chronometer (not a railway clock), whose error and rate are known. In this way a very fairly accurate determination could be effected, and one which would command respect as to its reliability.

The subject of time-keeping is one which might well occupy the attention of all racing bodies, but more especially of automobilists, and we trust that the Automobile Club will, in framing its rules for racing, take steps to ensure that the timekeepers are competent observers, and that the instruments employed comply at least with the Admiralty standard of excellence.

G. H. L., in the *Automobile Journal*.

AN INCREASING DEMAND

It is stated that the New York Central and Hudson River Railroad Company will replace its present cabs with the latest improved automobiles, just as soon as the new vehicles can be manufactured. The Lake Shore road, too, will discard its cabs in Chicago. It has placed orders for the new vehicles with the new $10,000,000 Woods Motor Vehicle Company, recently incorporated in New Jersey. Vice-President J. Wesley Allison, of the Woods Company, says he also has orders from Manager Boldt, of the Waldorf-Astoria Hotel, for automobiles for his hotel service.

Floral Parade of French Auto Club

One Year's Progress of Automobilism

By Félicien Michotte

I AM very much tempted to say, and should like to be able to say, that 1899 has been the year of triumph of automobilism in France; but if I did say so, I should be no longer able to say so after 1900. For, without being a great prophet, I foresee that what has been done in 1899 is but a trifling matter as compared with what will be accomplished in 1900, from the viewpoint of manufacture and sales, and especially from that of the general adoption of the automobile in France. The number of these vehicles in use at present in Paris alone is more than three thousand.

The automobile has now thoroughly entered into the French mode of life; timorous persons no longer exist; the sight of an automobile running in the streets no longer attracts any attention; and a break-down alone causes the gathering of a crowd, and that of much smaller size than one that is collected by the falling of a horse.

Odors no longer call forth protests, since people have grown used to them. Horses have been pleased to put themselves upon

The Automobile Magazine

a level with man (and for this the poor pedestrian should be thankful to them) and no longer take fright or even shy at mechanical or other street cars that come toward them at 15 miles an hour, and graze their nose at such speed. The noise made by petrolettes, motocycles, quadricycles, etc., leaves everybody (even dogs) indifferent, and it may be said that the horse, the dog and man are now trained.

Another progress made by automobilism in its entrance into our mode of life is seen in the theatre. For example, in a piece put upon the stage by the " Variétés," one of the principal incidents is an accident that happens to the automobile of the Prince, and which is followed by a happy denoument for the latter, the piece and the author.

Racing Van at Timing Station

In a lecture delivered three years ago upon the future of automobilism, I said: " Our fathers judged of the wealth of a person by the criterion, ' he has a horse and carriage '; in a near future, in the second year, he has an automobile." I did not know that I was saying a thing so true and so soon to become a reality; for it may be said that nearly every wealthy person now has an automobile and those who have not, have one ordered from some manufacturer, and, as the saying is, will have to wait months and months to have it delivered, so great is the demand for these vehicles.

The middle class, the bourgeoisie, has adopted the voiturette and the quadricycle, which are within easier reach as regards purchase price, and this latter, like the automobile, leaves the public absolutely indifferent, so many prices are there.

Is it necessary to say that the cycle has done nothing but grow and become handsomer; that new adepts are counted by thousands every year, to the great profit of the Treasury, which taxes them 10 francs; that the Monts de Pieté (pawn-broker establishments) have been obliged to provide vast installations for the housing of bicycles during winter, since it appears that it is cheaper to borrow from one's " uncle " than to put a bicycle in storage; and that the members of the Automobile Club now number more than two thousand (despite the distinctly " aristocratic " character of the association). and more than the Touring Club includes at present.

The Goal

MOTOCYCLETTES

The Werner Motocyclette, a bicycle which carries in front a diminutive motor of the De Dion system, has some adherents; and several manufacturers have been endeavoring to push it, but, so far, without much success. This type of vehicle does not appear to me to have much chance of coming into general favor.

MOTOCYCLES

The manufacturers of the motocycle have increased in number, but the general character of the vehicle has not changed. It is always the De Dion motor that holds supremacy. A few Gladiator motors and several derivatives from the De Dion have followed in its train, but very unassumingly.

The Automobile Magazine

Decorating a Peugeot Victoria

This year, we have seen the De Dion motor of 1¾ horse-power, which is merely that of 1¼ horse-power with a slight increase in internal diameter; then the 2 horse-power, and finally the 2¼ horse-power motor—always the same, with a slightly increased bore. Ere long, we shall have the 2¾ and also the 3¼ horse-power; but that will cost nothing but writing, since such powers exist only upon paper.

An exceptional motor that I tested, and which was called 1¾ horse-power, gave ¾ horse-power with difficulty, and a 2¼ horse-power gave one only, and even that was doing well.

Chainless tricycles or " acatenes " have been constructed; the De Dion surface carbureters have been replaced by certain others, such as the Longuemare vaporizing apparatus or analogous ones; the forms, complications and dimensions of lubricators have increased, and gasoline reservoirs have assumed gigantic proportions, which do not ornament the tricycle as seen from the rear, but far from it. Let us hope for the sake of æsthetics that a halt may be called here, lest the motocyclist have the appearance of having a manufactory at his back.

MOTOCYCLES AND VOITURETTES

The voiturette coupled to a motocycle came and had a certain run; but, unfortunately for it, the quadricycle, which is less costly, caused it to be abandoned. The latter is less cumbersome, and is stronger, and presents to the eye more solidity and stability.

One Year's Progress of Automobilism

QUADRICYCLES

The quadricycle presented itself in the form of a front hauler of the Chenaud system, with a seat upon two wheels, which were situated at the place of the front wheel. It was a success, and a well merited one for the modest inventor. Such success involved the construction of the quadricycle now so much employed. It is handsome and practical, and permits of a conversation being carried on between a lady and gentleman, and does not make it appear as if the lady were being driven by a servant. This vehicle is destined to be still further developed, for it is more within the reach of many purses and less cumbersome than the voiturette.

The " Victoria Combination " voiturette is a vehicle intermediate between the quadricycle and the voiturette. It is the back of a motocycle turned about and forming but one with a voiturette seating two persons. It is pleasing to the eye and has achieved a certain amount of success.

VOITURETTES

The voiturette has quite a number of new manufacturers, but, notwithstanding this, remains stationary, since the price of it is somewhat high; and fault is found with some of the types that they are poor hill climbers and do not afford sufficient speed.

Voiturettes, with a few rare exceptions, have all been actuated by De Dion motors, and are, for the most part, " carriage-finished " motocycles, if I may so express myself.

A Decorated Coupé

The Decauville Voiturette, pretty and graceful, is provided with two coupled De Dion motors; but it is feeble as to motive power, and lacks strength. Nevertheless, there is a certain demand for it.

The voiturette with front-hauler, heavy and ugly, and perhaps more powerful, has not had the success of the other types. On the contrary, the little two or three passenger " Voiture " De Dion (for such is the name that has been given to it) has had in its favor its pleasing aspect and the name of its manufacturer. The Bollée " Fameuse " Voiturette seems to be becoming more and more neglected.

The hit of the year is a voiturette which has received no puffing, but which has in its favor the possession of a motor of

De Dion Bouton Steam Stage in Operation

4 horse-power (at the brake), without a circulation of water, and yet is a vehicle that *climbs*. It is the Niullary Voiturette, light and handsome, which by the next year will find a ready sale.

CARRIAGES

The gasoline carriage has this year made a great advance. Its form has been improved, and it has become lighter, more elegant and nearly silent.

We now see running a few landaus, one or two hacks, several coupés, some dog-carts and phætons, and a large number of omnibuses. The vis-à-vis type has been nearly abandoned.

One Year's Progress of Automobilism

Motors have remained *in statu quo,* but have become of 6 and 8 more or less effective horse-power. It is always the Phenix, Panhard or Peugeot that take the lead.

Many small engine builders have been desirous of engaging in the construction of automobiles, but all have desisted after perceiving that the construction of a motor carriage is not so simple a matter as it might seem to be, and that if certain persons, such as Panhard and Peugeot, have the precedence it is because they have justly merited it. In their train follow Delahoye, Mors, Hunter, Rochet, etc.

Some attempts at importation have been made, and that by the American builders of the Duryea carriage, Locomobile and Columbia. Manifold (though fruitless) attempts have been made to extend the patents on these vehicles; but, despite their real value,

Replenishing Station for French Autocars

the purchase of foreign patents at a phenomenal price would be a mistake that our manufacturers would be slow to commit.

Steam Carriages

Among steam carriages, we have witnessed the development of " heavy weights "; the De Dion traction carriages have been used upon the Metropolitan Railway for drawing cars loaded with excavated material; and a Scotte steam truck has been seen running in the streets of Paris.

As regards pleasure carriages, the " Hawley," an American vehicle, has had some success as a matter of curiosity; but that is about all. It is very well arranged, but lacks strength and is too

Steam Truck During Heavy Weight Contest

complicated, and will not come into favor, since many persons will be afraid to sit over a steam boiler registered at 170 pounds to the square inch. I pity the person whose duty it will be to make the connections and keep the joints tight.

ELECTRIC CARRIAGES

After being successfully launched, so to speak, the electric carriage has finally started off. It has proved a great success in France, despite the general lack of charging stations.

A Steam Bus at Standstill

Mechanical Propulsion and Traction

By Prof. G. Forestier

Third Paper

For the transportation of heavy indivisible masses that exceed in weight the load that can be economically imposed upon the usual team of five horses in tandem, inventors, discarding the automobile truck, have for a long time been incited by traction upon rails, and have devised mechanical vehicles for hauling independent trucks.

The reasons for the final failure of all regular systems of hauling upon roads are twofold:

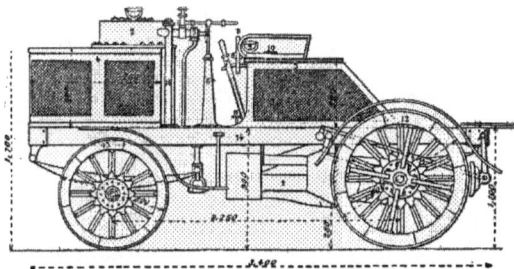

Fig. 12.   The de Dion and Bouton Steam-Power Carriage

(1) With a high speed there is a certain difficulty in maneuvering without hazarding the stability of the vehicle hauled, but especially a danger from the failure of the brakes to act in unison. Such are the causes for which, without speaking of a blinding dust, passenger trains upon roads are limited to quite a feeble speed, in order to prevent the inconveniences of attaining a dangerous one, but which is then inadequate to satisfy the requirement of going fast.

(2) For the carriage of merchandise at a slow speed the use of a vehicle for hauling loaded trucks would not constitute an economical application of mechanical traction.

In fact, let us see what takes place upon railways themselves. Although the hauled train is the rule upon plains, where the line presents only similar levels, the tendency, in a mountainous dis-

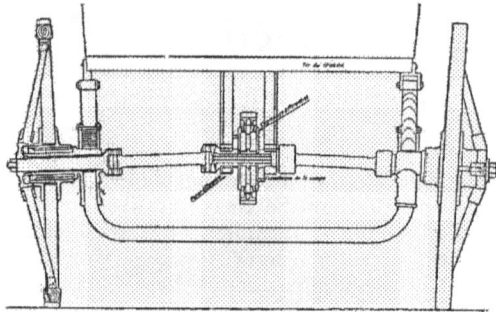

Fig. 13.   Arrangement of Driving-wheel Gear of the de Dion and Boulton
Power Carriage

trict, upon a line with a broken profile, where the gradients reach
.4 of an inch to the foot, is to have recourse to the automobile
vehicle.

The locomotive upon a line on which the maximum declivities
vary from .08 to .12 of an inch to the foot hauls from *ten to eleven
times* its weight.   If the declivities are from .12 to .16 of an inch
the load hauled descends to *eight or nine times* the weight of the
locomotive.   If they reach from .16 to .2 of an inch the load
hauled falls to five or six times the weight of the locomotive.   At
from .3 to .4 of an inch the weight of the train reaches scarcely
one and a half times that of the locomotive.   Beyond this the
latter would hardly be able to haul its tender.

Now, upon a road, although the coefficient of resistance to the
rolling of the vehicle varies from 44 to 88 pounds per ton, owing
to the fact that the coefficient of sliding friction is higher than it
is upon rails (0.35 instead of 0.14), a mechanical vehicle will
nevertheless be able, upon a level, to haul two or three times its
own weight at a low speed.   But upon many roads we meet with
declivities of .5, .6 and 1 inch to the foot, and upon these the
resistance of the vehicle hauled will be doubled or tripled while at
the same time the power of the traction engine will be diminished
by so much.†

† It results from experiments made with an Aveling and Porter road locomotive of 6-horse power,
weighing 13,410 pounds, that it was capable of hauling :
    On a level and on a good road..................five times its weight.
    Upon a 4% declivity and on a good road ....three   "
    Upon a 7% declivity and on a good road........two   "
    Upon a 10% declivity and on a good road .....one time its weight.
—but that, in regular service, it had to be made to haul only ⅗ of these experimental maximum
oads, that is to say :
    On a level...... ....3.33 times its weight.
    Upon a 4% declivity ....2.25   "
    Upon a 7% declivity ....1.31 time its weight.
    Upon a 10% declivity ....0.66   "

Mechanical Propulsion and Traction

At all events, the police regulations as to transportation do not permit a four-wheeled vehicle to exceed a weight greater than 16 tons. Hence, upon a somewhat broken road, the load hauled will not sensibly exceed that which a team of five horses may be made to draw economically.

In order to solve the problem of the mechanical traction of any load in bulk exceeding the normal and economical power of the draught animal without necessitating the construction of a special automobile truck in each particular case, we see only the following arrangement: The mechanical vehicle should be arranged in such a way that it can be used as a fore-carriage for the truck, upon which would be placed the indivisible load, as if animal traction were to be employed. This fore-carriage may evidently be provided with two, three or four wheels. In the first case the fore-carriage will be wholly for steering; in the two others the front wheel in the tricycle and the two front wheels in the bogie fore-carriage will be used for steering.

This type of mechanically propelled vehicle, in which a portion of the load to be hauled contributes toward giving the driving wheels the necessary adhesion, has been particularly studied and improved by MM. de Dion and Bouton.

It was a traction vehicle of this kind (Fig. 12) which, running isolatedly, was the first to arrive in the race organized between Paris and Rouen by the *Petit Journal* in 1894.

Further along we shall give the necessary details as to its boiler, motor and transmissions. For the moment we shall be content to state that the transmission of power to the driving

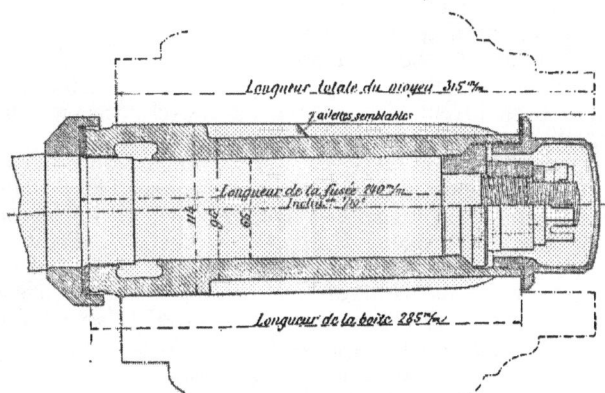

Fig. 14. Patent axle-journal used by the Compagnie Générale des Omnibus of Paris

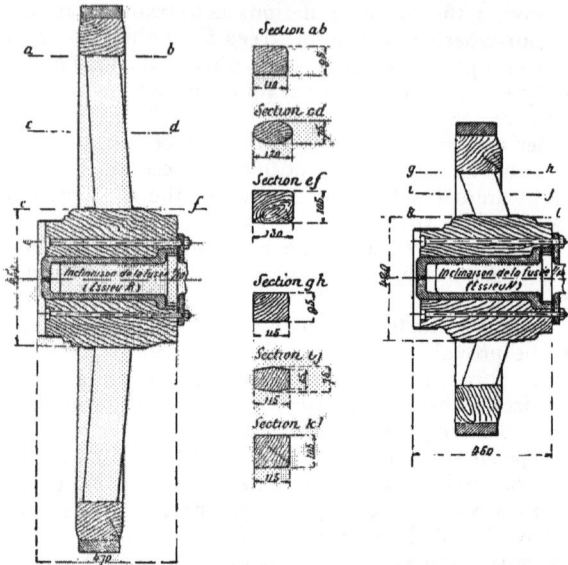

Fig. 15.   Patent axle-journals used on the Trucks of the Say Refinery

wheels is effected through the intermedium, not of a chain, but of a cardan joint placed between the axis of the differential gearing and the axle journal (Fig. 13).

We cannot refrain from saying a word as to the *gasoline carriage*, which, entering seriously into the lists with the Daimler motor, along about 1890, has more powerfully aided in the development of automobilism than the very persevering and interesting efforts of those inventors who have successively occupied themselves with the improvement of the steam carriage.

At the Exposition of 1889 MM. Panhard and Levassor exhibited an omnibus upon rails which was actuated by a Daimler motor.   In 1891 a Peugeot carriage, provided with a motor of the same type, participated, it appears, in the Paris-Brest race. Yet in 1892 an editor of *La Nature,* in giving an account of Prof. Unwin's experiments with gasoline motors, mentioned pleasure navigation only as an interesting application of this new motor.

Specially translated for the Automobile Magazine from *Le Génie Civil.*

(To be continued in our next issue.)

The Automobile in Local Transit

By Sylvester Baxter

ONE of the most vital problems relating to the modern municipality is that relating to local transit. The development of the automobile bears upon this question in two important ways. One of these replaces an existing form of animal-traction service and the other supplements a highly developed form of mechanical traction. Each has the advantage of a greater flexibility in operation than the corresponding methods in existence. The first, a motor-cab service, is already in a well advanced stage in our great American cities. This may be called the individual aspect of automobile local transit, giving to a person at a moderate charge all the advantage of a privately owned carriage; taking the passenger, as it does, from his own door to any spot that he may desire to reach. No investment is required on the part of the individual; the vehicle company assumes all the care, repairing, cost of charging, etc.; and with a telephone in his house or office a person may bring to his door at any moment what practically amounts to his own carriage. While, of course, there is a manifest economy in the possession of one's own automobile when it is wanted for regular use every day, for a very large number of persons—particularly those who live in cities—the hired vehicle will prove highly economical, and we may expect to see an enormous increase in the use of the motor-carriage for individual business and recreative purposes, such as errands to sections of a city not reached by the ordinary transit routes, pleasure-drives in the parks and suburbs, etc. The celerity of the motor-vehicle, its easy motion, its tirelessness, the free views on all sides, make it an ideal form of recreative conveyance, and with its growing availability we may expect to see a very great increase in the number of those who patronize livery carriages. An economic aspect of this form of service lies in the fact that the capacity of the plant is so much in excess of an aggregation of a corresponding number of privately owned carriages, each one of which is used, as a rule, only by the person or the family possessing it. But when a company owns the same number of vehicles for hire, they are naturally in use by a very much larger number of persons and consequently make a much greater return on the investment in the way of utility, which in the matter of the commercial company is equivalent to profit. This suggests the desirability of what may be called a modified form of private ownership in the shape of collective ownership on the part of groups of individuals. For instance, a number of persons might act coöperatively by clubbing together and purchasing any desired number of motor-vehicles to be kept for their common use as required. In this way there would be a material economy in expenses for stabling, care, charging of batteries, experienced drivers, etc., while a greater variety of vehicles would be at the disposal of the individual. Such a function might be assumed by automobile clubs to the advantage and convenience of their members, and we may expect to see a great growth of clubs for this very purpose, with membership as large or as small as may seem desirable. It is a common custom for boat and canoe clubs to own various kinds of water-craft for the use of their members, and there is no good reason why an automobile club should not follow the example. In this connection it may be suggested that a highly popular feature for a golf-club would be the ownership of automobile conveyances to take its members to and from the links. This was proved last year at Newport.

The public aspect of automobile local transit is represented by the motor-omnibus service, now in successful operation in various European capitals and in the near future to be installed in several great American cities. The possibilities of automobile omnibus lines is something enormous, and there is probably no more promising field for the profitable application of the principle than is offered in this. While it cannot be expected that this form of traction will to any degree supplant the tramway method, in all probability the motor-omnibus will rival the street-railway to a considerable extent, as it certainly will supplement the latter most conveniently. Moreover, it can easily be made a most valuable auxiliary to a street-railway service. A street-railway system, of course, has the advantage of larger vehicles than are practicable for motor-omnibuses, as well as that of a higher rate of speed. For in the former the element of guidance, so far as steering is concerned, does not appear, and the matter of control being therefore limited to fewer particulars, a higher speed in city streets is practicable, as a rule, than where the matter of direction enters very largely into consideration. But these advantages are counterbalanced by certain factors that tell in favor of the motor omnibus. The latter has a much greater flexibility than the tramway system possesses. If a street-car breaks down it brings to a stand every car behind it until the way is

The Automobile in Local Transit

cleared. The same interruption occurs in the case of a blockade of the street for any reason—from the collapse of a heavily loaded team—and loaded teams seem invariably to have the perversity to choose the car-tracks as the scene of misfortune—to the congestional effects of processions, circus-parades and gas-main explosions. But the breaking down of a motor-omnibus can have no effect on the other vehicles of the line, and in case of the blockade of any portion of the route the traffic can at once be diverted around through other streets for the time being. The motor-omnibus can also take the fullest advantage of improved forms of pavement. An asphalt pavement, for instance, converts the entire street-surface into what for the motor-vehicle is practically one broad rail of indefinite width—or what is the equivalent of a rail in every property except that of guidance. It can skillfully thread its way through the tangles of a crowded thoroughfare where the street-car is constantly brought to a halt by this or that interruption. So, even in the matter of speed through the busy streets of a city, the motor-omnibus will be found to have the advantage over the street-car. Another point in favor of the former is its greater directness. A street-car line between two cardinal points—such, for instance, as terminal railway stations on opposite sides of a city—makes its route as devious as possible —going the longest way around so as to pick up the greatest number of passengers practicable. An omnibus line, however, having for its special object the conveyance of passengers between two such points, would take the shortest cut across and get its patrons to their destination in the quickest possible time. As to elements of profit—while a motor-omnibus cannot carry anywhere near the number of passengers that a street-car can, and while a storage-battery system is much more costly in operation and maintenance than a trolley system, on the other hand the motor-omnibus is entirely free from the very costly factor of construction, maintenance and renewal of track—items that consume a very large proportion of the capital and the receipts of a street-railway company. For the motor-omnibus line the traction surface is provided by the public. And when we consider what enormous profits a street-railway would make were the element of rails, roadbed, trolley-wire, return-conduits, etc., eliminated, it may be imagined what a field for investment is offered in the establishment of motor-omnibus lines in the cities of this country.

It is safe to say that the prospective demand in this direction alone is sufficient to exceed enormously the capacity of all the existing manufacturing facilities in this country were they con-

The Automobile Magazine

fined solely to that special feature, instead of being overrun with orders for motor-vehicles of every description.

There are three cities in North America where lines of automobile omnibuses will be in full operation before the end of the present year, and each of these present exceptionally favorable opportunities for the purpose. These are New York, Boston and Mexico. In the City of New York, Fifth avenue, with its magnificent extent of asphalt pavement unbroken by car-tracks, offers a superb opportunity for the profitable operation of such a line. And this has been taken advantage of by the purchase of the Fifth Avenue Stage Line—so long celebrated for its decrepit horses, the butt of the caricaturists—by interests connected with the great Electric Vehicle Company. This will be the pioneer electric omnibus line of the continent, and the change will mark the beginning of an important epoch in local transit matters for this country.

In Boston there have lately been some remarkable transit developments. While fundamentally they have to do with broad principles of municipal administration, incidentally they have a very important bearing upon automobile interests. It is manifestly important that every city should have at least one great thoroughfare unobstructed by car-tracks. New York has such a street in Fifth avenue, and as a civic possession it is priceless. Boston is now assured the permanence of such a thoroughfare, or continuous line of thoroughfares, running into the very heart of the business section. It may be remembered that the celebrated Boston Subway, that carries the street-car traffic underground through the congested section of the city, was the result of an agitation to prevent the venerated Common from encroachment by street-railway tracks. When this great enterprise was authorized it was provided that upon its completion the car-tracks should be removed from Tremont and Boylston streets around the Common, and from the former throughout the rest of its surface to its terminus at Scollay square. The tracks had been removed but a few weeks when the Boston Elevated Railroad Company, which practically monopolizes the local transit service, instituted an agitation for their restoration. This agitation was ostensibly in the interest of the " masses "—and much was said about the poor shop-girls, who desired to be conveyed to their shop doors in inclement weather, etc. It was supposed to be a popular measure, and every daily newspaper but one, and the great majority of the politicians, were vociferous for the change. Legislation to restore the tracks was secured, but a threat of a veto from the Governor led to the incorporation of a provision that the question

The Automobile in Local Transit

be left to a vote of the citizens. When the time for the vote drew near, at the annual municipal election in December, a public-spirited citizen, Mr. B. F. Keith, the theatre manager, came forward and headed the movement against the restoration, contributing freely of his means and devoting his energies to the work. In his popular theatre the cinemetagraph presented the arguments in favor of keeping the tracks away from the streets with telling directness, showing the street in its former insufferably congested condition and in its present dignified aspect. The *Evening Transcript* also did yeoman service in the work. The result was a great civic victory for municipal development on correct lines. The longer the tracks were off the streets the better the public liked the effect. The people saw through the motives of the newspapers and of the politicians, and they buried the proposition under a great adverse vote of practically two to one—nearly 52,000 against and a little over 26,000 for, only two wards voting in the affirmative. So Tremont and Boylston streets remain stately thoroughfares, soon to be paved with asphalt and free to use by all freely moving vehicles.

This very notable demonstration of civic sense on the part of a great city's population has such an important bearing on the adaptation of our cities to the great reforms that are impending in their transit methods with the advent of the automobile, that the foregoing concise review has its place in these columns. And it will not be amiss to chronicle an amusing episode of the campaign. A certain Commoncouncilman, a ward politician, made a speech on the subject to the following effect: " I am in favor of putting the tracks back on Tremont street, because the people want them back! I am in favor of putting them back because B. F. Keith does not want them back. B. F. Keith is opposed to putting them back for the reason that he has on Washington street an entrance to his theatre to which the poor people and the common people are brought by the street-cars. He has an entrance on Mason street where his wealthy patrons come in their carriages. And now he wants Tremont street kept clear of the car-tracks in order that the aristocrats of the Back Bay may roll up to his door in their automobiles!"

The determination of the future of Tremont street has indeed given a great impetus to automobile development in the direction just considered. The Boston Transit Company, one of the subsidiary companies of the Electric Vehicle Company, had evidently been awaiting the action of the public on the question. No sooner had it been determined that Tremont and Boylston streets were to remain unobstructed, and therefore ideal thoroughfares for automobile use, than the company petitioned the Board of Adermen for the privilege of running lines of motor-omnibuses over three important routes, two of which included Tremont and Boylston streets. Electric omnibuses, of the same pattern as those designed for Fifth avenue in New York, are to be placed in service as soon as they can be made ready. One of these routes is intended chiefly for recreative purposes, although incidentally it will serve residents along the way. Whether this route will be operated depends upon the action of the Park Commissioners, for it lies chiefly within the jurisdiction of the Park Department. The Boston Park Board expressly disclaims any hostility to the automobile, but holds that the time has not yet come for a relaxation of the restrictions. In the interests of all concerned, however, the park board should look favorably on the proposition.

The third city which offers remarkable opportunities for automobile transit is the stately capital of Mexico. Conditions of climate, topography and population are alike favorable. The streets are level and smoothly paved, and, of course, are never obstructed by snow, while the public gives a most profitable patronage to local transit lines. The Broadway of Mexico, described by George Augustus Sala among his pen pictures of the famous streets of the world, comprises in a great central thoroughfare the successive streets known as *la calle de los Plateros,* the first, second and third *calles de San Francisco* and the *puente,* or bridge, of the same name, and the Avenida Juarez, to the magnificent pleasure drive of the Paseo de la Reforma, with its statues, monuments and splendid promenades extending out to the castle of Chapultepec. In the central portion of the city the streets included in this magnificent thoroughfare are too narrow and too thronged to admit occupation by street-cars, but the whole extent from the national Palace on the Plaza Mayor to Chapultepec, a distance of something like four miles, offers a superb route for the motor-omnibus line soon to be established by the Electric Vehicle Company of Mexico, a corporation allied with the great company of a similar designation in this country.

Couthon's Automobile
A Relique of the Last Century

IT is not known to everyone that Couthon, the French revolutionary fanatic who died on the scaffold with his friends and colleagues St. Just and Robespierre, was the possessor of a mechanical carriage in which some of the features of our modern automobiles have been anticipated. Like Scarron, the creator of French burlesque, who lived a century before him, Couthon was paralyzed in both legs. Both the revolutionist and the burlesquer, although physically disabled, were men of rare talent.

Nevertheless, the lives which they led were different. Scarron could devote himself to his literary and intellectual work in the privacy of his study, seated in a stationary chair, and could discourse at his ease with the brilliant men to whom he communicated his original ideas. Couthon, however, lived in more active times. The Convention, not the home, was the place for a gifted man during the Revolution; and in order to travel between the assembly hall and his house, he employed a vehicle, which to this day is preserved in the Musée Carnavalet.

Couthon's carriage, or chair, is a true automobile, although driven by hand. Its construction is as simple as it is ingenious, and is just as remarkable as that of the Ozanam carriage (called so, probably, because it was built by Elie Richard).

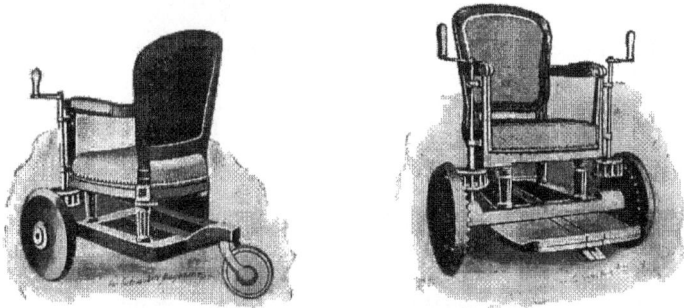

Couthon's Autochair

The carriage has two vertical shafts provided at their upper ends with hand-cranks, and at their lower ends with cog-wheels, which engage pins projecting laterally from the carriage-wheels.

Here we have the germ of the modern automobile driven by a
motor coupled with the front axle—the *avant-train moteur*.

The vehicle is easily steered by turning one crank more rapidly than the other. By turning both cranks in opposite directions, it can be swung around as upon a pivot.

One naturally wishes to know the name of the inventor of
this curious, old mechanical chair, and the date of its construction. Were similar vehicles in use, or is this the only one of its
kind? Unfortunately, no light can be thrown upon this point.
The chair, it is said, was lent to Couthon by the *Mobilier national*, and came from Versailles, where it had been used by
the Comtesse d'Artois.

Was it built especially for her? Was the inventor the Comte
d'Artois himself, or perhaps Louis XVI., who, as everyone
knows, was a skilful locksmith? Or was it merely an imitation
of a model already in use, of that, for example, which formed
part of the effects of Abbé D——, and which in 1785 was adver-
tised in the *Annonces-Affiches* as a " joli fauteil monté sur trois
roulettes, propre pour une personne infirme qui, au moyen, des
manivelles, peut se promener dans sa chambre ou dans son jar-
din " (a fine chair mounted on three small wheels, suitable for
an invalid who, by means of cranks, can ride about in his room
or in his garden). Chi lo sa?

THE OPEN ROAD

In a recent issue of the *London Spectator* there is a charming
article on the " Open Road," in which the delights of travelling
by road are dwelt upon. The article concludes as follows:

" We may be sure that before very long the roads of all countries will obtain, socially and commercially, an entirely new significance. The bicycle has restored the roads to the pleasure-
seeker; the motor car and motor wagon will restore them to
the non-athletic lover of the open air, and to the trader. The
new century will see the railways relegated to their proper place
as providers of very swift and very heavy transit, while the roads,
which have always this advantage that they pass all men's doors,
will once again be thronged by carriages and carts. In a few
years' time the man of business who lives six or seven miles out
of London will never dream of taking, as he does now, a fifteen-
minutes' drive in a carriage to the station, then twenty minutes
in a train, and then ten minutes more in a cab, to reach the
office. Instead, he will drive from door to door in a motor victoria, and save money, time, and temper."

Compressed Air in Europe

THE delivery-wagon represented herewith, constructed by Messrs. Molas, Lamielle and Tessier, was exhibited at the second Exposition des Tuileries, where it attracted considerable attention by reason of the relative simplicity of arrangement of its maneuvering devices and the limited amount of space occupied by the motive apparatus.

In this vehicle, the air-storage reservoirs employed consist of hammered steel tubes of 8-inch external diameter, the ratio of the weight of which to that of the air stored up is $\frac{1.36}{7.73} = \frac{1}{4.35}$ The heating is done directly by gasoline, instead of by steam from a boiler, as in the Mekarski system; the manufacturers being of the opinion that, since a direct heating of the air permits it to be raised to a temperature much higher than that which could be obtained by means of heating by steam, they obtain also a greater increase of volume and, consequently, of work that compensates for the heating during expansion obtained with the above-named system.

The stove, burners and gasoline reservoir used for heating have here a feeble weight as compared with the arrangement in which a boiler is employed. The ratio of the weight of air stored up to that of the reservoirs and heating apparatus does not descend below $\frac{1}{4.35}$.

The air reservoirs are eleven in number and are distributed in two groups, one of six forming the " battery," and the other of five constituting the " reserve." The capacity of these groups is, respectively, 11 and 7.5 cubic feet, or, altogether, 18.5 cubic feet.

With a charging pressure of 4,290 pounds to the square inch, the weight of air stored up, at a temperature of 13° C, is 1,100 x 2.7 x 638 = 392,370 pounds.

Before its admission to the cylinders of the motor, the air passes into a steel worm of .28-inch internal diameter, of .14-inch thickness, and of a length of 19.7 feet, which raises its temperature to a figure that varies with the discharge of air and the intensity of the burners that heat the worm. The temperature may thus reach 150° C. The air afterwards passes into an expander (maneuvered by the driver), which lowers its pressure by about from 11 to 44 pounds, according to the difficulties of the road, the load and the speed. This expansion of the air gives rise to a lowering of the temperature, in order to raise which, the air thus expanded is made to pass into a second worm concentric with the first and heated by the same burners. When the air makes its exit from this second worm, its temperature may be as high as 250°. It is claimed that this double heating is capable of more than doubling the volume, and, consequently, the work of the air stored up. This air is then admitted to the cylinders through an expansion distribution with a special change of speed. The admission takes place upon only one of the faces of each piston, and the motor is thus a single acting one. But the cylinders are four in number, and cast in pairs, and the connecting rods (jointed directly to the pistons) actuate the same shaft, the cranks of which are set at an angle of 180° per group, and at 90° from one group to the other. The motor thus has the same power as a two-cylinder double-acting engine. The arrangement adopted offers numerous advantages, and permits especially of suppressing the shocks at the joints of the connecting rods that occur in double-acting engines at the time of the change of direction of the piston; of doing away with piston-rod stuffing-boxes, which it is often difficult to keep tight, and which absorb a certain amount of work in friction; of removing and putting in place a connecting rod and its piston without having to dismount a cylinder bottom.

The space available for a motor under the seat of a carriage is sufficient widthwise, but not lengthwise; and so the motor must

Compressed Air in Europe

occupy more space in the former than in the latter direction. Now, two double-acting cylinders of the same bore, placed either longitudinally or vertically, always take up more space than the single-acting four-cylinder arrangement adopted; and from this fact, it has been possible to have a greater available space for occupancy by the reservoirs of compressed air.

The distribution is effected by means of cams and valves. The opening of the latter takes place very rapidly. Their closing, on the contrary, occurs gradually, and is hastened or retarded through the shifting of the reversing lever, according to the admission that is desired. The minimum admission employed is 10 per cent., but this may be increased to 60 per cent. for starting. The exhaust and the compression are fixed, and have a duration of 20 per cent. of the stroke of the pistons.

There is no special cut-off of fluid between the expander and the cylinder admission valves. These latter open wide for all admissions between 10 and 60 per cent. of the forward and backward running of the engine.

The burners, which are three in number, are arranged beneath the worms in an iron plate jacket 10.75 inches in diameter and 12 inches in height, surmounted by a chimney which leads the gases of combustion to the top of the roof of the vehicle. The burners are supplied by a gasoline reservoir of a capacity of about ten quarts in which an air pressure of from 45 to 70 pounds to the square inch is created. The pipe that leads the gasoline to the burners may be partially closed by a screw plug. Through such arrangements, it is possible for the driver of the wagon to proportion the intensity of the heating to the output of air, so as to obtain a temperature that is always sensibly the same.

The quantity of gasoline consumed by the burners is about 15 ounces per hour of running; but this might easily be increased if it were found that the operation of the motor became thereby more economical.

The motor rests upon the floor of the carriage, beneath the seat occupied by the driver, who can thus examine it while it is running. Its rotary velocity for a speed of five miles an hour made by the vehicle is 200 revolutions per minute. The transmission is effected by means of chains of the " Varietur " system, which connect two sprockets, keyed at the extremities of the driving-shaft, with two toothed wheels of six times larger diameter fixed through bolts to the spokes of the hind wheels. The power of the motor, upon the crank-shaft, is 20 horse at a velocity of 300 revolutions, and at a pressure of 285 pounds.

The differential consists of two friction-cones fixed near the

The Automobile Magazine

extremities of the driving-shaft and actuated by the steering apparatus. The control of these cones is such that, in a turning about, the adhesion of the one that is situated at the side of the internal wheel diminishes in such a way as to permit of a certain amount of sliding of the cone in its socket, and, consequently, of a retardation in the revolution of the corresponding wheel; while the adhesion of the cone situated at the side of the wheel that is to traverse the wide radius increases, so that no sliding in its socket can occur, despite the greater resistance of the wheel that is moving ahead.

The pivoting can thus be effected upon a driving wheel that is rendered absolutely immovable, from the view-point of revolution, and, consequently, in a circle having as a radius the distance of such wheel from the opposite wheel of the front axle, and that, too, in running forward as well as backward.

The steering is done by means of a hand wheel keyed upon the axis of an endless screw, which gears with a toothed wheel keyed upon the shaft of the reversing gearing situated beneath the vehicle. An indicator placed before the eyes of the driver exactly reproduces the changes in direction of the axis of the vehicle's path, and thus permits him to keep in a straight line.

The intermediate screw lengthens the maneuvering, but renders it sure, stable and gentle. Such an arrangement, however, offers advantages only for heavy vehicles designed for running always at a low speed.

Finally, let us say that the accumulators are placed longitudinally under the floor of the vehicle in two superposed rows, and are connected by easily accessible couplings placed in the rear. Movable panels permit of an inspection and of a tightening of the joints and couplings.

The motor and vehicle as a whole are well elaborated and denote upon the part of the manufacturers the possession of real ingenuity and a complete knowledge of the question with which they had to deal.

AN ENGLISH RACING MACHINE

According to the London *Daily Mail*, a committee has been organized among the members of the Automobile Club of Great Britain for the construction of a purely British racing machine. It is intended to enter this machine in the Paris races to be held in connection with the International Exposition of 1900.

A Front-driven—Front-steered Automobile

MOST automobiles are driven from the rear axle; they are *pushed*, in other words. Experience has shown that in pushing a vehicle, more effort is expended than in pulling it. A man can readily roll a four-wheeled carriage out of a stable with a slight pull upon the shafts; but in pushing it out, considerably more work is performed. Engineers have proven that the resultant of the forces applied in pushing a vehicle tends to press the front wheels on the ground; when the road is muddy,

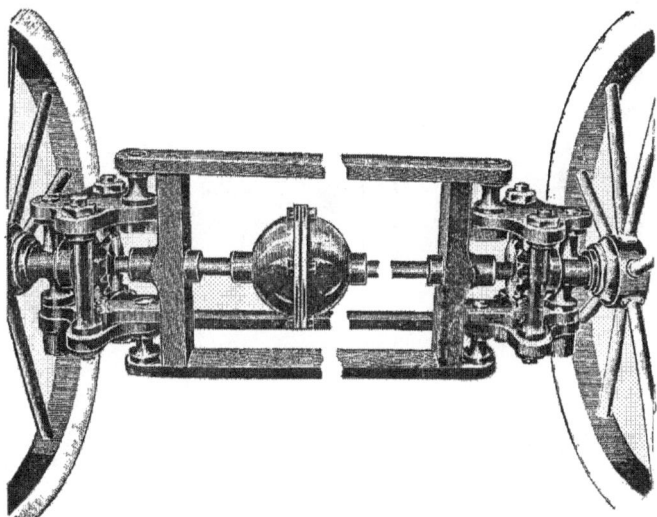

Fig. 1.   Perspective View of the Fore-Carriage with Driving and Steering Gear

the wheels are embedded. If, on the other hand, the vehicle be pulled, the forces are applied only in a horizontal direction, and the amount of work performed is frequently 20% less than in the first case.

Experience has furthermore demonstrated that a carriage which is pulled is more easily steered than a carriage which is pushed. When the fore-carriage both drives and steers, the

The Automobile Magazine

entire carriage immediately responds to the touch of the *chauffeur's* hand on the lever or wheel, the rear wheels naturally following in the track of the front wheels. But when the motive power is otherwise applied, the rear wheels do not begin to turn in the direction desired before the front wheels have, to a certain extent, been forced into the ground. It therefore follows that on muddy or sandy roads, if the front wheels be not sufficiently wedged in the ground, the rear wheels will not pursue the direction desired. The " waltzing " of automobiles—if it can be so

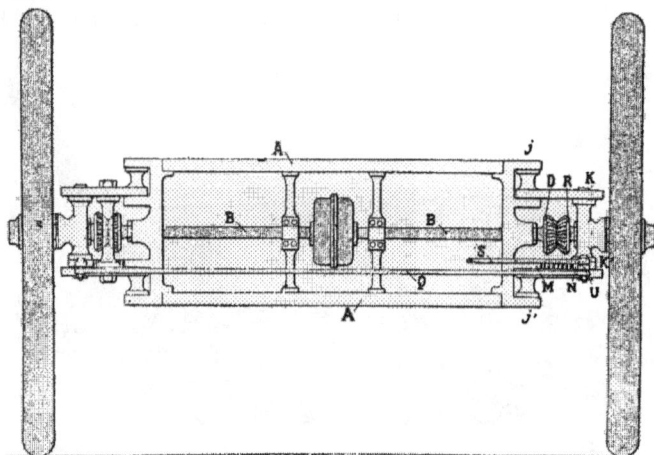

Fig. 2. Elevation of the Fore-Carriage. *AA*, Carriage-Frame *BB*, Motor-Shaft Carrying the Differential. *D*, Hemispherical Gear on the End of the Motor-Shaft. *R*, Hemispherical Gears on the Spindle. *KK'*, Bracket in which the Wheels are held and turning about the Pivot *jj'*. *MN*, Guiding Segment Gears. *U*, Steering Rod. *S*, Steering Lever. *Q*, Bar connecting the Two Steering Rods.

termed—is caused always by the inability of the fore-carriage to extricate the vehicle from its position. The proof of this is found in the only remedy known in such cases—by suddenly swinging the steering wheels into a position in which they will oppose the movement begun by the rear wheels.

Finally, experience has still further shown, that when the motor is mounted on the rear carriage the front wheels cannot be swung beyond a small angle (25°, approximately); for otherwise the vehicle could not be started, since the fore-carriage would block the rear carriage. On the road, the vehicle might even be

A Front-driven—Front-steered Automobile

overturned. The rear driven carriage can be steered out of a line of vehicles only when there is a clear road of several yards between it and the preceding carriage; it cannot be turned about on a straight road except by a succession of rearward and forward propulsions, or "backing and filling," as it were. The *avant-train moteur,* in which the engine is mounted on the forecarriage, can be started even though the front wheels be almost at right angles with the rest of the carriage. Exceedingly short turns can be made.

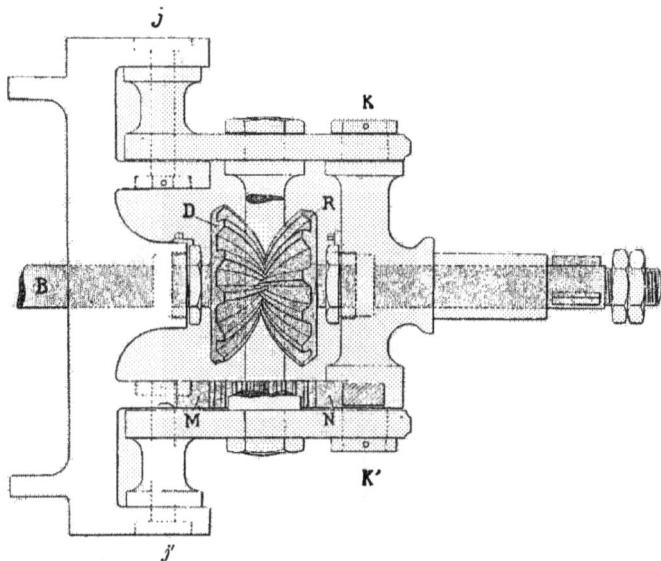

Fig. 3. Detail View of the Gears. *B*, Motor-Shaft carrying the Gear *D*. *R*, Gear carried by the Spindle. *M* and N, Guiding Segment-Gears. *KK'*, Pivoted Mounting of the Spindle. *jj'*, Pivots for permitting the Swinging of the Wheel relatively to the Motor-Shaft.

The superiority of the pulled to the pushed carriage being admitted, the question naturally arises: Why do not our automobile makers mount their motors on the fore-carriage? The answer is found in the mechanical difficulties which have hitherto not been overcome.

On *a priori* grounds it might be inferred that the problem could be most simply solved by reversing the functions of the front and rear wheels, or, in other words, by turning the carriage about and converting the rear driving wheels into front wheels, and the front or steering wheels into rear wheels. But it has been found in automobiles constructed on this principle that there is no more untrustworthy steering-gear, for speeds over 12 miles per hour, than that which is actuated from the rear wheels. To ride in such a vehicle at a speed of 24 miles per hour is suicidal.

The problem which confronted the inventor was, therefore, the construction of a carriage which would be both driven and steered from the front axle.

If there be but a single front wheel, the entire mechanism must perforce be mounted on that wheel. The difficulty of steering so heavy a wheel and the complexity of the construction are obstacles which are well-nigh insurmountable.

If the vehicle be a four wheeler, the two front wheels can be

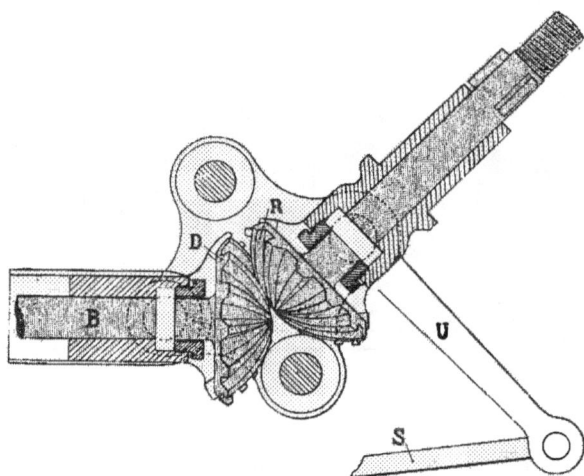

Fig. 4. Spindle-Controlling Mechanism *B*, the Motor-Shaft. *D*, Gear carried by the Shaft. *R*, Gear carried by the Spindle. *U*, Steering Rod for the Right Wheel. *S*, Steering Lever.

mounted on a rigid, centrally-pivoted axle. By an arrangement of gears, the power developed by the motor could be transmitted to the wheels in any position assumed by the front axle with respect to the carriage. But this method of steering by means of front wheels mounted on a pivoted front axle (as in the ordinary horse-drawn vehicle) has been criticized with some show of reason. It is argued that with fairly high speeds this method of steering is dangerous, because in order to turn ever so little to the right or to the left, the axle must be turned through a large angle, and in order to make a short turn, though it be comparatively slight, one of the wheels must be swung almost under the carriage, thereby considerably reducing the hold of the carriage on the ground. Finally, the system has the defect of being applicable only to high-built carriages, since sufficient room must be left for the front wheels to pass beneath the body.

The problem therefore narrows down to a carriage which is to be driven and steered from the fore-carriage and in which the front wheels are mounted on the divided axle usually met with in automobiles.

At first blush it seems an impossibility to transmit the movement of a rotating shaft to two wheels mounted at the ends of

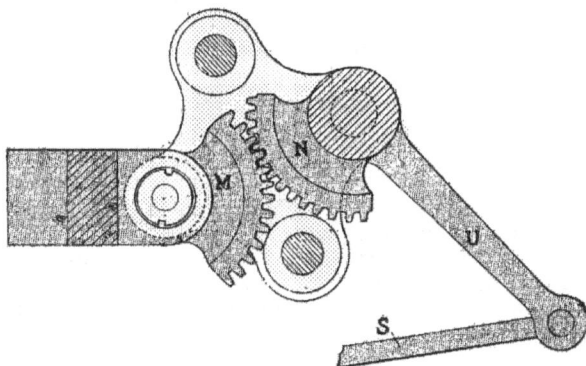

Fig. 5. The Guiding Segment-Gears. *M, N,* Safety Gear for guiding the Hemispherical Gears. *U,* Steering-Rod. *S,* Steering Lever.

the shaft, when these wheels, in order to steer the vehicle, will rarely be in a straight line with the shaft and will be constantly turned in ever-changing directions.

We shall see how, by a simple and novel arrangement of gears, an inventor has succeeded in solving this difficult problem. The system of this inventor is illustrated in the accompanying perspective and diagrammatic views, the casing within which the gearing is contained being removed.

In order that the driving-shaft of a motor may actuate a spindle with which it is not in alignment, but with which it forms a definite angle, bevel-gears must be employed. If the wheels were constantly turned in the same direction, if that be possible, the problem would have been solved long ago. But the wheels are ever changing their course, and the spindle forms an infinite number of angles with the driving-shaft of the motor. In order that steering-wheels could form any desired angle, a new form of gearing was evidently needed. The requirement was met by the spherical gear.

The driving-shaft of the motor, carrying the differential, is provided with a hemispherical gear at each of its ends, which is cut with tapering teeth. On each end of the spindle, opposite the wheel, a similar hemispherical gear is carried, which meshes with that of the driving-shaft of the motor.

The operation of these special gears is particularly effective, because when front and rear wheels are in alignment (as happens most frequently) there is no sliding friction between the gears of the driving shaft and the spindle, since under these conditions there is only an interlocking of the ends of the driving shaft and of the spindle. As the steering-angle increases, the thicker and stronger portions of the gears will be brought into engagement with one another; their resistance increases in proportion with the work which they have to transmit.

But a serious obstacle arises. The teeth of the spherical gears may have been originally badly cut or may have been worn away by constant use. The gears might then slide upon each other; and hence the jointing and pivots may bind.

The inventor has ingeniously and simply overcome the obstacle, by placing beneath the spherical gears two segment gears M and N which act as guides. The segment gear M is rigid; it is always parallel to the end of the driving shaft. The segment-gear N is movable; it turns with the wheel, but is always parallel to the end of the spindle.

Figs. 4 and 5 show clearly the relative positions of the two sets of gears. The manner in which the hemispherical gears adapt themselves to the steering angle is so plainly indicated that a verbal explanation is unnecessary.

Besides the advantages mentioned previously, this construction has other merits, among which may be mentioned its simplicity; the great strength of the spherical gears; and the total lack of all complicated mechanism. When the carriage is running in a straight line and the steering-angle is 0°, the gears engage each other at the ends. Power is transmitted with a minimum loss. The parts can all be inclosed in a casing, whereby they will be protected from dust and mud.

L. BAUDRY DE SAUNIER.

An Automatic Starting Gear for Hydrocarbon Motors

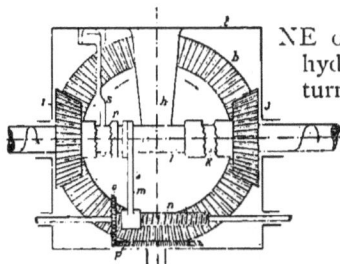

Fig. 1

ONE of the greatest objections to the hydrocarbon motor is the necessity of turning a crank-wheel in order to set the parts in motion. The objection has been overcome in an ingenious invention, illustrated in the annexed engravings, Fig. 1 being a front elevation, Fig. 2 a side elevation, Fig. 3 a plan view, Fig. 4 an end view of a spring-casing, and Figs. 5 and 6 detail views of parts which will be described further on.

The apparatus comprises essentially a box or barrel a (Figs. 2 to 4), carrying a bevel-gear b. Within the barrel a spirally-coiled spring C is arranged, one of the ends of which is secured to a lug d, integrally formed with the inner wall of the barrel a, and the other end of which is secured to a lug e, on a shaft f, held in position by two supports g, h.

When it is desired to wind up the spring, the barrel is turned in the direction of the arrow L; when it is desired to unwind the spring or reduce its tension, the barrel is turned in the opposite direction, or in the direction of the arrow B.

The bevel-gear b of the barrel (Figs. 1 to 6) meshes with two bevel-pinions i, j, loosely mounted on the shaft K, directly controlled by the motor. These bevel-pinions each carry a clutch-member which can be thrown into engagement with a corresponding clutch-member on the sleeve t. The sleeve can be shifted longitudinally along the shaft K. By means of a fork m, this coupling-sleeve t is automatically shifted along the shaft.

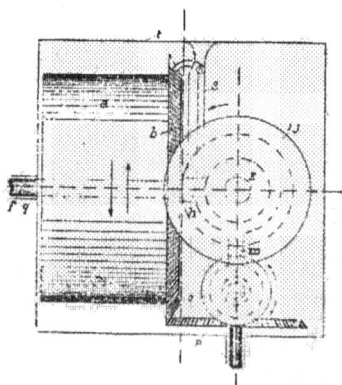

Fig. 2

The Automobile Magazine

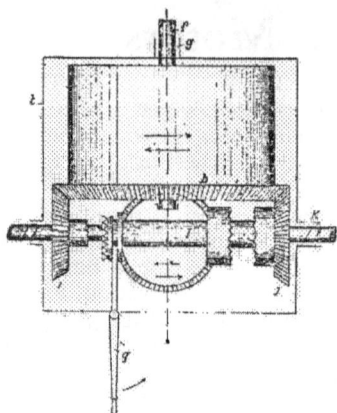

Fig. 3

The fork is operated by an endless screw n', which is rotated by means of bevel-gears o and p, connected with the bevel-gear b of the barrel a. As the screw turns, the fork m is shifted.

In winding up the spring, the sleeve t is thrown over into engagement with the clutch-member on the bevel-pinion j; and the shaft K turning in the direction of the arrow, communicates its motion to the barrel a. The barrel turns the endless screw m through the medium of the two bevel-gears o and p; and the fork m moves toward the left, sliding leftward of the sleeve t until it abuts against a shoulder, engages the sleeve, and throws the parts out of gear.

The movement of the fork is so regulated by the relative sizes of the gears and the pitch of the endless screw, that, when once the spring is wound, the parts are automatically disconnected.

The energy of the coiled spring is applied in starting the motor in the following manner:

By means of a lever q (Fig. 3) or pedal, the sleeve t is thrown over into engagement with the pinion i (Fig. 5). The flange r, on the end of the sleeve, then engages the arm of the detent S, which prevents the casing from turning in the opposite direction, by engaging the teeth of the bevel-gear b, as shown in Fig. 2.

When the barrel turns in a direction opposite to that followed in winding the spring, the direction of motion of the endless screw n is likewise reversed. The fork m is then shifted to the right (Fig. 5) until it engages the shoulder on the sleeve t, thereby forcing the sleeve into engagement with the pinion j (Fig. 6). When the sleeve is thus moved, the detent S falls away from the bevel-gear b; and the motor-shaft K starts the bar-

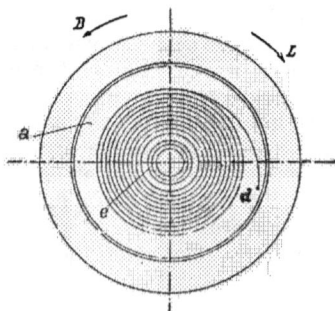

Fig. 4

Starting Gear for Hydrocarbon Motors

rel as explained above. The apparatus when once wound up, automatically, the parts are automatically disengaged and are ready to beging their operation anew. The entire mechanism is inclosed in a casing.

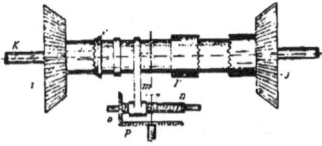

Fig. 5

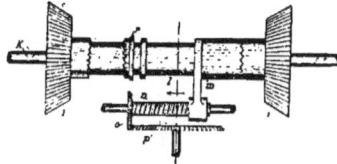

Fig. 6

— —

Gaillardet's Speed-changing Gear

WE illustrate herewith a speed-changing gear devised by M. Gaillardet for use upon light motor carriages. The apparatus comprises a differential gear, a gearing for obtaining two speeds at will, an automatic friction clutch, an

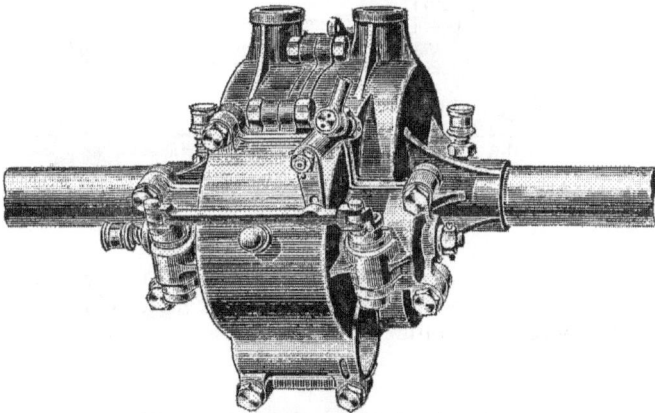

Fig. 1.   Gaillardet's Speed-changing Gear.
(External View.)

The Automobile Magazine

automatic brake for controlling the backward and forward running of the vehicle, and a chain-wheel with ratchet.

Fig. 2. Lower case, containing the speed-changing gearing, the clutch and the wheel that controls the motive pinion.

The advantages of this system are as follows: (1) It permits of throwing the motor out of gear; (2) the change from a low to a high speed, or inversely, during the running of the vehicle, is effected without any shock; (3) the chain is rendered immovable while the motor is in operation; (4) axles connected with the differential gear through joints assure the transmission of motion without resistance even though bearings are out of order or the axles are bent.

The axles are .95 of an inch in diameter and .88 of an inch at the collar. The entire mechanism is of cemented steel, refined and tempered. It is inclosed in a cast-steel case and operates in a bath of oil.

In addition to the mechanism, the complete apparatus com-

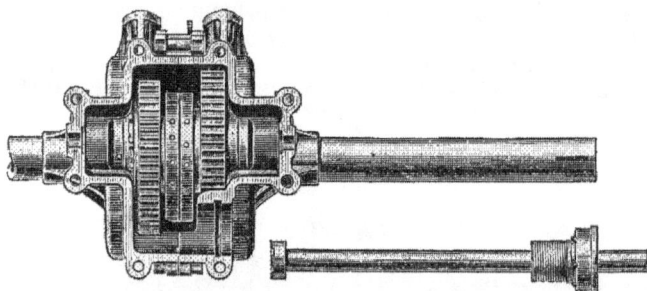

Fig. 3. Section showing the wheels that control the change of speed, the chain wheel and the brakes.

prises the hubs and the ball-bearings, that is to say, with the exception of the wheels, all the elements of a motor hind-carriage for a motocycle or a voiturette.

The Serpollet Lubricator

SO numerous are the parts to be lubricated in an automobile, that for the sake of simplicity and convenience, it is customary to employ a single oil-feeding device by which the moving parts of the driving mechanism are automatically oiled. Among the many devices of this type may be mentioned the Serpollet lubricator, noteworthy for the ingenious, simple means provided for feeding the lubricant.

The Serpollet lubricator is essentially a pump which forces oil from a reservoir into passages leading to the parts to be oiled, and which con-

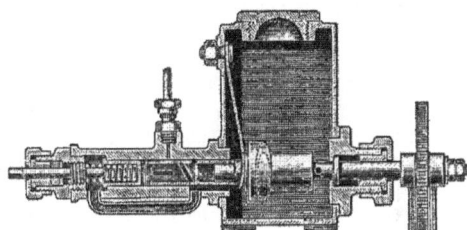

Cross Section

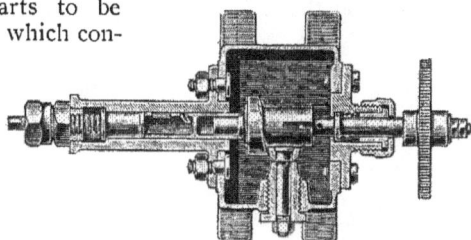

Plan

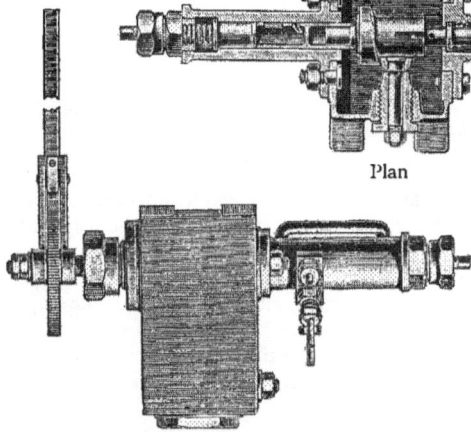

Side Elevation

tains no easily-deranged, delicate valves. The oil is constantly returned to the pump and used anew.

The lubricator consists of an oil-reservoir, through the lower part of which passes a shaft driven by the motor. Within the reservoir this shaft carries a piston, which, besides the rotary movement of the shaft, has also a reciprocating motion imparted by a cam spring-pressed on a roller rotating about a fixed axis. The piston at one end is provided with a hollow, perforated cylinder having a simultaneous rotary and reciprocating motion within another cylinder constituting the body of the pump, from which the various oil-ducts lead.

During the period of admission, that is, when the piston enters the cylinder, at the same time making a semi-revolution, the admission orifice of the piston-slide communicates with the reservoir; the return orifice at this period corresponds with the entire wall of the pump-body.

During the period in which the oil returns, that is, when the piston is on the return stroke and is making the second semi-revolution, the admission-opening is closed, while the return opening passes successively before the orifices of the return passages, discharging into them the oil pumped during the previous stroke.

The oil can be returned at a pressure varying between 2 and 50 pounds, by adjusting the tension of a spring pressing on the piston.

OBERAMMERGAU'S INNOVATION

The large iron theatre now under construction at Oberammergau for next year's Passion Play, is nearly finished. The stage will be in the open air. The auditorium will be 143 feet long and will accommodate 4,000 spectators. Two further improvements are also announced. There will be an office established for the purpose of assigning visitors suitable lodgings, and the tedious two-hours' drive to the village from the railway will be abolished, as well-equipped electric motor carriages will take visitors from Oberau in half an hour. Needless to say, the villagers do not like these departures. They say it will detract from the devotional attitude which all visitors to the Passions-spiel are supposed to assume. Apart from that, it is bound to interfere with the decennial prosperity of the village stage-driver and other rustic Jehus.

The Present Uses of Aluminium

By Leon Auscher

SOME recent discussions have attracted the attention of the public upon a new score to the use of aluminium. The commendations of its manifold advantages by some and the criticisms of its numerous shortcomings by others are certainly making of it at the present moment the most discussed of metals; and it might be supposed that the aluminium industry, bandied by such appreciations of opposite nature, was languishing and wavering, did not the increasing figures of production step in to prove a rapid development of this new branch of metallurgy.

Although the very marked increase in the production permits, on the one hand, of exactly answering the requirements of the moment, the number of new applications, on the other, continues increasing; and a new use of aluminium is rendered practical nearly every day.

Is this as much as to say that the present discussions are destined to be left out of consideration, and that we must deduct therefrom only what is favorable to this metal? Such is not the opinion of those who are endeavoring to derive profit, in favor of aluminium, from all the experiments made on both sides.

From these discussions, on the contrary, is evolved a moral, and that is that aluminium, a new metal, but slightly known primitively as regards methods of use, imperfectly refined, and mixed in improper proportions with other metals, began by being regarded by nearly everybody as the metal of all-work. It did not suffice that it was the lightest and one of the least oxidizable of metals, but it was seen fit to require of it all the other qualities of all the other metals to their highest degree. It was asked to be the most resistant to flexion and torsion, the most ductile, the most unalterable, etc. And, every time a check in detail was experienced in the accomplishment of this vast programme of research, a hue and cry arose against aluminium. The fact was not taken into consideration that it was not the metal that was to blame, but rather the injudicious employment of it, or, more plainly speaking, the wrong manner of using it.

Thus, it appears from an interesting note from M. Ditte to the Academy of Sciences (March 27, 1899), that the aluminium plates employed in Madagascar for canteens and platters, or as panel-material for the Lefèvre carriages, offered poor

The Automobile Magazine

resistance to the various ordeals of so hard a campaign. A certain number of the facts that were made known by M. Ditte are to be found in a note from M. Moissan, of the 10th of last April. The most remarkable of these relates to the proportion of impurities found in mixtures of aluminium from 1893 to 1897. M. Moissan calls our attention to the fact that the progress made in the manufacture has caused aluminium to pass from an average of 93 to 99 per cent. of pure metal, and has eliminated such important impurities as sodium, carbon, iron and silicium.

The 96 per cent. alloy of aluminium, however, seems to be the metal that is most widely employed at the present time under the most diverse forms. Moreover, as this metal is produced in bars and plates and in angle and T-form, it lends itself to all the processes of construction that require extreme lightness combined with great resistance.

The every-day applications of aluminium are manifold. The following are the principal ones:

> *Hardware*—Under almost all its forms.
> *Telegraphy*—Telephones, wires and cables.
> *Army*—Field and mountain gun-carriages, armor-plate, ammunition-chests, cartridge-shells, sabre-scabbards. lance-staffs, helmets, breast-plates, stirrups, and encampment and equipment material.
> *Cycling*—All detached pieces, fellies, frames, etc.
> *Lighting*—Tubes for coal gas and acetylene.
> *Music*—Wind instruments.
> *Surgery*—All kinds of instruments.
> *Kitchens*—Plates, knives, forks and kitchen utensils generally.
> *Watch-making*—Watch-cases.
> *Optics*—Opera-glasses, field-glasses, telescopes, etc.
> *Fancy Articles*—Confectionery boxes, bread-baskets, hand-mirrors, ash-pans, writing-table articles, etc.

Finally, among the new applications, it is pertinent to mention the uses created by the very prosperous, and essentially French, automobile industry.

For the accessory pieces of the frames, as well as for the principal parts of the carriage-work, a large proportion of aluminium, and especially of aluminium alloys, is used. Up to a certain point, these alloys permit of the choice of a metal possessing any such qualities as the use to which it is to be put may require. *Apropos* of this, it is impossible to omit a mention of the name

The Present Uses of Aluminium

of M. Henry Partin, a specialist in aluminium, if there ever was one, and who presented French industry with *partinium*—an alloy of aluminium (density 2.56) and tungsten (density 18), which, with the lightness of aluminium, combines a resistance that increases with the proportions of the alloyed metal.

Melted in sand, its density is 2.89. Its resistance to traction is from 17,000 to 24,000 pounds to the square inch, and its elongation is from 12 to 6 per cent., according to the proportions of the alloyed metals. When rolled, its density is 3.09, its resistance to traction from 46,000 to 53,000 pounds to the square inch, and its elongation from 8 to 6 per cent. Partinium is employed for making those casings of well-known aspect that inclose the motors of tricycles, and also the large casings employed on the De Dion 30 or 40 horse-power steam-carriages.

For all such uses, bronze and copper have, without any inconvenience, been replaced by a metal which weighs half less and possesses a third greater resistance. Let us add that the net cost of the raw material for such pieces is the same, whether it be bronze or partinium, and that the net cost of the finished piece is less for the latter, since the metal is more easily worked.

Finally, the rolled metal has been employed for a year past in carriage-work, for the bodies of automobiles. It lends itself to all shapes and is capable of receiving metallic mouldings. A body of this kind, mounted upon a frame of angle pieces, with a plate lining, constitutes a metallic combination that, with equal resistance, weighs from 50 to 60 per cent. less than wood, and that, provided certain precautions be employed, may receive the same coats of paint as a fancy carriage. Here again the industry has made gigantic strides.

The first partinium carriage-body dates back to the Paris-Bourdeaux race of 1898. At present, all *chauffeurs* have adopted this important improvement in automobile carriagework; racers, because lightness plays an important part in the results of a contest; and tourists, because it is better to substitute passengers or useful baggage for from 200 to 450 pounds of dead weight.

Finally, in the train of some experiments on crushing, in which this metal withstood a pressure of 54,600 pounds to the square inch without distortion, the manufacture of houses that can be taken apart and transported has been undertaken.

It will be seen that aluminium, in all its forms, is being further and further employed; and it will be still more extensively used when the price of it, which is already reasonable, shall permit of new industrial applications being thought of.

The following figures show the decreasing prices of the metal from the time of the discovery of the first processes of manufacture by Sainte-Claire Deville:

1854 .... $270.00 per pound.
1856 .... 135.00 " "
1859 .... 36.00 " "
1864 .... 9.00 " "
1889 .... 2.25 " "
1891 .... 1.80 " "
1892 .... 1.09 " "
1894 .... .45 " "
1896 .... .36 " "
1899 .... .32 " "

The consumption for the last three years in France has been about 600 tons per annum—a figure which it seems ought to be doubled in 1899.

If we consider that bauxite, an argillaceous ore of aluminium, is one of the most common ores of France, and that the metal is being extracted from it under more economical conditions from day to day, we may believe that the above decreasing scale is still far from its final limit. Although, then, we are perhaps wrong in emphatically denominating aluminium as *the* metal of the future, it cannot be denied that it is, to the highest degree, a metal *with* a future. Its fate will depend in either case upon the judicious use that will be made of it, and upon the chemical composition of its alloys.

Partinium Body for an Automobile

The Gobron and Brillié
Automobiles

By G. Chauveau

THE Gobron and Brillié automobiles, which, comparatively speaking, have but recently made their appearance, attracted the attention of specialists at the very outset by reason of their very interesting peculiarities.

Among the different types of carriages brought out by the above-named manufacturers may be mentioned an elegant wagonet, which is actuated by a six horse-power motor, and which we illustrate in perspective in Fig. 1, and in side elevation and plan in Figs. 2 and 3.

This automobile is provided with a frame of steel tubes to which are brazed cast-steel couplings. The longitudinal members of this frame have the form of trussed girders, that assure absolute rigidity to the whole and that rest, through springs, upon the front steering and hind driving axles.

The steering is done through a divided axle maneuvered in a very peculiar manner, which we shall describe further along.

The motor M (Figs. 3 and 4), through toothed wheels, actuates an intermediate shaft, G, by means of a cone-clutch, E. This

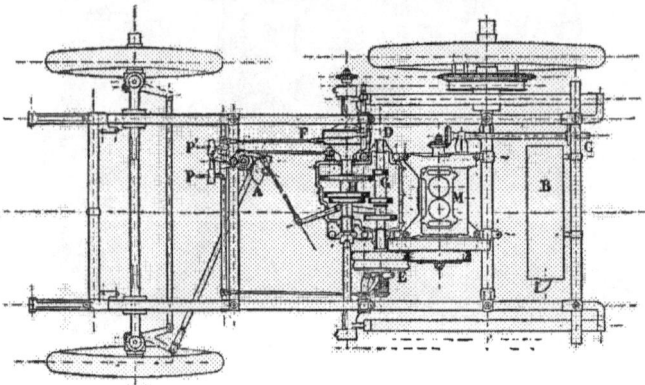

Plan

shaft carries a train of three pinions which slide longitudinally and are capable of being thrown into engagement with three wheels fixed to the box of the differential gear, *D,* so as to give three multiplications. A complementary pinion permits of reversing the motion of the carriage.

The differential shaft, *H,* at its extremities, controls the hind wheels through the intermedium of sprockets, chains and toothed wheels.

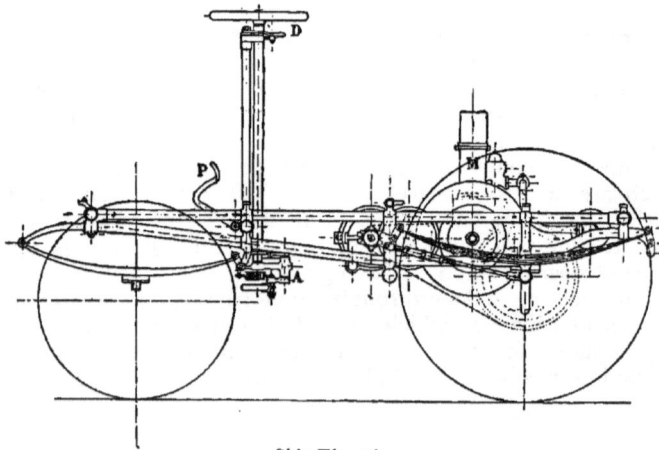

Side Elevation

The Gobron and Brillié Automobiles

The gearings are enclosed in a tight aluminum casing.

The changes of speed are effected through a lever mounted upon a bar near the steering-shaft. Of the two pedals P and P', one actuates a disengaging gear, and the other a band-brake, F, mounted upon the differential as well as upon the disengaging gear. The starting is effected at the rear through a winch at C.

The motor is so combined as to reduce the vibrations to a minimum through an equilibrium as perfect as possible of the parts in motion. It consists of two vertical cylinders placed side by side and open at their two extremities; and is of the four cycle type. The two lower pistons, $f f'$, through their connecting-rods, $d d'$, actuate the central crank of a three-throw crank-shaft. The cylinder bottoms are replaced by two other pistons, $e e'$, which move in a direction contrary to that of the first. These

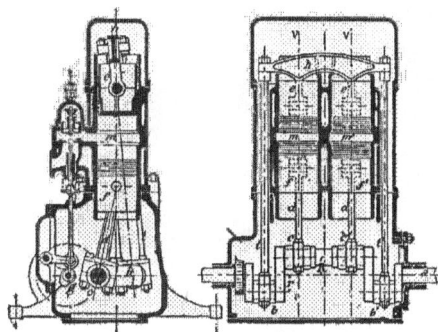

Sections of the Motor

pistons are connected through a cross-head, h, with the two connecting-rods, $t t'$, that act upon the lateral cranks, $b b'$, which are both set at an angle of 180% with respect to $c c'$.

The lower pistons have a stroke slightly longer than that of the upper ones, so as to compensate, by their difference in velocity, for the difference of the mass, and to obtain a perfect equilibrium.

The inside diameter of the cylinders is 3.2 inches; the stroke of the lower pistons is 3.2 inches, and, of the upper, 2.4; and the number of revolutions is 900.

The distribution is effected through vertical valves placed in boxes at the side of the cylinders. The admission is automatic. The exhaust-valves are controlled by rods lifted by cams placed upon a shaft parallel with the driving shaft. This cam-shaft, which is controlled by gearings, revolves with a velocity twice less than that of the driving-shaft. Finally, this entire mechanism is enclosed in a tight casing.

The electric ignition is effected through batteries, accumulators or induction coils. The cooling is done through a circulation of water, with a reservoir, pump and radiator.

As for the regulation, that is obtained through a suppression of the explosive mixture by modifying the supply of gasoline that enters. To this effect, the carbureter (Figs. 5 and 6) consists of a box, E, similar to that of a stop-cock, the key, A, of which is provided at its circumference with a series of cells. E is full of gasoline that enters at P. At a certain point of the box there debouch two ajutages, r and G, the former of which communi-

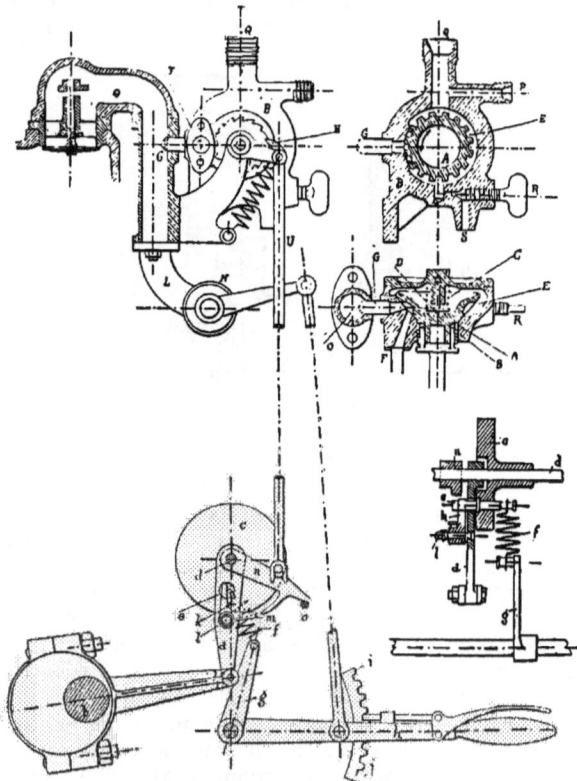

Elevation of Carbureter and Regulator

The Gobron and Brillié Automobiles

cates with the exterior, and the latter with a suction-conduit that extends to the admission-valves, o. The key A is moved by a ratchet-wheel controlled by an eccentric secured to the driving-shaft, between which and the distributer is interposed a regulator which we shall presently describe. In the revolution of the key, a cell full of gasoline presents itself in front of r and G at the moment at which an admission occurs; and the air drawn in through r vaporizes the gasoline which is carried along in the current of air.

It suffices to arrest the action of the eccentric upon the ratchet in order that (the key remaining stationary) no formation of

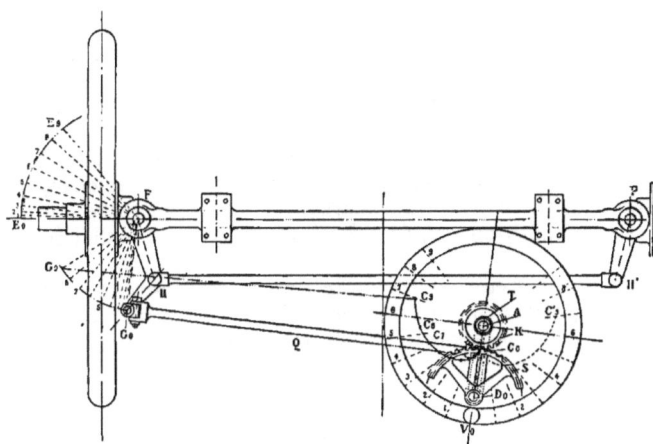

Steering Apparatus

explosive mixture may take place. To effect this is the office of the regulator (Figs. 7 and 8) just mentioned.

The eccentric, which through the intermedium of a system of levers, $d\ n$, a rod, u, and a click, controls the ratchet of the gasoline distributer, imparts an oscillatory motion to a mass C, held by the lever d through a tappet which passes with a certain amount of play into the said lever. Upon this lever d is pivoted a detent, k, one branch of which is held by a spring, f, against the tappet, while the other is provided with a notch, m, with which engages a finger projecting from the lever n. When an increase in the normal velocity occurs, the mass C receives an impulsion. the detent k oscillates and releases the finger of the lever n, and the consequence is a suppression of the motion of the ratchet.

The Automobile Magazine

A hand-lever, *j*, permits of varying at will the tension of the spring, and, consequently, the velocity, which may vary from 300 to 1,500 revolutions.

In the construction of the steering-apparatus, which is very peculiar, it has been the object of the manufacturers to obtain both safety and rapidity in turnings about. Speaking in a general manner, it may be said that safety is obtained by a great reduction, and rapidity by a feeble one. With the arrangement here adopted there is at once a great reduction in running in a straight line, and for slight deviations, and a feeble reduction for turning about. We have here, then, a variable reduction apparatus responding to the conditions indicated above.

To effect such a result, the steering hand-wheel, *V*, is so arranged as to carry along in its rotary motion a lever, *L*, which presents a vertical axis, *D*, upon which, through a bush, is mounted a toothed sector that rolls upon a stationary pinion, *K*. From such an arrangement it results that, during the rotary motion of the hand-wheel *V*, the stud *C*, jointly with the toothed sector *S*, describes an epicycloid. To the angular changes of position of the hand-wheel, marked 1, 2, 3, 8, 9, there therefore correspond positions C^1, C^2, C^3, C^8, C^9 of the stud *C* that controls the rod *Q*.

It will be seen that the more the rotary motion of the hand wheel is prolonged, the greater will be the angular changes of position of the journals of the steering wheels.

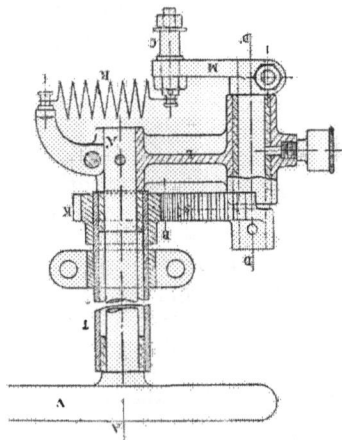

Longitudinal Section

The Bravo Gasoline Motors

By Maurice Chérie

AMONG the interesting novelties of the latter part of the year just passed must be mentioned the gasoline automobiles built by the Bravo establishment recently installed at Clichy.

The carriage represented above offers certain peculiarities, the most interesting of which we purpose to pass in review.

Let us say at once, that through the judicious arrangement of its parts and the form of its absolutely rigid frame, it cannot fail to satisfy the requirements of all those who are looking for a strong, elegant and comfortable vehicle.

All the pieces that enter into its construction are of great strength and are so combined as to afford a very lengthy service without necessitating troublesome repairs. All the transmitting parts, reservoirs, etc., are almost entirely concealed in the carriage-work, so that the vehicle bears a remarkable stamp of elegance.

The motor is vertical, with two convergent cylinders, and the pistons drive the shaft through the intermedium of a single crank. All the parts in motion are enclosed in an absolutely tight casing. The connecting-rod bearings, which are very wide, are of cemented steel, refined and tempered. The bearings of the shafts, which are of phosphor-bronze, are also very wide and thus assure the motor an operation of long duration without any appreciable wear. Through an ingenious arrangement, a continuous circulation of oil takes place in the casing, and this renders the consumption of the lubricant insignificant, and necessitates no surveillance.

535

The ignition is effected either through an electric spark or incandescence. With the sparking, an extremely simple device in which but a single cam is employed permits of obtaining all the velocities of the motor, say from 300 to 400 revolutions, through the simple maneuver of a button.

In the ignition by incandescence, a centrifugal governor provided with ball bearings acts upon the motor in such a way as to keep the exhaust-valves open during the change of speed. During this period, the admission-valve allows a small charge of fresh air to enter the cylinder, and the motion of the piston expels the products of combustion, the effect of which is to improve the work of the motor.

One very interesting point in the Bravo motors is the method

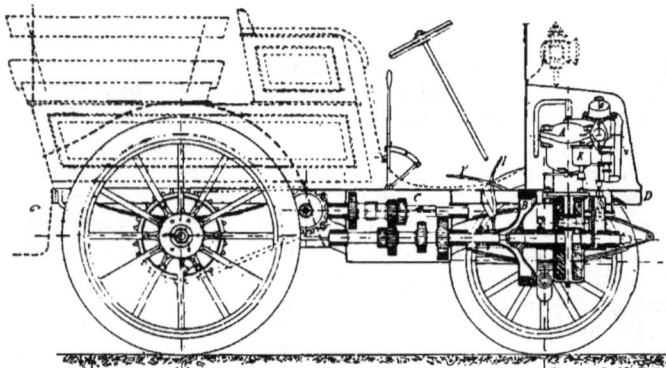

Fig. 1. Longitudinal Section

of starting them, this being done instantaneously by a single revolution of a winch.

The cooling is effected through a centrifugal pump set in motion by the fly-wheel of the motor. The tension of the pump's friction is regulated automatically through a spring.

The Bravo establishment has brought out types of motors of 5, 8, 10, 16 and 20 horse-power, which, according to requirements, may be mounted upon frames through couplings.

Fig. 1 represents a frame provided with an 8 horse-power motor with electric ignition. The motor, which is arranged in front, is put in connection with the speed-changing gear through a leather-faced cone-clutch. As may be seen from the figure, the gearings are enclosed in a casing, C, which receives also a reduction gearing, r, that reduces the velocity in the ratio of about

The Bravo Gasoline Motors

1 to 3. As the angular velocity of the gearings is thus relatively feeble with respect to that of the motor, it follows that the changes of speed are effected with the greatest facility, without noise, and in avoiding the frictions and shocks that are so prejudicial to the duration of the parts.

The reversal of motion is effected through the shifting of two bevel gearings, $p\,p'$, mounted upon the shaft of the differential gear M. For the purpose of avoiding the thrusts due to the changes of level of the frame in violent joltings, the differential shaft is divided into four parts (1, 2, 3, 4), the extreme junctions of which are made by universal joints.

The carriage-body is mounted upon the principal frame, D, while the motor and the mechanism are fixed upon the separate frame D'.

One interesting peculiarity of the Bravo motors consists in the dismountable steering-axle (Fig. 3) which is of a special system. The head of the tubular axle is formed of a vertical

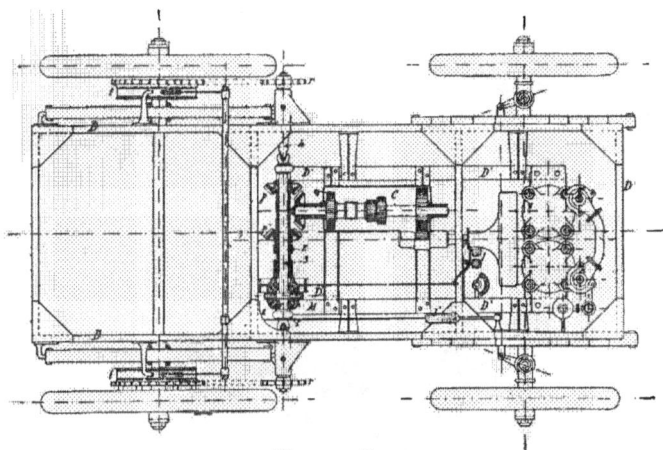

Fig. 2. Plan

socket, B, which receives an axis, A, upon which, through a socket-clutch, is mounted the journal-support, H, which is held simply by a nut, L. The weight supported by each journal is distributed over a horizontal row of balls that roll between two tempered steel plates. A large circular nut, R, that closes the ball-case permits of the regulation of the play in a perfect manner. The screw, V, prevents any loosening.

In this steering pivot, the pieces having a motion are entirely

The Automobile Magazine

immersed in the oil with which the box B is filled at the time of the first mounting. The oil afterwards lasts indefinitely. This process of mounting renders it very easy to do the steering, which is effected through an inclined handwheel with a transmission through bevel gearings and rigid levers with a universal joint.

The safety apparatus comprise a band-brake mounted upon the differential gear and controlled by a pedal; and a hand-brake that acts upon the tires of the wheels. An electric interrupter permits of the instantaneous stoppage of the motor by throwing it out of gear without the driver having to change his position.

All the controlling parts are grouped in such a way as to render the driving of the vehicle very easy. The maneuvering levers are straight, without any latch or connecting-rod. All the points of oscillation are of tempered steel, thus considerably diminishing the play and wear of such pieces.

As may be seen, the carriages built by the Clichy establishment present many well elaborated peculiarities of detail that render them very resistant and cause them to run with great regularity.

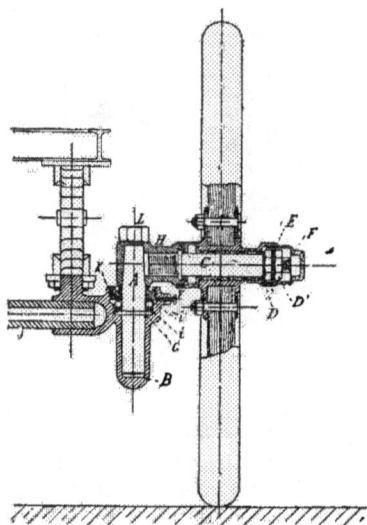

Fig. 3. Section of Steering Wheel and Axle

The Locomobile

SINCE the Locomobile was described in our January number, a number of modifications and improvements have been made in it. In view of the attention at present being devoted to steam automobiles generally, the following description and detail illustration of the latest may not be without interest at the eve of the spring trade opening.

The oil fuel is led from the fuel tank, where it is carried under 20 or 25 pounds air pressure through a vaporizing pipe, which passes through the boiler, and thence the vapor goes through a passage which may be closed by a regulator-valve to the burner. In front of the burner a special automatic petroleum supply regu-

Part Sectional View of Locomobile Showing Location of Engines and Boiler

lator is provided, by means of which as soon as the steam pressure in the boiler attains a certain degree the flame of the burners is automatically lowered, there being a by-pass of limited cross-section leading around the regulator valve which keeps the fire alight. So soon as the regulator valve acts to reduce the fire it also opens a large area of cold air entrance to the fire-box, which has the effect of instantly checking the steam production. This makes the action of the burner exceedingly prompt, and keeps the steam almost exactly at the regulator pressure in the boiler, no matter whether the carriage is traveling up-hill or coasting down-hill. The pump delivery is about constant; the regulator is set at about 150 pounds, and the safety valve at 160 pounds. No cylinder drainage cocks are used, so that the boiler feed is always about the same for, say, each ten miles run, and hence the whole regulation of the fire and water are thus automatically taken care of. The boiler carries 8 inches of water above the tube sheet, leaving 5 inches of steam space, but an inch or two either way in the water level makes no difference, the boiler acting perfectly down to 1 inch of water over the lower sheet. Thus the driver has nothing to do except to steer and handle the throttle valve. A glass gauge on the outside of the wagon body shows at a glance where the water level is. The burner gives an absolutely perfect noiseless combustion, the up-take discharge not having any odor, and being wholly invisible. The fire is also invisible, appearing through the fire-box peep-hole as a wavering, bluish haze when burning hard.

The transmission of the power from the engine crank-shaft is effected by a hard sprocket of twelve teeth on the engine-shaft, connected by a single light central-driving chain to a twenty-four tooth sprocket-wheel, on the compensating gear-box on the rear axle. The chain adjustment is obtained by a right and left threaded screw strut, jointed at one end to the yoke of the rear axle support which surrounds the compensating gear, and at the other end to the lower part of the engine frame, in about the plane of the crank-shaft. This permits the rise and fall of the engine and small chain-wheel sprocket without material change of chain length, and relieves the pneumatic tires of all weight not carried on springs.

Steering is controlled by a bar acting on the front wheels which are mounted on vertical pivots in the usual way. Three band brakes are provided; one controlled by a foot-pedal acting on the differential and one each, actuated by a hand lever, on the hubs of the rear wheels. The wheels are of the cycle type, shod with small pneumatic tires. It climbs any gradient, and can be got ready by an expert in less than five minutes, from five to ten minutes being required to get up to its fullest power.

The Automobile

MAGAZINE

An *ILLUSTRATED* Monthly

VOL. I   No. 5   NEW YORK FEBRUARY 1900   PRICE 25 CENTS

The AUTOMOBILE MAGAZINE is published monthly by the United States Industrial Publishing Company, at 31 State Street, New York. Cable Address : Induscode, New York. Subscription price, $3.00 a year, or, in foreign countries within the Postal Union, $4.00 (gold) in advance. Advertising rates may be had on application.

Copyrighted, 1900.

Entered in the Post Office at New York as second-class matter.

Editorial Comment

THE PRESENT STATUS

A S a matter of historical record it will not be amiss to note the actual condition of the automobile industry in this country at the opening of this last year of the century. Our sense of moral responsibility to the community, which supports us by its subscriptions and its generous commendation and words of encouragement, impels us to set forth a few facts that might well be overlooked by the casual reader who scans the pages of the AUTOMOBILE MAGAZINE and admires the numerous pictures of motor vehicles, many of which have been thought of, some of which have been built and a few of which have been sold and are supposed to be in use.   It may be true that the streets of Boston have been forsaken by the horse, as described by the "Listener " of the Boston *Transcript,* and reproduced elsewhere in our columns, but as far as the avenues of New York are concerned truth compels us to say that aside from the occasional cab or trap of the Electric Vehicle Company and the runabout of the Locomobile Company, to which the public has become somewhat accustomed, the appearance of a vehicle without a horse to draw it attracts a curious crowd with greater rapidity than a fire.

The Automobile Magazine

The same conditions, more or less, prevail in Chicago, Philadelphia, and Washington, while in the smaller cities it is safe to say that the reading of eloquent predictions for the future of the automobile and the still more glowing prospectuses of companies organizing for the production of twenty or more vehicles per day, is the only evidence of the new industry which is destined to work such beneficent changes. In fact the literature of the automobile is at present more voluminous and picturesque by far than the manufacture. There are more publications on this subject, some appearing regularly and others " every little while," than there are factories in actual operation, prepared to furnish automobile vehicles with reasonable promptitude and at a fair price. As for the prospectuses of companies which have a limited amount of treasury stock for sale to the hopeful investor, it is to be regretted that the same amount of air, whether liquefied or compressed, which the promoters have made use of in the flotation of their shares, could not have been successfully employed to propel their imaginary vehicles.

Meanwhile even the reputable persons who are at the head of millionaire companies, and whose pictures adorn the pages of the trade papers, are singularly slow in perfecting their organization and turning out work. We hear that they are " rushed with orders " and that it is impossible for them to supply their demand, but is it not time that we should see some tangible results of this enormous production? Yankee ingenuity should surely devise some means to gratify the desires of those who are ready to sell their horses and adopt the automobile.

It must be confessed, and as far as we are concerned, with the keenest regret, that the present status of the automobile industry would be more correctly defined as the present hiatus. It cannot be denied that the majority of the manufacturing enterprises in this line are still enveloped in darkness and slowly groping towards the light. Instead of being " rushed with orders " and producing hundreds of vehicles weekly they are, in reality, either looking for capital or new patents or improvements in their present methods which shall insure greater economy and practicability in their product. The would-be *chauffeur* often stands aghast when he views for the first time the inner workings of the vehicle which he is asked to buy. Some of the clumsy and intricate affairs so far devised could better be described as traveling machine shops than as vehicles. Naturally the manufacturers of these interesting but complicated mechanisms are not in a hurry to turn out a large number, fearing with good reason that simpler and cheaper ones will soon be in the field.   The manufacturer well knows that the ordinary man has little knowledge of machinery and less desire to be bothered with it.

A glance at the advertising pages of the AUTOMOBILE MAGAZINE shows the small number of American factories that are really in position to supply vehicles, in comparison with the many in Europe which are in successful operation.   We are at present in the educational period of automobilism on this side of the Atlantic—everybody is intensely interested in the subject, everybody wishes to learn.   The best tire, the best wheel, the best bearing, motor, steering device, battery—all these things are eagerly sought after by manufacturers and future users of the automobile. For this reason there is sure to be an attentive audience for all those who have new devices to offer, whether American or foreign, and while for the moment, we find much better material obtainable in Europe and consequently must devote to it greater space in the MAGAZINE than we should like to, there is little doubt that during the present year we shall witness such a step forward on the part of our inventors and manufacturers as will eventually place this country at the head of this industry, and in a position to sell automobile vehicles to all parts of the world, as is already the case with the bicycle.

THE COMING CONTEST

THE coming automobile contest for James Gordon Bennett's international challenge cup, the racing conditions for which were set forth in the January number of this magazine, is already widely engrossing the attention of automobile interests.   Mr. Bennett has taken up the racing aspect of the automobile with the same enthusiasm—youthful enthusiasm, it might be said of one whose interest in sport is so perennially keen—that he originally gave to yachting contests.   It seems singular that the racing possibilities of the great invention of mechanical locomotion should have had to wait for their development until the closing years of the century that witnessed its practical introduction.   The invention that in less than a century has effected the most marvellous transformations in civilization—although it has given us the swiftest means of transportation, attended by the most exciting form of motion—has hardly been the subject of racing-contests until the perfection of the automobile emancipated the

locomotive from the thralldom of the rail.   This has brought the
proverbial " iron horse "—in its contemporary form as a steed of
steel—into the same popularity that has made the centaur-like
combination of man and bicycle the rival of the horse in the world
of sport.   It has been a most extraordinary progression.   First
the animal organism putting forth its muscular energies in the
development of high speed under the guidance of human intelligence and skill.   Next, the lightest and most delicately adjusted
form of motor-mechanism—the bicycle—impelled by the trained
and specially developed muscular energy of man under the guidance of human intelligence and skill.   And finally, human intelligence and skill sitting supreme in the guidance of a most complex
and powerful mechanism impelled by its own motive power, the
element of animal intelligence that was an essential factor in the
first of these three forms here being combined with the human
organism.   In other words, we first have a highly developed
natural mechanism actuated by organic energy responding to
animal intelligence subordinate to human intelligence.   In the
second place we have a highly developed artificial instrument
actuated by organic energy directly responsive to human intelligence.   And in the third place we have a highly developed artificial instrument actuated by inorganic energy directly responsive
to human intelligence.   We thus at last have the locomotive given
the freedom of the road—all the dynamic potentiality of the
Transcontinental Express free to be exercised on the common
highway!   It is no wonder that the manifestation of this extraordinary power should appeal in the highest degree to the emotions
that delight in contests where energy, speed and skill are involved
in most exciting combinations.   For what has just been characterized as the emancipation of the locomotive is in reality the
transference of its subjection from the sure inevitable control of
the rail of steel to that of the subtile band of the immaterial rail
that exists solely in the human mind, where a momentary break
in its continuity, a slight relaxation in the tension maintained by
facile hand and perfectly balanced nerve, would bring disaster
and destruction.   All this means, of course, one of the highest
forms of athletic contest possible to man.

While the utility of such contests may not be directly apparent
—for the automobile is essentially a vehicle for the highway, and
the highway is no place for racing, as a rule, but the world's common instrumentality of intercourse between communities and
nations—nevertheless in the promotion of skill and courage that
comes with such contests, and the development of inventive
resources in consequence thereof, we inevitably have results of

Editorial

the greatest value to society at large, and civilization correspondingly profits thereby.

The American challengers for the contest, through the Automobile Club of America—Messrs. Winton and Riker and the third contestant—are thoroughly representative, and their entrance will be most encouraging for prospects of a successful issue. France is naturally the home of skilled and courageous *chauffeurs,* and it is notable that so many are eager to enter the contest that some of the best French racers may secure the coveted opportunity to enter only under the challenge of the Belgian club. The German and Austrian clubs have also sent in their challenges, and it is assured that the contest that is to mark the last year of the twentieth century will begin a famous series of international races most worthily.

TEMPORA MUTANTUR

The remarkable headway which has been attributed to the automobile in our great cities of late is indicated by the following observation on the situation in Boston made by the " Listener " in a recent number of the Boston *Transcript:*

" It is certainly marvellous to see the rapidity with which horseless vehicles are supplanting horses for all the lighter work. Horse-drawn cabs have almost entirely disappeared. Horse-drawn delivery wagons are out of date. Private carriages still go to a great extent with horses; I should suppose they always would, for it is pleasanter to travel with a horse than with a machine. And yet, when automobiles have become considerably cheaper than they are, a great many men will adopt them, though they might prefer horses, simply because less labor and expenditure are necessary to take care of them."

The " Listener " adds that for complete success something horseless needs to be devised to take the place of the ordinary comfortable carryall—a vehicle which, without being bulky, will carry from four to six persons. He also comments on the desirability of providing robes or blankets for automobile cabbies, having noticed that during the recent severe weather, while well clad as to their heads and bodies in good caps and overcoats, they had to depend on their trousers and boots to keep their legs warm.

The Automobile Magazine

The Automobile Club of America is rapidly becoming a large and influential association, who will undoubtedly carry out reforms affecting all users of highways, far more quickly than the Wheelmen's Association could ever expect to do. Cycling is merely a pastime, while automobilism is a new means of traction designed to have an enormous effect on commercial interests.

AUTOMOBILE APPAREL

The leather coat used so much in France does not seem to meet with American tastes. What we need, is a waterproof, windproof automobile driving coat, protecting the automobilist thoroughly and effectually against all inclemencies of the weather, presenting at the same time a stylish appearance. Something in the line of a box driving coat of waterproof material, lined with fibre chamois, in order to make it wind proof, and woolen lining. A broad collar and four or five pockets properly located, with perpendicular openings would be advisable. The sleeves should be provided with wind cuffs, and the coat had better be double breasted. Leggins and a wide brimmed felt hat fitting securely would complete the outfit.

TIT FOR TAT

As to automobiles in the parks—why not take the bull by the horns and settle the question for good and all simply by providing that no horse shall be allowed in a park until he has learned not to shy at an automobile? An efficacious test would be to station at each park entrance one of the most formidable kinds of automobile in action. Then at the first symptoms of fright the animal could be turned back. Better still, and in the line both of public utility and of enhanced pleasure, would be to provide that park carriage-services be on an automobile basis. The service would thus be more efficient and profitable, pleasanter for the public, and horses would become so familiarized with motor-vehicles that they would be no more likely to shy thereat than they do at bicycles.

Automobile Humors

AN ELUSIVE PRODUCT

Simplex—How is it we hear so much of automobiles but meet so few of them on the streets?

Duplex—Must be the manufacturers are turning them out so fast that you can't see them with the human eye.

RUS IN URBE

Aleck—Well, Uncle; this is very different from the farm, isn't it?
Uncle Rube—It feels strange, sure enough; but I guess the owners of them new-fangled things know all about watering their stock, just the same.
C. R. McAndrew, in *Collier's Weekly.*

A CRUEL JEST

"I wish we had a horseless carriage," said the farmer's son. "We have," replied the farmer; "and now that you speak of it, you might as well get it and bring a load of potatoes up to the house."

Automobile Humors

DOUBLE SPEED

UNCLE ABE—Dem automobiles go so fast it 'ud take two niggers to tell about 'em.

SAMBO—How's dat?

UNCLE ABE—One ter say " Here she comes " an' one ter say " Thar' she goes!"

A MAN OF PROGRESS

" What luxury! Have you grown too lazy to grind your own stone ? "
" Mon Dieu, Suzette, I have to keep pace with our automobile age."
(*O'Galop in La Charivarai.*)

THE CHAFFEUR'S APOSTROPHE

My lady ought to be inside—
Auto! Do thy endeavor!
Reinless through lampless streets we glide
While I hug on forever.

Press Notices and Book Reviews

A RECENT issue of the *Automotor and Horseless Vehicle Journal* contains an excellent paper on steam automobiles read by George A. Burls before the Civil and Mechanical Engineers' Society of England. Among other things, Mr. Burls has this to say:

The owners of this steamobile consider that it easily replaces three of their two-horse wagons; that is, six horses, three wagons, three drivers, and three lads. According to Mr. Kempe, one horse working eight hours daily may be expected to maintain a speed of two and a half miles an hour, and perform 14 net ton-miles, on ordinary turnpike roads; two horses harnessed to one wagon may achieve, say, 35 net ton-miles daily, in regular work; 152 net ton-miles per diem will therefore require $\frac{152}{35} = 4.3$ two-horse units; or, say, nine horses. This result suggests that in the estimate just mentioned, " three *two*-horse wagons " should be altered to " three *three*-horse wagons." At a very moderate estimate the prime cost of nine horses, with harness, and three suitable wagons, materially exceeds that of the steamobile, while the total cost of running is almost certainly more than 6*d.* per net ton-mile; there is therefore a very marked saving in favor of the steam-wagon.

————

The London *Speaker* in a recent issue, under the heading of " Pace on the Roads," has several remarks to make about brakes. Rightly enough, the great brake power of the wayside " electric trolley car " is alluded to, and also the brake-power of a bicyclist in relation to the weight of his vehicle. It may be that the electric brake is more efficient—somewhat more efficient—than brakes such as those used on railways and by bicyclists, which could quite, and do nearly, stop the wheels from revolving. It is not impossible to supply automobiles with brakes of similar principle and similar power. An active power greater " than that usually applied in propulsion " is certainly wanting, but an equally great power can be called into play; the wheel may be subdued, almost stopped, and Dunlop and MacAdam must fight it out between them. We may hesitate to make any ordinary tire and rim under the weight of an automobile do what steel may do upon steel on the railway, or rubber on macadam, under the light weight of the bicyclist. The thing can be worked out in theory and in practice —will be worked out—is being worked out. We have thought it as well to make our point clear. There are few bogies which terrify the British public so easily as " insufficient brake power."

The Automobile Magazine

The question of dress for a motorist is one of great importance. "I see no reason," says the "Man at the Wheel" in the London *Daily Mail*, "why the appearance of a man driving a motor carriage should differ from that of a man driving a horse-drawn vehicle. But, in consequence of the greater distance covered by the motor vehicle in a day and the more continuous and greater speed at which the motor travels, ordinary tweeds and cloths are unsuitable when there is a cold breath in the air. On the front seat of a motor it is found that the air will penetrate any ordinary cloth. Leather is indispensable, especially where the air pressure tells most—namely, in the front in the region of the shoulders, collar-bones, and breast generally. The black leather jackets worn in France are excellent for protection from wind, rain, and dust, and can be readily cleaned. But, except for long country drives through storm and rain, there is no reason, as I have before said, why the leather coats should be worn; for ordinary tweed suits, if lined with chamois leather, are efficient as protection against penetration of the wind, and the desire of most Englishmen is not to appear on the highways in any form of dress so uncommon as to cause remark. The secret of keeping warm is to wear something tight around or a gauntlet-glove to keep the wind from going up the sleeves."

In "The *Practical Engineer* Pocket-Book for 1900," issued by the Technical Publishing Company, of Manchester, we have a collection of engineering data, formulæ, tables and the like, that is of especial use to the mechanical engineer engaged upon such constructional work as boiler and engine building, hydraulic and automobile machinery, mining engineering, and so forth. The aim of the editor has been to embody his information in plain and concise language suitable for the average mechanic. Such important subjects as fuel, combustion, management of steam generators are treated at considerable length, and even the most experienced boiler attendant will find much new and useful information. The sections on injectors and feed pumps are very practical. There is also an excellent chapter on engine testing and on inertia. This somewhat recondite subject is well covered in the simplest language. There is also a good chapter on electrical engineering. To those who are engaged in mechanical engineering but whose technical knowledge is not extensive we can cordially recommend this pocket-book as being the best of its class, while the advanced student or draughtsman will find it a most serviceable reference book.

Press Notices and Book Reviews

The French publishing house of J. Fritsch has just issued " Le Manuel pratique du Conducteur d'automobiles," par Pierre Guédon, ingénieur, et Yves Guédon, ingénieur civil.

This is the second edition, revised and greatly enlarged, of a work of which the first edition has been reviewed in these pages. The success of this book is justified by the large amount of valuable information that it contains. It is divided into seven parts, the first of which is devoted to the subject of steam automobiles. Herein we find some interesting considerations upon the advantages of expansion and of the compound system for the motors of automobile carriages; upon the use of back-steam for brakes; and upon the calculation of the tractive stress of motors. The second part deals with gasoline automobile carriages, such as those of Panhard and Levassor, Peugeot, Lebrun, Gardner, Gauthier-Verhlé, Decauville, Dietrich, Pantz, Henriod, Koch and Roser.

The third part treats of electric automobiles, and contains some general considerations upon electric motors and accumulators. Herein we find descriptions of the Jenatzy, Jeanteaud, Krieger, B. G. S. and Mildé carriages. The fourth part, which is extremely interesting, considers alcohol motors, such as the Briest and Armand and the Martha. The fifth part comprises calculations of work. The sixth part gives descriptions of various accessories. Finally, the seventh part is made up of appendices giving numerous pieces of advice and many reports and regulations concerning automobilism; all very interesting.

An interesting work is Prof. T. O'Connor Sloane's book on " Liquid Air and the Liquefaction of Gases." This treatise gives the entire history of the liquefaction of gases, from the earliest times to the present, and also contains a description of all the experiments that have excited the wonder of audiences all over the country. It shows how liquid air, like water, is carried hundreds of miles and is handled in open buckets; also the Formation of frost on bulbs—Filtering liquid air—Dewar's bulbs—Liquid air in water—Tin made brittle as glass—India rubber made brittle—Descending cloud of vapor—A tumbler made of frozen whiskey—Alcohol icicle—Mercury frozen—Frozen mercury hammer—Liquid air as ammunition—Liquid air as basis of an explosive—Burning electric-light carbon in liquid air—Burning steel pen in liquid air—Carbon dioxide solidified—Atmospheric air liquefied—Magnetism of oxygen. Prof. Sloane also suggests what may be expected from liquid air in the near future. Though this is a work of scientific interest and authority, it is written in a popular style, and is therefore well within the grasp of any average reader.

The Automobile Index

Everything of permanent value published in the technical press of the world devoted to any branch of automobile industry will be found indexed in this department. Whenever it is possible a descriptive summary indicating the character and purpose of the leading articles of current automobile literature will be given, with the titles and dates of the publications.

Accumulators—

A serial article, by E. C. Rimington, on the construction of accumulators for automobiles. "The Automotor Journal," London, December 15, 1899.

A Fast American Run—

By Hiram Percy Maxim. "The Automobile Magazine," January, 1900.

An Artist's Appeal—

By Frances S. Carlin. "The Automobile Magazine," January, 1900.

An Improved Explosion-engine—

By Baudry de Saunier. "The Automobile Magazine," January, 1900.

An Improved Power-transmitting Mechanism—

By E. E. Schwarzkopf. "The Automobile Magazine," January, 1900.

A Simple Steering Device—

Patented by R. W. Jamieson. "The Automobile Magazine," January, 1900.

Automatic Starting—

Werhlé device for the automatic starting of any hydro-carbon motor. Description of same, with six illustrations. "La France Automobile," Paris, December 3, 1899.

Automobile Driving for Women—

An expression of views by Mrs. M. E. Kennard, the well-known English novelist, on the subject of automobile driving for women. "The Motor-Car World," London, December, 1899.

Automobile Management—

"How to manage a motor-car." Practical hints to users. A serial illustrated article, by Philanto. "The Motor-Car World," London, December, 1899.

Automobilism and Health—

"Are cycling and automobilism injurious to health?" First serial article of a study of this subject, by Dr. Raymond Sainton, Chief Surgeon of the Péan Hospital (Paris). "Le Chauffeur," Paris, November 25, December 11, 1899.

Automobilism in Italy—

Automobiles and their future in Italy (Gli Automobili ed il loro Avvenire in Italia). Guiseppe Spera. A report to the Commissioner of Public Works, discussing the relation of the automobile to the improvement of highways and the possible competition of automobile traction with railways. "Rivista Generale delle Ferrovie," October 29, 1899.

Carbureter—

Description and illustration of the "Abeille" carbureter. "La Locomotion Automobile," Paris, December 12, 1899.

Care of Storage Batteries—

J. K. Pumpelly's suggestions for the proper care of storage batteries, especially when they are employed in motocycle propulsion. "The Motocycle-Automobile," Chicago, December, 1899.

The Automobile Index

Charging Plants for Automobiles—

" A few practical suggestions on the installation and management of storage battery charging plant for electromobiles, etc." By J. M. Walsh. "Cycle and Automobile Trade Journal," Philadelphia, January, 1900.

Compressed Air Automobiles—

Full description of a compressed air delivery wagon devised and built by Molas, Lamielle et Tessier, of Paris. One illustration. "La Locomotion Automobile," Paris, December 7, 1899.

Electric Automobiles—

Two novel automobiles. Frank H. Mason. Describes two vehicles exhibited recently in Berlin, the Pieper double-motor carriage and the "Electra." "American Manufacturer and Iron World," November 2, 1899.

The Kriéger delivery wagon described and illustrated. "L'Avenir de l'Automobile et du Cycle," Paris, December, 1899.

Description and illustration of the Jenatzy delivery wagon. "L'Avenir de l'Automobile et du Cycle," Paris, December, 1899.

The new Kriéger "electrolette" illustrated and described. "La Vie au Grand Air," Paris, December 10, 1899.

The Crowdus electric automobile fully described. Two illustrations. "The Motor Age," Chicago, December 21, 1899.

A new Victoria built by the United States Automobile Co. Described with illustrations. "Cycle and Automobile Trade Journal," Philadelphia, January, 1900.

Electric Transformer—

Legros et Meynier's transformer used for charging the ignition accumulators on automobiles. An illustrated description by A. Delasalle. "La Locomotion Automobile," Paris, December 21, 1899.

Explosion Motors (Les Moteurs à explosion)—

A book by M. Georges Moreau. It is a study for the use of constructors and drivers of hydro-carbon automobiles. "La France Automobile," Paris, December 24, 1899.

Fire Engines—

Self-propelling steam fire-engines constructed by Messrs. Merryweather & Sons, with one illustration. "Engineering," London, September 15, 1899.

Self-propelled fire-engine constructed by Messrs. Cambier & Co., of Lille, France. Illustrated. "Engineering," London, November 24, 1899.

Fore-carriage System—

Serial illustrated articles on a new system for the propulsion and steering of automobiles by a special fore-carriage. L. Baudry de Saunier. "La France Automobile," Paris, December 3 and 10, 1899.

Gallery of American Automobiles—

" The Automobile Magazine," January, 1900.

Hydro-carbon Automobiles—

Full description, with six illustrations, of the Julien Bernard voiturette, called "L'Hirondelle Automobile." This vehicle is of novel design and is intended for public service. It has a double or reversible steering gear. "L'Avenir de l'Automobile et du Cycle," Paris, November, 1899.

Illustrated description of the Underberg voiturette. "L'Industrie Velocipédique et Automobile," Paris, November, 1899.

The improved Turgan-Foy voiturette, with divided axle for steering. An illustrated description. "L' Industrie Velocipédique et Automobile," Paris, November, 1899.

A brief description of the " Eole " voiturette, with one illustration. "Le Chauffeur," Paris, November 25, 1899.

The Gobron et Brillié wagonette fully described by Gustave Chauveau. Eleven illustrations. "Le Chauffeur," Paris, November 25, 1899.

The Pougnaud et Brothier gasoline carriage, called "La Charantaise." A full description of this new vehicle, with six illustrations. "La Locomotion Automobile," Paris, November 30, 1899.

The Panhard et Lavassor gasoline omnibus, described and illustrated. "L'Avenir de l'Automobile et du Cycle," Paris, December, 1899.

A brief description of the new friction-geared automobile built by F. Rough & Co., of Hereford, Eng. It has no chains, belts or toothed gears whatever. "The Motor-Car World," London. December, 1899.

Description of the "Ivel" automobile, built by Mr. D. Albone, of Biggleswade, Eng. One illustration. "The Motor-Car Journal," London, December 1, 1899.

The "Lady" voiturette, built by Mr. H. Cave, of Coventry, Eng. A brief description, with one illustration. "The Motor-Car Journal," London, December 8, 1899.

The N. Bravo gasoline automobile described by Maurice Chérié. With illustrations. "La France Automobile," Paris, December 10, 1899.

Illustrated description of the new two-seated quadricycle, built by the "Société anonyme des Voiturettes Automobiles." "La Locomotion Automobile," Paris, December 12, 1899.

Description of various new devices covered by the patent corresponding to Wm. Wallace Grant's gasoline carriage. "The Motor Age," Chicago, December 14. 1899.

Darracq's gasoline carriage described and illustrated. "The Automotor Journal," London, December 15, 1899.

Illustrated description of the Geo. Richard gasoline carriage. "The Automotor Journal," London, December 15, 1899.

The Pieper gasoline voiturette briefly described. One illustration. "The Motor-Car Journal," London, December 15, 1899.

The "Esculape" voiturette described and illustrated. "La France Automobile," Paris, December 24, 1899.

Description and illustration of the Bardon gasoline carriage. "La France Automobile," December 24, 1899.

The Bronhot wagonette described and illustrated. "The Motor-Car Journal," London, December 29, 1899.

Description and illustration of the Audibert-Lavirotte carriage. "The Motor-Car Journal," London, December 29, 1899.

The Old and New Bollée Voiturettes. "The Automobile Magazine," January, 1900.

Hydro-carbon Motocycles—

The Berthier gasoline bicycle described and illustrated. "L'Industrie Velocipédique et Automobile," Paris, November, 1899.

Illustrated description of the new gasoline tricycle made by Mr. H. Mayer, of Berlin, Germany. "The Motor-Car Journal," London, December 8, 1899.

The Shaw gasoline bicycle described and illustrated. "The Motor-Car Journal," London, November 24, 1899, and "The Autocar," Coventry, December 9, 1899.

Description of the Lamandière-Labre gasoline bicycle, with four illustrations. "The Motor-Car Journal," London, December 29, 1899.

Hydro-carbon Motors—

Description and illustration of the Rhéda motor. "L'Industrie Velocipédique et Automobile," Paris, November, 1899.

Description of the Lucas & Villain motor. Three illustrations. "La France Automobile," Paris, December 3, 1899.

The Secor oil motor. Hugh Dolnar. Reply to the criticisms of Mr. J. D. Lyon. "American Manufacturer and Iron World," November 9, 1899.

The "Abeille" gasoline motor described, with three illustrations. "La Locomotion Automobile," Paris, December 12, 1899.

The Automobile Index

Description and illustration of the Bennett and Thomas gasoline motor. "The Automotor Journal," London, December 15, 1899.

Illustrated description of the Partin motor. "La France Automobile," Paris, December 17, 1899.

A brief description (illustrated) of the Toepel motor for bicycles. It uses kerosene, gasoline or acetylene. "Cycle and Automobile Trade Journal," Philadelphia, January, 1900.

The Ravel "Intensif" motor. By Gustave Chauveau. "The Automobile Magazine," January, 1900.

J. W. Walters' device, consisting of a gasoline motor, a propelling wheel and a steering gear, all combined. The whole is arranged to work as a detachable fore-carriage. Two illustrations. "Scientific American," New York, January 6, 1900.

Igniters—

An article on the problem of ignition and the various kinds of igniters. By C. E. Lee. "The Motocycle-Automobile," Chicago, December, 1899.

The Apple igniting apparatus as made by the Dayton Electrical Mfg. Co. A brief description with one illustration. "Electrical World and Engineer," New York, December 16, 1899.

Several devices for electrical ignition as used on the Phœnix automobile. "La France Automobile," Paris, December 24, 1899.

Liquid Air—

By Harrington Emerson. "The Automobile Magazine," January, 1900.

Lubricating Device—

Serpollet's "multiple" lubricating device. With three illustrations. "La France Automobile," Paris, December 17, 1899.

Mechanical Traction and Propulsion—

By Prof. G. Forestier. "The Automobile Magazine," January, 1900.

Motocycle Management—

Serial articles, by Mr. A. J. Wilson, under the title of "Motor cycles and how to manage them." With illustrations. "The Autocar," Coventry, December 9, 23, 1899.

Motor Vehicles in the Stock Market—

Editorial discussion and warning, considering the four kinds of motive power used, and reviewing the standing of the industry. "Engineering News," November 2, 1899.

Motor Cycle Racing—

By Al Reeves. "The Automobile Magazine," January, 1900.

Recent Experiences with Steam on Common Roads—

Paper read by John J. Thornycroft, F. R. S., before Section G of the British Association. "Engineering," London, September 22, 1899.

Speed Changing Gear—

Description, with three illustrations, of the W. H. Newman system of speed changing gear for automobiles. "The Engineer," London, December 1, 1899.

The Buchet speed changing gear described. Two illustrations. "The Motor-Car Journal," London, December 1, 1899.

The two-speed gear made by Mr. A. Eldin, of Lyons, France. One illustration. "The Motor-Car Journal," London, December 29, 1899.

Speed Changing Mechanism—

The Delbruck speed changing mechanism described and illustrated. "The Automotor Journal," London, December 15, 1899.

Speed Reducing Device—

L. Brun's *demultiplicator* or speed reducing device, described and illustrated. "Le Chauffeur," Paris, December 11, 1899.

Starting Device—

Description of a new starting device invented by the English automobilist Mr. E. Estcourt. One illustration. "The Automotor Journal," London, December 15, 1899.

The Automobile Magazine

Steam Automobiles—

Steam motor-cars constructed by the Steam Carriage and Motor Co., Ltd., of Chiswick, Eng. With six illustrations. "Engineering," London, September 22, 1899.

Illustration and brief description of a new steam carriage built by Geo. A. and Peter Harris, of Manchester, Conn. This vehicle uses soft coal. "The Motor Vehicle Review," Cleveland, December 12, 1899.

Illustrated description of three vehicles of the Leach pattern. "Cycle and Automobile Trade Journal," Philadelphia, January, 1900.

The "Locomobile" described and illustrated. "The Automobile Magazine," January, 1900.

The Marsh carriage and its performances as a fast hill climber. One illustration. "The Motor Age," Chicago, January 4, 1900.

Steam Omnibus—

Illustrated description of the latest steam omnibus built by Dion et Bouton. "L'Avenir de l'Automobile et du Cycle," Paris, December, 1899.

Steam Wagon—

A short account of some modern steam wagons. George A. Burls. Excerpt of paper read before the Civil and Mechanical Engineers' Society. A general discussion of the efficiency, durability, power, speed, etc., of these vehicles, with table of data, and short discussion. "Automotor Journal," London, November, 1899.

Steam Heavy Weight Wagon—

Description and illustration of the new "Toward" heavy weight steam-wagon. "The Motor-Car Journal," London, December 1, 1899.

Steering—

Serial articles, fully illustrated, on steering by means of the divided axle. Dr. C. Bourlet. "La Locomotion Automobile," Paris, December 7, 1899.

Last of the serial articles on steering by means of the divided axle. Dr. C. Bourlet. "La Locomotion Automobile," Paris, December 21, 1899.

The Automobile Club of Great Britain—

By a founder member. "The Automobile Magazine," January, 1900.

The Automobile in Traction—

By Robert H. Thurston. Director of Sibley College, Cornell University. "The Automobile Magazine," January, 1900.

The Automobile Voiturette at the End of the Century—

A review of the various types of voiturettes produced up to the present time. "Le Chauffeur," Paris, December 25, 1899.

The Buchet Head for Working-chambers—

"The Automobile Magazine," January, 1900.

The Gordon Bennett Automobile Cup—

Official rules. "The Automobile Magazine," January, 1900.

The Truth About Motor Horse-Powers—

By Georgia Knap. "The Automobile Magazine," January, 1900.

Tires—

Illustrated description of Munger's tires. New patents. "The Motor Age," Chicago, December 14, 1899.

Traction Engines in South Africa—

Description, with five illustrations, of traction engines for common roads, sent to the Transvaal by Fowler & Co., of Leeds, England. "The Engineer," London, December 8, 1899.

Traction Engines in War—

Description and illustration of a military traction engine. "The Engineer," London, November 24, 1899.

Traffic Regulation and the Speed of Motor Vehicles on Highways—

A paper read before the Automobile Club of Great Britain by R. E. Crompton. "The Horseless Age," New York, January 3, 1900.

The Automobile Magazine

Vol. I   MARCH 1900   No. 6

CONTENTS

AGENCY FOR FOREIGN SUBSCRIPTIONS:

INTERNATIONAL NEWS COMPANY

BREAMS BUILDINGS, CHANCERY LANE   STEPHAN STRASSE, No. 18
LONDON, E. C.   LEIPSIC

Price 25 Cents a Number; $3.00 a Year
Foreign Subscription $4.00, Post-paid

A Midwinter Run

(Copyrighted by U. S. Industrial Publishing Co.)

The Automobile
MAGAZINE

VOL. I MARCH 1900 NO. 6

The Street Car of the Future
By Waldon Fawcett

IN business, as in warfare, contestants seldom evidence great care in estimating the strength of an opponent they do not fear. That the electric street railway interests of America are focusing upon automobilism a fierce light of investigation is, therefore, of considerable significance. It will be remembered that when the interurban electric lines a few years ago commenced their inroads upon the province which had up to that time been the exclusive property of the steam roads, the projectors of the former asserted that there was a place for both, and that by reason of its work as a " feeder " the electric interurban system would prove a valuable adjunct to its predecessor in the transportation field.

Now, a third factor is steadily and rapidly advancing to claim a place in the field, and also to aid in its development. All that was urged in favor of the trolley system in its relation to the steam road may be claimed for motor vehicles in their influence upon both. The new mode of locomotion will assuredly draw to some extent from the patronage of each of its competitors—principally, of course, from the electric lines—but, on the other hand, it will, quite as surely as did the interurban trolley, open up new lines of communication which will bring revenue to all transportation interests in greater or less degree.

To take up first the automobile in its relation to electric lines in the cities, it may be pointed out that its chief advantages will be found in speed and comfort. The displacement of the antiquated Fifth avenue stages in New York City by autocars of an approved type will serve to illustrate admirably one phase of the possibilities referred to. There will probably always remain in every principal American city at least one or two residence streets of considerable length from which street cars will be debarred. Some system of public conveyance for such thoroughfares is an essential, not only to meet the requirements of the residents, but also in view of the fact that such streets being favored with the handsomest residences are much sought by tourists.

Had an effort been made to design a vehicle for just such service nothing more completely filling the requirements than the

London Park Omnibus

automobile could be found. The absence of steel tracks and the minimum of noise constitute its chief qualifications, but there are also others which would cause it to be looked upon rather as an ornament than a detriment to exclusive residence districts.

While upon this subject mention must be made of another sphere of usefulness quite as important as any in which the horseless vehicles will come into more direct competition with the street cars. This is the service in parks and boulevards. Of late years American cities have broadened immensely the scope of their park systems, and there is an almost universal tendency to connect the various public playgrounds by boulevards, after the plan, the advantages of which have been so conclusively

The Street Car of the Future

demonstrated in Chicago and Cleveland. All considerations preclude the possibility of allowing street cars to gain entrance to any part of these pleasure highways, and yet the old carriage or omnibus system, long an eyesore in many of our American parks, is totally inadequate to meet the demands of that portion of the public which wishes to inspect a park system without devoting several days to the task. For such a service, on the other hand, the combination automobile omnibus and street car recently designed will have numerous qualifications and few disadvantages.

In the open field the automobile must rely for its share of the business of carrying the public between homes and shops and

Berlin Park Stage

offices on the advantage afforded by greater speed and comfort. The former must be primarily considered, inasmuch as any means of reducing the time of transit is bound to appeal strongly to mankind in general, a fact attested by the patronage accorded to the elevated railroads in the larger cities. It will also be found practicable to operate automobiles in sections of cities where excessive grades and other engineering obstacles would stand in the way of the establishment of an electric railway system.

In the suburban service there is a distinct niche for the motor vehicle. There are in the neighborhood of every city numerous districts rather too sparsely settled to make a suburban electric line a paying investment, and yet with a population sufficient to

The Automobile Magazine

net a handsome revenue to a carefully conducted automobile line Especially will this be true in the smaller cities ranging in population from thirty-five to one hundred thousand. In many such the suburban car service is limited in the extreme in scope, largely by reason of the fact that a single company controls the entire electric railway system of the place and has never been stimulated by competition. In such a field the schedule for automobile operation can be gauged very accurately to the volume of travel, and there will be, of course, no heavy first outlay for roadbed, rails and overhead wires.

Certainly the prediction may be made, without any danger of characterization as optimism, that the automobile will ultimately supplant every other form of conveyance on those routes to which a majority of the travelers who traverse them are attracted by the scenic possibilities. This is attested by the great demand for automobiles which has sprung up at all the leading summer resorts and by the project of American capitalists to establish an automobile line across the Island of Porto Rico, as well as numerous other similar propositions.

French Hotel Bus

The Street Car of the Future

American Hotel Bus

The branch of the service for which the automobile has yet to demonstrate its adaptability is the interurban. Many street railway magnates, while conceding that the motor vehicles are admirably adapted to all forms of public conveyance in the city, assert that they would prove impracticable for the runs of from twenty to fifty miles which would be required on interurban routes. The credence given to these representations is due largely to the lack of success which has attended some of the long distance automobile trips undertaken in the United States during the latter half of 1899. That the objections made are, however, unfounded, is proven by the splendid records for long tours made in England and France.

An exhibition which will further demonstrate the capability of the automobile for just this class of work will be found in the thousand mile tour for motors of every description to be held late in the spring, traversing a zig-zag route from London to Edinburgh. The award of gold, silver and bronze medals will serve to establish a relative rating of merit for the different machines from the standpoint of suitability as roadsters.

A type of vehicle which, with certain modifications, would seem to be admirably adapted to interurban service, is the combined street car and omnibus which has recently been introduced in several German cities. The vehicle is designed for use as an ordinary automobile on streets where there are no tracks and yet is gauged to run on the tracks of electric railways, where such

French Colonial Stage

exist. A distinguishing feature of the design is the location in front of two guide wheels. These are lifted up during the passage of the vehicle through streets without tracks, but serve to guide it on the tracks of electric lines. The automobile is so constructed as to be able to turn into very narrow streets.

In Berlin an Automobile Stage has been running during the last six months regularly over that stretch of the city lying between the Stettin R. R. Station and the Kreuzberg. The daily distance covered by this vehicle is 72 kilometres.

Reckoned by Watt hours the cost per kilometre is 0.6 pfennigs, which makes 2 cents per kilometre, counting a kilo-Watt hour at 16 pfennigs.

This omnibus weighs 7,100 kilograms, including the weight of 30 passengers, conductor and motorman. Without passengers the vehicle weighs 4,850 kilograms, including its electric batteries. The omnibus measures, according to continental standards, 3.6 in height, 4.3 in length, 3.2 axle distance. 1.7 in width, 1.3 wheel width. The height of the front wheels is 1.07.

The Street Car of the Future

and that of the hind wheels 1.2. The steering wheel has a diameter of 540 milimetres, and the gear is set at one to eighteen.

Power is transmitted through two motors of five horse-power each, at 160 volts and 30 amperes, giving approximately 450 revolutions per minute.

The results accomplished by this omnibus seem to have proved unusually satisfactory. Though the wagon has been fully laden with passengers day after day, the Berlin newspapers report that no serious breakdown has ever interfered with the regular schedule of the new automobile line.

Through the courtesy of the builders, Messrs. Lange & Gutzeit, of Berlin, we are able to give herewith an illustration of another large electrical omnibus they have lately constructed for the Neue Berliner Omnibus Gesellschaft. (See page 569.)

This vehicle has accommodation for twenty-eight persons, including the driver—two on the front platform, twelve inside, and fourteen outside. It is driven by two 4 horse-power electro-motors, geared by large spur wheels to the rear axle. The electrical energy is supplied by a battery of Tudor accumulators slung in a box underneath the body of the car. Unfortunately, we are unable to give the number or electrical capacity of the accumulators, but the makers inform us that a run of 12 kilome-

Martyn's Steam Omnibus

tres can be made on one charge and that the battery can be recharged in a very short space of time.

The vehicle is provided with both electrical and mechanical brakes, which can pull the vehicle up in a very short distance, even when traveling at full speed. The steering is effected by a vertical hand wheel connected with the front road wheels. The stage, we learn, has been in use in the German capital for over a month, and, according to the makers, has given every satisfaction.

Messrs. Lange & Gutzeit are at present engaged on the construction of a 'bus on similar mechanical lines, but having no outside seats. It will have accommodation for eighteen passengers, and is intended for service in small towns.

Coulthard's Colonial Steam Bus

An interesting comparison of the cost of operation of trolley cars and automobiles has been made by an authority on that subject. According to this estimate the total cost of installing and maintaining a five-mile trolley line over a fairly level road amounts to no less than fifteen thousand dollars a mile. This includes a small electric power plant and a single track with turn-outs, equipped with ordinary wooden telegraph poles. In addition to this there would be the cost for at least two cars carrying from 35 to 40 passengers each. This is $6,000 per car. The running expense of such cars amounts to sixteen dollars a day, or more in places where labor and fuel come high.

The Street Car of the Future

The installation and maintenance of such a small trolley road, in other words, would call for an investment of about $375,000, to which should be added the first cost of $75,000 for the tracks and $12,000 for the cars.

Traveling at the rate of thirty miles an hour a trolley car, in making twenty-minute trips over its five-mile course, would stand at rest for at least half of the time. The maximum carrying capacity under such conditions would be 2,800 passengers a day for every twelve hours. To meet its total running expense of $63 per day, therefore, each car would have to carry an ave-

Compressed Air Omnibus

rage of more than three-fourths of its maximum capacity, simply to meet expenses. Needless to state this proportion between income and outlay would never commend itself to any serious business man.

An automobile omnibus capable of carrying one-half the number of passengers of a trolley car would require a motor of at least fifteen horse-power. For a new vehicle equipped with such a motor $5,000 would be a good price. A motor of equal horse-power capable of adjustment to any vehicle, such as the Vollmer fore-carriage, or other similar devices now used in Europe, could be secured for less than $1,000. Taking the maximum price of

$5,000 as the basis, however, would make the total investment for an automobile omnibus line covering the same distance $30,000, as against $87,000 for the trolley line. This estimate provides for at least half a dozen omnibuses to carry the same number of passengers as two trolley cars. Thus only could the difference of carrying capacity in traveling time be fairly offset.

Allowing five years of service for each automobile, gives the maintenance cost of $6,000 a year, as against 12,370 points as against the trolley. With the initial cost of $30,000 the total investment for the motor omnibus system of equal carrying capacity and route mounts up to $164,000, as against $375,000 for the trolley line.

If other things are taken into consideration, such as the probable income from a light freight and package service and frequent economies in cost when superfluous automobiles are laid off at times when traffic is light, it may be safely assumed that such a motor omnibus system could be installed and operated at an outlay of about one-third of that required for a trolley line of corresponding value.

These figures show plainly that an automobile omnibus service designed as a feeder for some railway or street car already in existence would be a far cheaper investment than an extension of the older existing systems. Similarly, automobiles could be

German Electric Autocar

The Street Car of the Future

German Electric Stage

thus operated in country districts or as feeders to prosperous summer resorts and sanitariums, where a trolley line could only be a dismal failure.

Street railway men have long declared that the average American is in too much of a hurry to take time to climb to the second story of a double-deck car, but there is every reason to believe that a type of automobile, built after the fashion of a tallyho or an omnibus, with seats on top, will ultimately find favor with suburban and long distance passengers. The possibilities in the matter of speed may be realized from the recent record of an English machine which made the journey from Coventry to London, a distance of 92 miles, in four hours, an average of 23½ miles per hour.

The cheapening of the motor vehicle will naturally prove an important factor in the extension of its use as a competitor of the street car. Indeed, it will reach its fullest development in this direction only after the price of an autocar is approximately that of a street car fitted with an electric motor. Once started, the encroachments of the automobile on the field of municipal transportation will be rapid. The new vehicles may also be expected to displace street cars in many auxiliary services, such

The Automobile Magazine

as the conveyance of the mails and the transportation of farm products in the rural districts.

The following is a conservative view of one of the greatest railroad promoters of America:

" I do not think that any sort of traction that does not run on rails and that does produce, or carry in condensed form, its own motive power, will ever be a competitor for railways of any description. The production of power in quantities is the only economical way to produce it, and the resistance to traction on rails is so much less than on the best of roads, that, for regularly carrying large numbers of people, there can scarcely be a comparison.

" I can readily see, however, that there are cases where motor-vehicles would be far more economical than trolley cars. I can also see where they could be used to great advantage in connection with trolley lines. There are many trolley roads that feel the need of feeders and yet find that there is not enough traffic to justify the building of branch lines. The motor-omnibus is the very thing to fill the want. It would be particularly economical where storage batteries were used, as the company could charge the batteries at a minimum cost. In such cases, lines of motor-omnibuses could be used in connection with the regular trolley lines, to advantage, and, when the season was over, the buses could doubtless be utilized elsewhere, or, if not, they could be safely stored and there would be far less loss than with the trolley line.

" Yes; there is, no doubt, a great field for motor-omnibuses. They will prove a great factor in cities, too, but not in direct competition with the street railway lines. It will be their province to operate in the residence streets where it is impossible to get franchises to lay tracks."

TIRE PUNCTURES

According to a contemporary the greatest trouble with the pneumatic tire on heavy vehicles is not caused by puncturing, which accounts for only about 7 per cent. of the trouble, but results from the internal wear of the fibres of the tire, due to the weight. Some tires which have been examined show the fibres of the material reduced to a fine powder.

The New Sport Abroad

(By Our Own Correspondent)

THE new regulations for automobiles imposed by the municipal council of Paris have raised lively protests.   Particular objection is made to the rule restricting the speed in the Bois de Boulogne and Vincennes to twelve kilometres an hour, or seven and a half miles.   The Automobile Club has issued a circular on the subject, calling attention to the fact that at the meeting of November 28 the committee of the club censured the practice of driving at inordinate speed in crowded thoroughfares as compromising the interests of automobile locomotion by provoking regulations which by their rigor tend to impede the development of the industry.   In this censure were included those motorcyclists who, by not using their escape-valves, run the risk of frightening horses by the noise.   The committee also regretted the action of the municipal council limiting speed in the parks and expressed the hope that the regulation, if not withdrawn, would not be strictly enforced.

The President of the club, Baron de Zuylen, says the action of the council was due to certain madmen who go at the rate of 60 kilometres an hour simply to make themselves conspicuous.   This has caused such numerous complaints, he said, that the council imposed the new regulations.   He suggested that certain ways in the Bois be set apart for automobiles as others have been set apart for bicycles and others for riding.

President Loubet has recently become a convert to the automobile.   Accepting an invitation to join Count Potocki in a day's shooting at the latter's estate at Le Serray, the Count met the President at the railway station with his automobile.   They darted off at a good speed and by the time they had reached the Count's abode the President was enthusiastic over his experience.

A deal of interest has been excited by the reported bet of a young automobilist here in Paris that he can make 100 kilometres in an hour, staking 100,000 francs on the result.   It is said that the attempt will be made on the stretch of 70 kilometres between Evreux and Lisieux, where the only heavy grade lies near the middle portion of the road.

It appears that the automobile has not yet diminished the use of horses here in France.   Indeed, according to the report of the general inspector of horses, their number continues to increase.

The Automobile Magazine

In 1895 there were 2,881,226 horses in the country. In 1897 there were 3,005,541, and in 1898 the horse-population had increased to 3,025,502.

There does not yet appear to be any great boom in automobile manufacturing projects in Great Britain, although contrary to the experience with bicycle manufacturing, the demand is so enormous that there can be no danger of overstocking the market for years to come, and the industry must therefore prove highly profitable. There is considerable quiet activity, however, in the way of new enterprises. One of these is the Yorkshire Motor Car Manufacturing Company, Limited, incorporated with a capital of £50,000 to acquire and extend established concerns in Bradford, Halifax and Hipperholme. Various cycle companies are also going in for automobile manufacturing.

The London Electric Cab Company, which has already invested 1,500,000 pounds sterling in its plant, reports a great advance in storage-battery efficiency. Its latest accumulators comprise forty cells each, with a capacity of 150 ampere-hours. Two sets of batteries are provided for each vehicle. There are five rates of speed beside one for reversing. The motors, of 3½ horse-power, are of the Johnson-Lundell type. The company's station has room for 150 cabs, and there are 80 in service.

The automobile has proved itself useful in distributing war news on Sunday morning in London, when the regular local transit lines are at a standstill. The *Telegraph* and the *Westminster Gazette* place at the disposal of the *Weekly Dispatch* and the *Referee* all the war news received at their offices up to noon and these newspapers then send out, as soon as possible, their latest editions by automobile to all outlying districts, the vehicles being furnished by the Motor Van and Waggon Company, Limited.

A very useful novelty for ladies who fare either by automobile or bicycle is the Claxton Mask Veil, a recent device that is colorless, translucent and practically weightless. Invisible beneath an ordinary veil, it does not interfere with respiration and at the same time protects the face against cold winds as effectually as if the wearer were behind a screen.

The automobile omnibus service in Edinburgh is reported to be a great success, and steps have been taken to extend it by the addition of several new lines. The fare is the same as on the tramways—a penny a mile. The same rates are charged by the automobile lines in Glasgow, Dundee, Aberdeen, Falkirk and other Scottish cities.

Steps have been taken in Edinburgh to organize the Scottish

New Sport Abroad

Automobile Club, affiliated with the Automobile Club of Great Britain and Ireland. At the annual banquet of the latter it is interesting to note that Mr. Shaw Lefevre spoke of a belated movement against automobilism, manifest in the bombardment of the Local Government Board with petitions for further restrictions. Great progress had been made in the three years since the passage of the Light Locomotives Act, he said, though small compared with that made in France. The British demand for vehicles, however, was greater than the supply. He believed a great change was impending; automobilism would not only build up a new industry, but greatly improve conditions in the rural districts.

The Chinese ambassador to Great Britain, Chih-Chen-Io-Feng-Luh, who is making an inspection of the leading industrial centres of the country by order of the home government, to report on what he sees, particularly in relation to goods suitable to the Chinese markets, recently made a visit to Coventry. His party was met at the station by a fleet of automobiles belonging to the Daimler and Motor Manufacturing Companies and taken over the town to visit the various industrial establishments. He and his staff took a special interest in the process of automobile manufacturing. At a banquet in the evening he referred particularly to the automobile industry and said that one could scarcely realize the great future before the trade without seeing a factory. And he said that when the roads in China are restored these vehicles would be introduced in the far East as they are now in the West.

The English Automobile Club proposes to organize for next March a tour from London to Edinburgh and back.

" Ladymobile " is the English word proposed to express what is called a *chauffeuse* in French and *Automobilistin* in German.

An English company is to establish an automobile transit line between the Russian city of Memel and the port of Liebau on the Baltic. It is proposed to have a freight-van towed by the omnibus. The time for the trip will be from six to eight hours.

M. Orlovski has written for the *Vélo* an account of his trip from St. Petersburg to Paris by motor-tricycle. He says that the roads about St. Petersburg are miserable; the farther from that capital one gets the better they are. After passing Dvinsk, about 500 kilometres from St. Petersburg, the ways grow fine and one rolls along at the rate of 35 kilometres an hour. With a higher gear he could have made a much greater speed. From Kovno to Warsaw and from Warsaw to the boundary the roads are also good. His greatest trouble was with poor combustion caused by the damp weather and bad benzine. Orlovski undertook the trip to demonstrate the practicability of Russian roads for the automobile, the Parisian *chauffeurs,* Messrs. Thevin and Houry, sent by the Automobile Club of France to reconnoitre the route, having reported them impracticable at present.

The Russian, Paul Tidemann, has bought a motor-quadri-cycle here in Paris and intends to make the trip back to Moscow therewith.

Another Russian, Axel Monoinen, proposes to make an automobile trip throughout European Russia in the course of eight and a half months, going from St. Petersburg to Narva, Riga, Kovno, Ssuvalki, Plosk, Kalish, Lodz, Kamenez, Kishinev, Odessa, Nikolajev, Novotsherkask, Zarikyn, Novosdiensk, Uralsk, Perm, Vjatka, Makarjev, Kostroma and Jaroslau, back to St. Petersburg.

In Berlin, automobile liveries have been established with all types of vehicles for hire or on sale. The present charge is four

German Forecarriage Types

marks, or about a dollar, an hour. For a half day the charge is 15 marks; for a day, 25 marks; for a week, 125 marks, and for a month, 400 marks. A good business is done at these rates. Physicians, in particular, use the vehicles very extensively. Each establishment has a well patronized school for instruction in the use of the automobile.

The recent international automobile exhibition in Berlin was a gratifying financial success, although a deficit had been looked for. The receipts were 55,810 marks and the expenditures 43,702 marks, leaving a balance to the good of 12,108 marks. Deducting certain " gratifications " a surplus of at least 10,000 marks is left. The admission receipts, estimated in advance at 10,000 marks, amounted to 31,603 marks, and the rentals for space, also set at 10,000 marks in the preliminary estimate, reached 19,150 marks. It is proposed to hold the next Berlin exhibition in 1901.

In Berlin there are two electric automobiles in the service of

New Sport Abroad

the post-office. One of these, a *Güterspostwagen,* or postal freight wagon, has been in use for several months without serious interruption. The second, a *Paketbestellwagen,* or parcels-post wagon, has also been in successful operation for some time. But when the snow came their use had to be abandoned, together with the *Brief-carriol,* or letter carriage. The trouble was not with lack of power, but with the tires. It appears that, owing to the universal prevalence of asphalt pavement in Berlin, it was thought that steel tires would be sufficient. But when the snow came the traction proved inadequate. Rubber-tired automobiles, however, found no trouble in getting through the snow, and the postal authorities will naturally make the change from steel to rubber as soon as practicable.

The Forecarriage Detached

The German Automobile Club, organized with the Duke of Ratibor as president, has over 200 members. A handsome house on the Sommer-Strasse, Berlin, has become the home of the club.

Berlin Paketbestellwagen

The Automobile Magazine

Official and social relations with the automobile clubs of France, Austria and Switzerland have been entered into and the former has cordially opened its house to all members visiting Paris.

In Munich the number of automobiles licensed by the police authorities is 25, and this being the maximum allowed in the city under present regulations further official action will have to be taken before the number can be increased. The good Bavarians are so cautious in such matters that the Vienna *Fremdenblatt* suggests that for the protection of pedestrians a policeman be assigned to each person in the function of nursery maid!

Count Stefan Gyulai made a most interesting journey last September from Vienna to Venice and back. His vehicle was an eight-horse-power Daimler. From Vienna he journeyed through Lower Austria, Steiermark, Krain, Kaernten, Goerz-Gradiska to Venice and back by way of Upper Italy and the Southern Tyrol. The distance covered was 1,630 kilometres, the journey from Vienna to Mestre, opposite Venice, being 696 kilometres. This part of the journey was covered in 29 hours, an average of 24 kilometres an hour. The 179 kilometres from Goerz to Mestre occupied 6½ hours; an average of 25 kilometres an hour. From Bruck to Graz 53½ kilometres were covered in 1 hour and 47 minutes, an average of 29 kilometres an hour. From Graz to Marburg 66½ kilometres were made in 2 hours and 30 minutes, an average of 26 kilometres an hour. From Cilli to Laibach, 47½ kilometres, in 2 hours and 45 minutes, the average was 27 kilometres an hour. From Mestre to Bozen was 267 kilometres, from Bozen to Toblach 104½, Toblach to Klagenfurt 195, Klagenfurt to Semmering 217, Semmering to Vienna 89, an average for the whole of 22 kilometres an hour. The distance from Toblach to Vienna, 501 kilometres, was made in 23 hours. Imagine, if you can, a horse making such a distance at an average of nearly 14 miles an hour! The Count, who was accompanied by the director of his estates, Herr von Lukacsy of Arad, guided his carriage all the way himself. They were attended by one servant. They met with no accident or interruption beyond a slight delay caused by a hot bearing between Padua and Vincenza, due to insufficient oiling. No difficulty was encountered in surmounting the Semmering pass, 950 metres above sea-level, and the still higher Berhauer Sattel, 1,250 metres in altitude. The roads were good for most of the way, and were exceptionally adapted to the automobile in the Tyrol and Steiermark. The carriage was covered, sheltering the occupants well. The weather, however, was favorable most of the time.

The very mountainous Austrian province of Steiermark, which

The New Sport Abroad

Count Gyulai and Herr von Lukacsy Touring from Vienna to Venice
The trip extended over nearly 1,000 miles and lasted a fortnight.

is favored with excellent roads, is to have the benefit of an automobile service connecting various important parts of the province,
from East Steiermark to Hungary, with the city of Radlersburg.
A company has been organized for the purpose under the official
designation of *Erste steiermärksche Automobil-Verkehrs-Gesell-
schaft in Radlersburg*—the First Automobile Transportation
Company of Steiermark in Radlersburg. The following motor-
stage lines have been decided upon: Radlersburg, Purkla, and
Gleichenberg to Feldbach and return; Radlersburg, Kaltenbrunn,
St. Georgen, to Jennersdorf (Hungarian route) and return; Radlersburg, Muraszombad, Bellatinz to Alsolentva (Hungarian
route) and return; Radlersburg to Luttenberg and back. The
service will begin on May 1, 1900, with a ten horse-power omnibus on each route. The vehicles are to have a gauge of 1.16
metre, and will seat from eight to ten persons with standing-room
for two others. Benzine motors with water-cooling will be used.
The concession was given by the authorities with a promptness
that shows a most favorable disposition towards the undertaking.

The interest in the coming first international contest for the
Gordon Bennett cup is growing intense, and our French *chauf-
feurs* intend to strain every nerve to secure the trophy for France.
Various racing-machines are building expressly for the occasion.

It is estimated that one thousand new automobiles will be
marketed in Austria this year. Indeed the automobile, unlike
other things, seems to be enjoying a far greater boom in Austria
than in Germany.

Among all fashionable Automobile Clubs that have sprung
up like mushrooms abroad, the most energetic seems to be that
of Austria and Hungary, in Vienna, of which Count Pötting-
Persing is the president. The members of this club have
expressed their intention to take part in the following automobile events, to be conducted under the auspices of their own club
and that of Baden Baden. In the first week of May a motor
cycle race up the incline of the Exelberg, near Vienna. On
May 13 an automobile race to the Kahlenberger Berg. On May
20 a trial for the two kilometre record along the main avenue
of the Prater in Vienna. The last day of the month and the
opening week of June have been set aside for an elaborate automobile race from Munich to Salzburg and thence to Vienna,
open to all comers, with arrangements for public exhibitions of
the contesting vehicles at all three of these cities. After the
close of the projected Automobile Exhibition in Vienna will be
held another record run from Vienna to Graz and thence to
Baden-Baden. This will be followed by a series of races in

New Sport Abroad

Baden-Baden, from the end of June to the end of July, on the dates when the famous horse races of that popular watering place were wont to be held. Between the first week of August and the middle of that month another long-distance run will be made through the picturesque region of Saltzkammergut, covering the following points: Ischl-Strobl, Gilgen-Scharfling-See-Unterach-Weissenbach-Ischler Brücke. The most interesting point along this route, I dare say, will be found to be the

Count Pötting-Persing, President of Austrian Automobile Club

steep incline of the Scharflinger Berg. After these events the racing members of the club will betake themselves to France to try for new laurels there.

In the spring a motor-omnibus line is to be established for one of the most beautiful sections of the Thuringian Forest, running over the perfect roads between Ilmenau-Rammersberg, Stützerbach, Friedrichsroda, Oberhof, Schmucke and Schwarzburg.

In Nice a novel automobile party recently took place, the occasion being the baptism of the infant son of M. Leon Desjoyaux and the infant daughter of M. Noël Desjoyaux, both known as intrepid *chauffeurs*. They proceeded to the church in their automobiles, accompanied by various friends also in automobiles— among them the Prince Lubecki—forming a joyous procession. The ceremony over they returned to the residence of M. Noël Desjoyaux to honor the double baptism with appropriate festivities.

The Italian military authorities are taking an active interest in the possibilities of the automobile, and the ministry of war has directed inquiries to various foreign manufacturers as to what might be furnished, especially in the line of ambulances, field-post and munition wagons.

The sporting club of Baden, whose president is Prince Alexander Solms, has decided to establish an automobile section.

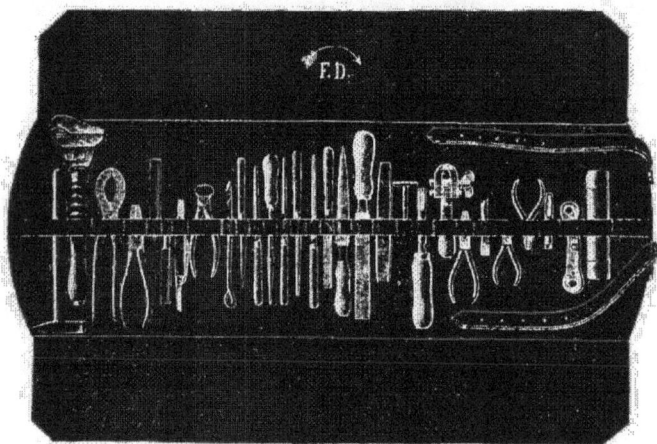

Automobile Tool Case

The Progress of Motor Vehicles

By W. N. Fitz-Gerald

Read before Twenty-seventh Annual Convention of the Carriage Builders

FOR several years past the carriage industry has been agitated over the advent of the motor vehicle to an extent that seems almost impossible of explanation, as its success means nothing more than the substitution of mechanical for horse power. In either event the carriage remains, and the carriage builders' vocation continues as in the past, while the effect upon buyers, if the new vehicle becomes a success, means increased interest in carriage riding and out-of-door exercise.

If science and skill succeed in producing a vehicle for pleasure or business that can take the place of the horse-drawn vehicles of to-day—one that will cost less to make and to keep; one that can be rendered equally available and has as great or greater elements of safety—then the automobile will become an important factor as an additional contribution to the needs of advanced civilization.

In New York City it is estimated that the number of horses in use is over 200,000. One-half of these are used for pleasure or light driving, and with three vehicles to every two horses, an estimate that is considered fair, we have 150,000 as the number of carriages in that city. On that basis, the total in our cities of 10,000 population and upward would be fully 1,500,000. There is, therefore, nothing in the advent of the motor vehicle to discourage the builders of high-priced carriages, while the builders of lower grades of work will not come in contact with it in any way.

Our people will either make it a success and encourage a large production, or the automobile will be a thing of the past in ten years. It is the duty of the carriage trade to give it every encouragement possible, consistent with safe business. Builders should bring their best skill to bear upon the carriage portion and be prepared to purchase reliable motors.

All carriage builders need not enter into the construction of the automobile, but none should antagonize it, for come it will, and antagonism will not check its progress.

Viewing the automobile industry from its favorable and unfavorable points, we can see nothing in it but good to the carriage industry, provided a conservative course is followed.

Gallery of Some Automobiles

Waverley Wagonette

Indiana Automobile Co. Brake

Shown at New York Exposition

Electric Stanhope

Vollmer Forecarriage Victoria

Automobile Tariffs

THE time is now come when the American automobile manufacturer, having established himself in his own country, looks about him for new worlds to conquer. No longer content with filling the wants of his own countrymen, he has cast his eye abroad, narrowly surveyed the unexplored field, and taken the necessary steps to secure a foothold in the new market. It is not the first time that the American manufacturer has thus peacefully invaded Europe. His locomotives are now running upon the railways of England, France and Russia; his machine tools are used in countless shops, from the Atlantic Ocean to the Ural Mountains, from the North Sea to the Mediterranean; his bicycles have supplanted the heavier and less graceful machines of his English and French rivals. His deep inroads in the various European countries have caused such consternation and alarm that the various European governments have been petitioned to check the invasion of the American upon their territory by the imposition of a restrictive tariff, which enabled the home industry at least to hold its own. The constantly increasing American export trade in automobiles has been regarded with little favor abroad. Some of the European governments have modified their old tariff laws, and rendered them more severe, in order to protect the automobile industry from the ravages of the encroacher. The reason for the protection of the native industry is to be found, not so much in the cheapness as in the better workmanship and more handsome appearance of the American automobile. In view of the gradually growing foreign trade of the United States in motor-carriages, we have deemed it advisable to indicate to the exporter exactly what obstacles must be overcome ere he can introduce his wares into foreign countries; what sums he must pay before Europe will open its portals to his automobiles. Undoubtedly the most important buyers in Europe are England, France and Germany; and to a discussion of their tariff laws the present article will therefore be confined.

England, ever true to her free trade traditions, welcomes the American automobile maker with open arms. No duties whatever are exacted from the importer of motor carriages. Automobiles are brought in and out without restrictions, whether they be of English or foreign make.

Automobile Tariffs

In France, on the other hand, an elaborate system of tariff regulations has been drawn up.

Automobiles are divided into two classes. In Class A are included motor carriages designed to carry persons. In Class B belong those vehicles used for carrying merchandise.

In both classes a general tariff and a minimum tariff is imposed. The general duty is applicable to wares made in those European and American countries from whom France has obtained no commercial concessions by treaty or otherwise. Certain portions of the United States are included under this head. The minimum duty is levied only upon goods imported from those countries with whom France has made commercial compacts for the common benefit of its own manufacturers and those of the countries in question. Those portions of the United State not included under the general tariff head receive this concession.

In Class A the duty is imposed on the carriage proper and on the driving mechanism. On carriages weighing 125 kilos. (275 lbs.) and more, the motor being included, a general duty of 60 francs per 100 kilos. or a minimum duty of 50 francs per 100 kilos. is imposed. For carriages weighing less than 125 kilos., motor included, a general duty of 150 francs and a minimum duty of 120 francs is collected. It follows, therefore, that a heavier tariff is levied upon the lighter than upon the heavier carriage, the French customs officials holding that, for an equal volume, the lighter carriage represents a greater value than the heavier vehicle.

In Class B a subdivision is made. A general duty of 30, 45, or 100 francs per 100 kilos., and a minimum duty of 20, 30, or 80 francs per 100 kilos. is levied upon electrically driven vehicles. On vehicles driven by petroleum, gas, air, etc., a general duty of 30 or 18 francs per 100 kilos. or a minimum duty of 20 or 12 francs per 100 kilos. is imposed.

A special tariff must be paid on imported electric accumulators, whether or not they be intended for automobile use. The general duty on a battery is 21 francs per 100 kilos., the minimum duty 16 francs, 50 centimes per 100 kilos. The importer of an electromobile is therefore compelled to pay two taxes, one on the carriage proper and driving mechanism (Class A), and another on the accumulators.

Motocycles, like ordinary bicycles, are taxed on their total weight (motor included). The general tariff is 250 francs per 100 kilos. and the minimum tariff 220 francs per 100 kilos.

Upon automobile parts imported into France, the same duty

is levied as upon the carriages themselves. When, however, the body and upholstered seat constitute 50 per cent., the body alone 40 per cent., and the wheels 28 per cent. of the total weight, the maximum tariff of 150 francs per 100 kilos. can be imposed; only, however, if the weights of the parts mentioned be respectively less than 62½ kilos., 50 kilos., and 35 kilos. When the wheels alone, tired or untired, are imported, the maximum tariff is imposed in all cases.

Besides the payment of a tax the importer must conform with certain requirements exacted by the Government. He must describe his vehicle in detail, state its weight, the country in which it was made and from which it was imported; give his name and address, and the distinctive marks of his vehicle, if there be any. Travelers temporarily residing in France, who import foreign motor carriages, must deposit the total amount of the tax levied upon their vehicles. This duty is returned when the traveler takes his automobile out of the country. In order to tour in other countries with his imported motor carriage the automobilist must obey certain formalities.

Germany, though she has ever jealously watched the development of American trade and endeavored in every way to protect her own industries, has been somewhat lax in restricting the importation of automobiles. The intention of the Electric Vehicle Company to establish a German branch house has raised a storm of protest among German manufacturers. For it is feared, and not without reason, that the keen competition of the American bicycle-maker may find a parallel in the wholesale importation of American motor-carriages. The German press has already taken up the cudgels in defence of the home industry and has vigorously advocated the erection of barriers which will effectually prevent the swarming-in of foreign automobiles.

In the existing German tariff schedule no special provision has been made for automobiles. Motor vehicles are therefore admitted as ordinary carriages upon the payment of the schedule duty of 150 marks for each vehicle. No restrictions are placed either upon size or weight.

On all imported articles the tariff rate is governed by the polish or outward finish of an article, in which respect the German system is unique. The distinction between rough and finished work often leads to curious decisions. Thus the rating of a fire-proof safe is changed, if, as in most American safes, one of the panels be decorated with a picture. The duty on carriages is increased when any portion of the vehicle is uphol-

Automobile Tariffs

stered. But since American manufacturers exercise sufficient care that their carriages be not unduly upholstered, the tariff usually remains at 150 marks per vehicle.

The German press has suggested the imposition of the usual machinery duty on automobiles—a duty based upon the material of which the machine is principally composed, modified, as already explained, by the polish, painting, finish, and outward decoration. These conditions, while theoretically simple, are in practice often exceedingly complex, so that it is frequently impossible, even for an expert German importer, to state precisely what will be the duty on a new article until a sample has been imported and subjected to actual classification and appraisal by the customs officials. An automobile is, however, essentially a vehicle, and the custom house will probably continue to impose the present imperial schedule rate until the tariff will have been modified in compliance with the demands of German manufacturers.

Thanks to the zeal, energy, knowledge, and thoroughness of our manufacturers, the American automobile industry has little to fear from European competition on our own soil. Although there be no need of a protective tariff, nevertheless foreign motor carriages are not admitted duty free. True it is that there is no special provision in the Dingley tariff schedule for automobiles, but under Section 193 " articles or wares not specially provided for * * * composed wholly or in part of iron, steel, lead, copper, nickel, pewter, zinc, gold, silver, platinum, aluminium, or other metal, and whether partly or wholly manufactured " are subjected to an *ad valorem* duty of 45 per cent. The Section is so broad in its scope that automobiles can readily be included. The Collectors of our ports accordingly levy a duty of 45 per cent. on all imported motor carriages, for the present.

At a time when the commercial struggle of the nations is fiercer than ever, when each country is trying to secure the largest possible share of the world's trade, the various governments are perhaps justified in establishing protective tariffs. But the best protection, as we have seen in our bicycle and machine tool industries, consists, not in the imposition of restrictive taxes and the enforcement of stern tariff laws, but in the production of wares which, by reason of their excellent workmanship, coupled with their greater cheapness, so completely out-class similar products of European manufacture that foreign competition is not to be feared.

The "Wartburg" Motocycle

THAT the Parisian types of motocycles have had a certain effect upon similar German automobiles goes without saying. The Werner system, perhaps more than any other, has influenced German makers, largely because it seems best adapted to the motocycle, because it does not require any change in the construction of the wheel, increases the weight but slightly, and enables the rider to propel the machine either by means of the pedals or by means of the motor.

The Werner system has been applied in the "Wartburg" motocycle with noteworthy success. The motor is mounted in front of the steering-head, and thus all the merits of the front-driven, front-steered automobile are obtained, the motocycle is rendered usually stable and is excellently balanced. The motor has several changes of speed, develops about $\frac{3}{4}$ H. P. and makes 1,300 revolutions per minute. By means of a belt and pulley, which constitutes one of the most silent and efficient transmission gears, the power developed by the motor is used in driving the front wheel.

The carbureter is secured to the upper brace of the bicycle and is composed of two parts—the carbureter proper and the benzine tank, communicating with each other by means of an aperture in the bottom of the tank. The tank has a capacity of one gallon—a quantity of benzine which will enable the cycle to run 75 miles. The proportions of the mixture of gas and air are regulated by means of a lever on the handle-bar. Through the medium of this lever alone, the motor can be started and stopped, and the speed changed at will. The explosive mixture is electrically ignited by a spark-coil energized by a small accumulator battery. When the motor is in operation the pedals can be held stationary. If it be desired to stop the motor the electric current is cut off and the lever so turned that the exhaust valve is closed, with the result that no gas is drawn into the cylinder. In order to drive the cycle at its highest speed the lever previously mentioned is operated to open the air valve as far as possible, and the gas-valve is regulated in accordance with the noise of the exhaust. The lubricator is mounted in front of the motor and automatically supplies oil to the moving parts. With all its accessories the motocycle weighs but 66 pounds.

A Typical French Voiturette

THE light, handsome carriage which forms the subject of the above illustration, is a type of *voiturette,* which has been successfully exhibited at the expositions of Paris, Berlin and Leipsic. As our perspective and plan views indicate, the vehicle is a small quadricycle, weighing about 450 pounds empty, and measuring 6 feet 7 inches in length and 4 feet 5 inches in maximum width; the distance between the wheels is 4 feet.

The front wheels are 25 inches in diameter, the rear wheels 29 inches. Each wheel is provided with steel tangent spokes and with a pneumatic tire $2\frac{1}{2}$ inches in diameter. The vehicle is steered by means of a divided axle, controlled by a tiller. The frame is made of cold-drawn steel and is firmly braced to secure the strength so necessary in automobiles. The frame is supported on the wheels by springs, the empty spaces being filled in by panels which constitute the carriage body and provide accommodation for two persons.

The vehicle is driven by a $2\frac{1}{4}$ H. P. motor of the De Dion-Bouton type, mounted in the rear portion of the carriage at *A.* Cooling of the cylinder is effected by means of a water-jacket, which takes the place of the usual head. The water is contained in a reservoir which has a capacity of 3.8 gallons and which is mounted above the motor. The thermo-siphon principle of circulation is used. The water evaporates at the rate of four to five pints per hour.

Power is transmitted to the rear wheels by two bevel-gears $B\,C$, controlling an intermediate shaft D, which drives another intermediate shaft H, by means of the gearing $E\,E'$ or $G\,G'$. The shaft H is connected with the shaft K, carrying the differential gear I, by the bevel-gears $L\,L'$. At the ends of the shaft K are two sprocket-wheels connected by chains $N\,N'$ with the sprocket-wheels $O\,O'$ secured respectively to the wheels $P\,P'$.

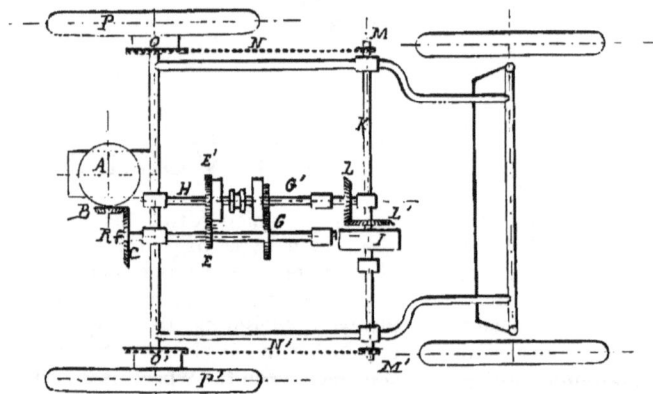

Plan View

The group of gears $E\,E'$ and $G\,G'$ provide two changes of speed. The gear-wheels E' and G' are loosely mounted on the shaft H, but can be thrown into gear by a clutch. By means of a lever controlling a sleeve sliding on the shaft H, either of the two speeds can be obtained. The special form of clutch used enables the *chauffeur* to pass from one speed to the other, silently and without any unnecessary shock.

The intermediate or higher speeds can be obtained by varying the proportions of the explosive mixture and by properly controlling the ignition. The motor is started by means of a crank connected with R, at the end of the shaft D. The highest speed which can be attained is 17 miles per hour. Grades can be ascended with the smallest speed-gear at a rate equal to one-half the maximum speed.

The Draulette Electromobile

BETWEEN the motor and the driving-wheels of most electromobiles an intermediate shaft is located which carries the differential. If this intermediate shaft could be dispensed with, the transmission gear would gain not only in simplicity but also in mechanical efficiency. Automobile-makers have, therefore, either connected the differential directly with the motor, or they have suppressed it altogether by employing

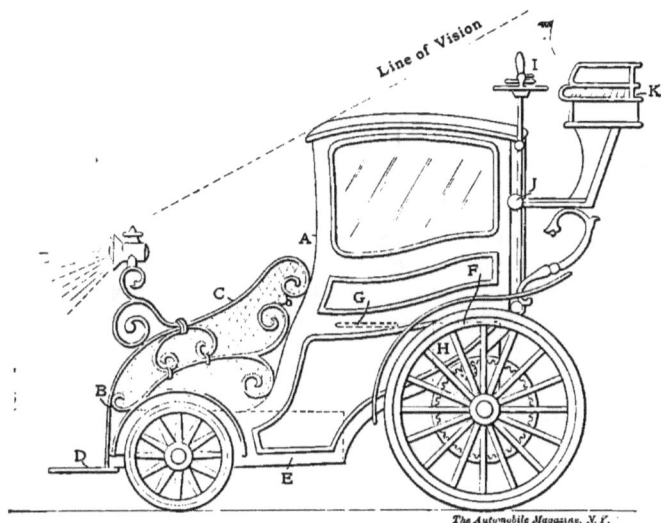

Fig. 1. Side Elevation of the Draulette Cab

two electric motors, one geared with each wheel. In the first method, a two-piece axle is employed which considerably impairs the strength and rigidity of the frame. Moreover, larger parts are required in the differential, since the gears must of necessity turn more slowly. The arrangement has been abandoned owing to its many disadvantages. The second construction is now in general use on electromobiles, but lacks simplicity and is both costly and heavy.

An improved arrangement of the driving mechanism has recently been patented by M. Draulette, in which the inconven-

The Automobile Magazine

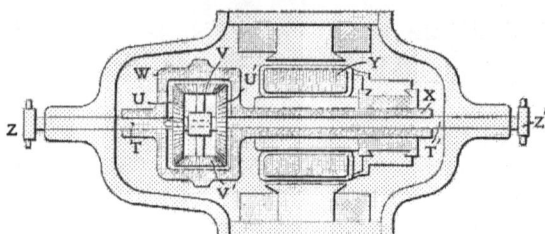

The Automobile Magazine, N.Y.

Fig. 4.   Section through Differential and Motor

iences of the first method and the complexity of the second have
been very ingeniously overcome.

The improvement in question is sectionally shown in Fig. 4,
in which T and T' are the two shafts of the differential upon
which the bevel-gears U and U' are secured, meshing with the
bevel-gears V and V', loosely mounted within the differential casing W, formed with an extension X, constituting a bearing for the
shaft T'.   Upon the bearing, the armature Y and the collector of
an electric motor are secured.   If it be so desired two collectors
can be used, one on the right and another on the left, each connected with the armature.   The arrangement is evidently applicable to all kinds of electric motors whatever may be the character
of the current by which they are driven.   The armature instead

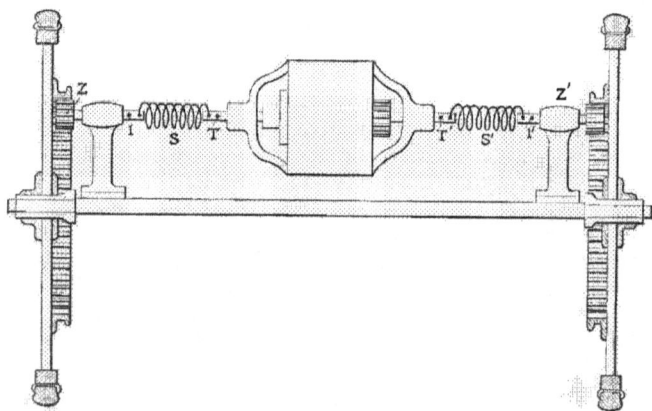

The Automobile Magazine, N.Y.

Fig. 5.   Flexible Connection between Motor-shaft and Wheels

The Draulette Electromobile

of being secured on the bearing X, can be mounted directly on the casing W of the differential.

The advantages of this improved construction over the methods already mentioned are its inexpensiveness, lightness, simplicity, and efficiency.

The driving-wheels can be directly actuated by pinions Z and Z' on the ends of the shaft T and T' of the motor (Fig. 4),

Fig. 2.   The Draulette Cab in Perspective

or by means of pinions Z and Z' (Fig. 5) on the ends of the shafts $L\ L'$, connected by flexible joints S and S' with the shafts T and T', or by a chain gear. In the second method (Fig. 5), in which the pinions Z and Z' are secured on the shafts $L\ L'$. mounted in supports bolted to the rear axle, the shocks due to obstructions and inequalities in the road are not trasmitted to the motor.

The Automobile Magazine

In order to facilitate the starting of the carriage and to avoid variations in the speed of the motor as well as strains on the transmission gear, the shaft carrying the bevel gears V and V' is connected with the casing W of the differential, not rigidly as in the usual construction, but elastically by means of springs R coiled around supports A A' secured to the casing W.

In most electromobiles, when the carriage is stopped by means of the brake-pedal, the current is cut off from the motor; but the switch remains in starting position and must be returned to zero by hand. If the driver forget to set the switch back, the carriage, when the circuit is completed, will start off at the speed which it had before cutting off the current, with results that are often disastrous to the motor. In the Draulette carriage the circuit-breaker is entirely discarded, and the mechanical brake-pedal and

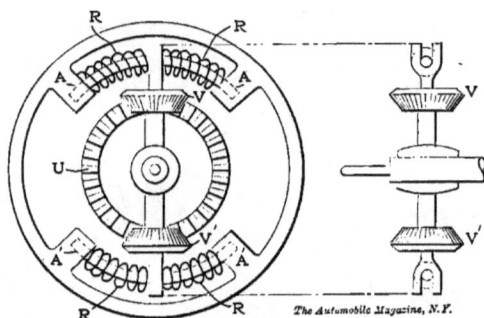

Fig. 6. Elastic Connection between the Differential Gearing and Its Casing

switch so connected that when the brakes are applied the switch is automatically returned to zero and the electric brakes operated. If it be so desired the electrical brakes can be actuated independently.

In Fig. 7, 1 is the switch-cylinder, turning with the shaft 2. The switch is operated by a wheel or hand lever. The pedal 3 is loosely mounted on the shaft 2 and actuates the mechanical brake by means of the link 5. The switch carries a lug 4 which raises the pedal 3 when, in order to increase the speed of the carriage, the cylinder is turned in the direction of the arrow. If pressure be applied to the pedal, the switch is turned in the opposite direction, set back to zero, and then moved in a position to operate the electrical brakes. It is evident that, when the switch is independently turned to apply the electrical brakes, the pedal remains stationary.

594

The Draulette Electromobile

Current is derived from an accumulator battery E (Fig. 1) inclosed in a special casing which can be slid in and out of the carriage through the front A, or lifted out by raising the removable platform B. The low position of the battery contributes much to the stability of the vehicle.

The steering-gear (Fig. 3) is of the irreversible type and is mounted upon a divided axle. The forward axle carries two lugs L and M, serving as bearings for a screw-shaft N, turned by a sprocket o and chain p. A nut Q is carried by the screw shaft

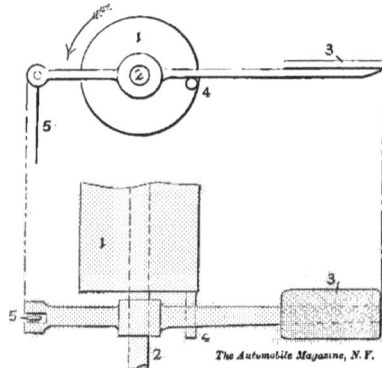

Fig. 7.   The Brake-Pedal and the Switch

and is connected with a link R and pivoted at Q' to one of two levers connected by a rod $S\ S'$. By turning the screw-shaft, the nut in moving will cause the levers to swing the steering-wheels in the desired direction.

From Figs. 1 and 2 it will be observed that passengers enter not from the side but from the front A, by means of the step D and the platform B, flanked by side members C. The Draulette cab differs from other electromobiles, not only in the nature of the driving mechanism, but also in general appearance. The construction possesses advantages too obvious to be dwelt upon.

AN AUTOMOBILE BRONCHO

In Texas a wealthy stockman will use an automobile for making inspection trips around the wire fence of his ranch. The country being level and free from brush and other obstructions, it is thought that the automobile will prove practical.

Lucas-Villain Explosion Engines

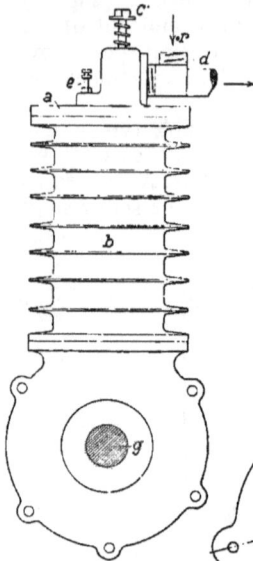

Fig. 1.

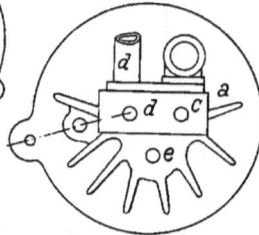

Fig. 2.

THE illustrations presented herewith represent a new type of explosion engine devised by Lucas and Villain.

The arrangement of valves and igniter directly on the cylinder-head necessitates but a single joint in the cylinder. The admission-valve c opens on the intake stroke of the piston and is reseated by means of the spring c'. The exhaust-valve is operated by a pinion f, secured to the motor-shaft g and engaged by a gear h, the sizes of pinion and gear being in the ratio of 1 to 5. The gear is provided with a cam i, with the peripheral surface of which a friction-roller j is always in contact. The roller is mounted in a cap k on a rod l, operatively connected with a lever m fulcrumed on a support n, to actuate the exhaust-valve stem. The exhaust-valve, like the admission-valve, is reseated by means of a coiled spring o; and a similar spring forces the roller j constantly into engagement with the cam i.

The explosive mixture of air and gas is led from the carbureter by means of the pipe r. The piston is provided with the usual rod t, connected by a crank u, with the shaft g.

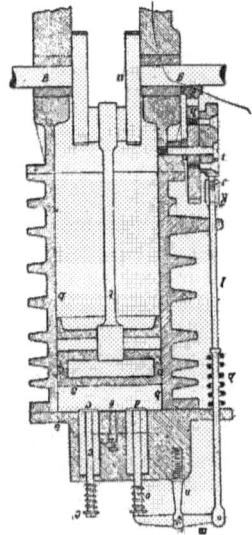

Fig. 3.

Accumulators

By Prof. Félicien Michotte

L ET us take two plates of lead and plunge them into water containing sulphuric acid; then let us connect one of them with the positive pole and the other with the negative pole of a source of electricity; and, finally, let us set the latter in operation. A current will be established through the liquid and the plates, and, under the influence thereof, there will be produced in the interior of the liquid a series of chemical phenomena, the result of which will be a chemical transformation of the lead of the plates; and such transformation will have absorbed the current sent into the latter.

If we suppress the source of electricity and connect the two plates, there will be produced in the wire that unites them an electric current that may be collected; and during such disengagement the plates will resume the chemical state that they possessed before receiving the action of the electric source.

The tangible result is therefore this: An electric current has been sent into the plates and has, as it were, *accumulated* therein, and, at a given moment, has disengaged itself therefrom. It is for this reason that such an apparatus is called an " accumulator."

Practical Accumulator.—In practice, things take place less simply. In order that an accumulator may operate, the lead must be oxidized and converted into a state of oxide, in order that there may be had what is called the formation of " active material."

The plates are therefore previously oxidized by an operation that constitutes the " formation " of the accumulator.

On another hand, the quantity of electricity accumulated in the two plates is feeble, and depends upon the quantity of active material that the accumulator contains. Now, since this material must be of relatively feeble thickness and be arranged as a superficies of the " working surface," we are obliged, in order to obtain a certain quantity of electricity, to employ plates of a number that varies according to the mode of construction of the accumulator and to what is required of it.

The assembling of several plates in the same box constitutes what is called an " element," and the quantity of electricity accumulated therein is called the " capacity " of the element. In order to compare accumulators with each other, we generally refer to them by weight, and speak of an accumulator of 6, 8, 10 or 12 kilogrammes, that is to say, of an accumulator that contains 6, 8, 10 or 12 kilogrammes of plates.

Boxes.—The boxes that contain the elements are made of ebonite or celluloid. Those made of the last-named material are the lighter and cheaper, but should not be used in automobilism, on account of their easy combustibility.

Voltage.—When an electric current is sent into an accumulator (an operation which has received the name of " charging "), it is found from the voltage that the electro-motive force of the apparatus gradually rises; but that, starting from the moment at which 2.4 or 2.45 volts are obtained, such force no longer increases. We say at this moment that the accumulator is *charged,* since, whatever be the current that is sent into it, the electro-motive force will not increase.

The maximum electro-motive force of an accumulator is therefore 2.45 volts; and whenever it is desired to obtain a greater voltage, several accumulators will have to be combined with one another.

Battery.—When several accumulators are combined, they constitute what is called a " battery." Each battery consists of a number of elements proportional to the voltage that it is desired to obtain—each element being reckoned at 2 volts. A battery to give 80 volts will require 40 elements; but, in practice, 2, or even 4, more are added, making 42 or 44 in all.

The weight of each element is proportional to the *quantity* of electricity that it is desired to obtain—such quantity being measured in ampere-hours.

We say that a battery gives 20 ampere-hours, and, if we wish to specify it more completely, we say that a battery of 6 kilogrammes gives 20 ampere-hours, this meaning that a battery having elements each containing 6 kilogrammes of plates, will, in discharging itself, give a quantity of electricity measured by 20 ampere-hours.

Tension: Quantity.—As well known, electric batteries are united in quantity or in tension according as it is desired to have a quantity of electricity or a sufficient electro-motive force. The same thing is done with accumulators.

The quantity that each element is capable of giving is increased by increasing its number of plates, or else by uniting several elements in quantity, that is to say, by uniting all the negative plates on the one hand and all the positive ones on the other. Then, in order to obtain electro-motive force, the elements are united in tension, that is to say, the positive pole of each of them is united with the negative pole of the following.

Rate of Charging.—When we are charging an accumulator we should not send into it any quantity of electricity whatsoever, but should limit the quantity, without which we might destroy the accumulator. The charging is generally done with a current of one ampere per kilogramme of plate. This is called the normal rate of charging. An accumulator of 6 kilogrammes will be charged by a current giving 6 amperes. The number of amperes per kilogramme of plates is the " rate of charging" of an accumulator.

Rate of Discharging.—Inversely, at the time of the discharging, the current produced should not exceed a normal rate of discharge, that is to say, we must give the accumulator only a limited number of amperes per kilogramme of plates, or else we shall destroy it. This rate generally varies between 1 and 1.5 ampere, for pasted oxide accumulators.

It must be remarked that, at the discharge, an accumulator does not render up all the electricity that it has received. When any element whatever is discharged, the voltage that was 2.4 at the beginning falls to 2.05 volts, remains at that figure for quite a long time, and then falls to 1.9 volt, and finally to 1.8. At this moment, the discharge must be arrested, since otherwise the accumulator would empty itself entirely and be destroyed. It is even prudent to stop at 1.9 volt.

1st Remark.—An accumulator allows its current to flow pretty nearly at the rate that is required of it, acting in the same way as would a reservoir of water provided with a cock, which would discharge a greater quantity of water in proportion as it was wider open; with the difference, however, that as the current is produced in measure as it is used, the accumulator will be destroyed if too large an amount be required of it. A certain rate of discharge per kilogramme must, therefore, not be exceeded. Such discharge depends upon the accumulator and its interior construction, and generally varies in practice from 1 to 1.5 ampere in oxide accumulators, and from 1 to 3 amperes per kilogramme in lead accumulators.

2d Remark.—In consequence of the production of the current by chemical phenomena, it follows that an accumulator renders so much more electricity in proportion as its discharge takes place more slowly. Consequently, the quantity of electricity rendered by an accumulator, that is to say, the *capacity* of the latter, varies with the rate of its discharge. The more slowly it is discharged, the more electricity it will produce. Thus, an accumulator of 12 kilogrammes will give 120 ampere-hours of capacity in discharging itself in one hour, and this will give a rate of 10 amperes per kilogramme; while in two hours, corresponding to a rate of 5 amperes, it will give a capacity of 125 ampere-hours. This accumulator will give in

1 hour,	at the rate of	10	amperes	capacity,	115	ampere hours.
2 hours,	" " " "	5	"	"	125	" "
3	" " " " "	4	"	"	135	" "
4	" " " " "	3	"	"	145	" "
5	" " " " "	2.5	"	"	155	" "

An account should therefore be taken of the rate of discharge, and, consequently, of the time that it has taken an accumulator to discharge itself, in order to estimate it. We ought to say that such an accumulator of 10 kilogrammes has a capacity of 145 ampere-hours at a rate of discharge of 3 amperes.

Measurement of Capacity.—The capacity of an element is estimated in ampere-hours, and the capacity of a battery, (1) in ampere-hours; (2) in watt-hours; and (3) in horse-hours.

1. In Amperes.—We say that a battery has a capacity of 135 ampere-hours (the word "hours" is often understood) at a rate of discharge of 25 or 30 amperes. If nothing is said about the rate, the normal one is understood. The normal capacity of the "Planté" is from 11 to 12 amperes to the kilogramme.

2. In Watt-hours.—As watts are the product of amperes by volts, the above-named battery would give:

135 amperes x 80 volts = 10,800 watts.

The capacity of the Planté accumulator is, per net weight of plates, from 16 to 24 watt-hours to the kilogramme. Per gross weight of the battery (that is to say, boxes and liquid included), it is from 11 to 16 watt-hours.

Accumulators

3. In Horse-hours.—The capacity of a battery is expressed also in horse-hours. This is easily done, since watts are the product of amperes by volts, and a horse is represented by 736 watts.

It suffices to multiply amperes by volts and divide the product by 736. A battery of 135 amperes and 80 volts will give:

$$\frac{135 \times 80}{736} = 14.67 \text{ horse-hours.}$$

As the amperes are ampere-hours, the product represents horse-hours, that is to say, the work produced by 14 horses during one hour, or by one horse during 14 hours.

Remarks.—(1) As the capacity varies with the output, the number of horse-hours varies in the same way. (2) A horse-hour is produced by a Planté with a weight of plates of from 30 to 40 kilogrammes, according to the rate of discharge, and per gross weight of battery (with box and liquid) of from 45 to 65 kilogrammes. (3) We often hear a person say: " My accumulator makes 100, 120, or 150 kilometres." This is a false expression that seems to say something and means nothing. It simply proves that those who use it know nothing themselves about accumulators, or else are endeavoring to deceive the public. In fact, the capacity necessary for making a determinate trip depends upon two essential things : the capacity of the accumulator, which is tantamount to its weight, and to the load that the motor which it actuates is to pull. An accumulator that will permit a carriage seating two or three persons to cover 60 kilometres, will be incapable of propelling an 8-passenger omnibus for 6 kilometres, while it will allow a tricycle to make a trip of 100 kilometres. The motor employed, its rendering, and the load drawn or propelled (factors often lost sight of) must also be taken into account.

Rapidly Charged or Discharged Accumulators.—The time required for charging or discharging an accumulator without injury thereto has caused manufacturers to employ the term " rapidly charged accumulator "—a denomination that has merely a relative signification and indicates nothing precise; for such or such an accumulator may be a rapidly charged one if it be compared with some type that is more slowly charged.

Rendering.—An accumulator receives a certain quantity of electricity at a charge, and, in emptying itself, renders a less amount of it. The ratio of these two quantities is the " rendering." An accumulator at a normal rate of discharge renders 80 per cent. of what it has received. Its rendering diminishes with the rapidity of its discharge.

The Automobile Magazine

Short-Circuit.—If two accumulator plates are in contact, there is produced what is called a " short-circuit," and the accumulator loses its electricity. It discharges itself upon itself and produces no current, and at the end of a certain length of time, it can no longer even be charged.

Dead Accumulators.—After a short-circuit has acted for a certain length of time, and the accumulator cannot be recharged, it is said that the latter is " dead."

Sulphatation.—If a battery remains for a long time out of service or has been too thoroughly discharged, the active material of the negative plates is converted into sulphate, and it is said that " sulphatation " has occurred. The accumulator is then out of service and cannot operate again until after its plates have been desulphated.

SYSTEMS OF ACCUMULATORS.

The systems of accumulators present a great variety as regards form and details of construction, but may all be referred to two different types—the pasted oxide and the lead.

Oxide Accumulators.—These apparatus, called also " pastille " accumulators, owe their name to their plates, which, instead of being " formed " with oxidized lead, consist of sheets of lead into which are punched holes for the reception of the active material, which is formed of oxides of lead.

Lead Accumulators.—When the plates consist of pure lead, the apparatus is called a lead or a Planté accumulator, or. simply a Planté, after the name of its inventor.

Accumulators to be Employed.—The only accumulators to be employed in automobilism are those of the Planté kind, since this type of apparatus is the only one that really has a long life and that permits at the same time of a rapid discharge without becoming deteriorated.' The pasted oxide accumulator has the drawback of being short lived. The oxides, whatever be their mode of fixation, swell up or become detached (if the current required be at a low rate of discharge), and falling, produce short-circuits. The result is that not only does the accumulator lose its capacity and discharge itself upon itself, but is very rapidly put out of service. Such destruction, moreover, is hastened by the incessant vibrations of the carriage; and there is not an oxide accumulator that can withstand these.

The lead accumulator has the advantage of being capable of being charged and discharged much more rapidly than the oxide accumulator, without being impaired.

Accumulators

An oxide accumulator charged at one ampere per kilogramme of plates will require for elements of 12 kilogrammes, with a capacity of 135 amperes,

$$\frac{135}{12} = 11 \text{ hours,}$$

while a lead accumulator will require:

12 kilogrammes x 3 amperes = 36 amperes.

$$\frac{135}{36} = 4 \text{ hours,}$$

and this rate of discharge of 3 amperes may be increased.

At its discharge, an oxide accumulator supports 2 amperes with difficulty, while with a lead one the discharge may reach 5 amperes without any inconvenience.

The lead accumulator has another advantage and that is that its capacity increases with its use, since it is "formed" a little more day by day. The contrary is the case in the oxide accumulator, the capacity of which is maximum at the beginning, but has a tendency to diminish with use. Certain parts of the carriage frame, entering into action, disturb its operation, and some of the active material becomes detached.

So, a test of a lead accumulator cannot be profitably made until after it has been in service for quite a long time, say for six months.

A test of an accumulator designed for automobilism can be made only upon a carriage, since, in a laboratory. although it is possible to re-

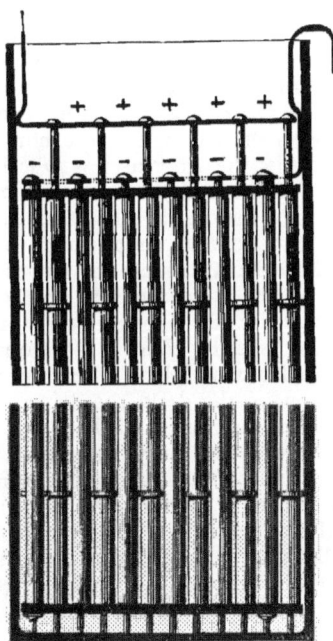

Phenix Accumulator

produce the vibrations approximately, it is impossible, whatever be the precautions taken, to bring about the fits and starts in the discharge that occur during its operation upon a carriage.

An Automobile for Land and Water

An invention which will enable a horseless carriage to travel with equal ease and facility on water as on land is the interesting creation that will be the next step forward in the perfecting of the automobile conveyance now sweeping so swiftly into popular favor. The perfection of locomotion will have been reached when this system of automobilism is in use, for the driver of the horseless carriage can send his conveyance speeding over the country or leave terra firma and take himself and his party for a trip on lake or river without the need of any one alighting, and with the same propelling power in use for both land and water.

Two long and hollow metallic cylinders placed parallel to each other give the buoyant properties to the floating vehicle. These cylinders are made of aluminum, as being the lightest metal. Their size depends upon the weight of the party and of the automobile. The heavier the freight the larger the displacement. The driver of the automobile sits in the front of the cylindrical float, on the same seat that is his perch while driving on a roadway. When he has driven the horseless carriage on to the float he fastens the now useless wheels to the deck by means of suitable clutches and throwing the machinery out of gear with the wheels of the carriage he attaches it to the gear of the float. This change consists simply of throwing off a chain from a sprocket wheel that transmits the power to the wheels of the horseless carriage and attaching it to another sprocket wheel that causes the propeller shaft between the two cylinders to revolve at any desired speed. When seated on the front of the horseless carriage in his usual place, after the wheels have been fastened into position on the deck, the driver, or, as he should now be called, the pilot, has close to his hand two upright handles, one to start and stop the machinery, the other to manipulate the rudder, the chain of which runs between the cylinders to the rear of the float. The rudder is manipulated by connection with the steering gear of the automobile. The starting and stopping is done in precisely the same way and with the same machinery as when the automobile is ashore. The only difference is that the machinery in this case moves a propeller instead of four wheels.

Mechanical Propulsion and Traction

By Prof. G. Forestier

Fourth Paper

Previous to 1895 automobilism occupied the attention of a very small number of informed and intelligent experts. After the triumph of Levassor everyone wanted to become a driver. The gasoline carriage, in fact, satisfied the demand that existed for a light and rapid private vehicle.

At present the carriage provided with an electric motor seems destined, from this point of view, to supplant the gasoline carriage in the favor of those interested; but, incapable for the moment of moving beyond a relatively narrow radius around electric works, it must abandon to its rival the privilege of making lengthy excursions.

IV.—RESISTANCES TO BE OVERCOME.—Since the very conditions under which the motive power must act in order to set a vehicle in motion involves a bad utilization of it, the first thing to be done is evidently to endeavor to reduce to a minimum the resistances to be overcome. To this effect, it is necessary to begin by well determining not only their nature, but also their value, so as to see whether it is possible to diminish the latter.

We shall therefore briefly recall the fact that the total resistant work results from the following factors: (1) The sliding friction of the journals in their boxes; (2) the force of translation upon the road; (3) the loss of live force resulting from the jarring of the different parts of the vehicle, produced by the roadway or by the motions of the motor, and transmitted to the ground or air; and (4) the pressure of the air.

The two first factors are proportional to the weight of the vehicle, and their numerical coefficients are sensibly constant within the limits of speed of the carriages considered.

The third is a function of the speed, of the flexibility of the springs, and of the elasticity of certain accessories, such as the dash-board, steps, lamps, etc.

The fourth is proportional to the square of the speed,* and to the surface of the projection of all the parts of the carriage upon

---

* This is conformable to usage. Nevertheless, it seems that for some time past certain railroad engineers have been inclined to adopt formulas in which the pressure of the air is simply proportional to the velocity beyond a certain value of the latter.

The Automobile Magazine

a plane at right angles with the direction of the motion; but its numerical coefficient varies with the form and the more or less partial overlap of the various parts of the vehicle.

This much said, we shall proceed to point out in succession how it is possible to express the values of the four above-mentioned factors for an automobile carriage.

1.—*Sliding Friction of the Journals in the Axle-boxes.*—This is given by the formula—

$$Tf = \varphi \frac{F}{D} p$$

in which F is the mean diameter of the journal; D the external diameter of the wheel; p the load supported by the journal in tons; and φ a coefficient expressed in pounds, and variable with the method of lubricating the journal, and with the nature of the rubbing surfaces.

2.—*The Force of Translation.*—If this consisted of the rolling friction alone it would be given by the classic formula—

$$Tr = (f \pm i) \frac{P}{D} \quad \dagger$$

in which P is the total weight of the vehicle in tons; i the declivity expressed in thousandths; f a coefficient expressed in pounds, and variable with the nature of the road and the width of the wheels.

This factor is more complicated. It comprises, in addition:

(*a*) The stress to be exerted in order to surmount the asperities of the road, and which has for value:

$$P \sqrt{\frac{h}{D}}$$

in which h is the height of the asperity. The work per corresponding foot will therefore be:

$$Tr' = \frac{n P}{l} \sqrt{\frac{h}{D}}$$

in which n is the number of the asperities of a height h that the road presents upon a length l.

(*b*) The stress due to depressions:

The surface of the road is not regular, but presents a series of concave and convex undulations. Dupuit has demonstrated

---

† Experimenters are not in accord as regards the influence of the radius of the wheel upon the tangential stress that opposes itself to the rolling thereof. According to Dupuit, such stress is inversely proportional to the square foot of the radius. On the contrary, Morin claims that it is inversely proportional to the radius. Others, like Baron Mauri, think that, independent of the radius for a certain degree of polish of the surfaces in contact, the stress is conformable to a certain relative size of the asperities.

that the drawing, v', of a two-wheeled cart over the concave parts, and v'' over the convex parts, are given, respectively, by the relations:

$$v' = v\sqrt{\frac{1}{1-\dfrac{R}{R_1}}}, \qquad v'' = v\sqrt{\frac{1}{1+\dfrac{R}{R_1}}}$$

where v represents the traction over a level upon the same even road; R the radius of the wheel; and R_1 the radius of curvature of the undulation.

For an automobile carriage, three cases must be considered: (1) The driving and steering wheels are placed upon undulations of the same nature—concave or convex; (2) the driving wheels rise upon a convexity, while the steering ones descend into a concavity; (3) the driving wheels descend into a concavity, while the steering ones rise upon a convexity.

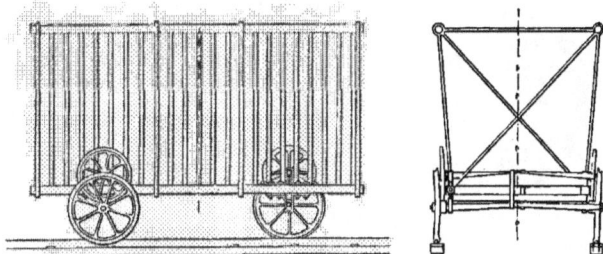

Fig. 16.   The Gerstner Arrangement

In the first case, the relations v' and v'' are applicable to the whole. In the second and third cases, since the load is not the same upon the driving and steering wheels, it will be necessary, in order to have the total draught that results from the depressions, to multiply v' and v'' by a coefficient proportional to the load before taking the difference of them.

(c) The stress that results from the alteration produced in the form of the subsoil and roadbed by the load of the wheels.

Since the time of Coriolis (1832), the following formula has been admitted:

$$Tr'' = \tfrac{3}{8}\sqrt[3]{\frac{12}{mb}\frac{P^4}{D^2}} \quad ,$$

in which m is a numerical coefficient variable with the compactness of the roadbed, and b is the width of the tires.

The Automobile Magazine

(*d*) The stress that results from curves.

In the traffic upon ordinary roads, as well as upon rails, the vehicle experiences an increase of resistance in sinuous passages.

In default of methodical experiments, it is impossible to give even an approximate value of the resistant work due to a curved motion as a function of the radius. All that we can say is that in some experiments made upon the total resistance with a Desdonit's dynamometric pendulum, it was found that there was a sudden increase as soon as the steering wheel was acted upon in order to change the line of travel.

Moreover, when the speed exceeds a certain limit, centrifugal force intervenes in an appreciable manner. The change of place laterally that it tends to produce is prevented by sliding friction

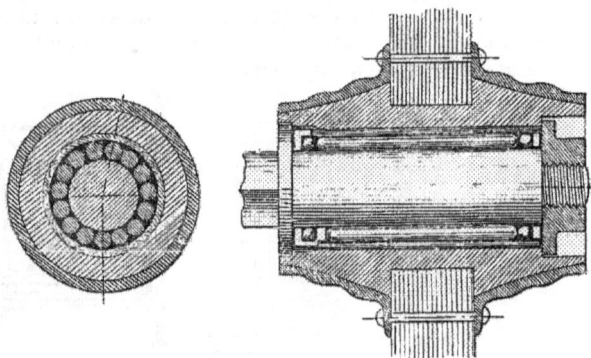

Fig. 17. Roller Bearings

or by an alteration in form of the surface of the road. If we examine a road upon which a vehicle has described a curve at a relatively high speed, we shall observe, provided the road be dusty or its materials be loose, that the groove made by the wheels presents on the external side of the trajectory a very pronounced ridge, a sure indication of the lateral reactions of the tire upon the roadbed.

3.—*Loss of Live Force Due to the Vibrations of the Vehicle.* —The loss of live force due to the vehicle's vibrations, which, produced by the jarring motions of the wheels upon the road, are transmitted and absorbed by the vibrations of the ground and air is a function of the speed, of the flexibility of the springs and of the elasticity of the other vibrating parts of the vehicle. The coefficient by which this factor is affected is such that it is very difficult to make an exact allowance for it in calculations.

Mechanical Propulsion and Traction

Two cases are to be examined according as the carriage is or is not provided with pneumatic tires. In the case of such tires, we can set aside the term function of the speed. If the tires are rigid, it will be well to increase the coefficient of the preceding term by a tenth.*

4.—*Pressure of the Air.*—For the pressure of the air (an elastic fluid) against a surface in motion, we admit the formula—

$$Ta = KSV^2,$$

in which S is the surface of the projection of all the parts of the vehicle upon a plane at right angles with the direction of the motion, that is to say, here, upon a vertical plane parallel with the axles; V^2 the square of the speed expressed in feet per second; and K a numerical coefficient variable within very wide limits with the form and degree of overlapping or the distance of the parts projected.

This formula seems to be theoretically true only for the pressure of water (an incompressible fluid).

For want of special experiments with automobile carriages, we may admit the coefficient 0.0288 for K.

5.—*Accelleration.*—If we wish to take into consideration the supplementary force to be required of the motor at the moment of starting and until the speed of the vehicle becomes constant, we must add to the preceding terms, the term—

$$\frac{P}{g}\frac{dV}{dt}.$$

which makes evident the advantage of a slow starting when the necessities of the traffic permit of it.

V.—Axle Journals.—From the moment that the resistant work due to the sliding of the journal upon the side of the axle-box has for expression—

$$Tf = \varphi\,\frac{F}{D}\,p,$$

it is necessary to try at once to diminish the relation $\frac{F}{D}$ and the coefficient φ.

The mean diameter, F, of the journal depends upon the quality of the metal used in it, and also upon the load. This quantity is therefore nearly definite for a vehicle of a given weight. On the contrary, we shall see further along that D (the diameter of the wheel) may vary considerably with the arrangements adopted for the vehicle. For the moment, we shall examine only the possible diminution of the coefficient φ.

---

* In the case of the bicycle, M. Bourlet, from a few experiments made in England, concludes that such increase should be 1/6.

The Automobile Magazine

In former times, two-wheeled carts, even when their axles were of iron, had journals that were conical and of a diameter notably less than that of the axle-box, an interval between the two being necessary for the reception of the lubricant with which the journal was surrounded. As the latter left an open space in the rear and front of the box, the lubricant had to be solid in order that it might not be lost.

At this epoch, the coefficient of friction of the journal upon its box was estimated at 220 pounds per ton.

In stage coaches, and still more so in private carriages, it has for a long time been regarded as necessary, in order to diminish φ, to substitute a liquid for a solid lubricant. In order to obtain such a desideratum, it has become necessary to place the journal in a space as perfectly closed as possible.

Such an arrangement is now employed even for the heaviest vehicles, as shown in Fig. 15 in our previous issue, which represents the patent journals of the trucks employed at the Say Refinery for the carriage of 22,000 pounds of sugar in bags. The coefficient of friction has thus been brought to 22 pounds per ton. This value corresponds to the case of the patent journal of a wheel revolving very regularly in a plane without lateral concussions.

In the case of wheels running upon a somewhat irregular pavement, we are led to think that the coefficient of the work absorbed by the friction of the journal upon its box is notably greater. We take as a basis a fact communicated to us by M. Jeantaud, the well-known manufacturer, who one day received from the manager of the Grands Magasins du Louvre an order to provide the house's small delivery hand-carts with patent journals, in hopes of diminishing the fatigue of the men who were obliged to draw them. Great was M. Jeanteaud's astonishment when the manager informed him that the delivery men had complained that they had experienced more fatigue than before. An experiment made by communicating a given speed to wheels provided with a patent journal and to others with an ordinary one, and then counting the number of revolutions that they made before stopping, demonstrated that the first were submitted to a friction ten times less than the second. Notwithstanding this, the delivery men continued to assert that their carts had become harder to draw.

Like M. Jeantaud, we can explain such a result only by the impossibility of relative displacements of the journal and the axle-box in cases of those lateral shocks that occur so frequently in street traffic.

Automobile Tires

By H. Falconnet

O NE of the most important problems in the construction of automobiles—a problem which concerns both the *chauffeur* and the manufacturer—is the provision of a suitable rubber tire. Up to the present time no attempt has been made to classify rubber tires, to describe them briefly, yet clearly, and to state their respective merits impartially. As a result, most automobile-owners merely utter the opinions which they find in the catalogues of tire-makers. Unfortunately, these catalogues are for the most part confined chiefly to a glowing descrip-

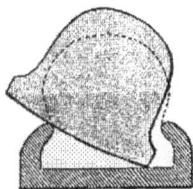

Fig. 1. Forcing the Tire in the Rim

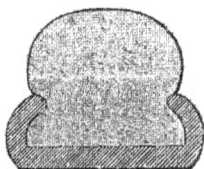

Fig. 2. Tire Fitted in an Inturned Round-flanged Rim

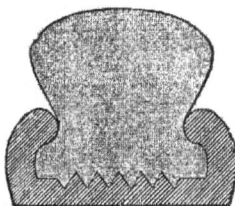

Fig. 3. Serrated Rim

tion of one particular type, and lead one to infer that all tires, with the single, noteworthy exception of the one under discussion, are absolutely worthless. Such flamboyant and extravagant descriptions are, of course, to be taken with a large pinch of salt; for there are many tires made both here and abroad which are very well adapted for motor-carriages. It is merely a question of selecting that system which is best suited for the purpose to which it is to be applied. In this spirit I have endeavored to bring together in the present article all those systems which have proven practicable, and to give an unprejudiced opinion of the advantages of each.

One of the simplest methods which has yet been devised for securing tires in place is the forcing of the rubber within the inwardly-turned flanges of the rim. The tire, as illustrated in Fig. 1, is held in position by the engagement of its recesses with the overhanging flanges, and is glued to the rim by means of an

Fig. 4.   Tire with Cross-piece   Fig. 5.   Tire with Single Lateral Support

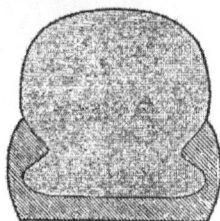

adhesive gutta-percha compound.   The rim-flanges can be made either straight or round, the latter being the more sightly.   The rim shown in Fig. 3 is longitudinally-serrated, for the purpose of preventing a displacement of the tire; but the serrations are not of much value.   In another method of holding the tire in place a steel cross-piece is employed (Fig. 4).   The rim shown in Fig. 5, formed with a lateral support, is an improvement on this type.

Tires secured by all these means are not very durable; they are soon destroyed by the sharp joints or the steel cross-ties.

By forming the rim with a double support the tire is more effectively secured and better able to withstand hard usage.   Figs. 7 and 8 respectively represent a hollow and a cellular tire, held in position by direct engagement with the rim.

Tires are also fastened to the wheel by vulcanization (Figs. 9 and 10).   An elastic connection of the tire with the rim is secured by employing four or five successive layers of gradually increasing hardness, the last of which is composed of ebonite. If the wheel be of wood, the tire must be shrunk on the rim, since the high temperature required in vulcanization would render it impracticable to follow the usual method.

Many manufacturers secure their tires to the wheel by mechanical means (Figs. 14 to 19).   In the system represented

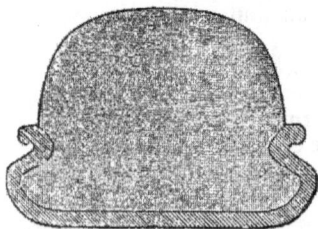

Fig. 6.   Steel Rim with Double Support

Fig. 7.   Cellular Tire

Automobile Tires

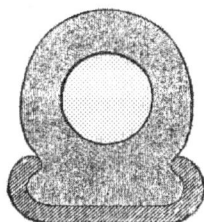

Fig. 8. Hollow Tire

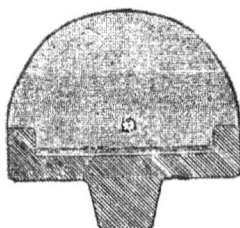

Fig. 9. Rim for Metal
Wheels

Fig. 10. Rim for
Wooden Wheels

in Fig. 14, for example, the tire is composed of a hard and an elastic portion, a pin being passed through the elastic portion so that its head will hold the hardened portion against the rim. In Fig. 15 a modification of this method is shown. The tire pictured in Fig. 16 is held in position by a tie composed of two curved metallic members, one of which rests on the wheel and the other of which serves as a protecting means for the rubber. Fig. 17 represents a tire secured to its rim by means of wires running longitudinally through the rubber. The same principle is shown in Figs. 18 and 19, the wire being superseded by an interior circular retaining member.

The tires which we have so far described have been either of the hollow or solid rubber types. We now come to pneumatic tires. In the month of June, 1896, a very exhaustive paper was read before the Society of Civil Engineers of France, from which paper I shall make excerpts of those portions in which the principles governing the construction of inflated tires is discussed.

The material employed comprises a rigid base of metal (or wood) on which the rubber is placed. The air-chamber is an

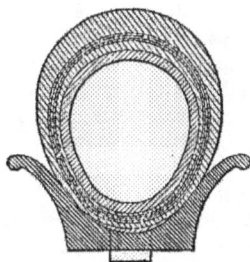

Fig. 11. Single Tube Tire

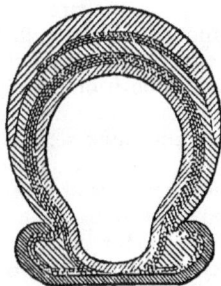

Fig. 12. Interlocking Pneumatic
Tire and Rim

endless tube of pure rubber. Canvas cannot be used; for all textile fabrics are permeable to air under pressure. The rubber tube, being unable to resist high pressure, is incased in an envelope composed of canvas, which imparts greater resistibility to the tube and likewise serves as a protecting jacket for the rubber. The canvas is coated with rubber to prevent its rotting, and is furthermore protected from wear by a thick outer covering of rubber. This exterior rubber covering is crescent-shaped in cross-section and is separated from the envelope proper, so that it can be readily removed and a new one substituted whenever it may be necessary.

The envelope is fastened to the rim by innumerable methods. As a general rule it is held in place by overhanging, inturned,

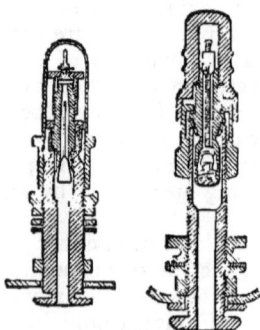

Fig. 13. Valves for Pneumatic Tires

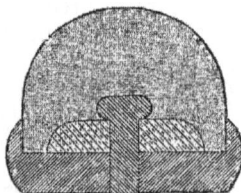

Fig 14. A Rubber Tire Pinned to the Rim

Fig. 15. Tire Secured by a Stud Embedded in the Rubber

locking flanges on the rim, which engage correspondingly-shaped portions on the tire. The tire is also pinned to the rim so that it cannot be dislodged. This method of fastening the envelope presents obvious advantages over those systems in which the tire is held in position merely by the air-pressure within the tube. If the pressure diminish, the envelope may become entangled in the spokes and cause serious accidents. Other means of securing the tire (by bolts and the like) have not been very successful.

Rubber tires are composed of Para gum, so prepared and disposed as to secure a maximum resistance at every point to all those strains which tend to destroy a tire. Evidently the forces which affect the central portion will not be the same as those to which the outer surface and the base are subjected. A substance may be capable of resisting friction, but may not have the resiliency

Automobile Tires

required; another substance may be capable of withstanding great pressure, but may be unable to resist tractive forces. It is natural, therefore, that so many inventors should have endeavored to devise some form of tire which would answer the multifarious requirements of the automobile. Some manufacturers have confined their efforts to the production of a perfectly seamless rubber tire; others have sought to lessen the inconveniences which attend the removal of the tire from the rim and its replacement.

Of all the tires which have been invented for the purpose of overcoming the obstacles mentioned, the compound tire seems to have given the best results. Experience has proven that it is admirably adapted to the automobile and that it is free from many objections to which the pneumatic tire is open.

The method of securing the compound tire to a steel rim is similar to that employed in pneumatic tires, laterally projecting

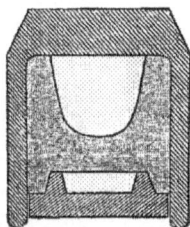

Fig. 16. Tire with Metal Shield

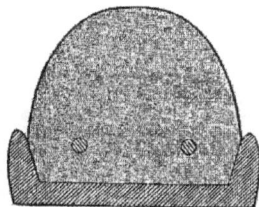

Fig. 17. Tire Secured by Longitudinal Wires

flanges and pins being used. In most systems the steel flanges of the rim have a tendency to cut the tire; in the compound system, on the other hand, the flanges serve as a double support for the rubber. Pins are chiefly employed for the purpose of preventing a displacement of the tire when suddenly wrenched.

When steel rims are used, care should be taken that there is no possibility of cutting the rubber. The form should be such that a permanent and an auxiliary support are provided, the former of which receives the tire under normal conditions and the latter of which serves to retain the tire when, by reason of a sudden wrench or other cause, it shows a tendency to spring from the rim. The rim flanges, moreover, should be sufficiently inturned to engage the recesses and laterally projecting portions of the tire. To prevent possible contact with the road, the rim should also be outwardly curved to form protecting wings of steel.

The firmness with which the compound tire can be held in position, its resiliency, the readiness with which it can be removed, the impossibility of its being cut by steel rims of proper shape, its freedom from joints and its ability to withstand long and hard usage render it particularly applicable to the automobile.

The considerations which should govern the automobilist in the selection of a tire are so many that it will be possible to mention only the most.

The first point to be considered is the type of motor used. If the motor comprise delicate mechanism, if it be incapable of withstanding shocks, then a tire of great elasticity should be selected.

Although the inherent disadvantages of the pneumatic tire are reduced to a minimum in the tricycle, they are often a source of trouble in light carriages. The weight of the vehicle should, therefore, receive some attention in the choice of a tire. In carriages weighing more than 600 pounds only seamless tires should

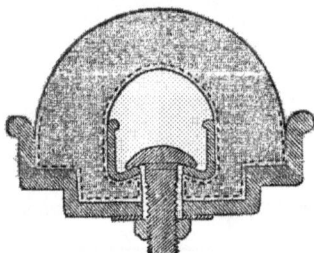

Fig. 18.   Hollow Tire with Interior Retaining Members   Fig. 19.   Solid Rubber Tire with Interior Retaining Members

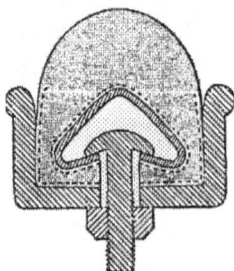

be used; and the rubber should be secured to the rim either by vulcanization or by some equally serviceable method.

A vehicle whose wheels are provided with pneumatic tires will hardly require springs. But if the carriage body be supported on the axle by springs, the difference in resiliency between the solid rubber and pneumatic tires is not so apparent, and is well-nigh inappreciable when the compound tire is employed.

Tires on vehicles which are used only in the limits of a city should be wide enough to prevent their binding in the tracks of street-car railways. When a carriage is used for touring and the automobilist does not desire to carry with him a repair kit, the pneumatic tire will not be of much use; for the best pneumatic tire is apt to be punctured, no matter how carefully it is made.

Automobile Tires

The statement holds good for delivery wagons, physicians' carriages, and for all vehicles in which absolute safety and certainty are required.

If the speed of a carriage be high, the centrifugal force will have a tendency to throw the tire from its rim; but engagement with the ground will tend to force it back into place. These two opposing forces acting on the tire destroy the rubber prematurely. For speeds of 15 to 18 miles per hour it is, hence, advisable to use tires composed of an endless ring of rubber without any transverse joint. The tires should be either vulcanized to the rim or held in place by mechanical means.

The advantages of vulcanizing a tire to the wheel are offset by certain disadvantages. If the wheel be of wood, the rim must be removed in order to substitute a new tire for the old. With

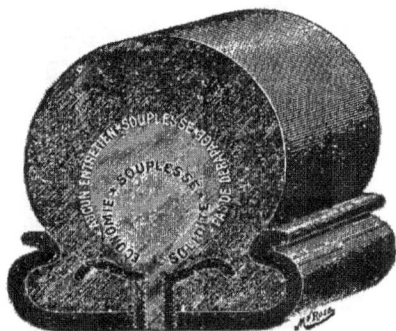

Fig. 20. A Compound Tire

all-metal wheels these difficulties disappear and vulcanization will be found to give excellent results.

A pneumatic tire, in the end, is a rather costly contrivance; for the constant repairs which are required by the outer covering, the canvas envelop, the air-tube, and the valves are often very expensive. Of solid rubber tires, the most economic and to a certain extent the most elastic are those having the greatest width and height. The wider the tire the smaller will be the pressure of the load per unit of surface, and likewise the strains which tend to wear away the rubber and destroy the tire. The higher the tire, the greater will be the surface provided for securing the rubber to the rim. The high tire, however, has the defect of being more readily torn from the rim under the influence of the lateral force exerted in withdrawing it from a rut or track. The more elastic the rubber, the smaller will be the danger of cutting the tire; for the material will yield under the pressure of an obstacle and thus possibly prevent an incision.

Except in those cases where the character of the motor requires a tire of unusual resiliency, those tires should be used which insure safety, comfort and economy.

Tricycles and light carriages without springs can be fitted with pneumatic, hollow or compound tires. For automobiles up to 1,000 pounds in weight (the load being equally distributed on the axles), almost any tire can be used; but the quality of rubber used and the means of attaching the tire to the wheel will affect the life of the tire. If great resiliency be necessary, the pneumatic, hollow, or compound tire should be employed. The compound is better than any solid rubber tire, because it always

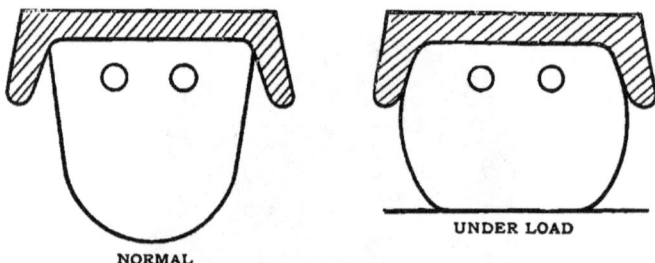

NORMAL   UNDER LOAD

retains its form and is not so readily chipped. For automobiles varying in weight from 1,000 to 10,000 pounds it is not advisable to employ tires with transverse joints, or those which are hooked, pinned, or wired to the rim. The disadvantages attending the use of such tires are obviated by employing very elastic compound, or solid rubber tires, or tires vulcanized on the rim. For vehicles weighing from 10,000 to 20,000 pounds, the tire should be vulcanized on the wheel to obtain the best results.

A few words regarding the care and preservation of tires will not be here out of place.

As a general rule, the manufacturer should be immediately informed of any signs of weakness and of unusual wear in a tire. Slight injuries can be almost always repaired as soon as they have occurred. When the transverse joint of an ordinary tire, for example, has parted, repairs can be made if the accident be of recent occurrence; otherwise the injury is irreparable.

With tires less than $1\frac{1}{2}$ inches in diameter care should be taken in traveling over streets covered with large paving-stones;

for the tire often enters the crevices between the stones, and the load being borne only by the sides, the rubber is often cut. A journey of 120 miles on a badly-paved street, where the vehicle is traveling at a high rate of speed, will produce as much wear on the tire as a journey ten times longer on an ordinary road, and at a reasonable speed.

If the tire be not of the endless ring type and the joint show a tendency to part, the lateral retaining devices should be readjusted and wire passed transversely around the tire to hold it to the rim.

The switch-tongues of railway tracks should be avoided, for they are a source of constant danger to the automobilist, especially when the tires are less than $1\frac{1}{2}$ inches in diameter.

The brakes should be gradually applied and the sudden stops produced by the rim-brake avoided. The friction-roller brake is perhaps the best for automobiles. If a rubber brake-shoe be employed, particular care should be exercised to avoid lateral strains.

The carriage should be steered only at reduced speeds.

If the rubber of the tire has been deeply cut, the incision should be cleaned with benzin and melted gutta percha poured in the cut. Even if the injury be not repaired, the cut should be thoroughly cleansed of sand.

The rubber should never be allowed to come into contact with lubricating-oil.

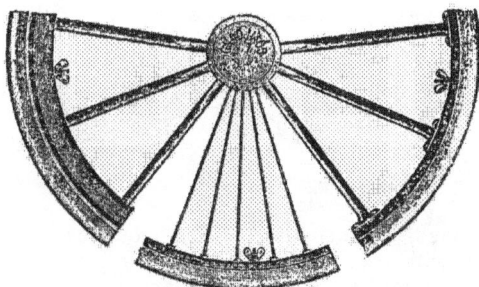

Fig. 21. Different Methods of Mounting a Compound Tire

Hertel's Voiturette

IN Hertel's Voiturette, named "The Impetus," the motor is
placed between the front wheels over the springs. The
motor is supplied with air-cooling flanges, and has a speed
change working by two brakes manipulated by a single lever.
These brakes also serve alternately to detach each of the two
gear chains.

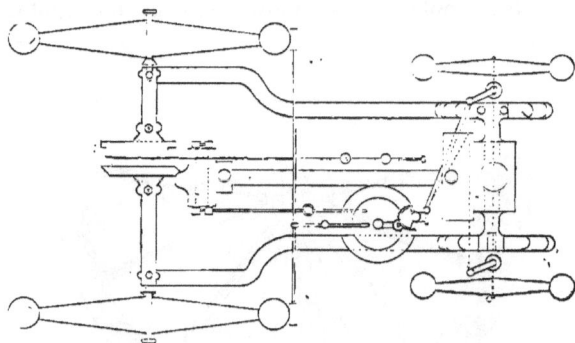

The carriage, when supplied with a motor of two and a half
horse-power with cooling flanges, or a three horse-power with
water jacket, is said to be capable of covering 33 kilometres in
one hour.

A Power Increasing Motor

By Henry Sturmey

EARLY this year the first trials upon the road of a new car
has been made by Messrs. Dawsons, a firm of engineers
of Canterbury, England, which possesses many novel
features, but more particularly relating to the engine construction. Hitherto the great advantage of steam over gasoline for
automobile traction has been its flexibility, and it is just this
flexibility which is possessed by the steam engine which Messrs
Dawsons have succeeded in giving to their oil engine, with a
result which is simply astonishing to those who know the limitations hitherto characteristic of the oil motor. We are not yet
in possession of minutely detailed particulars of the invention,
but we may say that its broad principle of construction consists
in the combination of compressed air with the gasoline charge.
Ordinarily the engine works as an autocycle explosion engine,
but the cylinders are extended in the direction of the crank
chamber in the form of an additional cylinder of larger diameter,
which is used as an air pump having an area equal to the difference of the two diameters. In ordinary action the valves connected with this portion of the motor remain open, and, with
free circulation of the air through the walls of the cylinder, it
moves without perceptible additional friction, and the motor is
in all respects identical with an ordinary autocycle explosion
motor. But this additional cylinder has three functions which
can be utilized at will to obtain with the motor all the flexibility
of power and action that is obtainable with a steam engine. In
the first place, the valves can be so shifted that the secondary
cylinder becomes an air force-pump, which compresses air into
a cylinder provided for that purpose under the driver's seat. By
reversing the position of the valves this compressed air can be
passed back again through the air pumps, which, by the reversal
of the valves, become air motors, and by this means the petroleum
action can be functioned at will, a touch of the valve lever causing the motor to be instantly started automatically, so that there
is no need for the driver to be getting down out of his carriage
and turning the usual handle. Not only is the power thus
obtained sufficient to start the motor when disconnected from
the carriage, but it will start and propel the carriage itself some
little distance, if need be, though, of course, this is not, as a rule,

The Automobile Magazine

necessary. Again, by an alteration of the valve gear the air motors can be started backwards, and thus the motor can be automatically set going reversed, so that no reversing gear in the car is required, this reversal, as well as the starting, being entirely automatic and effected by the merest touch of the finger upon the valve lever, and, of course, the motor, which is timed to fire at the dead point, will give equal power backwards as forwards; but the chief advantage obtained by the use of the secondary cylinder air system is the obtaining of additional power when required for hill work. This is not, as might at first be supposed, effected by assisting the gasoline system with the power of the compressed air system, but does not need the air storage tank at all, and Messrs. Dawsons have a second pattern of engine which dispenses with the automatic starting and reversing, and uses this automatic power accelerator device only. And the way it is done is this: By a suitable manipulation of the valves the air contained in the air cylinder is caused to pass into the forward end of the gasoline explosion cylinder at the moment of commencing its return stroke. This being the exhausting or emptying stroke, the exhaust is not only expelled by the backward passage of the piston but is swept out by a blast of fresh air, and thus the interior of the cylinder is more effectually cleared of the waste gases than could otherwise be effected; the charge which it then takes in is consequently of a purer and richer character, and this in itself is an advantage, but when the following return of the piston takes place on the compression stroke, this action is preceded by the admission into the cylinder of atmospheric air forced into it by the air pump at a pressure of about ten pounds to the square inch. This has the double effect of increasing the initial pressure of the charge and consequently vastly increasing its ultimate pressure at the moment of explosion, with a corresponding increase in the power of the stroke, and it also enriches the charge itself and thus gains a further advantage, and, if necessary, the amount of air so mixed with the charge can be still further increased and the amount of gasoline vapor admitted increased likewise to correspond, thus still further increasing the effectiveness of the engine, whilst the repeated admission of cold air into the cylinder between the strokes helps materially to keep the internal heat down. The result of this arrangement is that, when ascending long, steep hills, or meeting with other work necessitating increased power, the power of the motor can be virtually doubled.

The Bosch Magneto-Electric Igniter for Automobiles

A T the Berlin Automobile Exposition the firm of Robert Bosch, of Stuttgart, exhibited magneto-electric ignition apparatus of various sizes, remarkable chiefly for the simplicity of their construction and the astonishingly large number of sparks obtained per minute.

In the hydrocarbon motors so widely used in Europe for driving carriages the ignition devices constitute one of the most important features. Of igniters there are two kinds—ignition tubes and electric-spark apparatus. The latter may obtain their electric energy either from batteries or miniature magneto-electric machines. Despite the many disadvantages attending the

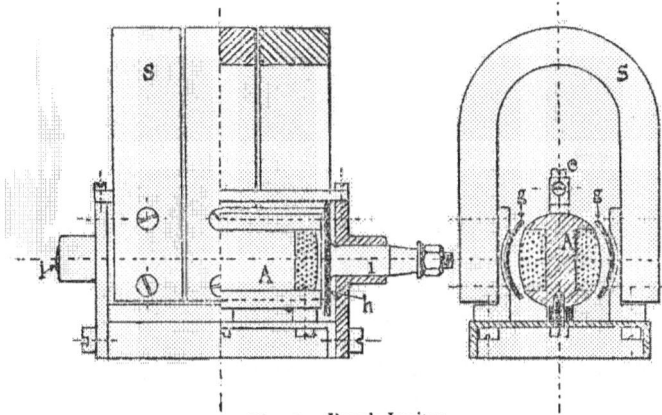

Fig. 1. Bosch Igniter

use of ignition-tubes, the only form of electric gas-exploding device which has been more or less widely used is the induction-coil energized by a battery current. But its sensitiveness to moisture, the care which it requires, and the frequent occurrence of short circuits have caused the battery to be abandoned by many *chauffeurs* for the magneto-electric machine.

For a number of years magneto-electric machines have been successfully used on stationary explosion engines making but a small number of revolutions per minute. But mechanical reasons prevented the application of this system of ignition to automobile motors. Finally Robert Bosch, of Stuttgart, succeeded in devising a magneto-electric machine which was applicable to fast and slow running motors alike, and thereby overcame an obstacle which had long baffled inventors.

Before describing the construction of this new apparatus, it should be remarked that its underlying principle is essentially the same as the magneto-electric igniter which for the past thirteen years has been used on stationary motors. Experience has proven that the magnetism of these machines is well-nigh permanent and that they are subject merely to wear. It should be remarked, furthermore, that the ignition is no wise affected by moisture and that, even though the apparatus be deluged with water, sparks will still pass between the terminals. The ignition is not influ-

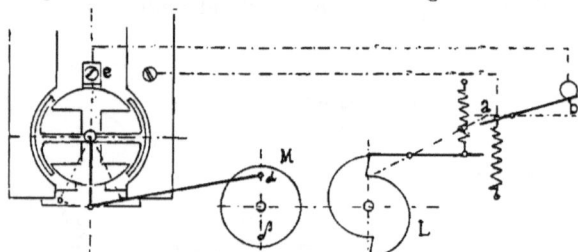

Fig. 2. Bosch Igniter

enced by atmospheric conditions; nor is it possible to ignite spilt benzin or any other hydrocarbon, since the sparks are produced only within the motor-cylinder.

If about the bar A (Fig. 1), insulated wire be wound, and a magnet S be brought in proximity with and removed from the ends of A, then an electric current will be generated in the wire, the intensity of which depends upon the rapidity with which the magnet approaches and recedes from A and upon the strength of the magnet. If the armature g of the apparatus be given an oscillating movement, the same result is obtained as when the magnet S approaches and recedes from the ends, i. e., the cylindrical surfaces of A. Electric currents are set up by this oscillation. If the circuit through which these currents are flowing be broken, sparks will be produced at the point of interruption, which sparks are admirably adapted for the ignition of explosive mixtures of air and gas.

An experience gained in constructing igniters for stationary motors proved to the inventor that the following conditions should be observed in making gas-exploding devices for automobiles:

The Bosch Igniter

The winding of the core should be secure in order that—

1. The connection of the one end of the winding with the insulated electrodes in the cylinder shall be constant, so that the employment of a collector is rendered unnecessary. (A collector is subject to much wear and cannot readily be kept clean.)

2. The current from the other end of the winding (which end, for the sake of simplicity, is directly connected, in such apparatus, with the iron of the core) shall not flow through oily and dirty bearings into the motor frame, but from the core over good, non-resistant surfaces to the electrodes.

3. The moving parts, which, for mechanical reasons as well as for the purpose of obtaining the speed necessary in ignition, are merely oscillated, shall not be too heavy, for otherwise the wear upon the driving mechanism would be inordinate.

These conditions are satisfied in the following manner:

The core A of the apparatus (Fig. 1 is firmly screwed to the base-plate; and between the hollowed poles of the magnet S and the core A, a cylindrical armature g is arranged, from the covering of which two longitudinal strips are cut. The armature gg is secured to the plates h, into which the ends of the shaft $i\,i$ are fastened.

The igniters can be arranged in any manner desired. In Fig. 2 such an arrangement is diagrammatically represented. To the driving shaft of the motor, a disk M is secured, with which an eccentric X is connected. To the shaft of the apparatus a lever-arm is fastened, which is in turn connected with the eccentric X by means of a pitman. When the driving shaft is rotated the armature is oscillated and a current is generated. To the binding-post a, with which the polished end of the winding is connected, a wire conductor is fastened, leading to an insulated electrode. By means of a coiled spring, the end b of the lever $a\,b$ is brought into contact with the electrode. A wire secured to the lever $a\,b$ runs to the igniting apparatus, which wire is really unnecessary, since the body of the motor could take its place. The end C with the insulated electrode is located in the cylinder (not shown). Upon the driving-shaft or any other shaft operatively connected therewith, a shouldered cam L is secured, the periphery of which is in contact with one end of a second lever, the other end of which is adapted to strike the end a of the first lever $a\,b$. When the parts are set in motion, sparks are produced between the terminal rods whenever the end b of the first lever is withdrawn from the insulated electrode and the circuit thereby broken.

The "Enomis" Odometer

THE question of odometers for automobile carriages has often been studied by manufacturers, but two great difficulties have always been encountered: (1) that of giving adequate strength to an apparatus that must resist all vibrations and shocks, and (2) that of regulation, which is different on every carriage, since a uniform diameter for wheels has never been adopted for automobiles, as it has been for bicycles.

An odometer, in order that it may be perfect, must therefore combine the two qualities of great strength and variable multiplication.

The "Enomis" odometer, which we illustrate herewith, may be placed indifferently upon all carriages, since it is capable of being regulated. The possibility of regulating these apparatus will be seen to be a very great advantage when we come to consider that the diameter of wheels often varies, either because of the escape of air from the pneumatic tire or the wear to which the solid tire has been submitted.

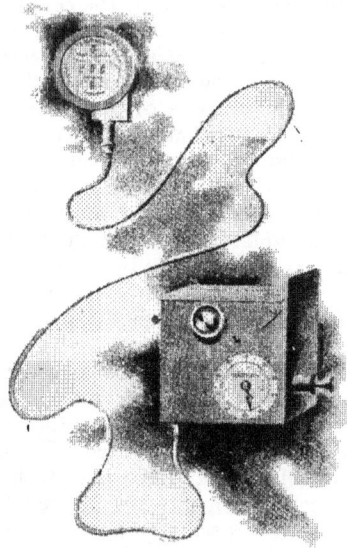

Fig. 1. The "Enomis" Odometer

Let us study in a mathematical manner the case of a wheel provided with a solid tire. Let us take a wheel 1 metre (3.28 feet) in diameter, the rubber tire of which is 3 cm. (1.18 inch) in thickness, which is normal. This wheel, when new, will cover a distance of 3.14 metres (10.3 feet) at each revolution of the crank. After the tire has become worn and is but a centimetre (.39 inch) in thickness, the diameter will have been reduced by 4 centimetres (1.57 inch), say to 96 centimetres (37.8 inches). We therefore have a loss of 13 centimetres (5 inches) in 3.14 metres (10.3 feet), say 41 metres (134.5 feet) per kilometre (.6 mile; i. e., 224 feet per mile). It

The "Enomis" Odometer

is therefore indispensable that there shall be a possibility of regulating the odometer in measure as the wheels wear. When the multiplication is fixed, such regulation is extremely difficult, since it is necessary to change the gearings. In the odometer under consideration, this operation is exceedingly simple, and may be performed along the route. It suffices to revolve a button in one direction or the other in order to set the multiplication back or ahead.

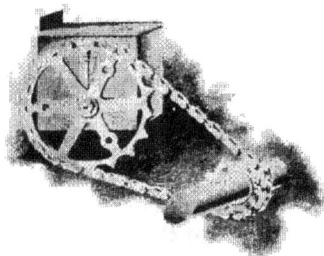

Fig. 2. Mounting of the Box Containing the Multiplying Gearing

The "Enomis" odometer (Fig. 1) consists of two distinct apparatus: (1) a box containing the multiplying-gearing (Fig. 3); and (2) a kilometric registering-apparatus (Fig. 4). These are connected with each other by a rubber tube that transmits the motion of the multiplying gearing to the registering-apparatus. Let us first examine the box enclosing the multiplying-gearing. This is placed near the differential shaft, with which it is connected by a chain actuated by a sprocket (Fig. 2). This method of control is infinitely superior to that effected by a trip placed directly upon the wheel. The trip, in fact, gives an abrupt shock at every revolution of the wheel, while the chain transmits a continuous and regular, and, consequently, a much smoother and more exact motion. Moreover, since the differential shaft revolves with greater velocity than the wheel, the regulation is on that account of much greater precision.

Fig. 3. Section of the Box Showing the Gearing

The variation in velocity is obtained through a disk (Fig. 4) over which moves a roller that may be made to recede from or approach the centre. Such regulation is of extreme sensitiveness:

and a half-revolution of the button suffices to cause the apparatus to gain or lose two or three metres per kilometre (about 11 or 16 feet to the mile). It therefore suffices to turn the external screw to the right when the odometer is running behind, and to the left when it is running ahead.

The transmission of motion is done pneumatically. At every hundred metres (328 feet), a lever strikes a rubber bulb, which, through a tube, transmits the motion to the kilometric registering apparatus. The pressure of the lever upon the bulb is obtained by means of a hammer situated at the exterior of the box. The degree of pressure upon the bulb is regulated by a screw at the extremity of the lever. Such a regulation, however, scarcely ever has to be effected, since the apparatus are delivered fully regulated.

Fig. 4. Regulating Apparatus

After the carriage has stopped, or when it is not desired to use the odometer, the collar of the pressure-spring must be screwed in such a way as to hold the hammer back. As the spring then no longer acts upon the lever, the odometer will be at rest.

As for the kilometric registering-apparatus, we have to consider:

(1) The Regulation.—The apparatus is provided with five dials, and is capable of marking distances up to 10,000 kilometres (about 6,200 miles). At every 100 metres, the figure of the hectometres changes and carries the other dials with it.

(2) Setting Back to Zero.—In order that the setting back to zero may be quickly done, it is effected through three axes—hectometres. kilometres and myriametres. In this way, the operation (which is performed by means of a key furnished with the apparatus) is effected in less than half a minute.

The Minerva Motor

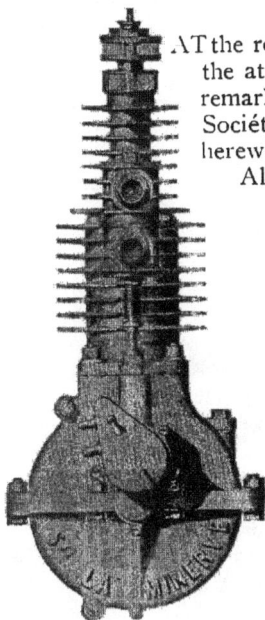

AT the recent Motor Competition at Aubervilliers, the attention of the public was attracted by a remarkable motor which was presented by the Société La Minerve, and which we illustrate herewith.

All explosion motors are subject to an excessive heating of the sides of the cylinder which interferes with the regularity of the explosions and renders the operation of the apparatus irregular just at the moment when there is most occasion to rely upon it; for example, when, in order to climb a long and steep acclivity, the explosions are hurried and the quantity of heat developed is thereby increased. Although flanges and circulations of water afford a means of averting such an inconvenience, they are merely inadequate palliatives, since, although the external sides of the cylinder allow their excessive heat to disengage itself, the internal sides and the piston do not get rid of it fast enough, and may be raised to such a temperature as to render lubrication impossible and thus cause the segments to gripe and injure the walls of the cylinder.

The originality of the " Minerva " motor consists in causing a current of air to pass into the interior of the piston so as to add a very efficient internal cooling to the external cooling produced by flanges. To this effect, the piston, which is hollow, is prolonged by a tube that forms a second piston which slides in a cylinder smaller than the explosion one. This tube, like the larger piston, is provided at its upper part with segments that assure the tightness of the apparatus and prevent the gases due to the explosion from escaping through the annular space left around the tube in order to prevent friction. In this way, the interior of the principal piston, and, consequently, the interior of the casing, are in constant communication with the external air, which is sucked in and forced out through the tube and assures the cooling of it to such a degree that stationary motors of this type have been able to withstand compressions of 650 pounds to the square inch without the maximum temperature exceeding 280° C, after a certain period of operation. When the motor operates upon a motocycle or a voiturette, a cooling by a circulation of air in the flanges is added to the internal cooling and prevents the temperature from rising to more than 150° C, even after several hours of running. This is an inappreciable advantage which, it is claimed, renders the " Minerve " motor infinitely superior to all the explosion apparatus that have made their appearance up to the present time; and, as the other parts are like those of motors of the same class, the addition that has been made to the " Minerve " may be applied to other motors so as to suppress the manifold inconveniences to which automobilists are exposed.

A lubricator, especially devised for the purpose, assures the perfect lubrication of the secondary piston.

The " Minerva " motor of the type M. A., which is the one now being constructed, and which has been submitted to experiment, is warranted to give, at the brake, an effective power of 957 foot-pounds, say an effective $1\frac{3}{4}$ horse-power. It weighs 748 pounds and occupies the same space as other motors of the same class. Other and more powerful types are in preparation and will soon be offered to the public.

A MOTOR VEHICLE DEFINED

At an English police court recently Captain Herve H. A. Errington Josse, of Norman Villa, Bargate, Grimsby, was summoned by the Inland Revenue for keeping a carriage without a license.

The question raised was whether a vehicle consisting of a car attached to a motor cycle was one carriage or two. If it was considered as one the maximum traveling speed allowed would be twelve miles, and if as two, six miles per hour.

The magistrates decided that it was one vehicle, and imposed a fine of one guinea.

The Loutzky Automobile

T HE Berlin automobile exposition gave German *chauffeurs* an opportunity of examining Loutzky carriages of various types. The Loutzky exhibit included two *voiturettes*, an automobile mail-wagon, two tricycles, and four traction-*voiturettes*. Of these nine vehicles, the two seated *voiturette* and the mail-wagon deserve more than a passing notice.

The *voiturette* in question is driven by a four-cycle, double-cylinder, hydrocarbon motor developing 3½ H. P. and making 1,200 to 1,500 revolutions per minute. The cylinders are cooled either by flanges or by water-jackets; and the carburetted gases are discharged by electric spark. The admission and exhaust valves are controlled by a single cam acting on two levers mounted in a common vertical plane, with their extremities separated by an arc of 90°. The gears, intermediate shaft and operating-

The Automobile Magazine

lever, as well as the motor-shaft and fly-wheel are inclosed in a casing which serves as a support for the cylinders.

The water-tank, in the jacketed type of engine, has a capacity of 4.4 gallons and is mounted in the forward portion of the carriage between the two steering-wheels. The water is circulated by means of a reciprocating pump.

The motor is arranged under the carriage-seat, between the two rear driving-wheels. Gearing directly connects the motor-shaft with the driving-wheels. Adjacent to the motor are mounted a petroleum-tank with a capacity of 4.5 to 6.6 gallons, the accumulators, and the spark-coil.

The carriage is provided with two changes of speed, the pinions and spur-wheels composing the gearing being respectively proportioned in the ratio of 1 to 9 and 1 to 17, corresponding in the former case with a speed of 15.5 to 21.7 miles per hour on a level road; the other change of speed is used on heavy grades. The intermediate speeds are obtaining by varying the proportions of the explosive mixture and by properly controlling the ignition. The speed-changing gear is thrown into and out of engagement by means of two friction clutches connected with a single lever mounted on a post upon which are also carried the steering-wheel and two hand-levers connected with the carbureter and igniters.

Steering is effected by means of a divided axle connected with the two front wheels, held in forks, as in bicycles.

In the remaining details of construction the carriage resembles most automobiles of the same type. The weight varies between 500 and 550 pounds.

MUNICIPAL ENTERPRISE ABROAD

The City Surveyor of Birmingham recommended his council to purchase a motor vehicle for heavy work. It has now for some time been in use, and he reports a saving of 50 per cent., as compared with the cost of performing similar work with horse-drawn vehicles. For some time past the Chiswick Vestry have employed motor dust-carts, and they report favorably on them; and the Mersey Docks and Harbor Board are now employing heavy motor vehicles.

The Pougnaud-Brothier Naphtolettes

IN the types of "naphtolette" brought out by Messrs. Pougnaud and Brothier, the motor has been mounted in the front of the carriage, for the reason that it can be more easily reached. Power is transmitted by gearing and by belts, whereby the inconveniences attending the use of the chain have been done away with.

In the carriage which is known as "Type C," two pinions on the motor-shaft engage two gear-wheels mounted on an intermediate shaft. Upon this intermediate shaft a pulley or drum is also carried, which receives a belt running beneath the carriage over a fast and loose pulley in the rear. The rear pulley shaft is geared with the differential carried on the rear axle.

In Type D the front pulley or drum is secured to the motor-shaft. The speed-changing gear is mounted between the rear pulley-shaft and the shaft of the differential. This type of carriage was designed to provide a means for increasing the adhesion of the belt and reducing the third speed as much as possible when the carriage was ascending heavy grades.

The motor used is the invention of Pougnaud and Brothier, has a stroke of 4 inches, makes 1,300 revolutions per minute, and

The Automobile Magazine

develops 2½ to 3 H. P. The cylinder is cooled by flanges, whereby the inconvenient water-jacket is dispensed with. The two valves, which hitherto have been arranged one above the other, have been placed at each side of the working-chamber, so that they can be easily reached. The exhaust-valve is not raised by the cam usually met with in automobile-engines, but by a lever.

The electric igniter used derives its energy from a powerful primary battery or from accumulators, the electric current being conducted to the usual spark-coil. The igniter is so mounted that it can be readily inspected.

In the carbureter, the intake-valve is reseated by means of a spring. The casing in which the valve is inclosed is so arranged that the air admitted can be either warmed or chilled, the proportions of the explosive mixture varied and the size of the valve-opening regulated, and the quantity of the explosive mixture admitted to the cylinder increased or diminished.

The carriage is steered from the front wheels, which are mounted on a divided axle. The steering-gear is perfectly trustworthy in operation and enables the driver to make very short turns.

Connected with the hollow steering bar are two concentric

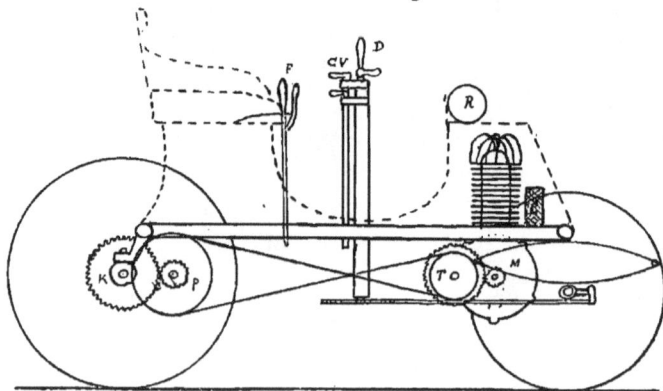

The Pougnaud-Brothier Naphtolettes

tubes, each terminating in a hand-lever. The one hand-lever controls the speed-changing gear and the other the mechanism for shifting the belt from the fast to the loose pulley. By shifting the belt only partially, the carriage can be gradually stopped and started without any shock whatever. By shifting the belt in this manner the speed can be changed without effort, shock, or noise. The steering-bar is also provided with two small levers controlling respectively the ignition and the admission of gas to the cylinder.

Two powerful band-brakes, one commanding the pulley-shaft, the other controlling the rear wheels, enable the driver to stop his vehicle almost instantly. The motor, if it be so desired, can be used as a third brake in descending steep grades.

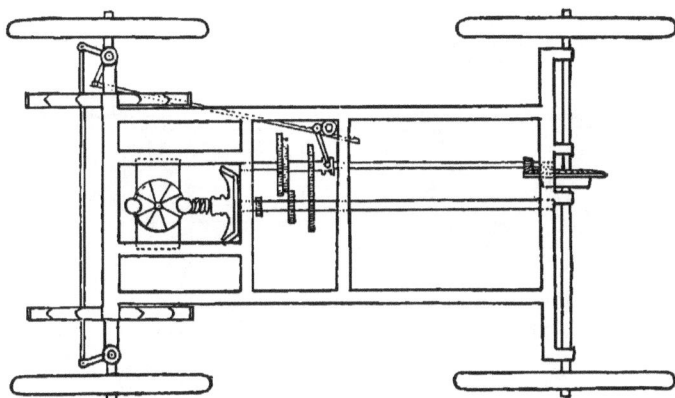

Under the driver's right foot is a belt-tightening pedal, which, however, is used but rarely.

The frame of the vehicle is composed of stout tubes which retain their shape permanently and which support a body that can be removed very readily. The body can be provided with a hood, a special top, a baggage-carrier, or a commercial traveler's sample box. The body is carried by springs, which, together with the pneumatic tires, break the force of all shocks and relieve the carriage of undue strains.

In other types the carriage is provided with two changes of speed, which, by means of the ignition, belt-shifter, and gas-admission cocks, can be varied from 0 to 19 miles per hour. A third speed of 4 to 6 miles per hour can also be obtained. By changing the proportions of the differential gear and coacting pinion, the speed can be raised or lowered.

635

The Automobile Magazine

For those *chauffeurs* who are opposed to the transmission of power by belt and pulley, Messrs. Pougnaud and Brothier have provided a friction-clutch mechanism which is used on the carriage known as " Type E." The exterior is exactly similar in appearance to that of the other type. As the plan view which forms the subject of Fig. 5 shows, it is essentially the same carriage, provided with the same frame, but with differently-arranged cross-bars, and with the motor mounted transversely in the front. The motor actuates the driving-shaft by means of a friction-cone. This shaft is longitudinally disposed as in the case of the single intermediate shaft.

What has been already said on the subject of speeds, steering-

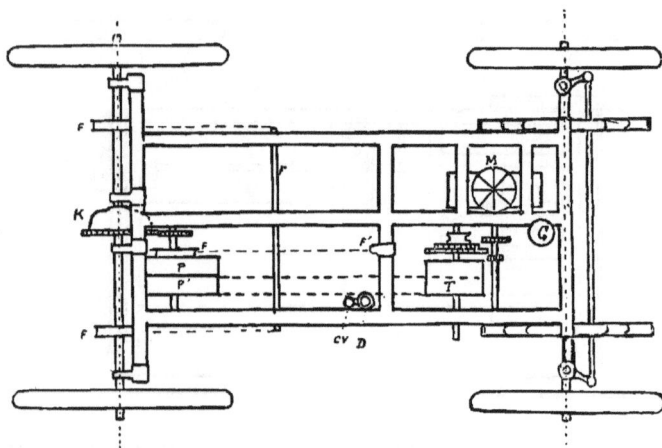

gear, and brakes, in connection with types C and D, applies equally well to type E, of course with the exception that there is no belt tightener or shifter.

In all types the operative mechanism is easily accessible. It is always placed beneath the frame so as to lower the center of gravity and render the carriage as stable as possible, even when running at high speeds or turning sharp corners. For similar reasons the seats have been placed very low, without, however, inconveniencing the occupants by reason of the dust.

All the parts have been arranged so that they can be easily examined, cleaned, and renewed. So simple is the construction of the carriage that the *chauffeur* requires little or no mechanical knowledge to understand his *teuf-teuf*.

The Automobile
MAGAZINE
An *ILLUSTRATED* Monthly

Vol. I No. 6	NEW YORK MARCH 1900	Price 25 Cents

The Automobile Magazine is published monthly by the United States Industrial Publishing Company, at 21 State Street, New York. Cable Address : Induscode, New York. Subscription price, $3.00 a year, or, in foreign countries within the Postal Union, $4.00 (gold) in advance. Advertising rates may be had on application.

Editorial Comment

THERE is a sort of grim humor in the idea of the exclusion of automobiles from cemeteries, as instanced in the recent action of the trustees of the Forest Hills Cemetery at Boston. As the Boston *Herald* remarked on the matter, " It is but fair to say that the automobiles are not headed that way."

It was the senior James Fiske, we believe, who declined to subscribe toward building a fence about the village cemetery, saying that there was no use for a fence; those who were inside could not get out, and nobody who was outside wanted to get in! The exclusion can hardly work any hardship to automobilists, for there is no motive for them to want to get into a cemetery. It is true that there are apt to be some stretches of good road there. But it is no place for pleasure-driving. And with the miles and miles of boulevards and other perfect roads about Boston—even though motor-vehicles are still excluded from the public parks after ten o'clock in the morning—there is nothing that should tempt an automobilist to explore the mortuary precincts of Forest Hills. The cemetery trustees, however, are said to have had visions of the horrible consequences of a stampeded funeral procession, with horses wildly cavorting across lots, demolishing costly marble tombstones, and attended by other incidents incompatible

The Automobile Magazine

with the solemnity of the occasion. A remedy suggests itself. But, while automobile weddings are getting to be the thing— and there was a very interesting one in Boston the other day— we have yet to hear of an automobile funeral. In the City of Mexico, to be sure, there will soon be nothing but trolley-car funerals. All funerals there have long been conducted on the street-cars, which run to all the cemeteries around the capital for the purpose. And now that the street-railway system in the City of Mexico is undergoing transformation to electric traction, the way to the grave will be entirely by that method. The auto-mobile-hearse may wait, however. It is doubtless on the way, but nobody appears to be in any hurry for it.

A Bright Future

" To the victor belong the spoils." It is to the emoluments of political office that this paraphrase of the ancient looting cry of " Væ Victis " is mostly applied here in America, but the developments in the mechanical and scientific world during the latter part of the century now closing would make it equally applicable to inventions. Machinery has at every stage of advancement put aside manual labor. At the same time it has provided for its ultimate employment. Machinery invented to do the work of hands has not worked injury to the man who owns them, on the contrary it has elevated and improved his opportunities. As with man and machinery, so it will turn out to be with the horse and machinery. Where the locomotive displaced the lumbering stage coach and the still clumsier freight wagon, it did not displace horses. It forced them to other uses. The horse is still " man's best friend." But the horse must now measure himself with another no less formidable rival. No matter what the difficulties or the impediments encountered, the evolution of the horseless carriage, the horseless truck, or the great horseless freight wagon is a certainty. In the immediate future our streets and our roads will teem with vehicles for every required purpose, driven by motor power of every desirable kind, and easily controlled and guided by one hand.

To-day hundreds of manufacturers here and abroad are battling with the problems confronting them. Many of them have already conquered and are producing most excellent and satisfactory vehicles. Others there may be that will drop by the way. Heavy, cumbersome, clumsy and lacking in vital essentials of construction, these curious errors will disappear before they have

Editorial

fairly come to our view, while newer and better devices will take their place and prove their value. Lightness, strength, simplicity and moderate cost in combination will win the battle, and the survival of the fittest will be the result. Some there are who halt when they see the superior work of their competitors, but it cannot deter the ultimate success of the best. The overstocked company with a brilliant prospectus but no money in its treasury, and no plant to manufacture, is doomed to an early death. Capital in plenty will continue to find investment in worthy concerns, and the close of the present year will see the automobile industry a child grown to vigorous manhood.

Automobile Traction

The present season will be likely to witness a wide extension of the automobile " sphere of influence." Probably the most notable direction in which mechanical traction will appear in evidence will be in the realization of the omnibus lines projected in various great cities—as described in the leading article of this issue. These will put the motor-vehicle on a nickel-basis and for the first time give the mass of the population in those cities the practical benefit of the innovation. Moreover the plying of the omnibuses at frequent and stated intervals along regular routes will contribute extensively toward familiarizing the public with the new form of locomotion and establish its popularity. In the way of the transportation of goods and parcels, we may look for a great increase in the use of the automobile for delivery service and postal purposes—uses that have been parallel in their development with that of the automobile for private use and for cab and livery service. We shall scarcely be likely to witness anything this year beyond the experimental stages in the very important field of heavy traffic, such as motor-trucks, carts, etc., and also the various municipal applications—beside the horseless steam-fire engine—like automobile street-sweepers and sprinklers. With these fields practically covered, as they must be in a comparatively near future, the automobile will proceed to enter completely into its own, and rapidly realize all the most confident expectations.

No Closed Season

Before electric traction for street railways had passed the experimental stage it was feared that the problem of snow in our Northern cities would prove an insuperable obstacle. Experience showed, however, that not only was electric traction immensely ahead of animal traction at times of heavy snow, but that electric cars found little trouble even from storms that pretty generally blockaded the steam railroads. This winter has not been of a nature to subject the automobile to severe tests, but there has been at least one snowstorm hard enough to indicate their qualities at such times. And it was found that the automobile, as a rule, did better than the horse. Its broad pneumatic tires gave it a good tread in the snow and enabled it to get through the drifts with little difficulty. Of course the work required much more power than for ordinary running, and electric vehicles had to make much more frequent trips to the stables for recharging. But in general, all classes of automobiles, whether driven by electricity, steam, or gasoline, had sufficient reserve power to meet the increased demands upon their strength. So there is not much to be feared on the score of snow. There is comparativly little call for pleasure-vehicles, either with animal or mechanical traction, in the cold months. Possibly some time we shall achieve an automobile sleigh, and until then a sleigh in tow of the wheeled motor-vehicle may be made to answer! And so far as ordinary traffic is concerned, the snow is now so quickly removed from city streets, and even the country roads are so soon broken out, that the automobile will be found to answer, on the average, all the winter demands, even better than does the horse.

CARRIAGE BUILDERS NEEDED

It is evident from the new automobile departments in the trade organs that the makers of carriages, as well as the dealers, are paying more and more attention to the possible inroads threatening their industry by this newcomer in their field—the automobile. In this connection we draw attention to the remarks of Mr. W. N. Fitzgerald at the last annual convention of carriage dealers, the substance of which we reproduce in this issue. It is undoubtedly a fact that much of the delay in this year's output of automobiles has been caused by the mistaken notion that the body of a carriage or wagon should be turned out by the hands of the same skillful mechanician that evolved the complicated machinery of the driving gear and motor. As a result we see powerful motors coupled with clumsy vehicles, which, at best, only bring down upon themselves that awkward circumlocution "the horseless carriage."

They do these things differently in France. There, after making the same mistake, the leading automobile constructors

Editorial

learned the wisdom of profiting by the experiences of others, and accordingly the most popular types of French automobiles are now constructed by the motor engineer and the carriage maker working in conjunction.

ACCESSORY MAKERS IN THE DARK

The disadvantages of our system of single-handed manufacture are obvious. Of three hundred and odd American manufacturers of automobiles, as enumerated in a recent English trade manual, barely half a dozen have their products in general evidence. These are the well-known Electric Vehicle Co., the Locomobile Co. of America, the Oakman Motor Vehicle Co. of New England, and two or three other equally well-known automobile makers. As a result of the protracted delay in the output of the other manufacturers, another industry which might be flourishing is floundering in the same slough of despond. This is the industry of automobile accessories, the possibilities of which may be judged from the rapid rise and profits of wheeling accessories which followed upon the general boom of the bicycle business a few years ago.

As matters now stand, those who would produce serviceable accessories for automobiles, such as gear-chains, driving belts, solid rubber rims and pneumatic tires, must needs resign themselves to await the adoption of some more or less general standard which will enable them to get some idea " where they are at." At present most of them, while valiantly trying this and that experiment, actually do not know what is really wanted.

HINTS FROM ABROAD

In the meanwhile some of the more enterprising manufacturers of automobiles abroad are improving the occasion by launching their best products in this country. In view of the all but prohibitive tariff levied on these importations this need not necessarily cause concern to our American builders of automobiles. On the contrary, a direct benefit to them must result from the searching tests to which all these foreign automobiles and their accessories are bound to be subjected.

A case in point was the series of trials conducted by various American experts with that recent German importation known as " Vollmer's Forecarriage." The American Express Co. had the Forecarriage hitched to one of their common delivery

The Automobile Magazine

wagons, and, with a load of one ton, drove the machine from New York City into Westchester County, taking all grades without any hitch and covering the distance of a little over seventeen miles in less than two hours. The officials of the American Express Co., under whose directions these trials were made, pronounced themselves fully satisfied with the results.

COMING SHOWS

The proposed automobile exhibitions in Amsterdam this March and in Paris next month should not be overlooked by American manufacturers. Some of them have achieved admirable results in their work of the past year, and in consequence the automobile has approached very considerably nearer the ideal. They should therefore not fail to make a showing of what they have accomplished. Though they may now be so rushed with work that they do not consider it worth while to attract more attention to the merits of their productions, the increased facilities for production that must come will sooner or later make judicious advertising of all kinds yield profitable returns. No business of any kind is so well established that it can afford to hide its light under a bushel.

Manufacturers who send their products to these expositions will probably have an opportunity to exhibit the same vehicles at the American Automobile Club's show projected for next fall. It has been definitely settled that this is to be an independent show, without reference to bicycles, carriages or other allied industries. What remains in doubt is whether it be held as an indoor exposition within the familiar precincts of New York's Madison Square Garden, or as an open air exposition with a more extensive track whereon to display the speeding capabilities of American automobiles. Those who favor the indoor project claim that the repeated failures of open air horse shows in New York have demonstrated that New Yorkers, and those who come to the city on these occasions, do not care for open air affairs. That such need not necessarily be the case, we think, has been abundantly proved by the unfailing popularity of Philadelphia's open air horse show—not to mention the great football games which draw such tremendous crowds, even in the worst of weather. So much is certain: New styles of automobiles, especially those not propelled by electricity, must be shown in the open air, and with plenty of elbow room, to be fully understood and appreciated.

Editorial

ANOTHER NEW COMPANY

Among the many new companies for the manufacture of automobiles which have lately sprung into existence, one of the most promising appears to be National Electric Automobile Co., of Indianapolis. L. S. Dow, the former manager of the Waverly branch of the American Bicycle Company or Bicycle Trust, and latterly the secretary and general manager of the Indiana Bicycle Co., is the promoter of the new enterprise. Associated with him are Arthur C. Newby, the leading spirit of the Indianapolis Chain and Stamping Factory, and Philip Goetz, the probable treasurer.

The company will manufacture electric automobiles and at the same time engage in the making of any other electrical appliances. After the concern is well under way the manufacture of gasoline automobiles will also be undertaken.

AUTOMOBILE IN A NEW ROLE

The automobile, which is becoming so familiar an object on our streets, has already figured in an elopement—so, at least, say the German despatches—and there is every probability that it will prove the vehicle of flight for runaway lovers of the incoming century. Its novelty, its exclusiveness, its speed, its silence, its simplicity in working, must all commend themselves to such as desire matrimonial bliss without the paternal consent. The railway carriage is so public an affair that runaway couples are in constant danger of recognition by friends or acquaintances who are shabby enough to happen on the same train; the horse and carriage affair is risky, because if it is kept too long there is danger of warrants for horse-stealing being issued; and the bicycle, tandem or single, is not a romantic form of locomotion, especially for the girl whose aim is speed. Walking, of course, is out of the question. The automobile, then, seems to be the ideal vehicle for elopers who have been forced to flight by stern parents. Side by side they can sit as they speed onward, the young man's attention only being distracted from his beloved by his being obliged to keep an eye upon the brake, and there is no need of a grinning, horrid coachman.

"Upon her lover's arm she leant,
And round her waist she felt it fold,
As far across the hills they went
Swift as the Valkyries of old."

AN IMPROVISED AUTO CLUB
From the *Deutscher Automobil Sport*

WHAT WE ARE COMING TO

From the Comic Section of the *New York Sunday World*

Press Notices and Book Reviews

Not a month passes now-a-days without the appearance of some new periodical publication devoted to the automobile and the new sports and industries that have arisen therewith. Last month brought us two such publications from Austria, one from Germany, and one from Nice, each of them profusely illustrated and full of bounding hopes for the future. Among the new arrivals this month the foremost is No. 1, Volume 1, of *The Irish Motor News*, published at Dame Court, in Dublin. This magazine, which is edited by R. Mecredy, the well-known editor of the *Irish Cyclist*, bids fair to rival some of the best publications that have already become well established in France and on this side of the Atlantic. The *Irish Motor News* indeed can be called a magazine in the American application of that term, which means that the character of its illustrations and letter press and quality of paper are unexceptionable. With these advantages the magazine shares the well-nigh universal defect of so many of our New World publications. to wit, triviality of subject matter and exaggerated attention given to topics of purely local import. Compared to such standard English publications as Mr. Sturmey's *Autocar* and the London *Automotor and Horseless Vehicle Journal,* their new rival across the Irish channel, however promising at first blush, appears somewhat provincial.

Lately the New York *Hub* has gone into automobilism. Specially noteworthy was what the editor of the *Hub* had to say on the subject of the so-called automobile exposition held in New York in January:

" There was a very general feeling of disappointment among those who attended the late show at Madison Square Garden, on seeing so small a number of exhibits of automobiles. The daily press had published glowing statements and promised the largest show of motor vehicles ever held in this country. and in addition stated that every prominent manufacturer of automobiles in the country would be present with his best products. The latter statement was true with those who did exhibit. but ten exhibitors could not be expected to make a great show. Then, too, the exhibits were so scattered as to make the show appear less important than it actually was. If automobile manufacturers are to have an exhibit that will do credit to them, they must cut loose from the bicycle. Go it alone. you who exhibit automobiles. Don't stifle the life of a new industry by hitching it to the carcass of a dead one."

Press Notices and Book Reviews

Les Moteurs à Explosion, Etude a l'usage des Constructeurs et Constructeurs d'Automobiles, par George Moreau, is a work issued by the publishing house of Béronger in Paris, a brief mention of which has already been made in the index of this magazine. M. Moreau's work is a study of explosion motors from the point of view of a theoretical mathematician. The author begins with a treatise of the physical and mathematical principles that govern explosion motors and then proceeds to the momentous subject of mechanical equivalents of heat. This is a subject that has been discussed so thoroughly during the latter part of this century that it would seem impossible to treat it from a really novel point of view. Yet the author seems to have succeeded in this, at least so far as it concerns explosion motors and the mechanical transmission of energies. In connection with his theories of the mechanical equivalents of heat, M. Moreau takes up the much-discussed theory of gases, in the study of which he goes back as far as Sadi Carnot, the famous author of the French law on the preservation of power. This is followed by general reflections on the accepted theories of motors and practical illustrations of the most recent diagrams and the loss of gases. An important chapter is that on power transmission and the theory of passive resistance, such as the resistance of air, effects of wind, axle friction and the resiliency of automobile tires. Of practical interest is the author's estimate of the various materials suitable for the manufacture of motors and the theoretic principles that should govern their selection. The work ends with a highly interesting contribution on the best fuels and gases that can be used for explosion motors.

A valuable publication is the " Automobile Pocket Book," issued by F. King & Co., the publishers of the *Automotor and Horseless Vehicle Journal,* in London. Mr. George Frederick Little, who has compiled this manual, is an experienced engineer and editor who has succeeded in grouping together all the most important subjects bearing on automobiles and everything connected with them in so lucid a manner as to make the book invaluable for quick reference. Besides an alphabetical index, giving all the subjects treated throughout the three hundred and odd text pages, there is a very serviceable glossary of all the most current automobile terms used in French, German and English.

The book is plentifully supplied with tables of comparative weights and measures, pressures of steam, metal expansions, speed tests, records and the powers and roots of numbers and

The Automobile Magazine

logarithms. Of special interest for English purchasers are the lists of regulations and laws relating to motor vehicles and those for securing patents on new inventions in the United Kingdom. More general in its interest is the complete list of British and foreign manufacturers and engineers, with directories of all automobile clubs and publications now existing in Austria, Belgium, France, Germany, Great Britain, Holland, Italy, Russia, Spain, Switzerland, and in the United States. American readers will be interested in the final ten pages of the book giving the names and local *habitat* of all the more prominent American firms of automobile manufacturers. They number nearly 300 in all.

Die Automobilen, ihr Wesen und ihre Behandlung, written by Dr. E. Möllendorf, with Capt. F. Kübel, and published by George Siemens in Berlin, is an excellent little hand book with many highly serviceable illustrations. The book is written for lay readers and is of interest to all who are concerned with the operation and manufacture of automobile vehicles.

After a brief preface and an historical review, the authors furnish a lucid treatise on the elementary principles of motors and on the general operation, cost and use of automobiles.

This is the first comprehensive work on this subject made intelligible to the general reader, that has appeared in Germany, and is bound to become something of a standard among German motor men who are not professional engineers. If certain parts of the text were adapted to American conditions, it might well be worth while to translate this excellent little hand book into English.

Das Automobil in Theorie und Praxis (Elementarbegriff des Fortbewegungsmittels mechanischer Motoren) is a German translation of the well-known French work by L. Baudry de Saunier. This authorized translation has been competently handled by Dr. R. von Stern and is published by A. Hartleben in Vienna.

As the original French edition has already been reviewed in these pages, it is scarcely necessary to enter into details of its German successor. Suffice it to state that the translation appears to be faithful and felicitous in many respects, while some of the illustrations appear even better than they did in the original.

648

The Automobile Index

Everything of permanent value published in the technical press of the world devoted to any branch of automobile industry will be found indexed in this department. Whenever it is possible a descriptive summary indicating the character and purpose of the leading articles of current automobile literature will be given, with the titles and dates of the publications.

Accumulators—
A serial article, by E. C. Bimington, on the construction of accumulators for automobiles. "The Automotor Journal," London, January, 1900.

Acetylene Bicycle—
Description and illustration of an acetylene motor applied to bicycle propulsion, devised by C. H. Offen. "La Locomotion Automobile," Paris, January 4, 1900.

Aluminium—
"The Present Uses of Aluminium." An article by Leon Auscher. "The Automobile Magazine, February, 1900.

Automobile Postal Service—
An article, with illustrations, by Perry S. Heath, First Assistant Postmaster General. "The Automobile Magazine, February, 1900.

Automobilism in 1899—
A review under the title of "The Progress of Automobilism in 1899." With illustrations. "The Motor-Car Journal," London, January 12 and 19, 1900.

Carbureters—
Illustrated description of the Lambert Carbureter. "La Locomotion Automobile," Paris, December 28, 1899.
Description of the new carbureter, patented by E. Georis, of Charleroi. "La Locomotion Automobile," Paris, January 4, 1900.
The Eldin carbureter described and illustrated. "La Locomotion Automobile," Paris, January 4, 1900.

Coils and Sparks—
A study by E. J. Stoddard. With diagrams. "The Horseless Age," New York, January 17, 1900.

Compressed Air in Europe—
A full description of the compressed air motor used for the propulsion of the Molas, Lainielle & Tessier delivery wagon "The Automobile Magazine," February, 1900.

Construction—
Recent Progress in Automobile Construction. W. Worby Beaumont. A review of the progress in details of motor vehicle construction during the past two years, showing the lines of advance in the removal of the mechanical imperfections revealed by experience. " Engineering Magazine," January, 1900.

Couthon's Automobile—
A relique of the last century briefly described. With illustrations. "The Automobile Magazine," February, 1900.

Electric Automobiles—
An article, by Frank B. Rae, on the design and control of electric-motors. With illustrations. "The Motocycle- Automobile." Chicago, January, 1900.

Description of the Rae carriage and its motor. With illustrations. "The Motocycle - Automobile," Chicago, January, 1900.

Illustrated description of a new electric carriage, built by Mr. Carl Opperman, of London. "The Automotor Journal," London, January, 1900.

The Automobile Magazine

Electricity on Common Roads—

A paper read by Mr. Thomas H. Parker, of Wolverhampton, before the Automobile Club of Great Britain. With three illustrations. "The Motor-Car Journal," London, January 12, 1900.

Electric Fore-carriage—

Description of the Solignac electric front hauler. With two illustrations. "La Nature," Paris, January 13, 1900.

Electric Motors—

An illustrated article, by F. B. Rae, on the design, construction and control of electric motors. "Electrical World and Engineer," New York, January 20, 1900.

The "Electromobile Ambulance," built by F. R. Wood & Son, of New York, for St. Vincent Hospital service. "Electrical Review," New York, January 24, 1900.

Fore-carriages—

"A Front-driven, Front-steered Automobile." Description and illustration of a new fore-carriage. "The Automobile Magazine," February, 1900.

Gas Motors—

Description of the reversible gas motor invented by Macdonald and Mackenzie, of England. "The Automotor Journal," London, January, 1900.

Gasoline and Gasoline M xtures—

An illustrated article by E. J. Stoddard. "The Horseless Age," New York, January 17, 1900.

Gasoline Vaporizers and Carbureters—

A study by Henry W. Struss. With illustrations. "The Horseless Age," New York, January 17, 1900.

Heavy Motor Wagons for Liverpool Traffic—

A paper read by Mr. Arthur Musker before the Liverpool Engineering Society. With 11 illustrations. "The Automotor Journal," London, January, 1900.

Hydro-carbon Automobiles—

The Plass voiturette described and illustrated. "La Locomotion Automobile," Paris, December 28, 1899.

The new model "Locomobile" carriage for 1900 described and illustrated. "The Motocycle-Automobile," Chicago, January, 1900.

The Wellington "Automobilette" described and illustrated. "The Automotor Journal," London, January, 1900.

A brief description of the new automobile built in Austria by the "Nesselsdorfer Wagenbau-fabriks Gesellschaft." "The Motor-Car Journal," London, January 5, 1900.

Brief description of latest improvements made on the Decanville carriages. Two illustrations. "La France Automobile," Paris, January 7, 1900.

The new Bollée voiturette described. With five illustrations. "La France Automobile," Paris, January 7, 1900.

A brief description of Luap-Legendre's "Mignonette." One illustration. "Le Chauffeur," Paris, January 11, 1900.

The Turgan-Foy carriage fully described and illustrated. "La Locomotion Automobile," Paris, January 11, 1900.

The voiturette called "La Princesse" (Barisien system) described and illustrated. "Le Chauffeur," Paris, January 11, 1900.

Illustrated description of the Levenn voiturette as built by Ernest & Co., of Paris. "The Motor-Car Journal," London, January 12, 1900.

Description of a two-seated phæton designed and built by Mr. Sutton, of Melbourne, Australia. "The Autocar," Coventry, England, January 13, 1900.

A full description and illustration of the Underberg voiturette. "La France Automobile," Paris, January 14, 1900.

Description of the "Eureka" carriage, built by Ough & Waltenbaugh, of San Francisco, Cal. With six illustrations. "The Horseless Age," New York, January 17, 1900.

The Automobile Index

Hydro-carbon Motors—

Illustrated description of the "Minerve" motor. "Le Chauffeur," Paris, January 11, 1900.

An illustrated description of the Bravo gasoline motors. "The Automobile Magazine," February, 1900.

Ignition and Ignition Troubles—

An illustrated article on this subject by P. M Heldt. "The Horseless Age," New York, January 17, 1900.

Ignition Verifier—

A device to verify electrical ignition on the hydro-carbon motors. Described by Gustave Hermite. One illustration. "La France Automobile," Paris, January 7, 1900.

Improvements in Internal Combustion Motors—

Description and illustration of the improved Crossley and Atkinson's motor. "The Automotor Journal," London, January, 1900.

Les Automobiles de Guerre—

An article pointing at the advantages of the Scotte traction engine as a hauler of trains of cars loaded with guns and other heavy materials of war. With four illustrations. "La Nature," Paris, January 6, 1900.

Lubricators—

Illustrated description of the Serpollet lubricator. "The Automobile Magazine," February, 1900.

Mechanical Propulsion and Traction—

A serial paper by Prof. G. Forestier. "The Automobile Magazine," February, 1900.

Motor Carriage Experiments—

An article on care and management of automobiles. By Mr. John Pope, of Liverpool, Eng. "The Autocar," Coventry, January 6, 1900.

Motor-Omnibus Lines—

Ideas on the possibilities of motor-vehicles in inter-urban passenger transportation compared with trolley lines. "The Motor Age," Chicago, January 18, 1900.

Motor Racing—

An article, with illustrations, by Edwin Emerson, Jr. "The Automobile Magazine," February 1900.

Multi-Cylinder Engines—

An article by P. M. Heldt. "The Horseless Age," New York, January 17, 1900.

Official Time-keeping Rules—

An article, by G. H. L. "The Automobile Magazine," February, 1900.

One Year's Progress of Automobilism—

An article by Prof. Félicien Michotte. With illustrations. "The Automobile Magazine," February, 1900.

Racing Rules—

The French racing rules (official). "The Automobile Magazine," February, 1900.

Radiators—

Illustrations and brief description of the Apprin radiators. "Le Chauffeur," Paris, January 11, 1900.

Speed Changing Gear—

Speed changing by means of extension pulleys. Lepillet's system described and illustrated. "La Locomotion Automobile," Paris, January 4, 1900.

Illustrated description of the "progressive" speed changing gear devised by Mr. H. Gérard. "La France Automobile," Paris, January 11, 1900.

A new speed changing gear for tricycles and voiturettes. Metz system described and illustrated. "L'Industrie Automobile," Paris, January 25, 1900.

Gaillardet's speed-changing gear, described and illustrated. "The Automobile Magazine," February, 1900.

The Automobile Magazine

Starting Device—

"An Automatic Starting Gear for Hydrocarbon Motors." With illustrations. "The Automobile Magazine," February, 1900.

Steam Automobiles—

"The Locomobile." An article describing the latest modifications and improvements made in the locomobile carriages. "The Automobile Magazine," February, 1900.

Steam Generators—

Illustrated description of the Blaxton steam generator. "The Automotor Journal," London, January, 1900.

Steam Lorry—

Full description of the Simpson-Bodman steam lorry. With seven illustrations. "The Motor - Car Journal," London, January 12, 1900.

Steering—

An article by R. D. on steering by means of the divided axle devised by Ph. Marot-Gardon et cie. "La Locomotion Automobile," Paris, January 18, 1900.

Steering Mechanism—

Illustrated description of the new Dion-Bouton steering mechanism. "La France Automobile," Paris, January 18, 1900.

Strength of Steel Balls—

A paper read before the December meeting of the American Society of Mechanical Engineers, by J. F. W. Harris. "The Motor Vehicle Review," Cleveland, O., January 23, 1900.

The Automobile in Local Transit—

An article by Sylvester Baxter. "The Automobile Magazine," February, 1900.

The Automobile Movement—

An article, by M. C. Krarup, on the styles of vehicles required and the usefulness of separate tractors. "American Machinist," New York, January 25, 1900.

The Construction of a Gasoline Motor Vehicle—

A serial technical article by Clarence C. Bramwell. "The Motor Vehicle Review," Cleveland, O., January 30, 1900.

The Gasoline Engine Indicator Diagram—

A study by E. C. Oliver, Engineering Department, University of Illinois. "The Horseless Age," New York, January 17, 1900.

The Use of Balls in Motor Construction—

An article by H. B. Adams. With illustrations. "The Motor Vehicle Review," Cleveland, O., January 23, 1900.

The Vibration of Explosion Motors—

An article by Herbert L. Towle. With illustrations. "The Horseless Age," New York, January 17, 1900.

Traffic Regulations and the Speed of Motor Vehicles on Highways—

A paper read before the Automobile Club of Great Britain, by R. E. Crompton. "The Horseless Age," New York, January 10, 1900.

Transmission Gear—

The transmission gear adopted by the "Société Française d'Automobiles." A brief description, with one illustration. "L'Industrie Velocipédique et Automobile," Paris, November, 1899.

Trials—

The Liverpool trials of motor vehicles. Judges' report, with illustrations. "The Automotor Journal," London, January, 1900.

Competitive trials of hydro-carbon motors in Paris, under the management of "La Locomotion Automobile." Described and illustrated. "La Locomotion Automobile," Paris, January 4 and 11, 1900.

Vaporizers and Carbureters—

An article, by Herbert L. Towle, on the requisites and action of vaporizers and carbureters. "The Horseless Age," New York, January 17, 1900.

"The Brown" Motor Quadricycle

Fitted with a two-speed gear, of extremely simple construction, by which means almost any hill can be ascended when the machine is carrying two persons. This gear also allows of the motor being entirely thrown out of gear when not required, and the machine can then be easily pedalled or moved about, an advantage that cannot be too highly recommended, as the motor can be kept running when in heavy traffic, or in any circumstance requiring a sudden stop. The motor is a genuine "De Dion," with electric ignition, and capable of travelling at a high rate of speed. The frame is very strongly constructed and is made of metal throughout. Both front and rear seats are extremely comfortable, and the steering very easy, and the cost of spirit for driving can be taken at about ¼d. per mile. The front seat is comfortably upholstered and provided with a leather apron. The front wheels are 26 in. diameter and the rear 28 in.

BROWN BROTHERS, Limited

Gesellschaft für Verkehrsunternehmungen

BERLIN (GERMANY)

Traffic Association of Berlin

Electric Wagonette with Interchangeable Body

Compagnie des Automobiles
et Moteurs Henriod

7-9 Rue de Sablonville, Neuilly
(Porte-Maillot) :: France

Carriages with 5 to 12 horse-power motors with flange-cooled cylinders.

Three-seat carriages with 4 horse power motors with flange-cooled cylinders.

The only carriage with an 8 horse-power motor not cooled by a water-jacket, which took part in the Paris-Bordeaux race, was a Henriod Carriage that made 321 miles in 19 hours.

These motors are vapor, petroleum, and alcohol engines.

Carburation is effected by means of a vaporizer.

INTERNATIONAL

: DOCTOR'S CAR :

Supplied with or without Hood, which is detachable

MORE doctors are users of this car than all other patterns put together. It is extremely reliable, very comfortable, and economical in use. It is a handsome carriage, and a credit to its owner and user. It will mount the steepest hills with ease, and can travel at a great speed on the level. It will carry enough oil for two hundred miles, and cannot catch on fire or explode. Over 1,600 of these motors are in use on the Continent and in England.

International Motor Car Co.

15, High Road, Kilburn, London, W.

DARRACQ & CIE.

Perfecta Works

Suresnes, France

LÉON BOLLÉE AUTOMOBILE—6 H.-P.—5 SPEEDS

MOTOCYCLES
TRICYCLES
QUADRICYCLES

WITH

DE DION, GAILLARDET, AND ASTER MOTORS
FITTED WITH PERFECTA PARTS

The record for the greatest distance covered in an hour was beaten
by a tricycle fitted with perfecta parts.

AMERICAN ROLLER BEARING CO.

27 STATE STREET, BOSTON, MASS.

SUBSCRIBER'S ORDER FORM TO

$3.00 a year
in the
United States | *The* AUTOMOBILE MAGAZINE | $4.00 a year
in Foreign
Countries

THE U. S. INDUSTRIAL PUBLISHING CO.

31 State Street, New York.

Gentlemen: Please enter my name as a Subscriber to The Automobile Magazine for one year, beginning with the.................... number, for which I remit...

STEAM VAN BUILDERS

London (England)
20, Abchurch Lane, E. C.

THE LIQUID FUEL ENGINEERING Cº EAST COWES. I. W.

Practical Directions for the Management of Petroleum and Electric Automobiles

BASED ON LECTURES DELIVERED
AT THE POLYTECHNIC INSTITUTE

By Professor FÉLICIEN MICHOTTE

Consulting-Engineer and Expert, Chevalier of
the Mérite Agricole, Officer of the Academy and
of the Nichau Iftikhar, Laureate of the Society
for the Encouragement of National Industry.

—   —

One Volume, 259 Pages, in French

PRICE, $1.50

Orders received at the

TECHNICAL OFFICE

...OF...

The Automobile Magazine, New York City

BUT IT DIDN'T!

CANADA: "Watch that new-fangled machine get a jar."

TO AUTOMOBILE USERS

Owners of automobiles will confer a favor upon the publishers of

THE AUTOMOBILE MAGAZINE

by communicating information as to results obtained, faults discovered, suggested improvements, etc. The make, style, and motive power of the vehicle should always be stated. Information so received will be used for the purpose of giving our readers the benefit of practical experience with the different classes of automobiles procurable, and when found of especial value will be published in the magazine with due credit to the correspondent.

ADDRESS

EDITOR, AUTOMOBILE MAGAZINE

U. S. INDUSTRIAL PUBLISHING CO., NEW YORK

www.ingramcontent.com/pod-product-compliance
Lightning Source LLC
Chambersburg PA
CBHW020850210326
41598CB00018B/1625